LONDON MATHEMATICAL SOCIETY LECTURE NOTE SERIES

Managing Editor: Professor J.W.S. Cassels, Department of Pure Mathematics and Mathematical Statistics, University of Cambridge, 16 Mill Lane, Cambridge CB2 1SB, England

The books in the series listed below are available from booksellers, or, in case of difficulty, from Cambridge University Press.

17 Differential germs and catastrophes, Th. BROCKER & L. LANDER
34 Representation theory of Lie groups, M.F. ATIYAH *et al*
36 Homological group theory, C.T.C. WALL (ed)
39 Affine sets and affine groups, D.G. NORTHCOTT
40 Introduction to H_p spaces, P.J. KOOSIS
43 Graphs, codes and designs, P.J. CAMERON & J.H. VAN LINT
45 Recursion theory: its generalisations and applications, F.R. DRAKE & S.S. WAINER (eds)
46 p-adic analysis: a short course on recent work, N. KOBLITZ
49 Finite geometries and designs, P. CAMERON, J.W.P. HIRSCHFELD & D.R. HUGHES (eds)
50 Commutator calculus and groups of homotopy classes, H.J. BAUES
51 Synthetic differential geometry, A. KOCK
54 Markov processes and related problems of analysis, E.B. DYNKIN
57 Techniques of geometric topology, R.A. FENN
58 Singularities of smooth functions and maps, J.A. MARTINET
59 Applicable differential geometry, M. CRAMPIN & F.A.E. PIRANI
60 Integrable systems, S.P. NOVIKOV *et al*
62 Economics for mathematicians, J.W.S. CASSELS
65 Several complex variables and complex manifolds I, M.J. FIELD
66 Several complex variables and complex manifolds II, M.J. FIELD
68 Complex algebraic surfaces, A. BEAUVILLE
69 Representation theory, I.M. GELFAND *et al*
74 Symmetric designs: an algebraic approach, E.S. LANDER
76 Spectral theory of linear differential operators and comparison algebras, H.O. CORDES
77 Isolated singular points on complete intersections, E.J.N. LOOIJENGA
78 A primer on Riemann surfaces, A.F. BEARDON
79 Probability, statistics and analysis, J.F.C. KINGMAN & G.E.H. REUTER (eds)
80 Introduction to the representation theory of compact and locally compact groups, A. ROBERT
81 Skew fields, P.K. DRAXL
82 Surveys in combinatorics, E.K. LLOYD (ed)
83 Homogeneous structures on Riemannian manifolds, F. TRICERRI & L. VANHECKE
85 Solitons, P.G. DRAZIN
86 Topological topics, I.M. JAMES (ed)
87 Surveys in set theory, A.R.D. MATHIAS (ed)
88 FPF ring theory, C. FAITH & S. PAGE
89 An F-space sampler, N.J. KALTON, N.T. PECK & J.W. ROBERTS
90 Polytopes and symmetry, S.A. ROBERTSON
91 Classgroups of group rings, M.J. TAYLOR
92 Representation of rings over skew fields, A.H. SCHOFIELD
93 Aspects of topology, I.M. JAMES & E.H. KRONHEIMER (eds)
94 Representations of general linear groups, G.D. JAMES
95 Low-dimensional topology 1982, R.A. FENN (ed)
96 Diophantine equations over function fields, R.C. MASON
97 Varieties of constructive mathematics, D.S. BRIDGES & F. RICHMAN
98 Localization in Noetherian rings, A.V. JATEGAONKAR
99 Methods of differential geometry in algebraic topology, M. KAROUBI & C. LERUSTE
100 Stopping time techniques for analysts and probabilists, L. EGGHE

101 Groups and geometry, ROGER C. LYNDON
103 Surveys in combinatorics 1985, I. ANDERSON (ed)
104 Elliptic structures on 3-manifolds, C.B. THOMAS
105 A local spectral theory for closed operators, I. ERDELYI & WANG SHENGWANG
106 Syzygies, E.G. EVANS & P. GRIFFITH
107 Compactification of Siegel moduli schemes, C-L. CHAI
108 Some topics in graph theory, H.P. YAP
109 Diophantine Analysis, J. LOXTON & A. VAN DER POORTEN (eds)
110 An introduction to surreal numbers, H. GONSHOR
111 Analytical and geometric aspects of hyperbolic space, D.B.A.EPSTEIN (ed)
112 Low-dimensional topology and Kleinian groups, D.B.A. EPSTEIN (ed)
113 Lectures on the asymptotic theory of ideals, D. REES
114 Lectures on Bochner-Riesz means, K.M. DAVIS & Y-C. CHANG
115 An introduction to independence for analysts, H.G. DALES & W.H. WOODIN
116 Representations of algebras, P.J. WEBB (ed)
117 Homotopy theory, E. REES & J.D.S. JONES (eds)
118 Skew linear groups, M. SHIRVANI & B. WEHRFRITZ
119 Triangulated categories in the representation theory of finite-dimensional algebras, D. HAPPEL
121 Proceedings of *Groups - St Andrews 1985*, E. ROBERTSON & C. CAMPBELL (eds)
122 Non-classical continuum mechanics, R.J. KNOPS & A.A. LACEY (eds)
123 Surveys in combinatorics 1987, C. WHITEHEAD (ed)
124 Lie groupoids and Lie algebroids in differential geometry, K. MACKENZIE
125 Commutator theory for congruence modular varieties, R. FREESE & R. MCKENZIE
126 Van der Corput's method for exponential sums, S.W. GRAHAM & G. KOLESNIK
127 New directions in dynamical systems, T.J. BEDFORD & J.W. SWIFT (eds)
128 Descriptive set theory and the structure of sets of uniqueness, A.S. KECHRIS & A. LOUVEAU
129 The subgroup structure of the finite classical groups, P.B. KLEIDMAN & M.W.LIEBECK
130 Model theory and modules, M. PREST
131 Algebraic, extremal & metric combinatorics, M-M. DEZA, P. FRANKL & I.G. ROSENBERG (eds)
132 Whitehead groups of finite groups, ROBERT OLIVER
133 Linear algebraic monoids, MOHAN S. PUTCHA
134 Number theory and dynamical systems, M. DODSON & J. VICKERS (eds)
135 Operator algebras and applications, 1, D. EVANS & M. TAKESAKI (eds)
136 Operator algebras and applications, 2, D. EVANS & M. TAKESAKI (eds)
137 Analysis at Urbana, I, E. BERKSON, T. PECK, & J. UHL (eds)
138 Analysis at Urbana, II, E. BERKSON, T. PECK, & J. UHL (eds)
139 Advances in homotopy theory, S. SALAMON, B. STEER & W. SUTHERLAND (eds)
140 Geometric aspects of Banach spaces, E.M. PEINADOR and A. RODES (eds)
141 Surveys in combinatorics 1989, J. SIEMONS (ed)
142 The geometry of jet bundles, D.J. SAUNDERS
143 The ergodic theory of discrete groups, PETER J. NICHOLLS
144 Uniform spaces, I.M. JAMES
145 Homological questions in local algebra, JAN R. STROOKER
146 Maximal Cohen-Macaulay modules over Henselian rings, Y. YOSHINO
147 Continuous and discrete modules, S.H. MOHAMED & B.J. MÜLLER
154 Number theory and cryptography, J. LOXTON (ed)

London Mathematical Society Lecture Note Series. 129

The Subgroup Structure of the Finite Classical Groups

Peter Kleidman,
Princeton University

Martin Liebeck
Imperial College, London

The right of the University of Cambridge to print and sell all manner of books was granted by Henry VIII in 1534. The University has printed and published continuously since 1584.

CAMBRIDGE UNIVERSITY PRESS
Cambridge
New York Port Chester Melbourne Sydney

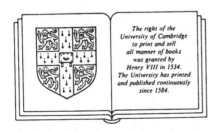

CAMBRIDGE UNIVERSITY PRESS
Cambridge, New York, Melbourne, Madrid, Cape Town, Singapore, São Paulo

Cambridge University Press
The Edinburgh Building, Cambridge CB2 8RU, UK

Published in the United States of America by Cambridge University Press, New York

www.cambridge.org
Information on this title: www.cambridge.org/9780521359498

First published 1990
Re-issued in this digitally printed version 2008

A catalogue record for this publication is available from the British Library

ISBN 978-0-521-35949-8 paperback

Preface

Following the classification of the finite simple groups, completed in 1980, one of the major areas of research in group theory today is the investigation of the subgroups of the finite simple groups, and in particular, the determination of their maximal subgroups. According to the classification theorem, the finite simple groups fall into four classes:

the alternating groups A_n ($n \geq 5$);

the finite classical groups — that is, the linear, symplectic, unitary and orthogonal groups on finite vector spaces;

the exceptional groups of Lie type;

the 26 sporadic groups.

In this book we concentrate on the classical groups, which we describe in detail in Chapter 2. Our work takes as its starting point the fundamental results of M. Aschbacher in [As$_1$]. Let G be a finite classical group. In [As$_1$], Aschbacher introduces a large collection $\mathcal{C}(G)$ of natural, geometrically defined subgroups of G, and shows that almost every subgroup of G is contained in a member of $\mathcal{C}(G)$ (the precise result is stated in Chapter 1). Thus the collection $\mathcal{C}(G)$ of subgroups plays a central role in the theory of classical groups. This book is intended to be a definitive investigation of the collection $\mathcal{C}(G)$. In it, we solve the three main problems concerning these subgroups — namely, we determine

(I) the group-theoretic structure of each member of $\mathcal{C}(G)$,

(II) the conjugacy among the members of $\mathcal{C}(G)$,

(III) precisely which members of $\mathcal{C}(G)$ are maximal in G and which are not — and, for non-maximal members H of $\mathcal{C}(G)$, we determine the maximal overgroups of H in G.

Some of these results have been obtained by Aschbacher and others, but many of them are new.

The layout of the book is as follows. In Chapter 1 we present a more detailed motivation and setting for the work in this book. In particular, we describe Aschbacher's theorem, and also survey recent developments in the study of the subgroup structure of the other (non-classical) simple groups.

Chapter 2 contains an introduction to the classical groups. This includes a detailed description of forms and standard bases for classical geometries, and of generators, automorphisms, simplicity and exceptional isomorphisms of classical groups. In §2.10 there is also a compendium of results concerning the representation of a classical group on its associated geometry.

In Chapter 3 we state the main results of the book. These are given in the form of nine tables (Tables 3.5.A-I). The chapter contains explanations of how to use the tables to read off the solutions to problems (I), (II) and (III) above.

Chapter 4 contains the proof of all the results concerning the structure and conjugacy of the subgroups in $\mathcal{C}(G)$. The subgroups in $\mathcal{C}(G)$ are divided into eight families $\mathcal{C}_i(G)$ ($1 \leq i \leq 8$), and the material in the chapter is accordingly presented in eight sections.

Each section begins with the definition and explicit description of the subgroups in $\mathcal{C}_i(G)$.

The remainder of the book is devoted to the solution of problem (III) — that is, to finding the maximal overgroups of subgroups in $\mathcal{C}(G)$. We handle this problem in two parts: (a) finding overgroups which themselves lie in $\mathcal{C}(G)$ and (b) finding the other overgroups. The overgroups occurring in (a) are constructed in Chapter 6, and in Chapter 7 we prove that these are the only such examples. As for (b), it follows from Aschbacher's theorem (described in Chapter 1) that it suffices to find just those overgroups which are almost simple and absolutely irreducible on the vector space associated with the classical group G. We find all such overgroups in Chapter 8. This chapter is the only one in the book in which the classification theorem is actually used. To employ the classification we require a good deal of detailed information about the simple groups, particularly concerning their permutation actions and their representations. This information is gathered in Chapter 5. One feature of the examples in Chapter 8 is the occurrence of spin modules for symplectic and orthogonal groups in characteristic 2. Some detailed calculations with these modules are required, and the groundwork for these is laid in §5.4 in Chapter 5.

We have aimed to be comprehensible to a first year graduate student with a background in algebra. We have also tried to present our material in such a way as to be useful to any research worker using simple groups, even if he or she is not especially interested in subgroup structure. In particular, we believe that Chapter 2 will serve as a useful introduction to the basic properties of classical groups, and Chapter 5 as a guide to the methods available when solving problems using the classification theorem.

Standard notation and terminology

Group-theoretic notation

Let G and H be finite groups, p a prime and n an integer.

$H \preceq G$	H is isomorphic to a subgroup of G
$H.G$	extension of H by G
$H{:}G$	split extension of H by G
$H \circ G$	central product of H and G
$\frac{1}{n}G$	normal subgroup of index n in G
G'	derived group of G
G^{∞}	last term of the derived series of G
$m_p(G)$	p-rank of G
$Syl_p(G)$	set of Sylow p-subgroups of G
$\lvert G\rvert_p$	power of p dividing $\lvert G\rvert$
$O_p(G)$	largest normal p-subgroup of G
$O^p(G)$	subgroup of G generated by all its p'-elements
$\lvert g\rvert$	order of element $g \in G$
$Z(G)$	centre of G
$\mathrm{soc}(G)$	subgroup of G generated by its minimal normal subgroups
$\mathbf{Z}_n$ or just n	cyclic group of order n
$\mathbf{Z}_p^n$ or just p^n	elementary abelian group of order p^n
$[n]$	arbitrary soluble group of order n
quasisimple group	perfect group G such that $G/Z(G)$ is non-abelian simple
component of G	quasisimple subnormal subgroup of G

Other notation

$\mathbf{F}_q$	field of q elements
$\mathbf{F}^*$	non-zero elements in the field $\mathbf{F}$
$(\mathbf{F}^*)^n$	n^{th} powers in $\mathbf{F}^*$
$T_{\mathbf{F}_o}^{\mathbf{F}}, N_{\mathbf{F}_o}^{\mathbf{F}}$	trace and norm maps $\mathbf{F} \mapsto \mathbf{F}_o$ where $\mathbf{F}_o$ is a subfield of $\mathbf{F}$
$\mathrm{Gal}(\mathbf{F}{:}\mathbf{F}_o)$	Galois group of $\mathbf{F}$ over $\mathbf{F}_o$
A^{t}	transpose of the matrix A
$(a_1, \ldots, a_n)$	highest common factor of the integers $a_1, \ldots, a_n$
$[a, b]$	lowest common multiple of the integers a, b
$[x]$, $\lfloor x\rfloor$	greatest integer $\leq x$
$\lceil x\rceil$	smallest integer $\geq x$
$x - \epsilon$	used for $x - \epsilon 1$ where $\epsilon = +$ or $-$

Contents

Chapter 1
MOTIVATION AND SETTING FOR THE RESULTS

§1.1 Introduction

In this chapter we provide the motivation and setting for this book by surveying recent developments in the study of the subgroup structure of finite groups. We also provide, in §1.2 below, a rough statement of the results which will be proved in the later chapters (see Theorem 1.2.3).

By the Jordan-Hölder Theorem, every finite group is 'built' out of simple groups, and so questions about arbitrary finite groups often reduce to questions about simple groups or *almost simple* groups — here, a group G is said to be *almost simple* if

$$G_o \trianglelefteq G \leq Aut(G_o) \tag{1.1.1}$$

for some non-abelian simple group G_o. Much of the information about a group can be gleaned from a study of its subgroups. For these reasons it is important to study the subgroup structure of the almost simple groups, and in particular their *maximal* subgroups. This subject has a rich history, dating back as far as Galois' famous letter to Chevalier, written on the eve of his ill-fated duel in 1832 [Ga_1]. In recent years there have been far-reaching developments in the area, which we now describe. Throughout this chapter, G_o denotes a finite non-abelian simple group and G a group satisfying (1.1.1). The maximal subgroups of G which contain G_o are not of interest to us, for they correspond merely to the subgroups of $Out(G_o) = Aut(G_o)/G_o$, which is a soluble group with a transparent structure. Thus throughout Chapter 1, the subgroups of G discussed are assumed not to contain G_o.

The non-abelian finite simple groups have been competely classified (for a discussion of this classification see [Go], for example), and they fall into four families:

(1) the alternating groups A_n ($n \geq 5$);

(2) the finite classical groups — that is, the linear, symplectic, unitary and orthogonal groups on finite vector spaces;

(3) the exceptional groups of Lie type;

(4) the 26 sporadic groups.

The families under (3) and (4) can be regarded as 'groups of bounded rank', and the study of their subgroups requires methods significantly different from those used for the groups in (1) and (2). We describe some recent work on the subgroups of groups in (3) and (4) briefly in §1.3 below. For the families (1) and (2) of 'unbounded rank', there are powerful *subgroup structure theorems* available, due to O'Nan and Scott (see [Sc, Appendix] and [A-S, Appendix]) for the alternating groups, and to Aschbacher [As_1] for the classical groups. In these structure theorems, a 'natural' collection of subgroups of G is defined (where G satisfies (1.1.1) and G_o is alternating or classical, and it is proved that for any subgroup H of G not containing G_o,

either H is contained in a member of the natural collection,

or $H \in \mathcal{S}$, where $\mathcal{S}$ is a set of almost simple subgroups of G satisfying certain 'irreducibility' conditions.

We call the natural collection $\mathcal{A}(G)$ in the alternating case and $\mathcal{C}(G)$ in the classical case, and these along with $\mathcal{S}$ are described in §§1.2-1.3.

As a consequence of these subgroup structure theorems, every maximal subgroup of G lies in $\mathcal{A}(G), \mathcal{C}(G)$ or $\mathcal{S}$. Thus to obtain a classification of the maximal subgroups of G, one seeks to determine when the subgroups in $\mathcal{A}(G), \mathcal{C}(G)$ and $\mathcal{S}$ are in fact maximal in G. This programme is complete in the alternating case [L-P-S$_1$], and those results are outlined in §1.3. It is the purpose of this book to contribute to this programme for the classical groups. In a strong sense, a large portion of the maximal subgroups of a classical group G lies in $\mathcal{C}(G)$, and we shall solve three main problems concerning $\mathcal{C}(G)$ — namely, we determine precisely which members of $\mathcal{C}(G)$ are maximal in G and which are not, the exact group-theoretic structure of each member of $\mathcal{C}(G)$, and the conjugacy amongst the groups in $\mathcal{C}(G)$. We now outline these results and some consequences. The precise results are stated in Chapter 3.

§1.2 The classical groups

The reader unfamiliar with the classical groups is referred to Chapter 2, where notation is established and definitions and basic properties of these groups are given. Let G_o be one of the following classical simple groups:

$$\begin{gathered} PSL_n(q),\ PSU_n(q),\ PSp_n(q)\ (n \text{ even}), \\ P\Omega_n^{\pm}(q)\ (n \text{ even}),\ \Omega_n(q)\ (nq \text{ odd}). \end{gathered} \tag{1.2.1}$$

Write V for the natural n-dimensional vector space over the finite field $\mathbf{F}$ associated with G_o (so that $\mathbf{F} = \mathbf{F}_q$ for the linear, symplectic and orthogonal groups and $\mathbf{F} = \mathbf{F}_{q^2}$ for the unitary groups), and write $q = p^f$, where p is prime. Also let Γ be the full semilinear classical group corresponding to G_o. For example, if $G_o = PSL_n(q)$, then $\Gamma = \Gamma L_n(q)$. The group Γ contains the scalars as a normal subgroup, and writing $\overline{}$ for reduction modulo scalars we see that $G_o \trianglelefteq \overline{\Gamma} \leq Aut(G_o)$, and $\overline{\Gamma} = Aut(G_o)$ except in the cases which we mention after Theorem 1.2.1, below.

The subgroup structure theorem for the classical groups is due to Aschbacher [As$_1$]. In [As$_1$], eight collections $\mathcal{C}_i(\Gamma)$ $(1 \leq i \leq 8)$ of natural subgroups of Γ are defined. The precise definitions and structures of the groups in $\mathcal{C}_i(\Gamma)$ are given in Chapter 4, so here we content ourselves with a 'rough' description in the left hand side of Table 1.2.A. In the right hand side we illustrate by giving the approximate structures of the subgroups $H \cap GL_n(q)$, where $H \in \mathcal{C}_i(\Gamma)$ and $G_o = PSL_n(q)$. Some of the notation used in the table will be discussed in Chapters 2 and 4.

Table 1.2.A

$\mathcal{C}_i$	rough description	rough structure in $GL_n(q)$
$\mathcal{C}_1$	stabilizers of totally singular or non-singular subspaces	maximal parabolic
$\mathcal{C}_2$	stabilizers of decompositions $V = \bigoplus_{i=1}^{t} V_i$, $\dim(V_i) = a$	$GL_a(q) \wr S_t$, $n = at$
$\mathcal{C}_3$	stabilizers of extension fields of $\mathbf{F}_q$ of prime index b	$GL_a(q^b).b$, $n = ab$, b prime
$\mathcal{C}_4$	stabilizers of tensor product decompositions $V = V_1 \otimes V_2$	$GL_a(q) \circ GL_b(q)$, $n = ab$
$\mathcal{C}_5$	stabilizers of subfields of $\mathbf{F}_q$ of prime index b	$GL_n(q_o)$, $q = q_o^b$, b prime
$\mathcal{C}_6$	normalizers of symplectic-type r-groups (r prime) in absolutely irreducible representations	$(\mathbf{Z}_{q-1} \circ r^{1+2a}).Sp_{2a}(r)$, $n = r^a$
$\mathcal{C}_7$	stabilizers of decompositions $V = \bigotimes_{i=1}^{t} V_i$, $\dim(V_i) = a$	$\overbrace{(GL_a(q) \circ \cdots \circ GL_a(q))}^{t} .S_t$, $n = a^t$
$\mathcal{C}_8$	classical subgroups	$Sp_n(q)$, n even $O_n^\epsilon(q)$, q odd $GU_n(q^{1/2})$, q a square

For any subgroup Y of Γ, define $\mathcal{C}_i(Y) = \{C \cap Y \mid C \in \mathcal{C}_i(\Gamma)\}$ for $1 \leq i \leq 8$, and let

$$\mathcal{C}(Y) = \bigcup_{i=1}^{8} \mathcal{C}_i(Y).$$

Further, put $\mathcal{C}_i(\overline{Y}) = \overline{\mathcal{C}_i(Y)}$ and $\mathcal{C}(\overline{Y}) = \overline{\mathcal{C}(Y)}$. Thus the right-hand column of Table 1.2.A gives the approximate structure of members of $\mathcal{C}_i(GL_n(q))$ when $\Gamma = \Gamma L_n(q)$.

Next we define the class $\mathcal{S}$ of subgroups of G, where G is as in (1.1.1) and G_o as in (1.2.1).

Definition of $\mathcal{S}$. The subgroup H of G lies in $\mathcal{S}$ if and only if the following hold.

(a) The socle S of H is a non-abelian simple group (i.e., H is almost simple).

(b) If L is the full covering group of S, and if $\rho : L \to GL(V)$ is a representation of L such that $\overline{\rho(L)} = S$, then ρ is absolutely irreducible.

(c) $\rho(L)$ cannot be realized over a proper subfield of $\mathbf{F}$.

(d) If $\rho(L)$ fixes a non-degenerate quadratic form on V, then $G_o = P\Omega_n^\epsilon(q)$.

(e) If $\rho(L)$ fixes a non-degenerate symplectic form on V, but no non-degenerate quadratic form, then $G_o = PSp_n(q)$.

(f) If $\rho(L)$ fixes a non-degenerate unitary form on V, then $G_o = PSU_n(q)$.

(g) If $\rho(L)$ does not satisfy the conditions in (d), (e) or (f), then $G_o = PSL_n(q)$.

Remark. Condition (b) in the definition of $\mathcal{S}$ is imposed to ensure that a group H in $\mathcal{S}$ is contained in no member of $\mathcal{C}_3(G)$. Similarly, (c)-(g) are imposed to ensure that H is contained in no member of $\mathcal{C}_5(G) \cup \mathcal{C}_8(G)$. In item (c), the phrase 'realized over a proper subfield' is defined in §2.10 (just before Lemma 2.10.7). As for (d)-(f), we refer the reader to §2.1. For a definition of 'covering group' in (b), see [As_8, §33].

We are now in a position to state the main subgroup structure theorem of [As_1].

Theorem 1.2.1 (Aschbacher [As_1]). *Let G be a group such that $G_o \trianglelefteq G \leq \overline{\Gamma}$, with G_o and Γ as in (1.2.1) above, and let H be a subgroup of G not containing G_o. Then either H is contained in a member of $\mathcal{C}(G)$ or $H \in \mathcal{S}$.*

Theorem 1.2.1 covers almost all groups whose socle is a classical simple group. However, there are three cases in which $\overline{\Gamma} \neq Aut(G_o)$, as follows.

(i) $G_o = PSL_n(q)$ and $n \geq 3$. Here G_o has an 'inverse-transpose' automorphism, called ι say, and $Aut(G_o) = \overline{\Gamma}\langle\iota\rangle$ (see §2.2). In [As_1, §13], an extra family $\mathcal{C}_1'(G)$ of subgroups of G is introduced when $G \not\leq \overline{\Gamma}$, and a version of Theorem 1.2.1 is proved using $\mathcal{C}_1'(G)$. As a convenience, we redefine $\mathcal{C}_1(G)$ so as to include $\mathcal{C}_1'(G)$.

(ii) $G_o = Sp_4(q)$ with q even. Here G_o admits a graph automorphism, which is discussed in [As_1 (§14),Ca_1 (Ch.12),Fl_2] and $|Aut(G_o) : \overline{\Gamma}| = 2$. When $G \not\leq \overline{\Gamma}$, the definition of $\mathcal{C}(G)$ is suitably modified in [As_1, §14] in such a way that a version of Theorem 1.2.1 still holds. This was also done in [Wi_6].

(iii) $G_o = P\Omega_8^+(q)$. Here G_o admits a 'triality' automorphism as described in [Ca_1 (Ch.12)] and $|Aut(G_o) : \overline{\Gamma}| = 3$. If $G \not\leq \overline{\Gamma}$, then no conclusive result is obtained in [As_1] — however [Kl_1] contains a full determination of the maximal subgroups of such groups G.

By Theorem 1.2.1, to obtain a classification of the maximal subgroups of the classical groups, one must consider the maximality of subgroups in $\mathcal{C}(G) \cup \mathcal{S}$. Cases where $n = \dim_{\mathbf{F}}(V)$ is small have been studied for a long time, dating as far back as the famous letter of Galois to Chevalier in 1832, which contains significant observations on the subgroups of $PSL_2(p)$. Many of the classical groups with $n \leq 7$ were dealt with prior to the classification of finite simple groups (see [K-L] for references). Using the classification and Theorem 1.2.1, the following result has now been proved.

Theorem 1.2.2 [Kl_1,Kl_2]. *Let G_o be as in (1.2.1) above with $n \leq 12$, and let $G_o \trianglelefteq G \leq Aut(G_o)$. Then the maximal subgroups of G are known.*

We now come to the main result of this book, dealing with the maximality of the subgroups in $\mathcal{C}(G)$ in general. In view of Theorem 1.2.2, we can restrict our attention to the case where $n \geq 13$. Put in relatively loose terms, the main result can be stated in the following way.

Main Theorem 1.2.3. *Let G_o be as in (1.2.1) and let $G_o \trianglelefteq G \leq Aut(G_o)$. Then*

(A) *the group-theoretic structure of each $H \in \mathcal{C}(G)$ is known;*

(B) *the conjugacy amongst the members of $\mathcal{C}(G)$ is known;*

(C) *for $H \in \mathcal{C}(G)$ all overgroups of H which lie in $\mathcal{C}(G) \cup \mathcal{S}$ are known (for $n \geq 13$).*

The precise version of Theorem 1.2.3 is stated as the Main Theorem in §3.1 and Tables 3.5.A-I. Sections 3.2-3.4 describe how these tables are to be read, and how one can obtain from them all the information advertised in Theorem 1.2.3, above. Parts (A) and (B) of Theorem 1.2.3 actually hold for all n, and not just for $n \geq 13$. Several portions of (A) and (B) are addressed in [As_1, Theorems A,B], but a number of significant questions are left unresolved there. As for (C), there is an extensive literature concerning various parts of the problem: for subgroups $H \in \mathcal{C}_1(G) \cup \mathcal{C}_2(G)$, the result has been obtained by geometric methods in [Dy_3,Ke_1,Ke_2] and in [Ki_1,Ki_2,Ki_3,Ki_4,Ki_5]. The maximality of subgroups in $\mathcal{C}_5(G) \cup \mathcal{C}_8(G)$ has been established in certain cases in [B-G-L,Dy_1,Dy_2,Ki_6,Ki_7,Mc,Po]. Some subgroups in $\mathcal{C}_3(G)$, and some in $\mathcal{C}_4(G)$, are treated in [Dy_4,Li_1,Li_2]. For the remaining cases — $\mathcal{C}_6(G), \mathcal{C}_7(G)$ and parts of $\mathcal{C}_3(G), \mathcal{C}_4(G), \mathcal{C}_5(G)$ — our result is new. We shall include complete proofs for all the groups in $\mathcal{C}(G)$, since our proofs are fairly uniform, and there is not much advantage to be gained by omitting various previously known special cases. However, it should be noted that we make use of the classification of finite simple groups in our proofs, whereas none of the above references does. (Note, however, that we use the classification only in Chapter 8 where we determine the overgroups in $\mathcal{S}$ of subgroups in $\mathcal{C}$.) In order to prove Theorem 1.2.3, we require a wealth of information about the structure and representations of the finite simple groups. We summarize this information in Chapter 5.

Consequences of Theorem 1.2.3

When studying the maximal subgroups of G (which do not contain G_o), one often distinguishes two general cases: the local case and the non-local case. More specifically, let H be such a maximal subgroup of G and set $H_o = H \cap G_o$. With a slight abuse of terminology, we call H *local* if $O_r(H_o) \neq 1$ for some prime r, and *non-local* otherwise. Since by Theorem 1.2.1 any local maximal subgroup of G obviously lies in $\mathcal{C}(G)$, we have the first main consequence of Theorem 1.2.3.

Corollary 1.2.4. *The local maximal subgroups of the finite classical groups are known.*

The precise list of local maximal subgroups can be read off from the Propositions in Chapter 4. This result, together with various results concerning the local maximal subgroups of the alternating, exceptional and sporadic groups discussed in §1.3, yields the following

Theorem 1.2.5. *The local maximal subgroups of all the finite simple groups are known, apart from the 2-locals of the Monster and the Baby Monster.*

Second, as mentioned in §1.1, Theorem 1.2.3 reduces a classification of the maximal subgroups of the classical groups to the problem of determining when groups in $\mathcal{S}$ are maximal in G. Work on this problem is in an interesting state at present, and we briefly describe some recent results. Let $H \in \mathcal{S}$, so that $S \trianglelefteq H \leq Aut(S)$ for some non-abelian simple group S, with S absolutely irreducible on V. The goal is to classify all such groups H which fail to be maximal in $HG_o = G$. Now if H is non-maximal in G, then $H < K < G$ for some maximal subgroup K of G. According to Theorem 1.2.1, we have $K \in \mathcal{C}(G) \cup \mathcal{S}$, and it is clear from the definitions that H is not contained in a member of $(\mathcal{C}_1 \cup \mathcal{C}_3 \cup \mathcal{C}_5 \cup \mathcal{C}_6 \cup \mathcal{C}_8)(G)$ (see the remark after the definition of $\mathcal{S}$). Thus two situations can arise:

(i) $K \in (\mathcal{C}_2 \cup \mathcal{C}_4 \cup \mathcal{C}_7)(G)$;

(ii) $K \in \mathcal{S}$.

If S is not of Lie type in characteristic p (here p is the defining characteristic of G_o), then first steps have been achieved in [Se$_3$]. However, for the most part, relatively little is known about the possibilities for H and K; for example classifying all instances of $H < K \in \mathcal{C}_4(G)$ is presently intractable, for it amounts to classifying all absolutely irreducible representations of the quasisimple groups which are tensor decomposable. So let us suppose that S is of Lie type in characteristic p. The p-modular representation theory of such groups S is sufficiently advanced to gain good control over situation (i), and so let us focus our attention on situation (ii). We may write $T \trianglelefteq K \leq Aut(T)$ with T non-abelian simple, and we have

$$S < T < G_o.$$

Consequently the aim here is to classify all such triples (S, T, G_o). The main theorem of [L-S-S] determines all but finitely many of the triples where S is of Lie type in characteristic p while T is not. In the case where both S and T are both of Lie type in characteristic p, significant advances have been achieved by Seitz and Testerman [Se$_1$,Se$_2$,Te].

§1.3 The alternating, sporadic and exceptional groups

We complete this survey by outlining the present state of knowledge on maximal subgroups of the alternating groups, the sporadic groups and the exceptional groups of Lie type.

We first consider the alternating groups A_n. As mentioned in §1.1, there is a subgroup structure theorem, due to O'Nan and Scott (see [Sc, Appendix]), which is analogous to Aschbacher's theorem for the classical groups. Let A_n and S_n act naturally on $I = \{1, \ldots, n\}$. As for the classical groups, we now describe five collections $\mathcal{A}_i$ $(1 \leq i \leq 5)$ of natural subgoups of S_n in Table 1.3.A, below.

Table 1.3.A			
$\mathcal{A}_i$	rough description	structure in S_n	comments
$\mathcal{A}_1$	stabilizers of subsets of I	$S_a \times S_b$	$n = a + b,\ a \neq b$
$\mathcal{A}_2$	stabilizers of partitions of I into subsets of equal size	$S_a \wr S_b$	$n = ab,\ a \geq 2,\ b \geq 2$
$\mathcal{A}_3$	stabilizers of affine structures on I	$AGL_d(p)$	$n = p^d$, p prime
$\mathcal{A}_4$	stabilizers of cartesian product structures on I	$S_a \wr S_b$	$n = a^b,\ a \geq 5,\ b \geq 2$
$\mathcal{A}_5$	the normalizer of T^k acting on the cosets of a diagonal subgroup, where T is a non-abelian simple group	$T^k.(Out(T) \times S_k)$	$n = \|T\|^{k-1},\ k \geq 2$

For a more detailed discussion of the families $\mathcal{A}_i$ consult [L-P-S$_2$].

For groups G satisfying $A_n \leq G \leq S_n$, define $\mathcal{A}_i(G) = \{A \cap G \mid A \in \mathcal{A}_i\}$ for $1 \leq i \leq 5$, and let

$$\mathcal{A}(G) = \bigcup_{i=1}^{5} \mathcal{A}_i(G).$$

The class $\mathcal{S}$ of subgroups of G consists of those almost simple groups acting primitively on I. With this notation, the subgroup structure theorem for the alternating groups may be stated as follows.

Theorem 1.3.1 (O'Nan-Scott [Sc]). *Let G be A_n or S_n and let H be a subgroup of G not containing A_n. Then either H is contained in a member of $\mathcal{A}(G)$ or $H \in \mathcal{S}$.*

Thus we may obtain a classification of the maximal subgroups of the alternating and symmetric groups by determining precisely when the subgroups H in $\mathcal{A}(G) \cup \mathcal{S}$ are maximal in A_nH. This problem is now completely solved:

Theorem 1.3.2 [L-P-S$_1$]. *Let G be A_n or S_n and let $H \in \mathcal{A}(G) \cup \mathcal{S}$. Then either H is maximal in A_nH, or $H < K < A_nH$, where (H, K, n) is given in an explicit list of triples.*

As far as the sporadic groups are concerned, the maximal subgroups of all but three of them are completely determined (the three exceptions are Fi'_{24}, B and M). Most of the lists of maximal subgroups can be found in [At] and a recent survey of the subject appears in [Wi$_1$].

For the exceptional groups of Lie type, we list in Table 1.3.B those groups G_o for which the maximal subgroups of G are known (where as usual $G_o \trianglelefteq G \leq Aut(G_o)$).

Table 1.3.B	
G_o	references
${}^2B_2(q)$	[Su$_1$]
${}^2G_2(q)$	[L-N,Kl$_3$]
$G_2(q)$	[As$_2$,Co$_2$,Kl$_3$,Mi]
${}^3D_4(q)$	[Kl$_4$]
${}^2F_4(2)'$	[Tc,Wi$_3$]
${}^2F_4(q)'$	[Ma]
$F_4(2)$	[N-W]
${}^2E_6(2)$	[No]
$E_6(2)$	[K-W$_3$]

In addition to the references appearing in Table 1.3.B, the maximal subgroups of $G_2(4)$ were obtained independently in [Bu,P-T,Wi$_2$], and those of ${}^3D_4(2)$ by Wilson, whose result appears in [At]. Recently Aschbacher [As$_3$,As$_4$,As$_5$,As$_6$,As$_7$] has obtained a wealth of information concerning the maximal subgroups of G when G_o is the simple group $E_6(q)$ and G does not induce a graph automorphism on G_o. Here, almost all the maximal subgroups have been determined, with the possibility of a few 'small' exceptions left open. These results have been extended to the case where G does induce a graph automorphism in [Kl$_5$], and this has led to a subgroup theorem for groups with socle ${}^2E_6(q)$.

As we stated in Theorem 1.2.5, it is a fact that all local maximal subgroups of the exceptional groups are known. Of course, each reference already mentioned in this section contributes to the proof of this fact. However, the complete list of the local maximal subgroups has been obtained in [C-L-S-S].

Finally, we remark that the recent paper [L-Se], following work of Borovik [Bor], reduces the study of maximal subgroups of the exceptional groups of Lie type to the study of almost simple maximal subgroups, and even handles these subgroups in some cases for large characteristics.

Chapter 2
BASIC PROPERTIES OF THE CLASSICAL GROUPS

§2.1 Introduction

In this chapter we construct the classical groups and describe their fundamental properties. We begin by introducing some basic definitions, terminology and notation.

Throughout, V will denote a vector space of finite dimension n over the field $\mathbf{F}$, where $\mathbf{F}$ is either a finite field or an algebraically closed field of characteristic p. We write $GL(V, \mathbf{F})$ for the *general linear group* of V over $\mathbf{F}$, which is the group of all non-singular $\mathbf{F}$-linear transformations of V. Also $SL(V, \mathbf{F})$ denotes the *special linear group* of V over $\mathbf{F}$, the group of elements in $GL(V, \mathbf{F})$ with determinant 1. Loosely speaking, the classical groups are the stabilizers in $GL(V, \mathbf{F})$ and $SL(V, \mathbf{F})$ of suitable forms on V, such as non-degenerate symmetric, skew-symmetric or sesquilinear forms. In the following sections we will provide a technical account of the classical groups. Other references for the classical groups include [Ar,As$_8$,Di$_1$,Ka$_2$].

We usually reserve the letter β for a basis of V. If $\beta = \{v_1, \ldots, v_n\}$ is such a basis, then each element of $GL(V, \mathbf{F})$ is determined by its action on β. If $g \in GL(V, \mathbf{F})$, then g_β will denote the $n \times n$ matrix which satisfies $v_i g = \sum_{j=1}^{n} (g_\beta)_{ij} v_j$ and if $X \subseteq GL(V, \mathbf{F})$ then $X_\beta = \{g_\beta \mid g \in X\}$. For $\lambda_i \in \mathbf{F}^*$ $(i = 1, \ldots, n)$, we denote by $\mathrm{diag}_\beta(\lambda_1, \ldots, \lambda_n)$ the *diagonal* linear transformation which satisfies $v_i \mathrm{diag}_\beta(\lambda_1, \ldots, \lambda_n) = \lambda_i v_i$. If the λ_i are all equal to λ say, then $\mathrm{diag}_\beta(\lambda, \ldots, \lambda)$ is called a *scalar linear transformation,* or simply a *scalar.* The centre of $GL(V, \mathbf{F})$ is the group of all non-zero scalars, which is isomorphic to $\mathbf{F}^*$. So with a slight abuse of notation we write $\mathbf{F}^* \leq GL(V, \mathbf{F})$, and the scalar $\mathrm{diag}_\beta(\lambda, \ldots, \lambda)$ is denoted simply by λ. We also write $PGL(V, \mathbf{F})$ for the *projective general linear group* $GL(V, \mathbf{F})/\mathbf{F}^*$. And if X is any subgroup of $GL(V, \mathbf{F})$, then we write PX for the corresponding projective group $X/X \cap \mathbf{F}^*$. Thus for example $PSL(V, \mathbf{F})$ is the *projective special linear group.* Along with P, the symbol $\overline{}$ will also denote reduction modulo scalars. Thus $\overline{GL(V, \mathbf{F})} = PGL(V, \mathbf{F})$, and if $g \in GL(V, \mathbf{F})$ then $\bar{g}$ denotes the image of g in $PGL(V, \mathbf{F})$.

If V' is another vector space over $\mathbf{F}$, then $\mathrm{Hom}_{\mathbf{F}}(V, V')$ denotes the set of $\mathbf{F}$-linear maps from V to V'. Moreover we put $\mathrm{End}_{\mathbf{F}}(V) = \mathrm{Hom}_{\mathbf{F}}(V, V)$. For any subset $X \subseteq \mathrm{End}_{\mathbf{F}}(V)$, we write $\mathrm{End}_{\mathbf{F}X}(V)$ for the subset of $\mathrm{End}_{\mathbf{F}}(V)$ commuting with all elements of X.

Semilinear transformations

A map g from V to V is called an $\mathbf{F}$-*semilinear transformation of* V if there is a field automorphism $\sigma(g) \in Aut(\mathbf{F})$ such that for all $v, w \in V$ and $\lambda \in \mathbf{F}$,

$$(v + w)g = vg + wg \quad \text{and} \quad (\lambda v)g = \lambda^{\sigma(g)}(vg). \tag{2.1.1}$$

If g is an $\mathbf{F}$-semilinear transformation, then g is *non-singular* if $\{v \in V \mid vg = 0\} = \{0\}$. Now define $\Gamma L(V, \mathbf{F})$ as the set of all non-singular $\mathbf{F}$-semilinear transformations of V. It is easy to verify that if $g, h \in \Gamma L(V, \mathbf{F})$, then their composition gh also lies in $\Gamma L(V, \mathbf{F})$ and $\sigma(gh) = \sigma(g)\sigma(h)$. Consequently $\Gamma L(V, \mathbf{F})$ forms a group, called the *general semilinear group* of V over $\mathbf{F}$, and the map σ from $\Gamma L(V, \mathbf{F})$ to $Aut(\mathbf{F})$ is a surjective homomorphism with kernel $GL(V, \mathbf{F})$.

Obviously $\mathbf{F}^* \trianglelefteq \Gamma L(V, \mathbf{F})$ and upon factoring out the scalars we obtain the *projective general semilinear group* $P\Gamma L(V, \mathbf{F}) = \Gamma L(V, \mathbf{F})/\mathbf{F}^*$. As before we may define PX and $\overline{X}$ for any subset X of $\Gamma L(V, \mathbf{F})$.

Let $\beta = \{v_1, \ldots, v_n\}$ be a basis for V over $\mathbf{F}$ as above. Just as each element of $GL(V, \mathbf{F})$ is determined by its action on β, so is each element $g \in \Gamma L(V, \mathbf{F})$ determined by its action on β along with $\sigma(g)$. If $\alpha \in Aut(\mathbf{F})$ then we define $\phi_\beta(\alpha)$ as the (unique) element of $\Gamma L(V, \mathbf{F})$ which lies in $\sigma^{-1}(\alpha)$ and which fixes each v_i. Thus

$$\left(\sum_{i=1}^{n} \lambda_i v_i\right) \phi_\beta(\alpha) = \sum_{i=1}^{n} \lambda_i^\alpha v_i. \tag{2.1.2}$$

Forms and isometry groups

A map $\mathbf{f} : V \times V \to \mathbf{F}$ is a *left-linear form* if for each $v \in V$, the map $V \to \mathbf{F}$ given by $x \mapsto \mathbf{f}(x, v)$ is a linear map. In other words, for all $x, y, v \in V$ and $\lambda \in \mathbf{F}$, we have

$$\mathbf{f}(x + y, v) = \mathbf{f}(x, v) + \mathbf{f}(y, v) \quad \text{and} \quad \mathbf{f}(\lambda x, v) = \lambda \mathbf{f}(x, v). \tag{2.1.3}$$

There is an analogous definition for *right-linear form*. A map which is both a left-linear and a right-linear form is called a *bilinear form*. For any map $Q : V \to \mathbf{F}$, define $\mathbf{f}_Q : V \times V \to \mathbf{F}$ by

$$\mathbf{f}_Q(v, w) = Q(v + w) - Q(v) - Q(w). \tag{2.1.4}$$

Then Q is called a *quadratic form* provided

$$\begin{gathered} Q(\lambda v) = \lambda^2 Q(v) \text{ for all } v \in V \text{ and } \lambda \in \mathbf{F} \\ \text{and} \\ \mathbf{f}_Q \text{ is a bilinear form.} \end{gathered} \tag{2.1.5}$$

When Q is a quadratic form, $\mathbf{f}_Q$ will be called the *associated bilinear form*.

If $\mathbf{f}$ is any map from $V \times V$ to $\mathbf{F}$ and if $\beta = \{v_1, \ldots, v_n\}$ is any basis of V, then we define $\mathbf{f}_\beta$ as the matrix satisfying

$$(\mathbf{f}_\beta)_{ij} = \mathbf{f}(v_i, v_j).$$

Furthermore, for $v \in V$ we call $\mathbf{f}(v, v)$ the $\mathbf{f}$-*norm of* v. In many circumstances we will omit the symbol $\mathbf{f}$ by writing simply (v, w) instead of $\mathbf{f}(v, w)$, and referring to the norm of v, rather than the $\mathbf{f}$-norm.

Often V will come equipped with a left-linear form $\mathbf{f}$ or quadratic form Q, and we sometimes write $(V, \mathbf{f})$ or (V, Q) instead of just V in order to emphasize this extra structure. Furthermore, we may write $(V, \mathbf{F}, \mathbf{f})$ or $(V, \mathbf{F}, Q)$ when we want to emphasize the role of the field $\mathbf{F}$, as well. To avoid treating the two structures $(V, \mathbf{f})$ and (V, Q) separately, we sometimes consider the structure (V, κ), where κ is *either* a left-linear form *or* a quadratic form. Thus κ is a map from $V^\ell = V \times \cdots \times V$ (ℓ times) to $\mathbf{F}$, where ℓ is 1 or 2. And in view of (2.1.3) and (2.1.5), we have

$$\kappa(\lambda v_1, \ldots, v_\ell) = \lambda^{3-\ell}\kappa(v_1, \ldots, v_\ell). \tag{2.1.6}$$

In the following discussion, $\mathbf{v} = (v_1, \ldots, v_\ell)$ denotes an ℓ-tuple in V^ℓ. And if $g \in \Gamma L(V, \mathbf{F})$ then $\mathbf{v}g = (v_1 g, \ldots, v_\ell g)$.

Assume that $(V, \mathbf{F}, \kappa)$ and $(V', \mathbf{F}, \kappa')$ are two spaces of dimension n over $\mathbf{F}$, where κ and κ' are either both left-linear or both quadratic forms. An *isometry* is an invertible element $g \in \mathrm{Hom}_{\mathbf{F}}(V, V')$ which satisfies

$$\kappa'(\mathbf{v}g) = \kappa(\mathbf{v})$$

for all $\mathbf{v} \in V^\ell$. If such an isometry exists, then $(V, \mathbf{F}, \kappa)$ and $(V', \mathbf{F}, \kappa')$ are said to be *isometric* and we write $(V, \mathbf{F}, \kappa) \cong (V', \mathbf{F}, \kappa')$. An invertible element $g \in \mathrm{Hom}_{\mathbf{F}}(V, V')$ is a *similarity* if there exists $\lambda \in \mathbf{F}^*$ such that

$$\kappa'(\mathbf{v}g) = \lambda\kappa(\mathbf{v})$$

for all $\mathbf{v} \in V^\ell$. If such a similarity exists, then $(V, \mathbf{F}, \kappa)$ and $(V', \mathbf{F}, \kappa')$ are said to be *similar*.

We are particularly interested in the case $(V, \mathbf{F}, \kappa) = (V', \mathbf{F}, \kappa')$. Here the isometries are called *κ-isometries*, and it is clear that the set of κ-isometries forms a subgroup of $GL(V, \mathbf{F})$. This is called the *κ-isometry group* and is denoted $I(V, \mathbf{F}, \kappa)$. A *special κ-isometry* is a κ-isometry lying in the special linear group (that is, an element of $I(V, \mathbf{F}, \kappa) \cap SL(V, \mathbf{F})$) and the group of special κ-isometries will be written $S(V, \mathbf{F}, \kappa)$. In the same fashion, a similarity will be referred to as a *κ-similarity*, and the set of κ-similarities forms the similarity group, denoted $\Delta(V, \mathbf{F}, \kappa)$. A *$\kappa$-semisimilarity* is an element $g \in \Gamma L(V, \mathbf{F})$ which both multiplies κ by a scalar and induces a field automorphism on κ — that is, there exist $\lambda \in \mathbf{F}^*$ and $\alpha \in Aut(\mathbf{F})$ such that

$$\kappa(\mathbf{v}g) = \lambda\kappa(\mathbf{v})^\alpha \tag{2.1.7}$$

for all $\mathbf{v} \in V^\ell$. It is easy to check that the set of κ-semisimilarlities forms a group, which we denote by $\Gamma(V, \mathbf{F}, \kappa)$. Clearly $\Delta(V, \mathbf{F}, \kappa) \leq \Gamma(V, \mathbf{F}, \kappa)$. The following Lemma is a direct consequence of the definitions.

Lemma 2.1.1. *If $(V, \mathbf{F}, \kappa)$ and $(V', \mathbf{F}, \kappa')$ are similar, then $X(V, \mathbf{F}, \kappa) \cong X(V', \mathbf{F}, \kappa')$ for $X \in \{S, I, \Delta, \Gamma\}$.*

Lemma 2.1.2. *Assume that κ is surjective and take $g \in \Gamma(V, \mathbf{F}, \kappa)$ satisfying (2.1.7).*

(i) *The field element λ appearing in (2.1.7) is uniquely determined by g. Thus there is a well-defined map $\tau : \Gamma(V, \mathbf{F}, \kappa) \to \mathbf{F}^*$ satisfying $\tau(g) = \lambda$.*

(ii) *The restriction of τ to $\Delta(V, \mathbf{F}, \kappa)$ is a homomorphism to $\mathbf{F}^*$ with kernel $I(V, \mathbf{F}, \kappa)$.*

(iii) *The field automorphism α appearing in (2.1.7) is also uniquely determined by g, and in fact $\alpha = \sigma(g)$, where σ is given in (2.1.1).*

(iv) *The restriction of σ to $\Gamma(V, \mathbf{F}, \kappa)$ is a homomorphism to $Aut(\mathbf{F})$ with kernel $\Delta(V, \mathbf{F}, \kappa)$, and hence $\Gamma(V, \mathbf{F}, \kappa) \cap GL(V, \mathbf{F}) = \Delta(V, \mathbf{F}, \kappa)$.*

Proof. (i) Suppose that $\lambda_1 \in \mathbf{F}$ and $\alpha_1 \in Aut(\mathbf{F})$ also satisfy

$$\kappa(\mathbf{v}g) = \lambda_1 \kappa(\mathbf{v})^{\alpha_1} \tag{2.1.8}$$

for all $\mathbf{v} \in V^\ell$. Since κ is surjective, we can choose $\mathbf{w} = (w_1, \ldots, w_\ell) \in V^\ell$ so that $\kappa(\mathbf{w}) = 1$. Consequently (2.1.7) and (2.1.8) imply $\lambda_1 = \lambda$, which shows λ is uniquely determined by g.

(ii) This is an easy calculation.

(iii) Let μ be a generator for $\mathbf{F}^*$ and let $\mathbf{w}$ be as in the proof of (i). Then on the one hand we have

$$\kappa((\mu w_1)g, w_2 g, \ldots, w_\ell g) = \kappa(\mu^{\sigma(g)}(w_1 g), w_2 g, \ldots, w_\ell g) = (\mu^{\sigma(g)})^{3-\ell}\lambda,$$

by (2.1.6). On the other hand,

$$\kappa((\mu w_1)g, w_2 g, \ldots, w_\ell g) = \lambda \kappa(\mu w_1, w_2, \ldots, w_\ell)^\alpha = \lambda(\mu^{3-\ell})^\alpha.$$

Thus $(\mu^{3-\ell})^\alpha = (\mu^{3-\ell})^{\sigma(g)}$, and so α and $\sigma(g)$ agree on $(\mathbf{F}^*)^2$, the set of squares in $\mathbf{F}^*$. Since $\mathbf{F}$ is either finite or algebraically closed, it follows that $\alpha = \sigma(g)$, as desired.

(iv) If $g \in \ker(\sigma)$, then $g \in GL(V, \mathbf{F})$ by an earlier remark. Moreover $\alpha = \sigma(g) = 1$, and so $g \in \Delta(V, \mathbf{F}, \kappa)$. Therefore $\ker(\sigma) = \Delta(V, \mathbf{F}, \kappa)$. ■

Remark. Although the restriction of τ to $\Delta(V, \mathbf{F}, \kappa)$ is a homomorphism, τ itself is not, in general. In fact, an easy calculation shows $\tau(g_1 g_2) = \tau(g_1)\tau(g_2)^{\sigma(g_1)}$ for all $g_1, g_2 \in \Gamma(V, \mathbf{F}, \kappa)$. Hence τ induces a homomorphism from $\Gamma(V, \mathbf{F}, \kappa)$ to $\mathbf{F}^*/(\mathbf{F}^*)^2$.

As a matter of convenience we shall write

$$X = X(V, \mathbf{F}, \kappa),$$

where X ranges over the symbols S, I, Δ, and Γ. Thus we obtain a chain of groups

$$S \leq I \leq \Delta \leq \Gamma. \tag{2.1.9}$$

Sometimes we need to express the dependence of τ on V, $\mathbf{F}$ or κ by writing $\tau = \tau_{(V,\mathbf{F},\kappa)}$, $\tau_{(V,\kappa)}$, τ_V or τ_κ. We use similar notation for σ.

In the situations that we shall encounter, either κ will be surjective or κ will be identically 0. In the former case, Lemma 2.1.2.i provides a homomorphism τ from Δ to $\mathbf{F}^*$, and in the latter case we shall define τ to be the trivial homomorphism from Δ to $\mathbf{F}^*$. Thus in both cases we have $I = \ker_\Delta(\tau)$. We now summarize some of the notation we have accumulated so far in the table below.

Table 2.1.A

Γ	$=$		$\{g \in \Gamma L(V,\mathbf{F}) \mid \kappa(\mathbf{v}g) = \tau(g)\kappa(\mathbf{v})^{\sigma(g)}$ for all $\mathbf{v} \in V^\ell\}$
Δ	$=$	$\ker_\Gamma(\sigma)$	$= \{g \in GL(V,\mathbf{F}) \mid \kappa(\mathbf{v}g) = \tau(g)\kappa(\mathbf{v})$ for all $\mathbf{v} \in V^\ell\}$
I	$=$	$\ker_\Delta(\tau)$	$= \{g \in GL(V,\mathbf{F}) \mid \kappa(\mathbf{v}g) = \kappa(\mathbf{v})$ for all $\mathbf{v} \in V^\ell\}$
S	$=$	$\ker_I(\det)$	$= \{g \in SL(V,\mathbf{F}) \mid \kappa(\mathbf{v}g) = \kappa(\mathbf{v})$ for all $\mathbf{v} \in V^\ell\}$

Special forms

Before we discuss the classical groups, we need to make some more definitions concerning forms. As before, assume that $\mathbf{f}$ is a map from $V \times V$ to $\mathbf{F}$. Then $\mathbf{f}$ is called *non-degenerate* if for each $v \in V\backslash 0$, the maps $V \to \mathbf{F}$ given by $x \mapsto \mathbf{f}(x,v)$ and $x \mapsto \mathbf{f}(v,x)$ are non-zero. A quadratic form Q is called *non-degenerate* if its associated bilinear form $\mathbf{f}_Q$ is non-degenerate. The map $\mathbf{f}$ is *symmetric* if

$$\mathbf{f}(v,w) = \mathbf{f}(w,v) \quad \text{for all } v,w \in V. \tag{2.1.10}$$

Note that $\mathbf{f}_Q$ is symmetric when Q is a quadratic form. Also $\mathbf{f}$ is *skew-symmetric* if

$$\mathbf{f}(v,w) = -\mathbf{f}(w,v) \quad \text{for all } v,w \in V. \tag{2.1.11}$$

When $\mathbf{F}$ has even characteristic, symmetric forms and skew-symmetric forms are one and the same. Also notice that in odd characteristic, skew-symmetric forms satisfy $\mathbf{f}(v,v) = 0$ for all $v \in V$. However this need not be the case in even characteristic. We define $\mathbf{f}$ to be *symplectic* if

$$\mathbf{f} \text{ is skew-symmetric, bilinear, and } \mathbf{f}(v,v) = 0 \text{ for all } v \in V. \tag{2.1.12}$$

Finally, $\mathbf{f}$ is said to be *unitary* if $\mathbf{F}$ has an involutory field automorphism α and

$$\mathbf{f} \text{ is left-linear and } \mathbf{f}(v,w) = \mathbf{f}(w,v)^\alpha \text{ for all } v,w \in V. \tag{2.1.13}$$

Definitions of the classical groups

We are now in a position to define the finite classical groups. Let $\mathbf{F}$ be a finite field of characteristic p, and let be κ a left-linear form $\mathbf{f}$ or a quadratic form Q appearing in one of the four cases below:

case **L**	κ is identically 0
case **S**	$\kappa = \mathbf{f}$, a non-degenerate symplectic form
case **O**	$\kappa = Q$, a non-degenerate quadratic form
case **U**	$\kappa = \mathbf{f}$, a non-degenerate unitary form

For each of these four cases we obtain the sequence of groups appearing in (2.1.9), and we define further groups $A = A(V, \mathbf{F}, \kappa)$ and $\Omega = \Omega(V, \mathbf{F}, \kappa)$ as follows. In case **L** with $n \geq 3$, the group $S = SL(V, \mathbf{F})$ possesses an inverse-transpose automorphism ι (see (2.2.4) below). And in case **O**, the group S contains a certain subgroup of index 2 (see §2.5). We define

$$A = \begin{cases} \Gamma\langle\iota\rangle & \text{in case } \mathbf{L} \text{ with } n \geq 3 \\ \Gamma & \text{otherwise} \end{cases}$$
$$\Omega = \begin{cases} \text{this subgroup of index 2 in } S & \text{in case } \mathbf{O} \\ S & \text{otherwise.} \end{cases} \tag{2.1.14}$$

Thus we obtain a chain of groups

$$\Omega \leq S \leq I \leq \Delta \leq \Gamma \leq A. \tag{2.1.15}$$

This chain is always A-invariant, that is, each group is normalized by A. We will also have $\mathbf{F}^* \trianglelefteq A$, and so as before we write P and $\overline{}$ to denote reduction modulo scalars for subsets of A. Thus there is a projective version of the chain in (2.1.15).

Remark. There is a slight difference between our notation here and that in [As$_1$]. We write I for the isometry group, whereas Aschbacher writes O; and Aschbacher defines $\Omega = O'$, which differs from our Ω in a few small cases.

We now define a *(finite) classical group* to be any group G satsifying

$$\Omega \leq G \leq A \quad \text{or} \quad \overline{\Omega} \leq G \leq \overline{A} \tag{2.1.16}$$

in one of the four cases **L**, **S**, **O** or **U**. If G is such a classical group, then G will be called a *linear group, symplectic group, orthogonal group* or *unitary group,* in the cases **L**, **S**, **O** or **U**, respectively. Apart from a few special cases, the groups $\overline{\Omega}$ are in fact non-abelian simple (see Theorem 2.1.3, below), and they comprise the *(finite) classical simple groups.* Thus the groups $\overline{\Omega}$ may be identified with the groups G_o discussed in §1.2. Furthermore, with only a few exceptions (see Theorem 2.1.4, below), $\overline{A} \cong Aut(\overline{\Omega})$, and so $\overline{A}$ may be identified with $Aut(G_o)$.

The goal in the rest of this chapter is to give a precise description of the chain in (2.1.15) and the corresponding projective chain for each of the four cases. We begin this description in Table 2.1.B, where we explain our notation and terminology for the relevant groups. In the first column we indicate that cases **L** and **U** will sometimes be referred to as cases $\mathbf{L}^+$ and $\mathbf{L}^-$, respectively. Moreover, it turns out that case **O** actually trifurcates into three subcases, which we call cases $\mathbf{O}^\circ$, $\mathbf{O}^+$ and $\mathbf{O}^-$ (see Proposition 2.5.3). These subcases will be described explicitly in §§2.6-2.8. The notation and terminology for the

Table 2.1.B

case	X	notation	terminology
$\mathbf{L} = \mathbf{L}^+$	$\Omega = S$	$SL_n(q) = SL_n^+(q)$	linear groups
	$I = \Delta$	$GL_n(q) = GL_n^+(q)$	
	Γ	$\Gamma L_n(q)$	
	$A = \Gamma\langle\iota\rangle$		
$\mathbf{U} = \mathbf{L}^-$	$\Omega = S$	$SU_n(q) = SL_n^-(q)$	unitary groups
	I	$GU_n(q) = GL_n^-(q)$	
	Δ		
	$A = \Gamma$	$\Gamma U_n(q)$	
$\mathbf{S}$	$\Omega = S = I$	$Sp_n(q)$	symplectic groups
	Δ	$GSp_n(q)$	
	$\Gamma = A$	$\Gamma Sp_n(q)$	
$\mathbf{O}^\epsilon$, $\epsilon \in \{\pm, \circ\}$	Ω	$\Omega_n^\epsilon(q)$ *	orthogonal groups
	S	$SO_n^\epsilon(q)$	
	I	$O_n^\epsilon(q)$	
	Δ	$GO_n^\epsilon(q)$	
	$\Gamma = A$	$\Gamma O_n^\epsilon(q)$	

* In case $\mathbf{O}^\circ$ we actually write $Y_n(q)$ instead of $Y_n^\circ(q)$ for $Y \in \{\Omega, SO, O, GO, \Gamma O\}$.

relevant groups is for the most part standard, except for the signs $\pm$ involved in GL_n and SL_n. This notation serves as a convenience in later arguments. Our convention for the field $\mathbf{F}$ is

$$\mathbf{F} = \mathbf{F}_{q^u}, \text{ where } q = p^f \text{ and } u = \begin{cases} 1 & \text{in cases } \mathbf{L}, \mathbf{S} \text{ and } \mathbf{O} \\ 2 & \text{in case } \mathbf{U}. \end{cases} \tag{2.1.17}$$

Also, we reserve the letter α to denote the automorphism of $\mathbf{F}$ given by

$$\lambda^\alpha = \lambda^q \quad (\lambda \in \mathbf{F}), \tag{2.1.18}$$

so that α has order u in $Aut(\mathbf{F})$.

Observe that we have not as yet proved the uniqueness of the various groups in Table 2.1.B. However uniqueness follows from results in the ensuing sections. For example, in case $\mathbf{U}$, Proposition 2.3.1 shows that every non-degenerate unitary form on V is isometric to $\mathbf{f}$, and so by Lemma 2.1.1 we may write $GU_n(q)$ for 'the' unitary group $I(V, \mathbf{F}, \mathbf{f})$.

To denote projective groups, we precede the symbol appearing in the third column by the symbol P. Thus for example $PGSp_n(q)$ denotes the group $GSp_n(q)$ modulo scalars. As an abbreviation however, we write

$$L_n^\pm(q) = PSL_n^\pm(q) = \begin{cases} L_n(q) & \text{if the sign is } + \\ U_n(q) & \text{if the sign is } -. \end{cases} \tag{2.1.19}$$

At this stage we have enough notation to assert the main theorems concerning simplicity and automorphism groups of the classical groups. Assume that Ω, A and the corresponding projective groups $\overline{\Omega}$ and $\overline{A}$ are as above.

Theorem 2.1.3. *Assume that $n = \dim_{\mathbf{F}}(V)$ is at least 2, 3, 4, 7 in cases* **L, U, S** *and* **O**, *respectively. Then $\overline{\Omega}$ is non-abelian simple, except for $L_2(2)$, $L_2(3)$, $U_3(2)$ and $Sp_4(2)$.*

Note that the conditions on n imposed in Theorem 2.1.3 do not exclude any classical groups, in view of the isomorphisms occurring in Proposition 2.9.1.i-vii, below.

A proof of Theorem 2.1.3 can be found in [Ar] and [Di$_1$], although it was known as far back as [Dic$_1$]. More detailed information about the structure of the classical groups will be given in §2.9. As for the automorphism groups of the classical simple groups, we state the following.

Theorem 2.1.4. *Assume that $\overline{\Omega}$ is simple and that n is as in Theorem 2.1.3. Then $\overline{A} = Aut(\overline{\Omega})$, except when $\Omega = Sp_4(q)$ with q even and when $\Omega = \Omega_8^+(q)$.*

Dieudonne [Di$_2$] devotes an entire memoir to the proof of this Theorem. More modern proofs, using the Lie theory, can be found in [Ca$_1$, Ch. 12] and [St$_1$, Chs.10,11]. When $\Omega = Sp_4(q)$ with q even, or $\Omega = \dot{\Omega}_8^+(q)$, then $\overline{A}$ has index 2 or 3 in $Aut(\overline{\Omega})$, respectively. These 'extra' automorphisms in $Aut(\overline{\Omega})\backslash\overline{A}$ are the graph automorphisms mentioned after Theorem 1.2.1.

Some goemetry with the classical groups

Assume here that $\mathbf{F}$, κ, $\mathbf{f}$, Q, σ and τ are as in the previous discussion, and let $X = X(V, \mathbf{F}, \kappa)$, where X ranges over the symbols in (2.1.15). We shall refer to $(V, \mathbf{F}, \kappa)$ as a *classical geometry.* More specifically, we call $(V, \mathbf{F}, \kappa)$ a linear, symplectic, orthogonal or unitary geometry in the respective cases **L**, **S**, **O** or **U**. Now let W be an m-dimensional subspace of V. Then we obtain a sub-geometry $(W, \mathbf{F}, \kappa_W)$, where κ_W denotes the restriction of κ to W (this is a minor abuse of terminology, for strictly speaking, κ_W is the restriction to $W \times W$ in cases **L**, **U** and **S**). Sometimes we drop the subscript W and simply write $(W, \mathbf{F}, \kappa)$ or just (W, κ) when there is no danger of ambiguity. Observe that $(W, \mathbf{F}, \kappa_W)$ may not be a classical geometry; however we will usually be interested in the cases when it is. Indeed, we will be primarily concerned with the cases where W is *non-degenerate*, which means κ_W is non-degenerate, and where W is *totally singular*, which means $\kappa_W = 0$. If case **S**, **O** or **U** holds and W is non-degenerate, then clearly $(W, \mathbf{F}, \kappa)$ is a symplectic, orthogonal or unitary geometry, respectively. To this geometry are associated the functions $\tau_W = \tau_{(W,\mathbf{F},\kappa)}$ and $\sigma_W = \sigma_{(W,\mathbf{F},\kappa)}$, and the relationships amongst τ, σ, τ_W and σ_W are considered in Lemma 2.1.9, below. If W is totally singular then of course $(W, \mathbf{F}, \kappa)$ is a linear geometry. We define $\mathcal{U}_m$ to be the set of totally singular m-subspaces of V. We also define W to be *totally isotropic* if $\mathbf{f}$ vanishes on W. In cases **L**, **U** and **S** the terms totally isotropic and totally singular coincide, since $\kappa = \mathbf{f}$. In case **O**, we see that if W is totally singular then W is isotropic; however, the converse does not hold in characteristic 2. More will be said about this later, in §2.5. For the rest of this discussion we shall dispense with reference to the field $\mathbf{F}$, so that (V, κ) is to be understood as $(V, \mathbf{F}, \kappa)$.

Now let v be a vector V. In cases **L**, **S** and **U** we say that v is *singular* or *isotropic* if $\mathbf{f}(v, v) = 0$. Thus all vectors are singular and isotropic in cases **L** and **S**. In case **O** however, we distinguish between the terms singular and isotropic — namely, we say that v is isotropic if $\mathbf{f}(v, v) = 0$ and that v is singular if $Q(v) = 0$. It follows immediately from (2.1.4) that every singular vector is isotropic. However when q is even, every vector is isotropic while some vectors are not singular. In all three cases, vectors which are not singular will be called *non-singular*, and those which are not isotropic will be called *non-isotropic*.

For any subset J of V define

$$J^{\perp} = \{v \in V \mid \mathbf{f}(v, x) = 0 \text{ for all } x \in J\}.$$

Obviously when case **L** holds we have $J^{\perp} = V$ for all J. Two subsets J and K are said to be *orthogonal* if $\mathbf{f}(x, y) = 0$ for all $x \in J$ and $y \in K$, or in other words $J \subseteq K^{\perp}$ or $K \subseteq J^{\perp}$. If U and W are subspaces of V such that $U + W = U \oplus W$ and U is orthogonal to W, then we write $U + W = U \perp W$. More generally, we write $W_1 \perp \cdots \perp W_t$ if the spaces W_i are mutually orthogonal and their sum is direct. Here we collect a few easy facts.

Lemma 2.1.5. *Assume that κ is non-degenerate, W is a subspace of V and J a subset of V.*

(i) $J^{\perp} = \langle J \rangle^{\perp}$.
(ii) $\dim(W) + \dim(W^{\perp}) = \dim(V)$.
(iii) $(W^{\perp})^{\perp} = W$.
(iv) *W is totally isotropic if and only if $W \leq W^{\perp}$.*
(v) *W is non-degenerate if and only if $V = W \perp W^{\perp}$.*

Proof. Exercise. ■

Assume in this paragraph that κ is non-degenerate and that W is a non-degenerate subspace of V. Thus $V = W \perp W^{\perp}$, and observe that an element of the isometry group $I(W, \kappa)$ on W extends naturally to an element of $I = I(V, \kappa)$ merely by taking it to fix every vector in $W^{\perp}$. Thus we can regard $I(W, \kappa)$ as a subgroup of I. When there is no danger of ambiguity, we sometimes write $I(W)$ to denote this subgroup of I. We obtain a similar inclusion $S(W) \leq S$, since an element having determinant 1 on W extends to an element of determinant 1 on V. Thus in cases **U** and **S**, we also have $\Omega(U) \leq \Omega$, for in these cases $\Omega = S$. Although we have not as yet defined the group Ω in case **O**, it turns out that $\Omega(W) \leq \Omega$ here, too; this fact will be demonstrated in Lemma 4.1.1, below. In general, there is no analogous way of embedding $\Delta(W, \kappa)$ in Δ, or $\Gamma(W, \kappa)$ in Γ.

Assume in this paragraph that κ is non-degenerate and that W is a totally singular subspace of V. In particular, W is totally isotropic, so $W \leq W^{\perp}$. Thus we may form the quotient space $U = W^{\perp}/W$. As in the proof of Lemma 2.1.2, let $\ell = 1$ in cases **U** and **S** and $\ell = 2$ in case **O**, and write $\mathbf{v}$ for a k-tuple $(v_1, \ldots, v_\ell) \in V^{\ell}$. Observe that if $\mathbf{v} \in (W^\perp)^\ell$ and $\mathbf{w} \in W^\ell$, then $\kappa(\mathbf{v}) = \kappa(\mathbf{v}+\mathbf{w})$. Consequently, $\kappa_{W^\perp}$ induces a well-defined map κ_U on U^ℓ given by

$$\kappa_U(\mathbf{v} + W^\ell) = \kappa(\mathbf{v}). \tag{2.1.20}$$

It follows immediately from the definitions that κ_U is a unitary, symplectic or quadratic form according to whether κ is unitary, symplectic or quadratic. Moreover, when κ is quadratic, we put $\mathbf{f}_U = \mathbf{f}_{\kappa_U}$ and check that $\mathbf{f}_U(v_1 + W, v_2 + W) = \mathbf{f}(v_1, v_2)$. In all three cases, it is evident that κ_U is non-degenerate; for if $v \in W^\perp \backslash W$, then Lemma 2.1.5.iii implies that there exists $x \in W^\perp$ such that $\mathbf{f}(v,x) \neq 0$, and hence $\mathbf{f}_U(v+W, x+W) \neq 0$. This non-degenerate form κ_U is called the *induced form on* U and we say that κ *induces* κ_U *on* U. The geometry (U, κ_U) has the associated functions τ_U and σ_U, and their relationship to τ and σ appears in Lemma 2.1.9, below.

One of the fundamental results concerning the classical geometries is Witt's Lemma. A proof can be found in [As_8, §20].

Proposition 2.1.6 (Witt's Lemma). *Assume that (V_1, κ_1), (V_2, κ_2) are two isometric classical geometries and that W_i is a subspace of V_i for $i = 1, 2$. Further assume that there is an isometry g from (W_1, κ_1) to (W_2, κ_2). Then g extends to an isometry from (V_1, κ_1) to (V_2, κ_2).*

A particular consequence of Witt's Lemma is the following.

Corollary 2.1.7. *All maximal totally singular subspaces of (V, κ) have the same dimension. And if κ is non-degenerate, this dimension is at most $\frac{n}{2}$.*

The upper bound $\frac{n}{2}$ in Corollary 2.1.7 is supplied by Lemma 2.1.5.iii.

The next Lemma is helpful in calculating with matrices in the classical groups. Recall that if $g \in GL(V)$ and $\beta = \{v_1, \ldots, v_n\}$ is a basis of V, then g_β denotes the matrix of g with respect to β and $\mathbf{f}_\beta$ denotes the matrix whose $(i,j)^{\text{th}}$ entry is $\mathbf{f}(v_i, v_j)$. Also recall the map τ introduced in Lemma 2.1.2. In part (ii) of the following Lemma, $\tau^{-1}(\lambda)$ denotes the set $\{g \in \Gamma(V,\kappa) \mid \tau(g) = \lambda\}$. The proof of the Lemma is left as an exercise.

Lemma 2.1.8. *Assume that κ is non-degenerate and let $\alpha \in Aut(\mathbf{F})$ be as in (2.1.18). Also let β be any basis of V.*

(i) *$I(V, \mathbf{f}) = \{g \in GL(V) \mid g_\beta \mathbf{f}_\beta g_\beta^{\alpha t} = \mathbf{f}_\beta\}$. Thus if $\alpha = 1$ then $\det(I) \leq \langle -1 \rangle$.*

(ii) *For $\lambda \in \mathbf{F}^*$, we have $\tau^{-1}(\lambda) \cap GL(V) = \{g \in GL(V) \mid g_\beta \mathbf{f}_\beta g_\beta^{\alpha t} = \lambda \mathbf{f}_\beta\}$.*

(iii) *If case* **O** *holds, so that $\kappa = Q$ is a quadratic form, then $I(V,Q) = \{g \in I(V, \mathbf{f}_Q) \mid Q(vg) = Q(v)$ for all $v \in \beta\}$.*

If $g \in \Gamma L(V)$ and W is a g-invariant section of V, then g_W denotes the image of g in $\Gamma L(W)$.

Lemma 2.1.9. *Assume that $W \leq V$.*

(i) *If W is non-degenerate, and $g \in \Gamma_W$, then $g_W \in \Gamma(W, \kappa_W)$, $\tau(g) = \tau_W(g_W)$ and $\sigma(g) = \sigma_W(g_W)$.*

(ii) *Assume that W is non-degenerate and that $g \in \Gamma L(V)$ fixes both W and $W^\perp$. Suppose further that $g_W \in \Gamma(W, \kappa_W)$ and $g_{W^\perp} \in \Gamma(W^\perp, \kappa_{W^\perp})$. Then $g \in \Gamma$ if and only if $\tau_W(g_W) = \tau_{W^\perp}(g_{W^\perp})$.*

(iii) *Assume that W is totally singular and that $g \in \Gamma_W$. Let $U = W^\perp/W$. Then $g_U \in \Gamma(U, \kappa_U)$, $\tau(g) = \tau_U(g_U)$ and $\sigma(g) = \sigma_U(g_U)$.*

Proof. We prove only (i). Parts (ii) and (iii) are just as easy, and we leave them to the reader. For all $\mathbf{v} \in W^\ell$ (here $\ell = 2$ in cases **U** and **S**, and $\ell = 1$ in case **O**) we have $\kappa_W(\mathbf{v}g_W) = \kappa(\mathbf{v}g) = \tau(g)\kappa(\mathbf{v})^{\sigma(g)}$. Thus $g_W \in \Gamma(W, \kappa_W)$ (by definition), and the equalities follow from the uniqueness assertions in Lemma 2.1.2. ∎

Orders of the classical groups

We now provide further information about the groups in (2.1.15) in the four cases. In Table 2.1.C we give the order of I and that of $I \cap \mathbf{F}^*$ and in Table 2.1.D we give the indices relating the groups in (2.1.15) and the corresponding projective groups. Thus the orders of all the groups in (2.1.15) and the corresponding projective groups can be read off from these two tables. The information given in these tables will be proved in the remainder of this chapter. Recall the definition of the integer f given in (2.1.17).

Table 2.1.C

case	$\lvert I\rvert$	$\lvert S:\Omega\rvert$	$\lvert I:S\rvert$	$\lvert\Delta:I\rvert$	$\lvert\Gamma:\Delta\rvert$	$\lvert A:\Gamma\rvert$
L	$q^{n(n-1)/2}\prod_{i=1}^{n}(q^i-1)$	1	$q-1$	1	f	2^*
U	$q^{n(n-1)/2}\prod_{i=1}^{n}(q^i-(-1)^i)$	1	$q+1$	$q-1$	$2f$	1
S	$q^{n^2/4}\prod_{i=1}^{n/2}(q^{2i}-1)$	1	1	$q-1$	f	1
O$^\circ$	$2q^{(n-1)^2/4}\prod_{i=1}^{(n-1)/2}(q^{2i}-1)$	$2^\dagger$	2	$\frac{1}{2}(q-1)$	f	1
O$^\pm$	$2q^{n(n-2)/4}(q^{n/2}\mp 1)\prod_{i=1}^{n/2-1}(q^{2i}-1)$	2	$(2,q-1)$	$q-1$	f	1

Table 2.1.D

case	$\lvert I\cap\mathbf{F}^*\rvert$	$\lvert\overline{S}:\overline{\Omega}\rvert$	$\lvert\overline{I}:\overline{S}\rvert$	$\lvert\overline{\Delta}:\overline{I}\rvert$	$\lvert\overline{\Gamma}:\overline{\Delta}\rvert$	$\lvert\overline{A}:\overline{\Gamma}\rvert$
L	$q-1$	1	$(q-1,n)$	1	f	2^*
U	$q+1$	1	$(q+1,n)$	1	$2f$	1
S	$(2,q-1)$	1	1	$(2,q-1)$	f	1
O$^\circ$	$2^\dagger$	2	1	1	f	1
O$^\pm$	$(2,q-1)$	$a_\pm^{**}$	$(2,q-1)$	$(2,q-1)$	f	1

* When $n = 2$ we have $A = \Gamma$, and so the 2 should be replaced by 1.

** Here $a_\pm \in \{1,2\}$, $a_+a_- = 2^{(2,q)}$, and when q is odd, $a_+ = 2$ if and only if $\frac{1}{4}n(q-1)$ is even.

† When $n = 1$ we have $S = \Omega$.

Remark. We have $\mathbf{F}^* \leq \Delta$ in all cases; indeed, for all $\lambda \in \mathbf{F}^*$ we have

$$\tau(\lambda) = \begin{cases} 1 & \text{in case } \mathbf{L} \\ \lambda^2 & \text{in cases } \mathbf{S} \text{ and } \mathbf{O} \\ \lambda^{1+q} & \text{in case } \mathbf{U}. \end{cases} \tag{2.1.21}$$

If X is Ω or S, then $|X \cap \mathbf{F}^*|$ can be read off from these two tables, for

$$|X \cap \mathbf{F}^*| = \frac{|I \cap \mathbf{F}^*||\overline{I} : \overline{X}|}{|I : X|}. \tag{2.1.22}$$

Most of our discussion involves finding generators for one group in the chain (2.1.15) modulo the preceding group. We write g_X for an element satisfying $\langle Y, g_X \rangle = X$, where Y is Ω, S, I, Δ or Γ, and X is S, I, Δ, Γ or A, respectively. We construct suitable elements g_X to exhibit generators and relations for $Out(\overline{\Omega}) = Aut(\overline{\Omega})/\overline{\Omega}$. For the rest of this chapter we will assume that

$$\langle \mu \rangle = \mathbf{F}^* \quad \text{and} \quad \langle \nu \rangle = Aut(\mathbf{F}). \tag{2.1.23}$$

Furthermore, we introduce the notation $\ddot{\ }$ to denote reduction modulo $\Omega\mathbf{F}^*$ in A. Thus along with the chain in (2.1.15) and the corresponding projective chain, we obtain

$$1 = \ddot{\Omega} \leq \ddot{S} \leq \ddot{I} \leq \ddot{\Delta} \leq \ddot{\Gamma} \leq \ddot{A}. \tag{2.1.24}$$

A fair amount of the following discussion is devoted to finding the precise structure of $\ddot{A}$. We do so in terms of the elements $\ddot{g}_X$, where g_X are the generators mentioned above. In general, $\ddot{A}$ is the outer automorphism group of $\overline{\Omega}$ (see Theorem 2.1.4), and our study of conjugacy in part (B) of the Main Theorem (see Theorem 1.2.3 and §3.1) depends upon detailed information concerning $\ddot{A}$. We consider the four cases **L**, **U**, **S** and **O** in the remaining sections of this chapter.

We conclude this section with a brief discussion of the map τ. We have already seen in the Remark after Lemma 2.1.2 that τ induces a homomorphism from Γ to $\mathbf{F}^*/(\mathbf{F}^*)^2$. Observe that (2.1.21) shows that $\tau(\mathbf{F}^*) \leq (\mathbf{F}^*)^2$, and hence τ induces a homomorphism $\overline{\tau}$ from $\overline{\Gamma}$ to $\mathbf{F}^*/(\mathbf{F}^*)^2$. We also obtain a homomorphism $\ddot{\tau}$ from $\ddot{\Gamma}$ to $\mathbf{F}^*/(\mathbf{F}^*)^2$. This notation will be used throughout the book.

§2.2 The linear groups

Here we study case **L**, so that $\kappa = \mathbf{f}$ is identically zero, and

$$\Gamma = \Gamma L(V, \mathbf{F}) \cong \Gamma L_n(q)$$
$$I = \Delta = GL(V, \mathbf{F}) \cong GL_n(q)$$
$$\Omega = S = SL(V, \mathbf{F}) \cong SL_n(q).$$

Proposition 2.2.1. *We have* $|I| = q^{n(n-1)/2}\prod_{i=1}^{n}(q^i - 1)$.

Proof. Clearly $|I|$ is the number of ordered bases $(v_1, \dots, v_n)$ of V. There are $q^n - 1$ choices for v_1, then $q^n - q$ choices for v_2, and so on. The result follows. ∎

Fix a basis β of V and for $\lambda \in \mathbf{F}^*$ define

$$\delta_\beta(\lambda) = \mathrm{diag}_\beta(\lambda, 1, \dots, 1). \tag{2.2.1}$$

Also put $\delta = \delta_\beta(\mu)$ (recall μ is a generator for $\mathbf{F}^*$). Then clearly $\det(\delta) = \mu$, and hence

$$I = S{:}\langle\delta\rangle. \tag{2.2.2}$$

(Recall from page iii that $S\langle\delta\rangle$ denotes a semidirect product of S with $\langle\delta\rangle$.) Now define $\phi = \phi_\beta(\nu)$ (see (2.1.2) and (2.1.23)), so that

$$\Gamma = I{:}\langle\phi\rangle. \tag{2.2.3}$$

Finally, assume that $n \geq 3$ and let $\iota = \iota_\beta$ be the *inverse transpose* map with respect to β, which is the automorphism of I given by

$$(g^\iota)_\beta = g_\beta^{-1t} \quad (g \in I). \tag{2.2.4}$$

We may now extend ι to a map from Γ to itself by defining

$$(g\phi^j)^\iota = g^\iota\phi^j \quad (g \in I,\ j \in \mathbf{Z}).$$

It is easy to show that ι is an automorphism of Γ of order 2, and so we may form the split extension $A = \Gamma\langle\iota\rangle \cong \Gamma.2$. When $n \leq 2$ we put $A = \Gamma$, and so

$$A = \begin{cases} \Gamma{:}\langle\iota\rangle \cong \Gamma.2 & \text{if } n \geq 3 \\ \Gamma & \text{if } n \leq 2. \end{cases}$$

Summarizing, we have the following:

Proposition 2.2.2. *We may take our generators* g_X *as follows:*

g_S	g_I	g_Δ	g_Γ	g_A
1	δ	1	ϕ	ι if $n \geq 3$
				1 if $n \leq 2$

We now present the structure theorem for the group $\ddot{A}$.

Proposition 2.2.3. *Let* $n \geq 2$. *Then*

$$\ddot{A} = \begin{cases} \langle\ddot{\delta}, \ddot{\phi}, \ddot{\iota}\rangle \cong \mathbf{Z}_{(n,q-1)}{:}\mathbf{Z}_f{:}\mathbf{Z}_2 & \text{if } n \geq 3 \\ \langle\ddot{\delta}\rangle \times \langle\ddot{\phi}\rangle \cong \mathbf{Z}_{(2,q-1)} \times \mathbf{Z}_f & \text{if } n = 2, \end{cases}$$

subject to the relations $|\ddot{\delta}| = (n, q-1)$, $|\ddot{\phi}| = f$, $|\ddot{\iota}| = 2$, $[\ddot{\phi}, \ddot{\iota}] = 1$, $\ddot{\delta}^{\ddot{\phi}} = \ddot{\delta}^p$, *and* $\ddot{\delta}^{\ddot{\iota}} = \ddot{\delta}^{-1}$.

Proof. Observe that $\det(\mathbf{F}^*) = (\mathbf{F}^*)^n$, which has order $(q-1)/(q-1, n)$. Thus

$$|\ddot{\delta}| = |\ddot{I}| = |I : \Omega\mathbf{F}^*| = \frac{q-1}{|\det(\Omega\mathbf{F}^*)|} = \frac{q-1}{|\det(\mathbf{F}^*)|} = (q-1, n), \tag{2.2.5}$$

as claimed. Furthermore, notice that ϕ has order f modulo $GL(V, \mathbf{F})$ and so $f \mid |\ddot{\phi}|$. On the other hand $\phi^f = 1$, and so $|\ddot{\phi}| = f$. Finally, we check that $\delta^\phi = \delta_\beta(\mu^p) = \delta^p$ and $\delta^\iota = \delta_\beta(\mu^{-1}) = \delta^{-1}$, and that $[\phi, \iota] = 1$. Thus all statements in the Proposition now follow. ■

Note that all assertions concerning case **L** in Tables 2.1.C and 2.1.D have now been established. For example, $|\overline{I} : \overline{S}| = |\overline{I} : \overline{\Omega}| = |\ddot{I}| = |\ddot{\delta}| = (q-1, n)$ (see (2.2.5)). And by (2.2.3) and Proposition 2.2.2, we have $|\Gamma : \Delta| = |\overline{\Gamma} : \overline{\Delta}| = |\ddot{\phi}| = f$.

§2.3 The unitary groups

Assume here case **U** holds, so that $\mathbf{F} = \mathbf{F}_{q^2}$ and $\kappa = \mathbf{f}$ is a non-degenerate unitary form. As a convenience we write (v, w) instead of $\mathbf{f}(v, w)$ for $v, w \in V$. Also, recall from (2.1.18) that α denotes the involutory field automorphism of $\mathbf{F}$. We set $\mathrm{T} = \mathrm{T}^{\mathbf{F}}_{\mathbf{F}_q}$ (the trace map $\lambda \mapsto \lambda + \lambda^q$) and $\mathrm{N} = \mathrm{N}^{\mathbf{F}}_{\mathbf{F}_q}$ (the norm map $\lambda \mapsto \lambda\lambda^q$). Here is one of the fundamental results concerning unitary geometries.

Proposition 2.3.1. *The space* $(V, \mathbf{f})$ *has an orthonormal basis. Hence all unitary geometries of dimension* n *are isometric.*

Proof. First, we claim there exists $v \in V$ with $(v, v) \neq 0$. For $(\ ,\ )$ is non-degenerate and so there exist $x, y \in V$ such that $(x, y) \neq 0$. If both x and y are singular, then for each $\eta \in \mathbf{F}$, we have $(\eta x + y, \eta x + y) = \mathrm{T}(\eta(x, y))$, which is non-zero for a suitable choice of η. Thus such a v exists, and we put $v_1 = \lambda v$, where $\lambda \in \mathbf{F}$ satisfies $\mathrm{N}(\lambda) = (v, v)^{-1}$. Then $(v_1, v_1) = 1$ and $V = \langle v_1 \rangle \perp v_1^\perp$. The result now follows by induction on n. ■

The next Proposition provides us with an equally useful basis, which we call a *standard* or *unitary* basis.

Proposition 2.3.2. *The space* $(V, \mathbf{f})$ *has a basis*

$$\begin{cases} \{e_1, \ldots, e_m, f_1, \ldots, f_m\} & \text{if } n = 2m \\ \{e_1, \ldots, e_m, f_1, \ldots, f_m, x\} & \text{if } n = 2m+1 \end{cases}$$

where $(e_i, e_j) = (f_i, f_j) = 0$, $(e_i, f_j) = \delta_{ij}$ *and* $(e_i, x) = (f_i, x) = 0$ *for all* i, j, *and* $(x, x) = 1$.

Proof. Let $\{v_1, \ldots, v_n\}$ be an orthonormal basis. The result is trivial if $n = 1$, so assume that $n \geq 2$. Let $\zeta \in \mathbf{F}$ be a root of the polynomial $\mathbf{x}^2 - \mathbf{x} - 1$, and set $e_1 = v_1 + \zeta v_2$ and $f_1 = \zeta v_1 + v_2$. Then $V = \langle e_1, f_1 \rangle \perp \langle v_3, \ldots, v_n \rangle$, and we proceed by induction. ■

Clearly $\langle e_1, \ldots, e_m \rangle$ is a maximal totally singular subspace of V, so by Witt's Lemma 2.1.6 all maximal totally singular subspaces have dimension $[\frac{n}{2}]$.

Proposition 2.3.3. *We have* $|I| = q^{n(n-1)/2}\prod_{i=1}^{n}(q^i - (-1)^i)$.

Proof. The result is obvious for $n = 1$, so assume $n \geq 2$. Clearly $|I|$ is the number of ordered unitary bases $(e_1, \ldots, e_m, f_1, \ldots, f_m)$ or $(e_1, \ldots, e_m, f_1, \ldots, f_m, x)$. Using an orthonormal basis, one sees that the number of non-zero isotropic vectors in V is equal to the number of non-zero solutions to the equation $\sum_{i=1}^{n} \lambda_i \lambda_i^q = 0$. An inductive argument shows that this number is $(q^n - (-1)^n)(q^{n-1} - (-1)^{n-1})$. Hence this is the number of choices for e_1. Since $f_1 \in V \backslash e_1^{\perp}$, and $(e_1, f_1) = 1$, the number of choices for f_1 is $(q^{2n} - q^{2n-2})/(q^2 - 1) = q^{2n-2}$. Since $V = \langle e_1, f_1 \rangle \perp \langle e_1, f_1 \rangle^{\perp}$, the result follows by induction. ■

We now find suitable generators g_X. Recall from (2.1.14) that $\Omega = S$ and $A = \Gamma$. Let $\beta = \{v_1, \ldots, v_n\}$ be an orthonormal basis. Then by Lemma 2.1.8.i

$$I = \{g \in GL(V, \mathbf{F}) \mid g_\beta g_\beta^{\alpha t} = \mathbf{I}_n\}.$$

Thus for all $g \in I$ we have $\det(g)^{q+1} = 1$, and so $|\det(I)| \mid q+1$. Recall from (2.2.1) that $\delta_\beta(\lambda) = \mathrm{diag}_\beta(\lambda, 1, \ldots, 1)$ and observe $(v_1\delta_\beta(\lambda), v_1\delta_\beta(\lambda)) = (\lambda v_1, \lambda v_1) = \lambda^{q+1}$. Therefore $\delta_\beta(\lambda)$ lies in I if and only if $\lambda^{q+1} = 1$. Thus putting $\delta = \delta_\beta(\omega)$, where $\omega = \mu^{q-1}$ is a primitive $(q+1)^{\text{th}}$ root of unity, we have

$$|\det(I)| = q + 1 \quad \text{and} \quad I = S{:}\langle\delta\rangle. \tag{2.3.1}$$

Take $g \in \Delta$ and suppose that $v \in V$ satisfies $(v, v) = 1$. Then $\tau(g) = (vg, vg) = (vg, vg)^\alpha = \tau(g)^\alpha$, and so $\tau(g) \in \mathbf{F}_q^*$. Consequently $\tau(\Delta) \leq \mathbf{F}_q^*$. Now take $\lambda \in \mathbf{F}^*$ and note that $\tau(\lambda\mathbf{I}) = (\lambda v, \lambda v) = \lambda^{q+1}$. Thus $\langle\tau(\mu)\rangle = \mathbf{F}_q^*$, and hence

$$\tau(\Delta) = \tau(\mathbf{F}^*) = \mathbf{F}_q^*. \tag{2.3.2}$$

Consequently

$$\Delta = I\mathbf{F}^*. \tag{2.3.3}$$

Evidently $I \cap \mathbf{F}^* = \langle\omega\rangle$, which has order $q + 1$. Moreover, defining $\phi = \phi_{\mathbf{f},\beta}$ to be $\phi_\beta(\nu)$, it is clear that ϕ satisfies $\tau(\phi) = 1$ and $\sigma(\phi) = \nu$, and so

$$\Gamma = \Delta{:}\langle\phi\rangle. \tag{2.3.4}$$

Thus we have

Proposition 2.3.4. *We may take our generators* g_X *as follows:*

g_S	g_I	g_Δ	g_Γ	g_A
1	δ	μ	ϕ	1

Proposition 2.3.5. *We have*

$$\ddot{A} = \ddot{\Gamma} = \langle\ddot{\delta}, \ddot{\phi}\rangle \cong \mathbf{Z}_{(n,q+1)}{:}\mathbf{Z}_{2f},$$

with the relations $|\ddot{\delta}| = (q+1, n)$, $|\ddot{\phi}| = 2f$ *and* $\ddot{\delta}^{\ddot{\phi}} = \ddot{\delta}^{p}$.

Proof. Argue as in the proof of Proposition 2.2.3. ■

We conclude this section, as we did before, by pointing out that all assertions concerning case **U** in Tables 2.1.C and 2.1.D have been established. For example, (2.3.3) shows that $|\overline{\Delta} : \overline{I}| = 1$. And $|\Gamma : \Delta| = |\overline{\Gamma} : \overline{\Delta}| = |\ddot{\phi}| = 2f$.

§2.4 The symplectic groups

We assume in this section that case **S** holds, so that $\mathbf{F} = \mathbf{F}_q$ and $\kappa = \mathbf{f}$ is a non-degenerate symplectic form (see (2.1.12)). As in §2.3, we write (v, w) instead of $\mathbf{f}(v, w)$. A fundamental result concerning symplectic geometries is given in the following Proposition.

Proposition 2.4.1. *The dimension n of V is even, and V has a basis $\beta = \{e_1, \ldots, e_m, f_1, \ldots, f_m\}$ (where $n = 2m$), such that for all i, j*

$$(e_i, e_j) = (f_i, f_j) = 0 \text{ and } (e_i, f_j) = \delta_{ij}. \tag{2.4.1}$$

Thus up to isometry, there is a unique symplectic geometry in dimension n over $\mathbf{F}$.

Proof. Let e_1 be any non-zero vector in V. Since $(\ ,\ )$ is non-degenerate, there exists $f_1 \in V$ such that $(e_1, f_1) = 1$. Clearly $W = \langle e_1, f_1 \rangle$ is non-degenerate, and hence $V = W \perp W^{\perp}$. The result now follows by induction on n. ■

A basis satisfying the description in Proposition 2.4.1 will be called a *standard basis* or a *symplectic basis.*

For the rest of this section assume that β is a symplectic basis as in Proposition 2.4.1. Observe that

$$\left(\sum_{i=1}^{m} \lambda_i e_i + \mu_i f_i, \sum_{i=1}^{m} \lambda_i' e_i + \mu_i' f_i \right) = \sum_{i=1}^{m} \lambda_i \mu_i' - \mu_i \lambda_i'. \tag{2.4.2}$$

Using this symplectic basis we investigate a few properties of the geometry $(V, \mathbf{f})$. Observe that all vectors in $(V, \mathbf{f})$ are isotropic. Note also that $\langle e_1, \ldots, e_m \rangle$ is a maximal totally singular subspace, and hence all maximal totally singular subspaces have dimension $\frac{n}{2}$.

Proposition 2.4.2. *We have* $|I| = q^{n^2/4} \prod_{i=1}^{n/2} (q^{2i} - 1)$.

Proof. Argue as in Proposition 2.3.3. ■

We now seek suitable generators g_X. First of all, recall from (2.1.14) that $\Omega = S$ and $A = \Gamma$. Furthermore, it may be shown (see [As$_8$, 22.4]) that $S = I$ (that is, all isometries have determinant 1). Therefore we may take $g_S = g_I = g_A = 1$. It remains to find generators g_Δ and g_Γ.

If q is even, then $\tau(\mathbf{F}^*) = \tau(\mathbf{F}^*)^2 = \mathbf{F}^*$ by (2.1.21), and hence $\Delta = I \times \mathbf{F}^*$. Assume now that q is odd and for $\lambda \in \mathbf{F}^*$, define $\delta_{\mathbf{f},\beta}(\lambda)$ by

$$\begin{aligned} e_i\, \delta_{\mathbf{f},\beta}(\lambda) &= \lambda e_i \\ f_i\, \delta_{\mathbf{f},\beta}(\lambda) &= f_i. \end{aligned} \tag{2.4.3}$$

Thus $\tau(\delta_{\mathbf{f},\beta}(\lambda)) = \lambda$, and so putting $\delta = \delta_{\mathbf{f},\beta}(\mu)$ we obtain

$$\Delta = \begin{cases} I \times \mathbf{F}^* & \text{if } q \text{ is even} \\ I{:}\langle\delta\rangle & \text{if } q \text{ is odd.} \end{cases} \tag{2.4.4}$$

Finally, define $\phi = \phi_{\mathbf{f},\beta}$ to be $\phi_\beta(\nu)$, and observe that $\phi \in \Gamma$, $\sigma(\phi) = \nu$ and $\tau(\phi) = 1$. Therefore

$$\Gamma = \Delta{:}\langle\phi\rangle. \tag{2.4.5}$$

Thus we have proved,

Proposition 2.4.3. *We may take our generators g_X as follows.*

g_S	g_I	g_Δ	g_Γ	g_A
1	1	μ if q even	ϕ	1
		δ if q odd		

Proposition 2.4.4. *We have*

$$\ddot{A} = \ddot{\Gamma} = \begin{cases} \langle\ddot{\phi}\rangle \cong \mathbf{Z}_f & \textit{if } q \textit{ is even} \\ \langle\ddot{\delta}\rangle \times \langle\ddot{\phi}\rangle \cong \mathbf{Z}_2 \times \mathbf{Z}_f & \textit{if } q \textit{ is odd.} \end{cases}$$

Proof. As in the proof of Proposition 2.2.3 we see that $|\ddot{\phi}| = f$, and so the result is clear when q is even. Now suppose that q is odd. Then $|\tau(I\mathbf{F}^*)| = |(\mathbf{F}^*)^2| = \frac{1}{2}(q-1)$. Therefore, arguing as in the proof of Proposition 2.2.3,

$$|\ddot{\delta}| = |\ddot{\Delta}| = |\Delta : I\mathbf{F}^*| = \frac{q-1}{|\tau(I\mathbf{F}^*)|} = 2. \tag{2.4.6}$$

The Proposition now follows from (2.4.4) and (2.4.5). ■

When $(p,n) = (2,4)$, the group $\Omega \cong Sp_4(2^f)$ admits a graph automorphism (see for example [Ca₁, Prop. 12.3.3]), and so $|Aut(\Omega){:}\overline{\Gamma}| = 2$. Indeed, $Out(\Omega) \cong \mathbf{Z}_{2f}$. Thus when f is odd, $Out(\Omega) \cong \mathbf{Z}_f \times \mathbf{Z}_2$, and so any involution in $Aut(\Omega)\backslash\Omega$ induces a graph automorphism on Ω. Taking a suitable such involution x and assuming $f \geq 3$, we have $C_\Omega(x) = Sz(q) = {}^2B_2(q)$, the simple Suzuki group, first discovered in [Su₁]. (Suzuki did not come upon the groups $Sz(q)$ in this context, however. Rather, he discovered them through his investigations of Zassenhaus groups — that is, doubly transitive permutation groups in which the identity is the only element leaving three distinct letters fixed.) More information about the graph automorphism and the Suzuki groups can be found in [Ca₁,Fl₁,Fl₂].

The following fact will be useful in Chapter 4.

Lemma 2.4.5. *If $g \in \Delta$, then* $\det(g) = \tau(g)^{n/2}$.

Proof. If q is odd, then by (2.4.4) $g = h\delta^k$ for some $h \in I$ and some integer k. Since $\det(h) = 1$, we have $\det(g) = \det(\delta)^k = \mu^{kn/2}$. Moreover $\tau(g) = \tau(\delta)^k = \mu^k$, and the result is proved for q odd. The case q even is just as easy. ■

As for the assertions in Tables 2.1.C and 2.1.D involving case **S**, note that (2.4.4) implies that when q is even, $\overline{\Delta} = \overline{I}$, and when q is odd, $|\overline{\Delta} : \overline{I}| = |\ddot{\delta}| = 2$ (see also (2.4.6)). The other statements are by now clear.

§2.5 The orthogonal groups

Assume in this section that case **O** holds, so that $\mathbf{F} = \mathbf{F}_q$ and $\kappa = Q$ is a non-degenerate quadratic form. We put $\mathbf{f} = \mathbf{f}_Q$, the associated bilinear form, so that $\mathbf{f}$ is a non-degenerate symmetric bilinear form on V. As usual, we set $(v, w) = \mathbf{f}(v, w)$. Putting $v = w$ in (2.1.4) we obtain

$$2Q(v) = (v, v). \tag{2.5.1}$$

Thus when q is odd, we can recover Q from $\mathbf{f}$ because $Q(v) = \frac{1}{2}(v, v)$, and so

$$I(V, Q) = I(V, \mathbf{f}) \text{ when } q \text{ is odd.} \tag{2.5.2}$$

However when q is even, (2.5.1) degenerates to $\mathbf{f}(v, v) = 0$. Indeed, $\mathbf{f}$ is a skew-symmetric bilinear form and so $I(V, \mathbf{f})$ is actually a symplectic group. Thus

$$I(V, Q) < I(V, \mathbf{f}) \cong Sp_n(q) \text{ when } q \text{ is even.} \tag{2.5.3}$$

Since every non-degenerate skew-symmetric bilinear form must be defined on a space of even dimension (see Proposition 2.4.1), we deduce

Proposition 2.5.1. *If q is even, then n is even.*

The key to obtaining standard bases in case **O** lies in the following definition and Lemma.

Definition. A subspace W of V is called *anisotropic* if $Q(w) \neq 0$ for all $w \in W \backslash 0$.

Lemma 2.5.2. *Suppose that W is a non-degenerate, anisotropic subspace of V.*
(i) $\dim(W) \leq 2$.
(ii) *If $\dim(W) = 2$, then W has a basis x, y such that $(Q(x), Q(y), (x, y)) = (1, \zeta, 1)$, where the polynomial $\mathbf{x}^2 + \mathbf{x} + \zeta$ is irreducible over $\mathbf{F}$.*
(iii) *All non-degenerate anisotropic 2-spaces are isometric.*

Proof. (i) and (ii). We may assume that $\dim(W) \geq 2$ and we take $x \in W \backslash 0$. Since W is non-degenerate, $x^\perp \cap W < W$, and so there exists $y \in W \backslash \langle x \rangle$ such that $(y, x) \neq 0$. Multiplying y by a suitable scalar, we may write $(y, x) = Q(x)$. Now put $\zeta = Q(y)/Q(x)$, so that for all $\lambda \in \mathbf{F}$ we have

$$0 \neq \frac{Q(\lambda x + y)}{Q(x)} = \lambda^2 + \lambda + \zeta,$$

and hence the polynomial $\mathbf{x}^2 + \mathbf{x} + \zeta$ is irreducible over $\mathbf{F}$. Therefore $\mathbf{x}^2 - \mathbf{x} + \zeta$ is also irreducible, and letting ω be a root of this polynomial in $\mathbf{E} = \mathbf{F}_{q^2}$, we have $Q(\lambda x + \xi y) = Q(x) \mathrm{N}_{\mathbf{F}}^{\mathbf{E}}(\lambda + \xi\omega)$ for all $\lambda, \xi \in \mathbf{F}$. Since $\mathrm{N}_{\mathbf{F}}^{\mathbf{E}}$ is surjective, we can rechoose x so that $Q(x) = 1$. It remains to show that $W = \langle x, y \rangle$, and since W is non-degenerate it suffices to show that $\langle x, y \rangle^\perp \cap W = 0$. So take $v \in \langle x, y \rangle^\perp \cap W$. By the surjectivity of Q, there exists $u \in \langle x, y \rangle$ such that $Q(u) = -Q(v)$. But then $Q(u + v) = 0$, which means $u + v = 0$.

Thus $v \in \langle x, y\rangle \cap \langle x, y\rangle^{\perp}$. In particular, $(v, v) = 0$. Consequently, when q is odd we have $Q(v) = 0$, which means $v = 0$ as W is anisotropic. Now take q even. Writing $v = \lambda x + \xi y$ and using the fact that $(v, x) = (v, y) = 0$, we deduce that $\xi = \lambda = 0$, proving $v = 0$ once again. Thus $W = \langle x, y\rangle$, as required.

(iii). Observe that $\mathbf{E}$ is a 2-dimensional vector space over $\mathbf{F}$ and that the norm $\mathrm{N}_{\mathbf{F}}^{\mathbf{E}}$ is a non-degenerate $\mathbf{F}$-quadratic form on $\mathbf{E}$. Moreover, the proof of (i) shows that map $W \to \mathbf{E}$ given by $\lambda x + \xi y \mapsto \lambda + \xi\omega$ is an isometry from $(W, \mathbf{F}, Q)$ to $(\mathbf{E}, \mathbf{F}, \mathrm{N}_{\mathbf{F}}^{\mathbf{E}})$. Thus all non-degenerate anisotropic 2-spaces are isometric to $(\mathbf{E}, \mathbf{F}, \mathrm{N}_{\mathbf{F}}^{\mathbf{E}})$, and hence (iii) holds. ■

We now produce ***standard bases*** for the orthogonal geometries.

Proposition 2.5.3. *The space (V, Q) has a basis of one of the following forms:*

(i) $\{e_1, \ldots, e_m, f_1, \ldots, f_m\}$ *(with $n = 2m$), where $Q(e_i) = Q(f_i) = 0$ and $(e_i, f_j) = \delta_{ij}$ for all i, j;*

(ii) $\{e_1, \ldots, e_{m-1}, f_1, \ldots, f_{m-1}, x, y\}$ *(with $n = 2m$), where $Q(e_i) = Q(f_i) = 0$, $(e_i, f_j) = \delta_{ij}$ and $(e_i, x) = (e_i, y) = (f_i, x) = (f_i, y) = 0$ for all i, j, and $\{x, y\}$ is as in Lemma 2.5.2.ii;*

(iii) $\{e_1, \ldots, e_m, f_1, \ldots, f_m, x\}$ *(with $n = 2m+1$), where $Q(e_i) = Q(f_i) = 0$, $(e_i, f_j) = \delta_{ij}$ and $(e_i, x) = (f_i, x) = 0$ for all i, j, and x is non-singular.*

Proof. If V is anisotropic, then $\dim(V) \leq 2$ and (ii) or (iii) holds according as $\dim(V)$ is 1 or 2 (see Lemma 2.5.2). Thus we can assume that V contains a non-zero singular vector e_1. As in the proof of Proposition 2.4.1, there exists $f \in V$ such that $(e_1, f) = 1$. Then $f_1 = f - Q(f)e_1$ satisfies $Q(f_1) = 0$ and $(e_1, f_1) = 1$. Evidently $\langle e_1, f_1\rangle$ is non-degenerate, and so we can proceed by induction. ■

Remark. Notice that if e_1, f_1 are as in Proposition 2.5.3, then $Q(e_1 + \lambda f_1) = \lambda$. And if x, y are as in Proposition 2.5.3.ii, then the proof of Lemma 2.5.2 shows that the map $Q : \langle x, y\rangle \to \mathbf{F}$ is surjective. So provided $n \geq 2$, the map $Q : V \to \mathbf{F}$ is always surjective. Thus by (2.5.1), V has vectors of all $\mathbf{f}$-norms when q is odd. We also point out that for $n \geq 3$, the geometry (V, Q) always contains non-zero singular vectors.

In view of the trichotomy given in Proposition 2.5.3, we define

$$\mathrm{sgn}(Q) = \begin{cases} \circ & \text{if } n \text{ is odd} \\ + & \text{if } n \text{ is even and } (V,Q) \text{ has a basis of type 2.5.3.i} \\ - & \text{if } n \text{ is even and } (V,Q) \text{ has a basis of type 2.5.3.ii.} \end{cases} \tag{2.5.4}$$

We divide case $\mathbf{O}$ into cases $\mathbf{O}^{\circ}$, $\mathbf{O}^{+}$ and $\mathbf{O}^{-}$, accordingly. Furthermore we write

$$\mathrm{sgn}(\mathbf{f}) = \mathrm{sgn}(Q) \text{ when } q \text{ is odd} \tag{2.5.5}$$

and

$$X(V, \mathbf{F}, Q) \cong \begin{cases} \Omega_n^{\mathrm{sgn}(Q)}(q) & X = \Omega \\ SO_n^{\mathrm{sgn}(Q)}(q) & X = S \\ O_n^{\mathrm{sgn}(Q)}(q) & X = I \\ GO_n^{\mathrm{sgn}(Q)}(q) & X = \Delta \\ \Gamma O_n^{\mathrm{sgn}(Q)}(q) & X = \Gamma, A \end{cases} \tag{2.5.6}$$

Furthermore when n is odd we write

$$Y_n(q) = Y_n^\circ(q),$$

where Y ranges over the symbols Ω, SO, O, GO and ΓO. Although the group $\Omega = \Omega(V, \mathbf{F}, Q)$ has been referred to in (2.5.6) it has not as yet been discussed. This group will be introduced in Proposition 2.5.7, and will be described in the paragraphs following that Proposition.

Suppose here that n is even. Then according to Proposition 2.5.3.i,ii and Lemma 2.5.2.iii, all geometries with $\text{sgn}(Q) = +$ are isometric, as are all geometries with $\text{sgn}(Q) = -$. If e_i, f_j are as in Proposition 2.5.3, then $\langle e_1, \ldots, e_m \rangle$ is a maximal totally singular subspace when $\text{sgn}(Q) = +$, while $\langle e_1, \ldots, e_{m-1} \rangle$ is a maximal totally singular subspace when $\text{sgn}(Q) = -$. Thus the geometry with $\text{sgn}(Q) = +$ is non-isometric to the geometry with $\text{sgn}(Q) = -$.

Now assume that n is odd, so that q is also odd by Proposition 2.5.1. For each $\lambda \in \mathbf{F}^*$ let us write Q_λ for the non-degenerate quadratic form on V which has the standard basis given in Proposition 2.5.3.iii and which satisfies $Q_\lambda(x) = \lambda$. For $\lambda, \eta \in \mathbf{F}^*$, consider the map $h \in GL(V, \mathbf{F})$ defined by

$$e_i h = \lambda\eta^{-1} e_i, \quad f_i h = f_i, \quad xh = x.$$

Elementary calculations show that $Q_\lambda(vh) = \lambda\eta^{-1} Q_\eta(v)$ for all $v \in V$, and hence the geometries (V, Q_λ) $(\lambda \in \mathbf{F}^*)$ are all similar. Suppose for the moment that (V, Q_λ) is isometric to (V, Q_η). Then by Witt's Lemma 2.1.6, the map which acts as the identity on $W = \langle e_1, \ldots, f_m \rangle$ must extend to an isometry g from (V, Q_λ) to (V, Q_η). However, as $\langle x \rangle = W^\perp$ in both geometries, $xg = \zeta x$ for some $\zeta \in \mathbf{F}^*$. This forces $\zeta^2\eta = Q_\eta(xg) = Q_\lambda(x) = \lambda$, which is possible if and only if $\lambda \equiv \eta \pmod{(\mathbf{F}^*)^2}$. Finally we note that $\langle e_1, \ldots, e_m \rangle$ is a maximal totally singular subspace in (V, Q_λ) and thus we have proved

Proposition 2.5.4. *For each n, there are precisely two isometry classes of orthogonal geometries in dimension n.*

(i) *If n is even, then the two isometry types are distinguished by the dimension of their maximal totally singular subspaces. Indeed, the maximal totally singular subspaces have dimension $\frac{n}{2}$ or $\frac{n}{2} - 1$ according as $\text{sgn}(Q) = +$ or $\text{sgn}(Q) = -$.*

(ii) *If n is odd, then the two isometry types are distinguished by the value of $Q(x)$ modulo $(\mathbf{F}^*)^2$, where x is given in Proposition 2.5.3.iii. The two geometries are similar, and all maximal totally singular subspaces have dimension $\frac{1}{2}(n-1)$.*

When n is even and $\text{sgn}(Q) = +$, the orthogonal geometry (V, Q) is said to be *hyperbolic* or of *maximal Witt index* or *Witt defect 0.* When $\text{sgn}(Q) = -$, the geometry is said to be *elliptic* or of *non-maximal Witt index* or *Witt defect 1.* When n is odd, it follows from Proposition 2.5.4.ii and Lemma 2.1.1 that the isomorphism type of $X(V, \mathbf{F}, Q)$ is independent of the value of $Q(x)$.

As in the proof of Proposition 2.3.3, one may count the number of standard bases to derive the order of $|I|$. The proof is omitted. (In the statement of Proposition 2.5.5, we use the slightly more aesthetic notation $x - \epsilon$ instead of $x - \epsilon 1$ when ϵ is the sign $\pm$. Such notation will be used throughout this book — see page vii.)

Proposition 2.5.5. *We have*

$$|I| = \begin{cases} 2q^{(n-1)^2/4}\prod_{i=1}^{(n-1)/2}(q^{2i}-1) & \textit{in case } \mathbf{O}^{\circ} \\ 2q^{n(n-2)/4}(q^{n/2}-\epsilon)\prod_{i=1}^{n/2-1}(q^{2i}-1) & \textit{in case } \mathbf{O}^{\epsilon},\ \epsilon = \pm. \end{cases}$$

In the following three sections we shall investigate each of these three types of orthogonal geometries individually. For the rest of this section, however, we will amass information which pertains to all orthogonal geometries simultaneously.

Let $v \in V$ be non-singular and define the *reflection in* v as the map r_v from V to V given by

$$x\, r_v = x - \frac{(v,x)}{Q(v)}v \quad (x \in V). \tag{2.5.7}$$

It is straightforward to check that $r_v \in I$ and that $\det(r_v) = -1$. Moreover, it follows from Lemma 2.1.8.i that $\det(I) \leq \langle -1 \rangle$, and hence

$$\det(I) = \langle -1 \rangle, \tag{2.5.8}$$

where in characteristic 2 it is understood that $-1 = 1$. In almost all cases, the reflections generate the full orthogonal group I. Indeed, according to [As$_8$, 22.7]:

Proposition 2.5.6. *We have* $I = \langle r_v \mid Q(v) \neq 0 \rangle$, *provided* $I \neq O_4^+(2)$.

It follows directly from Proposition 2.5.6 that when q is odd, S is the subgroup of I consisting of elments which are a product of an even number of reflections. In the following discussion, we aim to give a description of the group Ω, first mentioned in (2.1.14). The most important result in this direction is Proposition 2.5.7. A proof can be found in [As$_8$, 22.9].

Proposition 2.5.7. *Provided* $I \neq O_4^+(2)$ *and* $n \geq 2$, *the group* S *has a unique subgroup of index 2, which we call* $\Omega = \Omega(V, \mathbf{F}, Q)$.

This result lies fairly deep and its proof uses the theory of Clifford algebras. It is important for our purposes to determine whether a given element of S lies in Ω or in $S \backslash \Omega$. Thus we now offer four descriptions of the subgroup Ω, each of which is useful in this respect. The final description (Description 4) also includes a definition of $\Omega_4^+(2)$.

Description 1. First suppose that $n \geq 2$ and q is odd, so that $\mathbf{F}^*/(\mathbf{F}^*)^2$ has order 2. It follows from Proposition 2.5.6 that any element $g \in S$ can be written as a product of an even number of reflections $g = r_{v_1} \ldots r_{v_k}$, for some non-singular vectors v_i. Define the *spinor norm* of g, written $\theta(g)$, as follows:

$$\theta(g) \equiv \prod_{i=1}^{k}(v_i, v_i) \pmod{(\mathbf{F}^*)^2} \in \mathbf{F}^*/(\mathbf{F}^*)^2. \tag{2.5.9}$$

It turns out (see [As$_8$, 22.11]) that θ is a well-defined homomorphism from S to $\mathbf{F}^*/(\mathbf{F}^*)^2$. Now V has vectors of all $\mathbf{f}$-norms (see the Remark after Proposition 2.5.3). So if $r_\square$ and $r_\boxtimes$ are reflections in vectors of square and non-square $\mathbf{f}$-norm, respectively, then $\theta(r_\square r_\boxtimes)$ is non-trivial. Therefore θ is surjective and so $|S : \ker(\theta)| = 2$. Thus by Proposition 2.5.7 we have

$$\Omega = \ker(\theta) \text{ when } q \text{ is odd.} \tag{2.5.10}$$

Description 2. Next suppose that q is even, so that n is also even and $S = I$. Assume moreover that $I \neq O_4^+(2)$. Then according to Proposition 2.5.6, every element of S can be written as a product of reflections. According to [Di$_1$, p.65], the subgroup of S consisting of products of an even number of reflections has index 2 in S, and this is the group Ω.

Description 3. Suppose that case $\mathbf{O}^+$ holds with q even. Consider the group I acting on the set $\mathcal{U}_m$ of maximal totally singular subspaces, which are of dimension $m = \frac{n}{2}$. It can be seen that any reflection acts as an odd permutation on $\mathcal{U}_m$ (this is an exercise for the reader, but a proof can be found in [Ka$_2$, p.17]). Hence the subgroup inducing even permutations has index 2 in I, and by Proposition 2.5.7 this is the group Ω, provided $I \neq O_4^+(2)$.

Description 4. Assume that case $\mathbf{O}^+$ holds. Define a relation $\sim$ on $\mathcal{U}_m$ by $U \sim W$ if and only if $m - \dim(W \cap U)$ is even. Then it can be shown (see [As$_8$, 22.13]) that $\sim$ is an equivalence relation and there are precisely two equivalence classes, called $\mathcal{U}_m^1$ and $\mathcal{U}_m^2$. Obviously Γ preserves this equivalence relation, and hence we obtain a homomorphsim $\gamma : \Gamma \to Sym\{\mathcal{U}_m^1, \mathcal{U}_m^2\} \cong \mathbf{Z}_2$. By Witt's Lemma 2.1.6, the group I is transitive on $\mathcal{U}_m$, and so $\gamma(I)$ is surjective. Thus when q is even and $I \neq O_4^+(2)$, it follows from Proposition 2.5.7 that $\ker_I(\gamma) = \Omega$. When $I = O_4^+(2)$, we define Ω as $\ker_I(\gamma)$. It is obvious from the definition that Ω has at least two orbits on $\mathcal{U}_m$, and the next result asserts that there are only two, namely $\mathcal{U}_m^i$, $i = 1, 2$.

Lemma 2.5.8. *Assume that case* $\mathbf{O}^+$ *holds.*

(i) $\ker_I(\gamma)$ *is either* S *or* Ω, *according as* q *is odd or even.*

(ii) $\mathcal{U}_m^1$ *and* $\mathcal{U}_m^2$ *are the two* S*-orbits on* $\mathcal{U}_m$ *when* q *is odd, and are the two* Ω*-orbits on* $\mathcal{U}_m$ *for all* q.

(iii) *If* $U \in \mathcal{U}_{m-1}$, *then* U *is contained in precisely two members of* $\mathcal{U}_m$, *one in each* $\mathcal{U}_m^i$.

Proof. (i) This holds by Description 4 when q is even. So assume that q is odd and let $r = r_{e_1 - f_1}$, the reflection in the non-singular vector $e_1 - f_1$ (where e_1, f_1 are as in Proposition 2.5.3). Evidently r interchanges e_1 and f_1, and hence interchanges the the two totally singular m-spaces $\langle e_1, \ldots, e_m \rangle$ and $\langle f_1, e_2, \ldots, e_m \rangle$. Thus $r \in I \backslash \ker_I(\gamma)$. It follows that any product of an even number of reflection lies in $\ker_I(\gamma)$. But the group generated by an even number of reflections is just S (see Description 1), and so $S \leq \ker_I(\gamma)$. Since $|I : S| = |I : \ker_I(\gamma)| = 2$, the result follows.

(ii) Suppose that $U, W \in \mathcal{U}_m$ and that $U \sim W$. Now by Witt's Lemma 2.1.6, there exists $g \in I$ taking U to W. Clearly $g \in \ker_I(\gamma)$, and hence $\ker_I(\gamma)$ is transitive on $\mathcal{U}_m^i$

for $i = 1, 2$. Thus using (i), to complete the proof of (ii) it remains to show that Ω is transitive on each $\mathcal{U}_m^i$ when q is odd. To do so, it suffices to show that $S_W\Omega = S$ (recall S_W is the stabilizer in S of the space W). With no loss of generality, $W = \langle e_1, \ldots, e_m \rangle$. Now observe that $s = r_{e_1 - \mu f_1} r_{e_1 - f_1}$ fixes W and has spinor norm μ (mod $(\mathbf{F}^*)^2$), which is non-trivial as μ is a generator of $\mathbf{F}^*$ (see (2.1.23)). Thus $s \in S_W \backslash \Omega$, and hence (ii) holds.

(iii) Without loss of generality, $U = \langle e_1, \ldots, e_{m-1} \rangle$, which is contained in $W_1 = \langle e_1, \ldots, e_m \rangle$ and $W_2 = \langle e_1, \ldots, e_{m-1}, f_m \rangle$. Obviously we can choose the indices so that $W_i \in \mathcal{U}_m^i$. Now if $U \leq W \in \mathcal{U}_m^i$, then $W \cap W_i$ must have even codimension in W. But this codimension is at most 1 (as they both contain U). Thus $W = W_i$, and (iii) is proved. ■

For the sake of clearing up any mysteries concerning $O_4^+(2)$, we state the following.

Proposition 2.5.9. *If $I = O_4^+(2)$, then $I \cong S_3 \wr S_2$ and $I/O^2(I) \cong D_8$. The three subgroups of index 2 in I are:*
(i) *the subgroup of I generated by reflections;*
(ii) *the subgroup of I inducing even permutations on $\mathcal{U}_2$;*
(iii) *the subgroup Ω, as given in Description 4, above.*

At this point it is important to express the relationship between the groups Ω, S and I. Continue to assume that $r_\square$ and $r_\boxtimes$ are as in Description 1 when q is odd. As a convenience, we also write $r_\square$ for any reflection in I when q is even. It follows directly from (2.5.8) that

$$I = \begin{cases} S & \text{if } q \text{ is even} \\ S\langle r_\square \rangle = S.2 & \text{if } q \text{ is odd.} \end{cases} \tag{2.5.11}$$

Now we claim that

$$S = \begin{cases} I = \Omega\langle r_\square \rangle = \Omega.2 & \text{if } q \text{ is even} \\ \Omega\langle r_\square r_\boxtimes \rangle = \Omega.2 & \text{if } q \text{ is odd.} \end{cases} \tag{2.5.12}$$

To verify (2.5.12), we use Description 1 when q is odd, and we may use Description 2 when q is even and $I \neq O_4^+(2)$. Description 4 can be used to establish (2.5.12) for all geometries in case $\mathbf{O}^+$ with q even. For as we saw in the proof of Lemma 2.5.8, all reflections lie in $I \backslash \ker_I(\gamma) = I \backslash \Omega$, and hence $I = \Omega\langle r_\square \rangle$. We will use assertions (2.5.11) and (2.5.12) in the following sections in order to find suitable generators g_S and g_I. Also, we point out that

$$\{r_1 \ldots r_k \mid r_i \text{ a reflection, } k \text{ even}\} = \begin{cases} S & q \text{ odd} \\ \Omega & q \text{ even and } I \neq O_4^+(2) \\ \frac{1}{2}\Omega & I = O_4^+(2). \end{cases} \tag{2.5.13}$$

Recall that $\frac{1}{2}\Omega$ denotes a subgroup of index 2 in Ω, and in fact this group $\frac{1}{2}\Omega$ is the intersection of the three subgroups appearing in Proposition 2.5.9. The second statement in (2.5.13) is immediate from Description 2 above, and the first follows from Proposition 2.5.6 along with the obervation that each reflection has determinant -1.

We conclude this section with a discussion of the discriminant of Q. Fix any two bases $\beta = \{v_1, \ldots, v_n\}$ and $\beta' = \{v_1', \ldots, v_n'\}$ of V, and let $g \in GL(V, \mathbf{F})$ be the linear transformation taking v_i to v_i'. Then $\mathbf{f}_{\beta'} = g_\beta \mathbf{f}_\beta g_\beta^t$, and hence $\det(\mathbf{f}_\beta) = \det(g)^2 \det(\mathbf{f}_{\beta'})$. Thus modulo $(\mathbf{F}^*)^2$, we see that $\det(\mathbf{f}_\beta)$ is independent of the choice of basis β. Therefore we may define the *discriminant of Q*, written $D = D(Q)$ as

$$D(Q) \equiv \det(\mathbf{f}_\beta) \pmod{(\mathbf{F}^*)^2} \in \mathbf{F}^*/(\mathbf{F}^*)^2. \tag{2.5.14}$$

Sometimes we will write $D(Q) = \square$ or $\boxtimes$, according as $D(Q)$ is a square or non-square. (Obviously, the discriminant is a useful object only when q is odd.) Notice that when n is odd, Proposition 2.5.4.ii shows that the two isometry types of orthogonal geometries are distinguished by their discriminants. The next result shows that the same holds when n is even.

Proposition 2.5.10. *Assume that $n = 2m$ is even and that q is odd.*
(i) *If $\mathrm{sgn}(Q) = +$, then $D(Q) = \square$ if and only if $\frac{1}{2}m(q-1)$ is even.*
(ii) *If $\mathrm{sgn}(Q) = -$, then $D(Q) = \square$ if and only if $\frac{1}{2}m(q-1)$ is odd.*

Proof. (i) With respect to the basis β given in Proposition 2.5.3.i, we see that $\det(\mathbf{f}_\beta) = (-1)^m$, which is a square if and only if $\frac{1}{2}m(q-1)$ is even.

(ii) Let β be a basis as given in Proposition 2.5.3.ii, and write W for the anisotropic 2-space $\langle x, y\rangle$, so that $\det(\mathbf{f}_\beta) = (-1)^{m-1}\lambda$, where λ is the determinant of the 2×2 matrix $(\mathbf{f}_W)_{\{x,y\}}$. As we mentioned in the Remark after Proposition 2.5.3, $Q(W) = \mathbf{F}$, and so we may choose $w \in W$ such that $(w, w) = 1$. Now let u span $w^\perp \cap W$, so that $W = \langle w\rangle \perp \langle u\rangle$. Since $\langle \mu\rangle = \mathbf{F}^*$, we know that $(u, u) = \mu^j$ for some j. Moreover, by replacing u by a suitable scalar multiple of itself we can assume that $j \in \{0, 1\}$. We now let $\beta_1 = \{e_1, \ldots, e_{m-1}, f_1, \ldots, f_{m-1}, w, u\}$, so that $D(Q) = \det(\mathbf{f}_{\beta_1}) = (-1)^{m-1}\mu^j$. If $q \equiv 1 \pmod 4$, then we must have $j = 1$ (for if $j = 0$ then $Q(w + iu) = 0$, where $i = \sqrt{-1}$, a contradiction); consequently $D(Q)$ is a non-square. And if $q \equiv 3 \pmod 4$ then we must have $j = 0$ (for if $j = 1$, then $-\mu = \lambda^2$ for some λ and $Q(\lambda w + u) = 0$, a contradiction); consequently $D(Q) = (-1)^{m-1}$, which is a square if and only if m is odd. This completes the proof of (ii). ∎

If W is a non-degenerate subspace of V, then we sometimes write

$$D(W) = D(Q_W) \quad \text{and} \quad \mathrm{sgn}(W) = \mathrm{sgn}(Q_W).$$

If, in addition, $\dim(W) = k$, we say that

$$W \text{ is a } \begin{cases} +k\text{-space} & \text{if } \mathrm{sgn}(W) = + \\ -k\text{-space} & \text{if } \mathrm{sgn}(W) = - \\ \square k\text{-space} & \text{if } \mathrm{sgn}(W) = \circ \text{ and } D(W) = \square \\ \boxtimes k\text{-space} & \text{if } \mathrm{sgn}(W) = \circ \text{ and } D(W) = \boxtimes. \end{cases} \tag{2.5.15}$$

Here are a few easy results concerning $D(Q)$.

Proposition 2.5.11. *Assume that $V = V_1 \perp \cdots \perp V_t$, where each V_i is a non-degenerate subspace of V.*

(i) $D(Q) = \prod_{i=1}^{t} D(Q_{V_i})$.
(ii) *If* $\dim(V_i)$ *is even for all* i, *then* $\mathrm{sgn}(Q) = \prod_{i=1}^{t} \mathrm{sgn}(Q_{V_i})$.

Proof. Part (i) is clear from the definition, and when q is odd (ii) is immediate from (i) and Proposition 2.5.10. So it remains to prove (ii) when q is even, and to do so we may assume by induction that $t = 2$. The cases in which $(\mathrm{sgn}(Q_{V_1}), \mathrm{sgn}(Q_{V_2})) = (+,+)$, $(+,-)$ or $(-,+)$ are straightforward and are left to the reader. So we treat only the case $(-,-)$, and by considering the standard bases described in Proposition 2.5.3, we quickly reduce to the case where $\dim(V_i) = 2$ for $i = 1, 2$. But now according to Lemma 2.5.2.iii we can choose bases x_i, y_i of V_i such that the map $x_1 \mapsto x_2$, $y_1 \mapsto y_2$ induces an isometry. Thus $\langle x_1 + x_2, y_1 + y_2 \rangle$ is a totally singular 2-space in $V_1 \perp V_2$, and hence the result follows. ■

Proposition 2.5.12. *If* q *is odd, then* V *has a basis* β *such that* $\mathbf{f}_\beta$ *is either* $\mathbf{I}_n$ *or* $\mathrm{diag}_\beta(\mu, 1, \ldots, 1)$, *according as* $D(Q) = \square$ *or* $D(Q) = \boxtimes$.

Proof. Take $v_1 \in V$ such that v_1 is non-singular. Multiplying v_1 by a suitable scalar we may assume that $(v_1, v_1) \in \{1, \mu\}$. Now $V = \langle v_1 \rangle \perp v_1^\perp$, and by induction $v_1^\perp$ has the desired basis $\beta' = \{v_2, \ldots, v_n\}$. If β' is orthonormal, then clearly $\{v_1, \ldots, v_n\}$ satisfies the conclusion of the Proposition. And if $(v_1, v_1) = 1$, then $\{v_2, v_1, v_3, \ldots, v_n\}$ is the required basis. Assume therefore that $(v_1, v_1) = (v_2, v_2) = \mu$. Choose $\lambda_1, \lambda_2 \in \mathbf{F}$ such that $\lambda_1^2 + \lambda_2^2 = \mu^{-1}$. Then $w_1 = \lambda_1 v_1 + \lambda_2 v_2$ has norm 1. Now choose w_2 so that $\langle v_1, v_2 \rangle = \langle w_1 \rangle \perp \langle w_2 \rangle$. Then using discriminants we have $(w_2, w_2) = (w_1, w_1)(w_2, w_2) \equiv \mu^2 \pmod{(\mathbf{F}^*)^2}$, and so (w_2, w_2) is a square. Thus λw_2 has norm 1 for some λ and $\beta = \{w_1, \lambda w_2, v_3, \ldots, v_n\}$ is orthonormal, as desired. ■

When q is odd, the information we have obtained so far allows us to describe where the scalar -1 lies in relation to Ω, S and I. Of course, when n is odd, $-1 \in I \backslash S$, which means $I = S \times \langle -1 \rangle$. When n is even, $-1 \in S$ and some care is required to decide whether $-1 \in \Omega$ or $-1 \in S \backslash \Omega$.

Proposition 2.5.13. *Assume that* q *is odd and* n *is even.*
(i) *If* $-1 \notin \Omega$ *then* $S = \Omega \times \langle -1 \rangle$.
(ii) *The following are equivalent:*
 (a) $-1 \in \Omega$
 (b) $D(Q) = \square$
 (c) $\mathrm{sgn(Q)} = (-1)^{(q-1)n/4}$.

Proof. Part (i) is trivial. To prove (ii), let β be a basis provided by Proposition 2.5.12. If $-1 \in \Omega$, then the spinor norm (see (2.5.9)) of -1 is a square, and so β must be orthonormal. Therefore $D(Q) = \square$, which gives (b). Similarly, if $-1 \notin \Omega$, then the spinor norm is a non-square and so β cannot be orthonormal; thus $D(Q) = \boxtimes$. We have now shown that (a) and (b) are equivalent. Their equivalence with (c) is immediate from Proposition 2.5.10. ■

In §§2.6-2.8 we study the three cases $\mathbf{O}^\circ$, $\mathbf{O}^+$ and $\mathbf{O}^-$ in greater detail.

§2.6 Orthogonal groups in odd dimension

In this section we assume that case $\mathbf{O}^\circ$ holds, so that $\mathbf{F} = \mathbf{F}_q$, $\kappa = Q$ is a non-degenerate quadratic form, and $n = 2m + 1$ is odd. Thus

$$I = O_n(q) = O_{2m+1}(q).$$

We write the associated bilinear form as $\mathbf{f} = \mathbf{f}_Q$, and $(v, w) = \mathbf{f}(v, w)$. Also note that by Proposition 2.5.1 we know that q is odd. Recall $D = D(Q)$ is the discriminant of Q. This first result is proved analogously to Proposition 2.5.12, and its proof is left to the reader.

Proposition 2.6.1. *There exists a basis β of (V, Q) such that $\mathbf{f}_\beta = \lambda \mathbf{I}_n$, where $D \equiv \lambda \pmod{(\mathbf{F}^*)^2}$.*

For the rest of this section we shall assume that β is a basis which satisfies the condition in the preceding Proposition. We now seek to find suitable generators g_X. Recall from (2.1.14) that $A = \Gamma$, so we may take $g_A = 1$. Also, we may put $g_S = r_\square r_\boxtimes$ by (2.5.12).

As we observed just before Corollary 2.5.13,

$$I = S \times \langle -1 \rangle. \tag{2.6.1}$$

We now claim that

$$\Delta = S \times \mathbf{F}^* = I\mathbf{F}^*. \tag{2.6.2}$$

To prove (2.6.2) we first prove

$$\tau(\Delta) = (\mathbf{F}^*)^2. \tag{2.6.3}$$

We proceed by induction on n. Note that (2.6.3) is trivial in dimension 1, so we take $n > 1$. Now fix $g \in \Delta$. In order to show that $\tau(g) \in (\mathbf{F}^*)^2$, it is clearly sufficient to assume that g is a 2-element. Let W be an irreducible faithful $\langle g \rangle$-submodule of V. We argue that $W \neq V$. To prove this, we may clearly suppose that $\dim(W)$ is odd. By Lemma 2.10.2 below, $\langle g \rangle$ has an absolutely irreducible faithful representation in $GL_a(q^b)$, for some a, b, with $\dim(W) = ab$. But an absolutely irreducible representation of a cyclic group has dimension 1, and so $a = 1$ and $\dim(W) = b$. Therefore $|g|$ divides $|GL_1(q^b)| = q^b - 1$, and since b is odd and g is a 2-element, $|g| \mid q - 1$. But then $\mathbf{F}$ is a splitting field for $\langle g \rangle$, and hence $\dim(W) = 1$. Therefore $W \neq V$, as desired. Since $W \cap W^\perp$ is also g-invariant, it follows that W is either non-degenerate or totally singular. Suppose first that $V = W \perp W^\perp$. Then $\tau(g) = \tau_W(g_W) = \tau_{W^\perp}(g_{W^\perp})$ by Lemma 2.1.9.i, and since either W or $W^\perp$ has odd dimension, (2.6.3) follows by induction. Suppose therefore that W is totally singular. Then if U denotes the quotient space $W^\perp/W$, we see that $\tau(g) = \tau_U(g_U)$ by Lemma 2.1.9.iii, and so (2.6.3) follows by induction once again. Assertion (2.6.2) now follows from (2.6.3). For $\tau(\mathbf{F}^*) = (\mathbf{F}^*)^2$, and hence $\Delta = I\mathbf{F}^* = (S \times \langle -1 \rangle)\mathbf{F}^* = S \times \mathbf{F}^*$.

Now define $\phi = \phi_{Q,\beta}$ to be $\phi_\beta(\nu)\lambda^{(p-1)/2}$, where λ is given by Proposition 2.6.1. Then it is easy to verify that $\phi \in \Gamma$ and that $\tau(\phi) = 1$ and $\sigma(\phi) = \nu$. Thus

$$\Gamma = \Delta{:}\langle\phi\rangle. \tag{2.6.4}$$

Recall from (2.5.12) that $S = \Omega\langle r_\square r_\boxtimes\rangle$, where $r_\square$ (respectively, $r_\boxtimes$) is a reflection in a vector whose norm is a square (respectively, non-square). Assertions (2.5.12), (2.6.1), (2.6.2) and (2.6.4) yield

Proposition 2.6.2. *We may take our generators g_X to be the following:*

g_S	g_I	g_Δ	g_Γ	g_A
$r_\square r_\boxtimes$	-1	μ	ϕ	1

Proposition 2.6.3. *Assume that case* $\mathbf{O}^\circ$ *holds, so that q is odd, and take $n \geq 3$. Then*

$$\ddot{A} = \ddot{\Gamma} = \langle\ddot{r}_\square\ddot{r}_\boxtimes\rangle \times \langle\ddot{\phi}\rangle \cong \mathbf{Z}_2 \times \mathbf{Z}_f.$$

Proof. Proposition 2.6.2 implies that $\ddot{\Gamma} = \langle\ddot{r}_\square\ddot{r}_\boxtimes, \ddot{\phi}\rangle$. Due to the facts that $r_\square r_\boxtimes \in S\backslash\Omega$ and $\mathbf{F}^* \cap S = 1$, it follows that $r_\square r_\boxtimes \notin \Omega\mathbf{F}^*$, and hence $|\ddot{r}_\square\ddot{r}_\boxtimes| = 2$. And just as in the proof of Proposition 2.2.3 we see that $|\ddot{\phi}| = f$. Since $S \trianglelefteq \Gamma$, we have $\langle\ddot{r}_\square\ddot{r}_\boxtimes\rangle \trianglelefteq \ddot{\Gamma}$, and hence the structure of $\ddot{\Gamma}$ is now clear. Since $A = \Gamma$ (by definition), the Proposition is proved. ■

Note that from (2.5.11) and (2.5.12) we have $|I : S| = |S : \Omega| = 2$, and that (2.6.1) and (2.6.2) imply $|\overline{\Delta} : \overline{I}| = |\overline{I} : \overline{S}| = 1$ and $|\Delta : I| = \frac{1}{2}(q-1)$. The other assertions in Tables 2.1.C and 2.1.D follow from Proposition 2.6.3, above.

We conclude this section with the observation that $\tau(\Gamma) = (\mathbf{F}^*)^2$, and hence

Lemma 2.6.4. *In case* $\mathbf{O}^\circ$, *the homomorphisms $\overline{\tau}$ and $\ddot{\tau}$ are trivial.*

Recall that $\overline{\tau}$ and $\ddot{\tau}$ are defined at the end of §2.1.

§2.7 Orthogonal Groups with Witt defect 0

In this section assume that case $\mathbf{O}^+$ holds, so that $\mathbf{F} = \mathbf{F}_q$, $\kappa = Q$ and $\mathbf{f} = \mathbf{f}_Q$ as before, and $n = 2m$ is even with $\mathrm{sgn}(Q) = +$. Thus we have

$$I = O_n^+(q) = O_{2m}^+(q).$$

Also $\mathbf{f}(v, w)$ is written (v, w), as usual. We let

$$\beta = \{e_1, \ldots, e_m, f_1, \ldots, f_m\}$$

be a standard basis as given in Proposition 2.5.3.i. We also write $D = D(Q)$ for the discriminant of Q when q is odd.

As usual, we seek to find suitable generators g_X. We retain the notation $r_\square$ and $r_\boxtimes$ introduced before. Note that according to (2.5.12) and Lemma 2.5.13, we have

$$S = \begin{cases} I = \Omega\langle r_\square\rangle \cong \Omega.2 & \text{if } q \text{ is even} \\ \Omega\langle r_\square r_\boxtimes\rangle \cong \Omega.2 & \text{if } q \text{ is odd} \\ \Omega \times \langle -1\rangle \cong \Omega \times 2 & \text{if } q \text{ is odd and } D = \boxtimes. \end{cases} \tag{2.7.1}$$

Recall from Proposition 2.5.10 that when q is odd, $D = \square$ if and only if $\frac{1}{2}m(q-1)$ is even.

For $\lambda \in \mathbf{F}^*$ define $\delta_{Q,\beta}(\lambda)$ just as in (2.4.3), and set $\delta = \delta_{Q,\beta}(\mu)$. Then the same reasoning as for the symplectic groups shows that $\delta \in \Delta$, $\tau(\delta) = \mu$ and that

$$\Delta = \begin{cases} I \times \mathbf{F}^* & \text{if } q \text{ is even} \\ I{:}\langle\delta\rangle & \text{if } q \text{ is odd.} \end{cases} \tag{2.7.2}$$

Next, define $\phi = \phi_{Q,\beta}$ as $\phi_\beta(\nu)$. Then $\sigma(\phi) = \nu$ and $\tau(\phi) = 1$ and hence

$$\Gamma = \Delta{:}\langle\phi\rangle. \tag{2.7.3}$$

Combining the facts in the preceding discussion with (2.5.11) yields

Proposition 2.7.1. *We may take our generators g_X as follows:*

g_S	g_I	g_Δ	g_Γ	g_A
$r_\square$ if q even $r_\square r_\boxtimes$ if $D = \square$ -1 if $D = \boxtimes$	1 if q even $r_\square$ if q odd	μ if q even δ if q odd	ϕ	1

Before proving the structure theorem for $\ddot{A}$, we need a lemma which provides, in addition to the four Descriptions in §2.5, yet another method of detecting whether an element in S lies in Ω or in $S\backslash\Omega$. We do not prove this lemma here, for it is a special case of the more general result Lemma 4.1.9.iii.

Lemma 2.7.2. *Assume that $g \in I$ fixes both $W = \langle e_1, \ldots, e_m\rangle$ and $U = \langle f_1, \ldots, f_m\rangle$. Then $g \in S$, and moreover $g \in \Omega$ if and only if $\det_W(g)$ is a square.*

Proposition 2.7.3.

(i) *We have*

$$\ddot{A} = \ddot{\Gamma} = \begin{cases} \langle\ddot{r}_\square\rangle \times \langle\ddot{\phi}\rangle \cong \mathbf{Z}_2 \times \mathbf{Z}_f & \textit{if } q \textit{ is even} \\ \langle\ddot{r}_\square\rangle \times \langle\ddot{\delta}\rangle \times \langle\ddot{\phi}\rangle \cong \mathbf{Z}_2 \times \mathbf{Z}_2 \times \mathbf{Z}_f & \textit{if } q \textit{ is odd and } D = \boxtimes \\ \langle\ddot{r}_\square, \ddot{r}_\boxtimes, \ddot{\delta}, \ddot{\phi}\rangle \cong D_8 \times \mathbf{Z}_f & \textit{if } q \textit{ is odd and } D = \square. \end{cases}$$

(ii) *If q is odd and $D = \boxtimes$, then $\ddot{r}_\square = \ddot{r}_\boxtimes$.*

(iii) *If q is odd and $D = \square$, then $|\ddot{r}_\square| = |\ddot{r}_\boxtimes| = 2$, $|\ddot{\delta}| = 2(2, m-1)$, $\ddot{r}_\square^{\ddot{\delta}} = \ddot{r}_\boxtimes$, $\ddot{r}_\boxtimes^{\ddot{\delta}} = \ddot{r}_\square$, $|\ddot{\phi}| = f$, and $[\ddot{\phi}, \ddot{r}_\square] = [\ddot{\phi}, \ddot{r}_\boxtimes] = 1$ and $\ddot{\delta}^{\ddot{\phi}} = \ddot{\delta}^{(-1)^{m(p-1)/2}}$. Thus*

$$\ddot{\Gamma} = \begin{cases} \ddot{\Delta} \times \langle\ddot{\phi}\rangle & \textit{if } \frac{1}{2}m(p-1) \textit{ is even} \\ \ddot{\Delta} \times \langle\ddot{r}_\square\ddot{\phi}\rangle & \textit{if } \frac{1}{2}m(p-1) \textit{ is odd.} \end{cases}$$

Proof. According to Proposition 2.7.1,

$$\ddot{\Gamma} = \langle \ddot{r}_\square, \ddot{r}_\boxtimes, \ddot{\delta}, \ddot{\phi} \rangle.$$

It is clear that $|\ddot{r}_\square| = |\ddot{r}_\boxtimes| = 2$ and $|\ddot{\phi}| = f$. Also note that $\ddot{\Gamma} = \ddot{\Delta}{:}\langle \ddot{\phi} \rangle$. We now argue that

$$[\ddot{r}_\square, \ddot{\phi}] = [\ddot{r}_\boxtimes, \ddot{\phi}] = 1. \tag{2.7.4}$$

For if $r_\square$ is the reflection in $v \in V$, then $[\phi, r_\square] = r_\square^\phi r_\square = r_{v^\phi} r_v$ and since $\tau(\phi) = 1$, we have $(v^\phi, v^\phi) = (v, v)^p$. Therefore the spinor norm of $[r_\square, \phi]$ is $(v,v)^{1+p} \in (\mathbf{F}^*)^2$. Thus $[r_\square, \phi] \in \Omega$, and similarly $[r_\boxtimes, \phi] \in \Omega$, proving (2.7.4). We have now accounted for all those relations involving only $\ddot{r}_\square$, $\ddot{r}_\boxtimes$ and $\ddot{\phi}$. We next consider those which involve $\ddot{\delta}$. In particular, we may assume for the rest of this proof that q is odd.

First we determine the order of $\ddot{\delta}$. As $\tau(\Omega\mathbf{F}^*) = (\mathbf{F}^*)^2 \neq \mathbf{F}^*$, we know that $\delta \notin \Omega\mathbf{F}^*$, and hence $|\ddot{\delta}| > 1$. Now $\delta^2\mu^{-1} \in S$, and so $\overline{\delta}^2 \in \overline{S}$. Thus $\ddot{\delta}$ has order 2 or 4. Now if $D = \boxtimes$, then $\overline{S} = \overline{\Omega}$, and so $|\ddot{\delta}| = 2$. Similarly, if m is even then $\det_W(\delta^2\mu^{-1}) = \mu^m \in (\mathbf{F}^*)^2$, where $W = \langle e_1, \ldots, e_m \rangle$, and so again $|\ddot{\delta}| = 2$ by Lemma 2.7.2. So consider the case where m is odd and $D = \square$, so that $q \equiv 1 \pmod 4$ by Proposition 2.5.10. We claim that $|\ddot{\delta}| = 4$. Assume for a contradiction that $\overline{\delta}^2 \in \overline{\Omega}$. Then $\delta^2\lambda \in \Omega$ for some $\lambda \in \mathbf{F}^*$, and since $\tau(\delta^2) = \mu^2$ and $\tau(\lambda) = \lambda^2$, we have $\lambda = \pm\mu^{-1}$. Therefore $(\pm\mu)^m = \det_W(\delta^2\lambda) \in (\mathbf{F}^*)^2$. But this is impossible as $-1 \in (\mathbf{F}^*)^2$, $\mu \notin (\mathbf{F}^*)^2$ and m is odd. Therefore we have shown

$$|\ddot{\delta}| = \begin{cases} 2 & \text{if } D = \boxtimes \\ 2(2, m-1) & \text{if } q \text{ is odd and } D = \square. \end{cases} \tag{2.7.5}$$

We now claim that

$$\ddot{r}_\square^{\ddot{\delta}} = \ddot{r}_\boxtimes, \quad \ddot{r}_\boxtimes^{\ddot{\delta}} = \ddot{r}_\square. \tag{2.7.6}$$

For if $r_\square = r_v$ as above and $r_\boxtimes = r_w$, then $(v\delta, v\delta) = \mu(v,v) \notin (\mathbf{F}^*)^2$, and so $(v\delta, v\delta)(w,w) \in (\mathbf{F}^*)^2$. Therefore $r_\square^\delta r_\boxtimes \in \Omega$, which shows $r_\square^\delta \equiv r_\boxtimes \pmod \Omega$. Similarly $r_\boxtimes^\delta \equiv r_\square \pmod \Omega$, and so (2.7.6) holds.

Finally we prove

$$\ddot{\delta}^{\ddot{\phi}} = \ddot{\delta}^{(-1)^{m(p-1)/2}}. \tag{2.7.7}$$

For $[\phi, \delta] = \delta_\beta(\mu^{-p})\delta = \delta_\beta(\mu^{1-p})$. Thus $g = [\phi, \delta]\mu^{(p-1)/2} \in N_I(W)$ and $\det_W(g) = \mu^{m(1-p)/2}$. Thus by by Lemma 2.7.2 we see that g lies in Ω if $\frac{1}{2}m(p-1)$ is even. Now if both $\frac{1}{2}m(p-1)$ and f are odd, then $D = \boxtimes$ and so $\ddot{g} \in \ddot{S} = \ddot{\Omega}$. Therefore $\ddot{\delta}^{\ddot{\phi}} = \ddot{\delta}$. But $|\ddot{\delta}| = 2$, and so $\ddot{\delta} = \ddot{\delta}^{-1}$, which shows that (2.7.7) holds once again. It remains to treat the case where $\frac{1}{2}m(p-1)$ is odd and f is even. Here $D = \square$ and $|\ddot{\delta}| = 4$ according to (2.7.5). Therefore $\ddot{\delta}^{\ddot{\phi}}$ is either $\ddot{\delta}$ or $\ddot{\delta}^{-1}$. Suppose for a contradiction that $[\ddot{\phi}, \ddot{\delta}] = 1$. Then $[\phi, \delta]\lambda \in \Omega$ for some $\lambda \in \mathbf{F}^*$. Clearly $\tau([\phi, \delta]) = \tau(\delta_\beta(\mu^{1-p})) = \mu^{(1-p)}$, and so $\lambda = \pm\mu^{(p-1)/2}$. Therefore $[\phi, \delta]\lambda = \pm g$, and so as before we have $\det_W(\pm g) = (\pm\mu)^{m(1-p)/2}$. But now $q \equiv 1 \pmod 4$ as f is even, and so $-1 \in (\mathbf{F}^*)^2$. Therefore $\pm\mu \notin (\mathbf{F}^*)^2$, and since $\frac{1}{2}m(1-p)$ is odd, we deduce that $\det_W([\phi,\delta]\lambda) \notin (\mathbf{F}^*)^2$, violating Lemma 2.7.2. This contradiction completes the proof of (2.7.7).

All the statements in the Proposition are straightforward consequences of the discussion above. To prove the final line in the statement of the Proposition, note that both $\ddot{r}_\square$ and $\ddot{\phi}$ invert $\ddot{\delta}$ (by conjugation) and so $\ddot{r}_\square\ddot{\phi}$ is central in $\ddot{\Gamma}$. Moreover $q \equiv 1 \pmod 4$ as $D = \square$ and m is odd, and so f must be even. Therefore $|\ddot{r}_\square\ddot{\phi}| = [2,f] = f$. ■

Remark. As stated in Theorem 2.1.4, $Aut(\overline{\Omega}) = \overline{A}$, so long as $n \geq 10$. When $n = 8$ however, $P\Omega_8^+(q)$ admits a triality automorphism of order 3 (see [Ca_1, Ch. 12]). In fact, in this case $|\mathrm{Aut}(\overline{\Omega}) : \overline{\Gamma}| = 3$ and

$$Aut(P\Omega_8^+(q)) \cong \begin{cases} S_3 \times \mathbf{Z}_f & \text{if } q \text{ is even} \\ S_4 \times \mathbf{Z}_f & \text{if } q \text{ is odd.} \end{cases}$$

Just as the graph automorphism of $Sp_4(2^{2a+1})$ gives rise to the exceptional Suzuki group $Sz(2^{2a+1})$, so does a triality automorphism of $P\Omega_8^+(q^3)$ give rise to the exceptional Steinberg triality group ${}^3D_4(q)$. These were first by discovered by Steinberg in [St_4]. Further details concerning $Out(P\Omega_8^+(q))$ appear in [Kl_1].

Recall from Description 4 in §2.5 that the set $\mathcal{U}_m$ of totally singular m-spaces splits into two families $\mathcal{U}_m^i$, $i = 1, 2$, and that there is a corresponding surjective homomorphsim $\gamma : \Gamma \to \mathbf{Z}_2$. We write $\overline{\gamma}$ and $\ddot{\gamma}$ for the induced maps on $\overline{\Gamma}$ and $\ddot{\Gamma}$, respectively.

Proposition 2.7.4.

(i) *Each reflection lies in* $I \backslash \ker_I(\gamma)$.

(ii) $|I : \ker_I(\gamma)| = 2$ *and*

$$\ker_I(\gamma) = \begin{cases} S & \text{if } q \text{ is odd} \\ \Omega & \text{if } q \text{ is even.} \end{cases}$$

(iii)

$$\ker_{\ddot{\Gamma}}(\ddot{\gamma}) = \begin{cases} \langle \ddot{\phi} \rangle & \text{if } q \text{ is even} \\ \langle \ddot{\delta} \rangle \times \langle \ddot{\phi} \rangle & \text{if } q \text{ is odd and } D = \boxtimes \\ \langle \ddot{r}_\square \ddot{r}_\boxtimes \rangle \times \langle \ddot{\delta} \rangle \times \langle \ddot{\phi} \rangle & \text{if } q \text{ is odd and } D = \square. \end{cases}$$

Proof. Assertion (i) has been established already in the proof of Lemma 2.5.8.i, and (ii) is a restatement of Lemma 2.5.8.i,ii. If $W = \langle e_1, \ldots, e_m \rangle$ as before, then clearly $\langle \delta, \phi \rangle \leq \Gamma_W \leq \ker(\gamma)$, and so $\ker(\gamma) = \ker_I(\gamma)\langle \delta, \phi \rangle = \Omega\langle r_\square r_\boxtimes, \delta, \phi \rangle$. Part (iii) now follows. ■

The next result is entirely analogous to Lemma 2.4.5, and its proof is identical.

Lemma 2.7.5. *If* $g \in \Delta$, *then* $\det(g) = \pm\tau(g)^{n/2}$.

By now the reader should be well versed in verifying all the assertions in Tables 2.1.C and 2.1.D. There is one subtlety which arises in case $\mathbf{O}^+$ however — namely, the entry appearing under $|\overline{S} : \overline{\Omega}|$ in Table 2.1.D. To check this entry, note first that $|S : \Omega| = 2$ (see (2.5.12)), and hence $|\overline{S} : \overline{\Omega}| \leq 2$. Now when q is even, we have $|\overline{S} : \overline{\Omega}| = |S : \Omega| = 2$; on the other hand, we indicate just below Table 2.1.D that $a_+a_- = 4$, and since $a_\pm \leq 2$, it follows that $a_+ = a_- = 2$. Thus $|\overline{S} : \overline{\Omega}| = a_+$ when q is even. When $\hat{q}$ is odd, observe that $|\overline{S} : \overline{\Omega}| = 1$ or 2, according as $-1 \in S\backslash\Omega$ or $-1 \in \Omega$. Hence by Proposition 2.5.13.iii, $|\overline{S} : \overline{\Omega}| = 1$ or 2, according as $\frac{1}{4}(q-1)n$ is odd or even. This is precisely the condition stated below Table 2.1.D.

§2.8 Orthogonal groups with Witt defect 1

In this section assume that case $\mathbf{O}^-$ holds, so that $\mathbf{F} = \mathbf{F}_q$ and $\kappa = Q$ with $n = 2m$ and $\mathrm{sgn}(Q) = -$. As always, $D = D(Q)$, $\mathbf{f} = \mathbf{f}_Q$ and $(v, w) = \mathbf{f}(v, w)$. Here

$$I = O_n^-(q) = O_{2m}^-(q).$$

Let

$$\beta = \begin{cases} \{e_1, \dots, f_{m-1}, x, y\} & \text{if } q \text{ is even} \\ \{v_1, \dots, v_n\} & \text{if } q \text{ is odd,} \end{cases} \tag{2.8.1}$$

where β is a standard basis as described in Proposition 2.5.3.ii when q is even, and is a basis described in Proposition 2.5.12 when q is odd.

To find suitable generators g_X we need only consider g_Δ and g_Γ, for (2.7.1) holds in case $\mathbf{O}^-$ as well as case $\mathbf{O}^+$. And of course we may take $g_A = 1$.

Assume here that q is odd, so that $\beta = \{v_1, \dots, v_n\}$ and $\mathbf{f}_\beta = \mathbf{I}_n$ or $\mathrm{diag}(\mu, 1, \dots, 1)$, according as D is a square or non-square. Choose $a, b \in \mathbf{F}^*$ such that $a^2 + b^2 = \mu$ and define $\delta = \delta_{Q,\beta}(\mu)$ as the element which satisfies

$$\delta_{\{v_{2i-1}, v_{2i}\}} = \begin{cases} \left(\begin{smallmatrix} a & b \\ b & -a \end{smallmatrix}\right) & \text{if } \{v_{2i-1}, v_{2i}\} \text{ is orthonormal} \\ \left(\begin{smallmatrix} 0 & \mu \\ 1 & 0 \end{smallmatrix}\right) & \text{otherwise.} \end{cases} \tag{2.8.2}$$

Note that $\{v_{2i-1}, v_{2i}\}$ is usually orthonormal, the only exception being when $i = 1$ and D is a non-square. One checks that $\delta \in \Delta$ and that $\tau(\delta) = \mu$. Hence as before we obtain

$$\Delta = \begin{cases} I \times \mathbf{F}^* & \text{if } q \text{ is even} \\ I\langle\delta\rangle & \text{if } q \text{ is odd.} \end{cases} \tag{2.8.3}$$

When q is even, define $\phi = \phi_{Q,\beta}$ as the unique element of $\sigma^{-1}(\nu)$ which fixes $e_1, \dots, f_{m-1}, x$ and which satisfies $y\phi = \zeta x + y$, where ζ is given in Lemma 2.5.2.ii. When q is odd, let

$$\phi = \phi_{Q,\beta} = \begin{cases} \phi_\beta(\nu)\mathrm{diag}_\beta(\mu^{(p-1)/2}, 1, \dots, 1) & \text{if } D = \boxtimes \\ \phi_\beta(\nu) & \text{if } D = \square. \end{cases} \tag{2.8.4}$$

Thus for all q we have $\phi \in \Gamma$, with $\tau(\phi) = 1$ and $\sigma(\phi) = \nu$. Therefore

$$\Gamma = \Delta\langle\phi\rangle. \tag{2.8.5}$$

Note that we do not use the symbol ':' here, for in general $\Delta \cap \langle\phi\rangle$ can be non-trivial. It now follows as in §2.7 that Proposition 2.7.1 holds, although δ and ϕ have different meanings here.

Before we state the main structure theorem for $\ddot{A}$, we first introduce the homomorphism $\gamma = \gamma_Q$ which is analogous to that defined in §2.5.

Let $\mathbf{F}_\sharp$ be a quadratic field extension of $\mathbf{F}$ and form the tensor product $V_\sharp = V \otimes \mathbf{F}_\sharp$, so that $V_\sharp$ is an n-dimensional space over $\mathbf{F}_\sharp$. We regard V as a subset of $V_\sharp$ by writing $V = \{v \otimes 1 \mid v \in V\}$. Thus β is a basis for $(V_\sharp, \mathbf{F}_\sharp)$, and we let $\mathbf{f}_\sharp$ be the bilinear form on $V_\sharp$ satisfying $(\mathbf{f}_\sharp)_\beta = \mathbf{f}_\beta$. It is easy to check that there is a unique non-degenerate quadratic form $Q_\sharp$ on $(V_\sharp, \mathbf{F}_\sharp)$ such that $\mathbf{f}_{Q_\sharp} = \mathbf{f}_\sharp$ and such that the restriction of $Q_\sharp$ to V is equal to Q. We claim that $\mathrm{sgn}(Q_\sharp) = +$. To see this, work with a standard basis given in Proposition 2.5.3.ii and note that the polynomial $\mathbf{x}^2 + \mathbf{x} + \zeta$ is reducible over $\mathbf{F}_\sharp$. Hence one can exhibit a totally singular m-space in $V_\sharp$. Now let $X_\sharp = X(V_\sharp, \mathbf{F}_\sharp, Q_\sharp)$ as X ranges over the symbols in (2.1.15) and let $\sigma_\sharp = \sigma_{Q_\sharp}$, $\tau_\sharp = \tau_{Q_\sharp}$. Define $N = N_{\Gamma_\sharp}(V)$ and for $g \in N$, let $g\varrho$ denote the restriction of g to V.

Proposition 2.8.1.
(i) *ϱ is a surjective homomorphism from N to Γ.*
(ii) $\ker(\varrho)$ *is generated by an involution in $\Gamma_\sharp \backslash \Delta_\sharp$.*
(iii) $(N_{X_\sharp}(V))\varrho = X$, *where X ranges over S, I and Δ.*
(iv) *We have*

$$(N_{\Omega_\sharp}(V))\varrho = \begin{cases} \Omega & \text{if } q \text{ is even} \\ S & \text{if } q \text{ is odd.} \end{cases}$$

Proof. (i) Take $g \in N$, $\lambda \in \mathbf{F}$ and $v \in V$. Evidently $(\lambda v)(g\varrho) = \lambda^{\sigma_1(g)}(v(g\varrho))$, where $\sigma_1(g)$ is the restriction of $\sigma_\sharp(g)$ to $\mathbf{F}$. Consequently $g\varrho \in \Gamma(V, \mathbf{F}) \cong \Gamma L_n(q)$. Furthermore, $Q(v(g\varrho)) = Q_\sharp(vg) = \tau_\sharp(g)Q(v)^{\sigma_\sharp(g)}$. Since both $Q(v(g\varrho))$ and $Q(v)$ lie in $\mathbf{F}$ (and since $Q(v) \neq 0$ for some $v \in V$), it follows that $\tau_\sharp(g) \in \mathbf{F}^*$. Consequently $g\varrho \in \Gamma$, with $\sigma(g\varrho) = \sigma_1(g)$ and $\tau(g\varrho) = \tau_\sharp(g)$. Thus we have shown that $N\varrho \leq \Gamma$. We must now show that ϱ is surjective. To do so, choose $h \in \Delta$, and observe that h extends naturally to an element $h_\sharp \in GL(V_\sharp, \mathbf{F}_\sharp)$, where $h_\sharp$ satisfies $(v \otimes \lambda)h_\sharp = vh \otimes \lambda$ ($v \in V$, $\lambda \in \mathbf{F}_\sharp$). Now

$$\begin{aligned} Q_\sharp((v \otimes \lambda)h_\sharp) &= Q_\sharp(vh \otimes \lambda) = \lambda^2 Q_\sharp(vh \otimes 1) = \lambda^2 Q(vh) = \lambda^2 \tau(h) Q(v) \\ &= \lambda^2 \tau(h) Q_\sharp(v \otimes 1) = \tau(h) Q_\sharp(v \otimes \lambda). \end{aligned}$$

Consequently $h_\sharp \in \Delta_\sharp$, and since $h_\sharp\varrho = h$, we see that $\Delta \leq N\varrho$. Next take $\phi = \phi_{Q,\beta}$ as given above, and extend ν to a generator $\nu_\sharp$ of $Aut(\mathbf{F}_\sharp)$. Define $\phi_\sharp$ as the (unique) element of $\Gamma(V_\sharp, \mathbf{F}_\sharp)$ which agrees with ϕ on the basis β of V. Then one checks that $\phi_\sharp \in N$ and that $\phi_\sharp\varrho = \phi$. Thus we have shown that $N\varrho$ contains $\Delta\langle\phi\rangle = \Gamma$, as desired.

(ii) Take $g \in \ker(\varrho) = C_{\Gamma_\sharp}(V)$. If $g \in \Delta_\sharp$, then $g = 1$, for g fixes each basis vector in β. So assume that $g \in \Gamma_\sharp \backslash \Delta_\sharp$, so that $\sigma_\sharp(g) \neq 1$. In the proof of (i) we saw that $\sigma(g\varrho)$ is the restriction of $\sigma_\sharp(g)$ to $\mathbf{F}$, and this restriction must be trivial. Therefore $|g| = |\sigma_\sharp(g)| \leq 2$. On the other hand, observe that $\phi_\sharp^f$ is an element of order at least 2 in $\ker(\varrho)$, and hence $\ker(\varrho) = \langle {\phi_\sharp}^f \rangle \cong \mathbf{Z}_2$.

(iii) Parts (i) and (ii) imply that ϱ induces an isomorphism between $N_{\Delta_\sharp}(V)$ and Δ. Moreover, the proof of (i) shows that $\tau(g\varrho) = \tau_\sharp(g)$ for all $g \in N_{\Delta_\sharp}(V)$, and so ϱ

also induces an isomorphism between $N_{I_\sharp}(V)$ and I. And since ϱ preserves determinants, there is a similar statement for S.

(iv) Suppose first that q is odd. According to the argument so far, any element in $N_{S_\sharp}(V)$ is of the form $h_\sharp$, where $h \in S$ and the map $h \mapsto h_\sharp$ is described in (ii). Now by (2.5.13), $h = r_{v_1} \dots r_{v_k}$, where v_i is a non-singular vector in V and k is even. Clearly $v_i \otimes 1$ is a non-singular vector in $V_\sharp$, and it follows that $h_\sharp = r_{v_1 \otimes 1} \dots r_{v_k \otimes 1}$. Moreover, $\mathbf{f}_\sharp(v_i \otimes 1, v_i \otimes 1) = \mathbf{f}(v_i, v_i) \in \mathbf{F}^* \leq (\mathbf{F}_\sharp^*)^2$. Therefore the spinor norm of $h_\sharp$ is a square, which proves $h_\sharp \in \Omega_\sharp$. Consequently $N_{S_\sharp}(V) = N_{\Omega_\sharp}(V)$, as desired. Now assume that q is even. This time we find that for a non-singular vector $v \in V$, the element $r_{v \otimes 1}$ lies in $S_\sharp \backslash \Omega_\sharp$. Therefore $|N_{S_\sharp}(V) : N_{\Omega_\sharp}(V)| = 2$. Since Ω is the unique subgroup of index 2 in S by Proposition 2.5.7, $N_{\Omega_\sharp}(V)$ must be the unique subgroup of index 2 in $N_{S_\sharp}(V)$. Part (iv) now follows. ■

Proposition 2.8.1 shows that the $O_n^-(q)$ geometry embeds naturally in the $O_n^+(q^2)$ geometry, and this embedding often provides a useful point of view for studying the $O_n^-(q)$ geometry. In several instances we will regard the groups Ω, S, I and Δ as contained in the corresponding groups $\Omega_\sharp$, $S_\sharp$, $I_\sharp$ and $\Delta_\sharp$ in $O_n^+(q^2)$. However, one must take care to note that the full semilinear group Γ does not in general embed in $\Gamma_\sharp$, for ϱ is not faithful (see Proposition 2.8.1.ii).

We are now in position to define $\gamma = \gamma_Q$. It will be defined not on all of Γ, but only on Δ, as follows. Take $g \in \Delta$ and let $g_\sharp$ be the unique element of $\Delta_\sharp$ to which it extends (see the proof of Proposition 2.8.1.i). Then define $\gamma(g) = \gamma_{Q_\sharp}(g_\sharp)$. We write $\overline{\gamma}$ and $\ddot{\gamma}$ for the induced homomorphisms on $\overline{\Delta}$ and $\ddot{\Delta}$, respectively. Since both $\Delta_\sharp$ and $\ker(\gamma_{Q_\sharp})$ are normal in $\Gamma_\sharp$, it follows that

$$\ker_\Delta(\gamma) \trianglelefteq \Gamma \quad \text{and} \quad \ker_{\ddot{\Delta}}(\ddot{\gamma}) \trianglelefteq \ddot{\Gamma}. \tag{2.8.6}$$

We will have more to say about $\ker(\gamma)$ in Proposition 2.8.3, below.

Proposition 2.8.2.

(i) *We have*

$$\ddot{A} = \ddot{\Gamma} = \begin{cases} \langle \ddot{\phi} \rangle \cong \mathbf{Z}_{2f} & q \text{ even} \\ \langle \ddot{\delta} \rangle \times \langle \ddot{\phi} \rangle \cong \mathbf{Z}_2 \times \mathbf{Z}_{2f} & q \text{ odd, } D = \boxtimes \\ \langle \ddot{r}_\square, \ddot{r}_\boxtimes, \ddot{\delta} \rangle \times \langle \ddot{\phi} \rangle \cong D_8 \times \mathbf{Z}_f & q \text{ odd, } D = \square. \end{cases}$$

(ii) *If q is odd and D is a square (so that $\frac{1}{4}n(q-1)$ is odd), then* $|\ddot{r}_\square| = |\ddot{r}_\boxtimes| = |\ddot{\delta}| = 2$, $|\ddot{\phi}| = f$, $[\ddot{\phi}, \ddot{r}_\square] = [\ddot{\phi}, \ddot{r}_\boxtimes] = [\ddot{\phi}, \ddot{\delta}] = 1$ *and* $\ddot{r}_\square^{\ddot{\delta}} = \ddot{r}_\boxtimes$.

Proof. To prove (i), assume first that q is even. Thus $\ddot{\Gamma} = \langle \ddot{r}_\square, \ddot{\phi} \rangle$ by Proposition 2.7.1 (which, as we have pointed out, still holds in this section). One easily verifies that $y\phi^f = \mathrm{T}^{\mathbf{F}}_{\mathbf{F}_2}(\zeta)x + y$, and since $\mathbf{x}^2 + \mathbf{x} + \zeta$ is irreducible over $\mathbf{F}$, it follows that $\mathrm{T}^{\mathbf{F}}_{\mathbf{F}_2}(\zeta) = 1$. Since ϕ^f centralizes $e_1, \dots, f_{m-1}, x$, we see that ϕ^f is the involution $r_x \in I \backslash \Omega$. It now follows that $\ddot{\Gamma} = \langle \ddot{\phi} \rangle$ and that $|\ddot{\phi}| = 2f$.

Thus for the remainder of this proof we may assume that q is odd. Now an easy calculation shows that $\delta^2 \in \mathbf{F}^*$, and since $\delta \notin \Omega\mathbf{F}^*$ (see the proof of Proposition 2.7.3) we have

$$|\ddot{\delta}| = 2 \text{ when } q \text{ is odd.} \tag{2.8.7}$$

We also see that (2.7.4) and (2.7.6) hold from §2.7.

Consider here the case where q is odd and D is a square. Evidently $\ddot{\Gamma} = \ddot{\Delta}\langle\ddot{\phi}\rangle = \langle \ddot{r}_\square, \ddot{r}_\boxtimes, \ddot{\delta}, \ddot{\phi}\rangle$ by Proposition 2.7.1. As in the proof of Proposition 2.7.3, $\ddot{r}_\square, \ddot{r}_\boxtimes$ and $\ddot{\delta}$ satisfy the relations in (i) above, generating a group D_8. Moreover $\phi^f = 1$, and so $|\ddot{\phi}| = f$, which is odd since $q \equiv 3 \pmod 4$. Because $Aut(D_8)$ has no non-trival elements of odd order, $\ddot{\phi}$ must centralize $\ddot{\Delta}$, and hence we obtain the desired result.

Finally, take the case where D is a non-square. Then $\ddot{\Gamma} = \langle \ddot{r}_\square, \ddot{\delta}, \ddot{\phi}\rangle$ and $\ddot{\Delta} = \langle \ddot{r}_\square\rangle \times \langle\ddot{\delta}\rangle \cong \mathbf{Z}_2 \times \mathbf{Z}_2$. Since $r_\square \notin \ker(\gamma)$ (see Proposition 2.8.3, below) we have in fact $\ddot{\Delta} = \ker_{\ddot{\Delta}}(\ddot{\gamma}) \times \langle \ddot{r}_\square\rangle$, and in view of (2.8.6) we deduce $\ker_{\ddot{\Delta}}(\ddot{\gamma}) \leq Z(\ddot{\Gamma})$. Moreover $\phi^f = r_{v_1}$, and hence $\ddot{\phi}^f = \ddot{r}_\square$. Therefore $\ddot{\phi}$ centralizes both $\ddot{r}_\square$ and $\ker_{\ddot{\Delta}}(\ddot{\gamma})$, and hence $[\ddot{\phi}, \ddot{\Delta}] = 1$. All statements in the Proposition have now been proved. ■

Proposition 2.8.3.
(i) *Each reflection lies in* $I \backslash \ker_I(\gamma)$.
(ii) $|I : \ker_I(\gamma)| = 2$ *and*

$$\ker_I(\gamma) = \begin{cases} S & \text{if } q \text{ is odd} \\ \Omega & \text{if } q \text{ is even.} \end{cases}$$

(iii)

$$\ker_{\ddot{\Delta}}(\ddot{\gamma}) = \begin{cases} 1 & \text{if } q \text{ is even} \\ \langle \ddot{r}_\square \ddot{\delta}\rangle \cong \mathbf{Z}_2 & \text{if } qm \text{ is odd and } D = \boxtimes \\ \langle \ddot{r}_\square \ddot{\delta}\rangle \cong \mathbf{Z}_4 & \text{if } q \text{ is odd and } D = \square \\ \langle \ddot{\delta}\rangle \cong \mathbf{Z}_2 & \text{if } q \text{ is odd and } m \text{ is even.} \end{cases}$$

Proof. We retain the notation introduced in the discussion before Proposition 2.8.1, so that $V \subset V_\sharp = V \otimes \mathbf{F}_\sharp$. If v is a non-singular vector in V, then it is also non-singular in $V_\sharp$. Moreover the reflection $r_v \in I$ extends to a reflection $r_{v\otimes 1} \in I_\sharp$, and so $r_v \notin \ker_I(\gamma)$ by Proposition 2.7.4. Thus (i) holds, and (ii) is now immediate in view of (2.5.13).

We now prove (iii). When q is even, $\ddot{\Delta} = \ddot{I} = \langle \ddot{r}_v\rangle$, and so by (i) $\ker_{\ddot{\Delta}}(\ddot{\gamma}) = 1$. For the rest of this proof take q odd, so that $\beta = \{v_1, \ldots, v_n\}$ is as described in Proposition 2.5.12. and δ is as given in (2.8.2). Set $V_j = \langle v_{2j-1}, v_{2j}\rangle \otimes \mathbf{F}_\sharp$, a +2-space in $V_\sharp$, so that $\delta_\sharp$ fixes each V_j, where $\delta_\sharp$ is the extension of δ to $V_\sharp$ (see the proof of Proposition 2.8.1.i). As a convenience we set $w = v_{2j-1}$ and $v = v_{2j}$. If $\mathbf{f}(w,w) = \mathbf{f}(v,v) = 1$, then the two singular 1-spaces in V_j are spanned by $w \pm v \otimes i$, where $i = \sqrt{-1}$. But $(w + v\otimes i)\delta_\sharp = (a+ib)(w - v\otimes i)$, and hence $\delta_\sharp$ interchanges the singular 1-spaces in V_j. And if $\mathbf{f}(w,w) = \mu$, then the two singular 1-spaces in V_i are spanned by $w \pm v \otimes \eta$, where $\eta \in \mathbf{F}_\sharp$ satisfies $\eta^2 = -\mu$. Then $(w + v\otimes\eta)\delta_\sharp = \eta(w - v\otimes\eta)$, and so once again $\delta_\sharp$ interchanges the singular 1-spaces in V_j. Therefore $\delta \in \ker(\gamma)$ if and only if m is even and $r_\square\delta \in \ker(\gamma)$ if and only if m is odd. Thus when D is a non-square, Proposition 2.7.1 and the proof of Proposition 2.8.2 yield $\ddot{\Delta} = \langle \ddot{r}_\square\rangle \times \langle\ddot{\delta}\rangle \cong \mathbf{Z}_2 \times \mathbf{Z}_2$, and hence $\ker_{\ddot{\Delta}}(\ddot{\gamma}) = \langle\ddot{\delta}\rangle$ or $\langle \ddot{r}_\square \ddot{\delta}\rangle$

according as m is even or odd. Finally, assume that D is a square, so that m is odd. Then according to Proposition 2.8.2.ii, $\ddot{\Delta} \cong D_8$ and $|\ddot{r}_\square \ddot{\delta}| = 4$. Therefore $\langle \ddot{r}_\square \ddot{\delta} \rangle$ has index 2 in $\ddot{\Delta}$, and so it must be all of $\ker_{\ddot{\Delta}}(\ddot{\gamma})$. The proof is now complete. ■

Just like Lemmas 2.4.5 and 2.7.5, we have

Lemma 2.8.4. *If $g \in \Delta$, then* $\det(g) = \pm\tau(g)^{n/2}$.

The statements in Tables 2.1.C and 2.1.D concerning case $\mathbf{O}^-$ are checked in the same way as for case $\mathbf{O}^+$. All this is left to the reader.

§2.9 Structure and isomorphisms

In this section we summarize information concerning the structures of the classical groups introduced in the previous sections, and the isomorphisms between them. We continue to assume that $\Omega = \Omega(V, \mathbf{F}, \kappa)$ in one of the cases $\mathbf{L}$, $\mathbf{U}$, $\mathbf{S}$ or $\mathbf{O}$.

As we mentioned in §2.1, the groups $\overline{\Omega}$ give rise to the finite simple classical groups (see Theorem 2.1.3). When $n \geq 7$, $\overline{\Omega}$ is always simple, and distinct geometries give rise to non-isomorphic simple groups. For small n, however, there are exceptional isomorphisms given in the following Proposition.

Proposition 2.9.1. *We have the following isomorphisms.*

(i) $SL_2(q) \cong Sp_2(q) \cong SU_2(q)$.
(ii) *For q odd,* $L_2(q) \cong \Omega_3(q)$.
(iii) $O_2^\pm(q) \cong D_{2(q\mp1)}$, $SO_2^\pm(q) \cong \mathbf{Z}_{q\mp1}.(2,q)$ *and* $\Omega_2^\pm(q) \cong \mathbf{Z}_{(q\mp1)/(2,q-1)}$.
(iv) $\Omega_4^+(q) \cong SL_2(q) \circ SL_2(q) \cong (2, q-1).(L_2(q) \times L_2(q))$.
(v) $\Omega_4^-(q) \cong L_2(q^2)$.
(vi) *For q odd,* $PSp_4(q) \cong \Omega_5(q)$.
(vii) $P\Omega_6^\pm(q) \cong L_4^\pm(q)$.
(viii) $L_2(2) \cong S_3$.
(ix) $L_2(3) \cong A_4$.
(x) $L_2(4) \cong L_2(5) \cong A_5$.
(xi) $L_2(7) \cong L_3(2)$.
(xii) $L_2(9) \cong A_6$.
(xiii) $L_4(2) \cong A_8$.
(xiv) $U_3(2) \cong 3^2.Q_8$.
(xv) $U_4(2) \cong PSp_4(3)$.
(xvi) $Sp_4(2) \cong S_6$.

This list in fact includes all isomorphisms amongst the classical groups and between the classical groups and the alternating groups. There are of course further isomorphisms which can be drawn by combining several parts of the Proposition. For example, $\Omega_6^-(2) \cong U_4(2)$ by (vii) and $\Omega_5(3) \cong PSp_4(3)$ by (vi), and hence $\Omega_6^-(2) \cong \Omega_5(3)$ by (xv). Also, one obtains $\Omega_4^-(3) \cong L_2(9) \cong A_6 \cong Sp_4(2)'$ by combining (v), (xii) and (xvi). Most of the isomorphisms occurring in the Proposition involve either of the groups Ω or $P\Omega$. There are of course related isomorphisms when we consider S, I, Δ, etc. and the corresponding projective groups. For example, $SO_3(3) \cong S_4$, $O_4^-(2) \cong S_5 \cong P\Gamma L_2(4) \cong PGL_2(5)$, $O_6^+(2) \cong S_8$ and $P\Gamma O_4^-(3) \cong P\Gamma L_2(9) \cong Aut(A_6)$. We now proceed with the proof of the Proposition. These facts are essentially well known, so we provide only a sketch proof.

Proof. Parts (viii) and (ix) are trivial, and (xiv) is also easy. Parts (x)-(xiii), (xv) and (xvi) can be found in [Di$_3$]. We sketch some proofs for parts (i)-(vii); details can be found in [Ar].

First consider (i). According to Lemma 2.1.8, we have

$$Sp_2(q) = \left\{ g \in GL_2(q) \mid g \begin{pmatrix} 0 & 1 \\ -1 & 0 \end{pmatrix} g^t = \begin{pmatrix} 0 & 1 \\ -1 & 0 \end{pmatrix} \right\},$$

and one checks that this group is precisely $SL_2(q)$. In (4.5.7) of Chapter 4 we exhibit the embedding $Sp_n(q) \leq GU_n(q)$, and since all elements in $Sp_n(q)$ have determinant 1, we have $Sp_n(q) \leq SU_n(q)$. Putting $n = 2$ and comparing orders exhibits the isomorphism $Sp_2(q) \cong SU_2(q)$, and so (i) holds.

As for (ii), let v_1, v_2 be a basis for the natural 2-dimensional module for $SL_2(q)$. Then $v_1 \otimes v_1, v_2 \otimes v_2, v_1 \otimes v_2 + v_2 \otimes v_1$ is a basis for the symmetric square, and with respect to this basis $SL_2(q)$ fixes the non-degenerate symmetric bilinear form given by $(v_i \otimes v_i, v_j \otimes v_j) = 1 - \delta_{ij}$, $(v_i \otimes v_i, v_1 \otimes v_2 + v_2 \otimes v_1) = 0$ and $(v_1 \otimes v_2 + v_2 \otimes v_1, v_1 \otimes v_2 + v_2 \otimes v_1) = -2$ $(1 \leq i, j \leq 2)$. This yields an embedding $L_2(q) \leq O_3(q)$, and since $O^2(L_2(q)) = L_2(q)$, it follows that $L_2(q) \leq \Omega_3(q)$. The isomorphism follows by comparing orders.

Now we justify (iii). Let e_1, f_1 be a standard basis for $O_2^+(q)$, as given in Proposition 2.5.3.i, and for $\lambda \in \mathbf{F}^*$ define

$$M(\lambda) = \begin{pmatrix} \lambda & 0 \\ 0 & \lambda^{-1} \end{pmatrix}, \quad N = \begin{pmatrix} 0 & 1 \\ 1 & 0 \end{pmatrix}.$$

Letting μ be a generator for $\mathbf{F}^*$ as before, we see that $\langle M(\mu), N \rangle \cong D_{2(q-1)}$. One easily checks using Lemma 2.1.8 that $\langle M(\mu), N \rangle \leq O_2^+(q)$, and since $|O_2^+(q)| = 2(q-1)$ by Proposition 2.5.5, we obtain $O_2^+(q) \cong D_{2(q-1)}$. Now consider $O_2^-(q)$. As demonstrated in the discussion preceding Proposition 2.8.1, $O_2^-(q)$ is contained in $O_2^+(q^2)$. Moreover, using the basis $\{x, y\}$ provided by Lemma 2.5.2, one checks that r_x and r_y do not commute; since any non-abelian subgroup of a dihedral group is again dihedral, we conclude using Proposition 2.5.5 that $O_2^-(q) \cong D_{2(q+1)}$. When q is even, we have $O_2^{\pm}(q) = SO_2^{\pm}(q)$, and by Proposition 2.5.7 $\Omega_2^{\pm}(q)$ is the unique subgroup of index 2 in $SO_2^{\pm}(q)$, which must be the cyclic subgroup of order $q \mp 1$. When q is odd, $\langle M(\mu) \rangle \leq SO_2^+(q)$, and so equality holds by comparing orders. Therefore $SO_2^+(q) \cong \mathbf{Z}_{q-1}$. Moreover $SO_2^-(q) \leq SO_2^+(q^2) \cong \mathbf{Z}_{q^2-1}$, and hence $SO_2^-(q) \cong \mathbf{Z}_{q+1}$. It now follows that $\Omega_2^{\pm}(q) \cong \mathbf{Z}_{(q \mp 1)/2}$ by Proposition 2.5.7.

Next we deal with (iv). Let W be a 2-space over $\mathbf{F} = \mathbf{F}_q$ on which $SL_2(q)$ acts naturally. By (i), W admits an $SL_2(q)$-invariant non-degenerate symplectic form $\mathbf{f}$. Now $SL_2(q) \times SL_2(q)$ acts naturally on $W \otimes W$ as follows: $(u \otimes w)(g, h) = ug \otimes wh$

($u, w \in W$, $g, h \in SL_2(q)$). Obviously $(-1,-1)$ acts trivially on $W \otimes W$, and so this is a faithful action of a central product $SL_2(q) \circ SL_2(q)$. If we let $\mathbf{g}$ be the bilinear form on $W \otimes W$ given by $\mathbf{g}(u_1 \otimes w_1, u_2 \otimes w_2) = \mathbf{f}(u_1, u_2)\mathbf{f}(w_1, w_2)$ ($u_i, w_i \in W$), then $\mathbf{g}$ is a non-degenerate symmetric bilinear form preserved by $SL_2(q) \circ SL_2(q)$. Thus we obtain an embedding $SL_2(q) \circ SL_2(q) \le I(W \otimes W, \mathbf{g})$. When q is odd, $I(W \otimes W, \mathbf{g}) \cong O_4^\epsilon(q)$, and in fact $\epsilon = +$ since $W \otimes w$ is a totally singular 2-space. Since $O^2(SL_2(q)) = SL_2(q)$, we have $SL_2(q) \circ SL_2(q) \le O^2(O_4^+(q)) \le \Omega_4^+(q)$, and the result now holds by comparing orders. When q is even, we let w_1, w_2 be a basis for W, and we leave it to the reader to check that there is a unique quadratic form Q on $W \otimes W$ satisfying $Q(w_i \otimes w_j) = 0$ ($1 \le i, j \le 2$) and such that $\mathbf{g}$ is the associated bilinear form. One also checks that $SL_2(q) \circ SL_2(q)$ preserves Q, and hence $SL_2(q) \circ SL_2(q) \le I(W \otimes W, Q)$. As before, we obtain $SL_2(q) \circ SL_2(q) \le O_4^+(q)$, and since each factor $SL_2(q)$ fixes either $W \otimes w_1$ or $w_1 \otimes W$, both of which are totally singular 2-spaces, it follows from Description 4 in §2.5 that $SL_2(q) \circ SL_2(q) \le \Omega_4^+(q)$. These groups have the same order, so the result is proved.

Now consider (v). Let W be the natural 2-dimensional module over $\mathbf{F}_{q^2}$ for $SL_2(q^2)$, and let v_1, v_2 be a basis. Let α denote the involutory field automorphism of $\mathbf{F}_{q^2}$, and put $\psi = \phi_{\{v_1,v_2\}}(\alpha)$. Thus ψ induces a field automorphism on $SL_2(q^2)$ (see §2.2). Define $\overline{W}$ to be the $SL_2(q^2)$-module with underlying space W and with action $w * g = wg^\psi$. Take $SL_2(q^2)$ to act on the 4-dimensional space $W \otimes \overline{W}$ over $\mathbf{F}_{q^2}$ as follows: $(w_1 \otimes w_2)g = (w_1 g) \otimes (w_2 * g)$. Clearly $-1 \in SL_2(q^2)$ acts trivially, and so we obtain an embedding $L_2(q^2) \le GL_4(q^2)$. Now $\beta = \{v_1 \otimes v_1, v_2 \otimes v_2, v_1 \otimes v_2 + v_2 \otimes v_1, \lambda v_1 \otimes v_2 + \lambda^\alpha v_2 \otimes v_1\}$ is a basis for $W \otimes \overline{W}$ over $\mathbf{F}_{q^2}$, where λ is any element in $\mathbf{F}_{q^2} \backslash \mathbf{F}_q$. Observe however that $L_2(q^2)$ actually fixes the 4-dimensional $\mathbf{F}_q$-space $W_o = \mathrm{span}_{\mathbf{F}_q}(\beta)$. This yields an embedding $L_2(q^2) \le GL_4(q)$. Since $SL_2(q^2)$ fixes a non-degenerate symplectic form on W and on $\overline{W}$, it follows as in the previous paragraph that $L_2(q^2)$ fixes a non-degenerate quadratic form on $W \otimes \overline{W}$. Restricting this form to W_o yields $L_2(q^2) \le O_4^\epsilon(q)$. Since $L_2(q^2)$ does not embed in $SL_2(q) \circ SL_2(q)$, it follows that $\epsilon = -$, and we deduce the desired result.

For (vii), let W be the natural 4-dimensional module over $\mathbf{F} = \mathbf{F}_q$ for $SL_4(q)$ or over $\mathbf{F} = \mathbf{F}_{q^2}$ for $SU_4(q) \cong SL_4^-(q)$. Then let $V = \Lambda^2(W)$, the exterior square of W of dimension 6. Then V admits a non-degenerate quadratic form Q satisfying $Q(w_1 \wedge w_2) = 0$ and $\mathbf{f}_Q(w_1 \wedge w_2, w_3 \wedge w_4) = w_1 \wedge w_2 \wedge w_3 \wedge w_4$ (here $w_1 \wedge w_2 \wedge w_3 \wedge w_4$ is an element of $\Lambda^4(W)$, which may be identified with $\mathbf{F}$ — see [La, p.426]). This can be used to embed $L_4^\pm(q)$ in $P\Omega_6^\pm(q)$, and the isomorphism follows by comparing orders. A similar idea gives (vi) — here the exterior square $\Lambda^2(W)$ (where W is the natural 4-dimensional module) is reducible for $Sp_4(q)$, having a 5-dimensional submodule.

Finally we consider (xvi). Let the symmetric group S_{2n} act on a vector space V of dimension $2n$ over $\mathbf{F}_2$ by naturally permuting a basis $\{v_1, \ldots, v_{2n}\}$. Evidently S_{2n} preserves the non-degenerate symplectic form on V given by $(v_i, v_j) = 1 + \delta_{ij}$. Moreover S_{2n} fixes $v = v_1 + \cdots + v_{2n}$ and acts on the symplectic space $v^\perp/\langle v \rangle$, giving an embedding

$$S_{2n} \le Sp_{2n-2}(2).$$

When $n = 3$, a comparison of orders shows $S_6 \cong Sp_4(2)$. ■

In light of the isomorphisms in Proposition 2.9.1, two classical groups can be isomorphic and yet have different natural geometries. For example, there is a classical group X which is isomorphic to $PSp_4(3)$, $\Omega_5(3)$, $U_4(2)$ and $\Omega_6^-(2)$, and hence there are four classical geometries associated to the same group X. Often we wish to use the symbol Ω to specify both the relevant classical group *and* its associated geometry. Thus if $C\ell_n(q)$ denotes a classical group with associated geometry $(V, \mathbf{F}, \kappa)$ of dimension n over $\mathbf{F} = \mathbf{F}_{q^u}$, then we write

$$\Omega \approx C\ell_n(q) \tag{2.9.1}$$

to emphasize that Ω is this classical group with this particular geometry $(V, \mathbf{F}, \kappa)$. To illustrate, $\Omega \approx \Omega_5(3)$ means that $\Omega \cong \Omega_5(3)$ *and* that Ω is to be regarded as a 5-dimensional orthogonal group over $\mathbf{F}_3$; whereas $\Omega \approx U_4(2)$ means that $\Omega \cong U_4(2)$ *and* that Ω is to be regarded as a 4-dimensional unitary group over $\mathbf{F}_4$. Thus the two expressions $\Omega \approx \Omega_5(3)$ and $\Omega \approx U_4(2)$ give different meanings to the symbol Ω. We adopt similar notation for the other symbols in (2.1.15), as well as for the corresponding projective groups.

Using Proposition 2.9.1 and Theorem 2.1.3, one can read off precisely which classical groups $P\Omega$ are simple, without the restriction on n appearing in Theorem 2.1.3. For example, $P\Omega_4^-(q)$ is always simple since $L_2(q^2)$ is; and for odd q, $P\Omega_3(q)$ is simple if and only if $q \geq 5$. In fact, we can deduce the following result, which conveys rather precise information concerning the structures of Ω and $P\Omega$. We shall be quoting this Proposition on many occasions in subsequent chapters.

Proposition 2.9.2. *Let $\Omega = \Omega(V, \mathbf{F}, \kappa)$ be as above.*

(i) *Ω is soluble if and only if either $n = 1$ or $\Omega \approx SL_2(2)$, $SL_2(3)$, $Sp_2(2)$, $Sp_2(3)$, $SU_2(2)$, $SU_2(3)$, $\Omega_2^\pm(q)$, $SU_3(2)$, $\Omega_3(3)$, $\Omega_4^+(2)$ or $\Omega_4^+(3)$. Moreover, Ω is trivial if and only if either $n = 1$ or $\Omega \approx \Omega_2^+(q)$ with $q \leq 3$.*

(ii) *If Ω is insoluble, then Ω' is a central product of quasisimple groups and exactly one of the following holds:*
   (a) *$P\Omega$ is simple;*
   (b) *$\Omega \approx Sp_4(2)$, which is almost simple, and $P\Omega'$ is simple;*
   (c) *$\Omega \approx \Omega_4^+(q)$ with $q \geq 4$ and $P\Omega \cong L_2(q) \times L_2(q)$.*

Here is a further observation concerning the structure of Ω. To prove this result, use Proposition 2.9.2 and (2.1.22).

Proposition 2.9.3. *Let $\Omega = \Omega(V, \mathbf{F}, \kappa)$ as above and assume that Ω is insoluble. Then $Z(\Omega) = \Omega \cap \mathbf{F}^*$ is the largest normal soluble subgroup of Ω, and its order is as follows:*

$$\begin{cases} (q \pm 1, n) & \text{if } \Omega \approx SL_n^\pm(q) \\ (q - 1, 2) & \text{if } \Omega \approx Sp_n(q) \\ 2 & \text{if } \Omega \approx \Omega_n^+(q) \text{ with } q \text{ odd and } \tfrac{1}{4}n(q-1) \text{ even} \\ 2 & \text{if } \Omega \approx \Omega_n^-(q) \text{ with } q \text{ odd and } \tfrac{1}{4}n(q-1) \text{ odd} \\ 1 & \text{otherwise.} \end{cases}$$

We conclude this section with a Lemma which relates the two components of $\Omega_4^+(q)$ (see Proposition 2.9.1.iv) with the two equivalence classes of totally singular 2-spaces described in Description 4 in §2.5.

Lemma 2.9.4. *Assume that $I \approx O_4^+(q)$, and write $\Omega = L_1 \circ L_2$, with $L_i \cong SL_2(q)$. Then Γ acts on the set $\{L_1, L_2\}$, and its action here is isomorphic to its action on $\{\mathcal{U}_2^1, \mathcal{U}_2^2\}$.*

Proof. Obviously L_i $(i = 1, 2)$ satisfies $L_i \cong SL_2(q) \cong C_\Omega(L_i)$, and elementary group theory shows that the L_i are the only two subgroups of Ω which have this property. Therefore Γ does indeed act on the set $\{L_1, L_2\}$. Let W_i be the natural 2-dimensional module for L_i, and identify V with the tensor product $W_1 \otimes W_2$. The quadratic form on V is described in the proof of Proposition 2.9.1.iv, and we see that $Q(w_1 \otimes w_2) = 0$ for all $w_i \in W_i$. Thus the subspaces $W_1 \otimes w_2$ $(0 \neq w_2 \in W_2)$ lie in $\mathcal{U}_2$, and since two of these subspaces intersect trivially, it follows from Description 4 in §2.5 that they all lie in the same equivalence class $\mathcal{U}_2^1$, say. In addition, Lemma 4.4.3.iv in Chapter 4 below implies that these subspaces are precisely the irreducible L_1-submodules of V. Similarly, the irreducible L_2-submodules of V are the subspaces $w_1 \otimes W_2$ $(0 \neq w_1 \in W_1)$, all of which lie in $\mathcal{U}_2$. As $(W_1 \otimes w_2) \cap (w_1 \otimes W_2) = \langle w_1 \otimes w_2 \rangle$, the irreducible L_2-submodules of V are contained in the other equivalence class $\mathcal{U}_2^2$. Since Γ acts on $\{L_1, L_2\}$ and since the irreducible L_i-submodules of V lie in $\mathcal{U}_2^i$, we may now conclude that an element in Γ interchanges L_1 with L_2 if and only if it interchanges $\mathcal{U}_2^1$ with $\mathcal{U}_2^2$. This completes the proof. ■

§2.10 Classical groups acting on their associated geometries

In this section we prove various results concerning the representation of a classical group on its associated vector space. We also make a few more general representation theoretic observations which will be useful in future chapters. We assume in this section that $(V, \mathbf{F}, \kappa)$ is a classical geometry with $n = \dim_{\mathbf{F}}(V)$, as described before, and that $X = X(V, \mathbf{F}, \kappa)$ as X ranges over the symbols in (2.1.15). Thus $\kappa = \mathbf{f}$ in cases **L**, **U**, **S** and $\kappa = Q$ in case **O**. As usual, set $\mathbf{f} = \mathbf{f}_Q$ in case **O** and write $(v, w) = \mathbf{f}(v, w)$ for all $v, w \in V$. Further, assume that G is any subgroup of $GL(V) = GL(V, \mathbf{F})$.

Suppose for the moment that G is irreducible in $GL(V, \mathbf{F})$. For any extension field $\mathbf{E}$ of $\mathbf{F}$, we can form the tensor product $V \otimes \mathbf{E}$, which is an n-dimensional space over $\mathbf{E}$. Moreover G acts on $V \otimes \mathbf{E}$ via $(v \otimes \lambda)g = vg \otimes \lambda$ $(v \in V, \lambda \in \mathbf{E}, g \in G)$, and thus we may regard $G \leq GL(V \otimes \mathbf{E}, \mathbf{E})$. However, just because G is irreducible on the $\mathbf{F}$-space V, it does not follow that G is irreducible on the $\mathbf{E}$-space $V \otimes \mathbf{E}$. We say that G is *absolutely irreducible* in $GL(V, \mathbf{F})$ if G remains irreducible in $GL(V \otimes \mathbf{E}, \mathbf{E})$ for all extension fields $\mathbf{E}$ of $\mathbf{F}$. It is clear that if $\mathbf{F} \subseteq \mathbf{E} \subseteq \mathbf{E}_1$ and G is irreducible in $GL(V \otimes \mathbf{E}_1, \mathbf{E}_1)$, then G is irreducible in $GL(V \otimes \mathbf{E}, \mathbf{E})$. The following result is a restatement of Theorem 29.13 and Corollary 29.15 in [C-R].

Lemma 2.10.1. *Let $G \leq GL(V, \mathbf{F})$, and assume that G is irreducible on V. Then the* *following are equivalent.*

(i) *G is absolutely irreducible.*

(ii) *G is irreducible in $GL(V \otimes \mathbf{E}, \mathbf{E})$, where $\mathbf{E}$ is the algebraic closure of $\mathbf{F}$.*

(iii) $C_{GL(V,\mathbf{F})}(G) = \mathbf{F}^*$.

(iv) $\mathrm{End}_{\mathbf{F}G}(V) = \mathbf{F}$.

One can define a subgroup of $PGL(V, \mathbf{F})$ to be absolutely irreducible if its preimage in $GL(V, \mathbf{F})$ is absolutely irreducible. In Lemma 4.0.5.i in Chapter 4, we will show that if G is perfect, then $C_{PGL(V,\mathbf{F})}(\overline{G}) = \overline{C_{GL(V,\mathbf{F})}(G)}$, where $\overline{}$ denotes reduction modulo scalars. Thus it follows from Lemma 2.10.1 that if G is perfect and irreducible, then $\overline{G}$ is absolutely irreducible in $PGL(V, \mathbf{F})$ if and only if $C_{PGL(V,\mathbf{F})}(\overline{G}) = 1$.

While the preceding Lemma describes properties of absolutely irreducible groups, the following result describes properties of those which are not absolutely irreducible. A proof can be found in [Is, 9.21]. In the statement of Lemma 2.10.2 we use the following notation. Assume that $\rho : G \to GL(W, \mathbf{E})$ is a representation of G. Fix a basis β of W and let $\psi \in Aut(\mathbf{E})$. Then we obtain a new representation ρ^ψ of G given by

$$g\rho^\psi = (g\rho)^{\phi_\beta(\psi)}.$$

(Recall $\phi_\beta(\psi)$ is defined in (2.1.2).) We write W^ψ for the module corresponding to the representation ρ^ψ. Observe that the notation ρ^ψ and W^ψ does not involve the basis β; this is justified because a different choice of basis gives rise to an equivalent representation.

Lemma 2.10.2. *Assume that $G \le GL(V, \mathbf{F})$ and that G is irreducible but not absolutely irreducible. Then the following hold.*

(i) *$\mathbf{E} = \mathrm{End}_{\mathbf{F}G}(V)$ is a field extension of $\mathbf{F}$ of degree r, where $r \mid n$.*

(ii) *As an $\mathbf{E}G$-module, $V \otimes \mathbf{E} \cong \bigoplus_{i=1}^{r} W^{\psi^i}$, where W is an absolutely irreducible $\mathbf{E}G$-module and $\langle \psi \rangle = \mathrm{Gal}(\mathbf{E}{:}\mathbf{F})$.*

(iii) *G acts faithfully on each W^{ψ^i}.*

The following Lemma shows that absolutely irreducible groups can fix only one form, up to scalar multiples.

Lemma 2.10.3. *Let $(V, \mathbf{F}, \kappa_i)$ ($i = 1, 2$) be two classical geometries. Assume that κ_1 and κ_2 are non-degenerate and that they are both unitary, both symplectic or both quadratic. Further suppose that $G \le I(V, \mathbf{F}, \kappa_1) \cap I(V, \mathbf{F}, \kappa_2)$ and that G is absolutely irreducible. Then $\kappa_2 = \lambda\kappa_1$ for some $\lambda \in \mathbf{F}^*$.*

Proof. Set $I_i = I(V, \mathbf{F}, \kappa_i)$, and we write $\kappa_i = \mathbf{f}_i$ in cases $\mathbf{U}$ and $\mathbf{S}$ and $\kappa_i = Q_i$ in case $\mathbf{O}$. Moreover in case $\mathbf{O}$ we set $\mathbf{f}_i = \mathbf{f}_{Q_i}$.

Assume first that κ_1, κ_2 are not quadratic forms in characteristic 2. Thus if β is a basis of V, Lemma 2.1.8.i and (2.5.2) imply that $I_i = \{g \in GL(V, \mathbf{F}) \mid g_\beta(\mathbf{f}_i)_\beta g_\beta^{\alpha t} = (\mathbf{f}_i)_\beta\}$. Thus for all $g \in G$, we have $(\mathbf{f}_1)^{-1} g \mathbf{f}_1 = (\mathbf{f}_2)^{-1} g \mathbf{f}_2$ (suppressing the subscripts β), which is to say $\mathbf{f}_1\mathbf{f}_2^{-1} \in C_{GL(V,\mathbf{F})}(G)$. The result now follows from Lemma 2.10.1.

Next consider the case where $\kappa_i = Q_i$ are quadratic forms in characteristic 2. In view of the preceding paragraph, we may scalar multiply κ_2 to ensure that $\mathbf{f}_1 = \mathbf{f}_2$. Thus by Lemma 2.1.8.iii, it suffices to prove that Q_1 and Q_2 agree on some basis of V. Choose $v \in V$ and $g \in G$ such that $vg \neq v$. Then $w = v + vg \neq 0$. Obviously $Q_i(v) = Q_i(vg)$, and so $Q_i(w) = Q_i(v) + Q_i(vg) + \mathbf{f}_i(v, vg) = \mathbf{f}_i(v, vg)$. Since $\mathbf{f}_1 = \mathbf{f}_2$, we conclude $Q_1(w) = Q_2(w)$. As G is irreducible, V has a basis $\{wg_1, \ldots, wg_n\}$ for some $g_i \in G$. Clearly Q_1 and Q_2 agree on this basis, and so the result is proved. ∎

Corollary 2.10.4. *Let $G \leq I$ and suppose that G is absolutely irreducible on V.*
(i) $N_{GL(V,\mathbf{F})}(G) \leq \Delta$.
(ii) *If Ω is absolutely irreducible on V, then $N_{GL(V,\mathbf{F})}(\Omega) = \Delta$.*
(iii) *If $G^x \leq I$ for some $x \in GL(V, \mathbf{F})$, then $x \in \Delta$.*

Proof. These assertions are trivial in case **L**, so we assume that κ is non-degenerate. Furthermore, it suffices to prove (iii), as (i) follows from (iii) and (ii) follows from (i). So take $x \in GL(V, \mathbf{F})$ and assume that $G^x \leq I$. Then G fixes the form κ' given by $\kappa'(\mathbf{v}) = \kappa(\mathbf{v}x)$, where $\mathbf{v}$ lies in V in case **O** and lies in $V \times V$ in cases **U** and **S**. Therefore by Lemma 2.10.3, there exists $\lambda \in \mathbf{F}^*$ such that $\kappa' = \lambda\kappa$, and consequently $x \in \Delta$, as required. ∎

In order to use Corollary 2.10.4 in future chapters, we shall need to know precisely when the groups Ω, S and I are absolutely irreducible on V. This is established in Proposition 2.10.6 below, the proof of which depends on Lemma 2.10.5, which describes the orbits of Ω on V. For $\lambda \in \mathbf{F}$, define

$$V_\lambda = \begin{cases} \{v \in V\backslash 0 \mid \mathbf{f}(v,v) = \lambda\} & \text{in cases } \mathbf{L},\ \mathbf{U},\ \mathbf{S} \\ \{v \in V\backslash 0 \mid Q(v) = \lambda\} & \text{in case } \mathbf{O} \end{cases}$$

In particular, V_o is the set of non-zero singular vectors in V. Clearly I preserves the sets V_λ. Indeed, I is transitive on V_λ by Witt's Lemma 2.1.6.

Lemma 2.10.5. *Assume that $n \geq 2$.*
(i) *In cases **L** and **S**, the group Ω is transitive on $V\backslash 0$.*
(ii) *Suppose that either case **U** holds with $n \geq 3$, or case **O** holds with $n \geq 4$. Then Ω is transitive on V_λ for any λ.*
(iii) *$SU_2(q)$ has $q+1$ orbits on V_o (each of size q^2-1), and is transitive on V_λ for $\lambda \neq 0$.*
(iv) *$\Omega_3(q)$ has two orbits on V_o (each of size $\frac{1}{2}(q^2-1)$), and is transitive on V_λ for $\lambda \neq 0$.*
(v) *$\Omega_2^+(q)$ has $2(2, q-1)$ orbits on V_o, and $\Omega_2^\pm(q)$ has $(2, q-1)$ orbits on V_λ for $\lambda \neq 0$.*

Proof. Part (i) is clear for case **L**, and is true by Witt's Lemma 2.1.6 in case **S** (recall $\Omega = I$ in case **S**). Since I is transitive on V_λ, and $\Omega \trianglelefteq I$, it follows that Ω has $|I : I_v\Omega|$ orbits on V_λ, where v is any vector in V_λ. Moreover, all Ω-orbits have the same size.

To prove (ii), we must show that for $v \in V_\lambda$,

$$I_v\Omega = I. \tag{2.10.1}$$

Consider first case **U**, with $n \geq 3$.. If $\lambda \neq 0$, observe that I_v contains $I(v^\perp, \kappa) \approx GU_{n-1}(q)$, whence $\det(I_v) = \det(I)$, proving (2.10.1). And if $\lambda = 0$, choose $w \in V$ such that $W = \langle v, w\rangle$ is a non-degenerate 2-space. Then I_v contains $I(W^\perp, \kappa) \approx GU_{n-2}(q)$, so (2.10.1) again follows. Now consider case **O**, with $n \geq 4$. Here we can find a non-degenerate space $W \leq v^\perp$ such that $\dim(W) \geq 2$ and $I(W, \kappa) \leq I_v$. Therefore (2.10.1) holds by Lemma 4.1.5.iii, below.

As for (iii), (2.10.1) holds if $\lambda \neq 0$ by the same argument as in (ii). But now suppose $v \in V_o$. By Witt's Lemma 2.1.6, there exists $w \in V$ such that v, w is a unitary basis as described in Proposition 2.3.2. Take $g \in I_v$ and write $wg = \zeta v + \eta w$. Then $1 = (v, w) = (v, \zeta v + \eta w) = \eta^q$, and hence $\eta = 1$. Thus $I_v \leq \Omega$, whence $|I : I_v\Omega| = |I : \Omega| = q + 1$. Thus (iii) holds.

Next consider (iv). If $\lambda \neq 0$, then $I(v^\perp, \kappa) \leq I_v$ and (2.10.1) holds as before. Now assume that $\lambda = 0$. By Witt's Lemma 2.1.6 we can assume that $v = e_1, f_1, x$ is a standard basis as given in Proposition 2.5.3. Rearranging this basis as $\beta = \{e_1, x, f_1\}$, a direct calculation using Lemma 2.1.8.i shows that if $g \in I_v$, then g_β is the product of a lower triangular matrix with r_x^i, where i is 1 or 2. Consequently $I_v = O_p(I_v)\langle r_x\rangle = O_p(I_v).2$. Since p is odd (see Proposition 2.5.1) and since $|I : \Omega| = 4$ (see Table 2.1.C), we have $O_p(I_v) \leq \Omega$. Therefore $|I : I_v\Omega| = |I : \langle r_x\rangle\Omega| = 2$, as required.

Finally, we prove (v). One checks that if $\lambda = 0$ then $I_v = 1$, and if $\lambda \neq 0$ then I_v is just the reflection in a non-singular vector in $v^\perp$. Thus part (v) is clear. ∎

Proposition 2.10.6.

(i) *Ω is absolutely irreducible on V if and only if $\Omega \not\approx \Omega_2^\pm(q)$.*

(ii) *S is absolutely irreducible on V if and only if $S \not\approx SO_2^\pm(q)$ with q odd.*

(iii) *I is absolutely irreducible if and only if I is irreducible, which occurs if and only if $I \not\approx O_2^+(2), O_2^+(3)$.*

Proof. Parts (ii) and (iii) follow readily from (i) and the description of the groups $O_2^\pm(q)$ given in §2.9, so we prove only (i). Since $\Omega_2^\pm(q)$ is cyclic (see Proposition 2.9.1.iii), it is not absolutely irreducible on V, so we exclude this case. We also assume that $n \geq 2$, and show that under these assumptions, Ω is absolutely irreducible.

First, using Lemma 2.10.5, we check that any orbit of Ω on $V \backslash 0$ spans V. This is obvious for cases **L** and **S**. Now consider $\Omega \approx SU_n(q)$ and take $v \in V_\lambda$. Assume first that $\lambda \neq 0$. It follows from (2.1.13) that $\lambda \in \mathbf{F}_q$, and hence there exists $\eta \in \mathbf{F} = \mathbf{F}_{q^2}$ such that $\eta^{1+q} = \lambda$. Now V has an orthonormal basis, and multiplying each vector in this basis by η yields a basis contained in V_λ. As Ω is transitive on V_λ, the claim follows. Assume next that $\lambda = 0$. If n is even, then the unitary basis given in Proposition 2.3.2 lies in V_o, and so again the claim follows provided $n \geq 4$, for here Ω is transitive on V_o. If $n = 2$, then v extends to a unitary basis v, w of V, and the element $\left(\begin{smallmatrix}0 & i\\ i & 0\end{smallmatrix}\right)$ (with $i = \sqrt{-1}$) lies in Ω, whence $\mathrm{span}(v\Omega) = V$. And if n is odd, then v extends to a unitary basis $v = e_1, \ldots, f_m, x$. But clearly there are singular vectors in $V \backslash \langle e_1, \ldots, f_m\rangle$, and so we can appeal to the fact that Ω is transitive on V_o when $n \geq 3$. Similar observations will also handle case **O**.

It follows that Ω is irreducible on V. Now by Lemma 2.10.1, it suffices to prove that

$$C_{GL(V)}(\Omega) = \mathbf{F}^*. \tag{2.10.2}$$

As a first step in an inductive proof of this, we consider the cases where Ω is one of the groups

$$SL_2(q),\ Sp_2(q),\ SU_2(q),\ \Omega_3(q),\ \Omega_4^{\pm}(q)\,(q \text{ even}),\ Sp_4(q). \tag{2.10.3}$$

In the first case, we calculate that only $\mathbf{F}^*$ centralizes both $\left(\begin{smallmatrix} 0 & 1 \\ -1 & 0 \end{smallmatrix}\right)$ and $\left(\begin{smallmatrix} 1 & 1 \\ -1 & 0 \end{smallmatrix}\right)$, so (2.10.2) holds. Thus (2.10.2) also holds for $Sp_2(q)$ (see the proof of Proposition 2.9.1.i). Furthermore, $SU_2(q)$ contains $\left(\begin{smallmatrix} \lambda & 0 \\ 0 & \lambda^{-1} \end{smallmatrix}\right)$ and $\left(\begin{smallmatrix} 0 & -1 \\ 1 & 0 \end{smallmatrix}\right)$ (with respect to an orthonormal basis), where λ is a primitive $(q+1)^{\text{th}}$ root of unity in $\mathbf{F}_{q^2}$. Thus (2.10.2) is true for $SU_2(q)$. If $\Omega \approx \Omega_3(q)$, let v_1, v_2, v_3 be a basis as described in Proposition 2.6.1. Then Ω contains $r_{ij} = r_{v_i} r_{v_j}$ for $1 \leq i, j \leq 3$ and the element g which acts as $v_1 \mapsto v_2 \mapsto v_3 \mapsto v_1$. Moreover $C_{GL(V)}(\langle r_{12}, r_{23}, r_{13}, g \rangle) = \mathbf{F}^*$, and so (2.10.2) holds once again. Now let $\Omega \approx \Omega_4^{\epsilon}(q)$ with q even. If $\epsilon = +$, let $\beta = \{e_1, e_2, f_1, f_2\}$ be a standard basis as given in Proposition 2.5.3.i. Then Ω contains the involution g acting as $e_1 \leftrightarrow e_2$, $f_1 \leftrightarrow f_2$ (it is obvious that $g \in O_4^+(2)$, and thus $g \in \Omega$ since g fixes $\langle e_1, e_2 \rangle$ — see Description 4 in §2.5). Also Ω contains the involution h acting as $e_i \leftrightarrow f_i$ $(i = 1, 2)$, along with the element k of order 3 acting as $e_1 \mapsto e_1 + f_1 + e_2 + f_2$, $e_2 \mapsto e_1 + f_2$, $f_1 \mapsto e_1$, $f_2 \mapsto e_1 + e_2$. Then we leave it to the reader to verify that $C_{GL(V)}(\langle g, h, k \rangle) = \mathbf{F}^*$. When $\epsilon = -$, let e_1, f_1, x, y be a standard basis as given in Proposition 2.5.3.ii and notice that $C_{GL(V)}(\langle r_x r_y, r_{e_1+f_1} r_x, r_{e_1+x} r_x \rangle) = \mathbf{F}^*$, and so we can appeal to (2.5.13). The case $\Omega \approx Sp_4(q)$ is left as an exercise.

Now suppose that Ω is not one of the groups in (2.10.3). We complete the proof by induction on n. Observe that V contains a subspace W of codimension 1 or 2 such that $\dim(W) > \frac{n}{2}$ and W is non-degenerate in cases **U, S, O**. By what we have already proved, Ω_W is irreducible on W; moreover, $\dim(W) > \frac{n}{2}$, so W is the unique Ω_W-invariant space of its dimension. Consequently $C := C_{GL(V)}(\Omega)$ fixes W. By induction, $C^W = \mathbf{F}^*$. Choosing another space W' with the same properties as W and satisfying $V = W + W'$, we conclude that $C^V = \mathbf{F}^*$, and so (2.10.2) holds. This completes the proof. ∎

Observe that if Ω is insoluble, then Ω is perfect, except when $\Omega \approx Sp_4(2)$. However $Sp_4(2)'$ is absolutely irreducible on the natural 4-dimensional module. Consequently Propositions 2.9.1 and 2.9.2 now yield the following Corollary.

Corollary 2.10.7. *If Ω is insoluble, then Ω' is absolutely irreducible on V.*

Now we turn our attention to the subfields of $\mathbf{F}$ over which the representation of Ω on V can be defined. In general, there are no proper such subfields, but there are a few exceptions which we record below. For any subset $X \subseteq \mathbf{F}$ define $\mathbf{F}_p[X]$ as the subfield of $\mathbf{F}$ generated by the elements in X. And for any subgroup H of $GL(V, \mathbf{F})$, define

$$\mathbf{F}_p(H) = \mathbf{F}_p[\text{tr}(h) \mid h \in H], \tag{2.10.4}$$

where tr(h) is the trace of h. If $\mathbf{F}_o$ is a subfield of $\mathbf{F}$, then we say that H *can be realized over* $\mathbf{F}_o$ if there is a basis of V such that H acts on the $\mathbf{F}_o$-span of the basis vectors. In other words, H can be realized over $\mathbf{F}_o$ if, with respect to some basis, the matrix of each element of H has entries in $\mathbf{F}_o$. If H can be realized over $\mathbf{F}_o$ then obviously $\mathbf{F}_p(H) \subseteq \mathbf{F}_o$. It turns out that the converse to this assertion is also true. This is stated in the following Proposition, a proof of which can be found in [Is, 9.14].

Proposition 2.10.8. *If $H \leq GL(V, \mathbf{F})$, then H can be realized over $\mathbf{F}_p(H)$.*

We are now ready to prove

Proposition 2.10.9.

(i) *Provided $n \geq 2$, we have $\mathbf{F}_p(\Omega) = \mathbf{F}$, except when $\Omega \approx SU_2(q)$, $\Omega_2^+(4)$ or $\Omega_2^+(9)$. In these exceptional cases, $\mathbf{F}_p(\Omega)$ is $\mathbf{F}_q$, $\mathbf{F}_2$ or $\mathbf{F}_3$, respectively.*

(ii) *$\mathbf{F}_p(I) = \mathbf{F}$, except when $I \approx O_1(q)$ or $O_2^+(4)$. In the exceptional cases, $\mathbf{F}_p(I) = \mathbf{F}_p$ or $\mathbf{F}_2$, respectively.*

(iii) *If $\Omega \not\approx \Omega_2^\pm(q)$, then Ω contains an element whose trace lies in $\mathbf{F}_p^*$.*

Proof. (i) and (ii). The result is clear when $n = 1$. Note that if $I \approx O_1(q)$, then $I = \langle r_v \rangle$ for $v \in V \backslash 0$, and hence $I = \langle -1 \rangle$. Hereafter we take $n \geq 2$.

First consider $\Omega \approx SU_2(q)$, so that $\mathbf{F} = \mathbf{F}_{q^2}$. Relative to an orthonormal basis β, we have

$$\Omega_\beta = \left\{ \begin{pmatrix} \lambda & \eta \\ -\eta^q & \lambda^q \end{pmatrix} \;\middle|\; \lambda,\, \eta \in \mathbf{F}_{q^2}^*,\ \lambda^{1+q} + \eta^{1+q} = 1 \right\}.$$

Thus as the trace $\mathrm{T}_{\mathbf{F}_q}^{\mathbf{F}}$ and norm $\mathrm{N}_{\mathbf{F}_q}^{\mathbf{F}}$ are surjective, we have $\mathbf{F}_p(\Omega) = \mathbf{F}_q$. Clearly $\mathbf{F}_q \subseteq \mathbf{F}_p(I) \subseteq \mathbf{F}$, and $GU_2(q)$ contains $g = \mathrm{diag}_\beta(\lambda, 1)$, where λ is a primitive $(q+1)^{\mathrm{th}}$ root of unity. As the trace of g does not lie in $\mathbf{F}_q$, we deduce $\mathbf{F}_p(I) = \mathbf{F}$, as claimed.

Next consider $\Omega \approx \Omega_2^+(q)$. Relative to a standard basis β (as given in Proposition 2.5.3.i) we have

$$\Omega_\beta = \langle \mathrm{diag}_\beta(\mu^2, \mu^{-2}) \rangle, \text{ where } \langle \mu \rangle = \mathbf{F}^*.$$

Now take $\rho \in Aut(\mathbf{F})$, and observe that ρ fixes $\mu^2 + \mu^{-2}$ if and only if $\mu^2(\mu^2)^\rho(\mu^2 - (\mu^2)^\rho) = \mu^2 - (\mu^2)^\rho$, which occurs if and only if $(\mu^2)^\rho = \mu^{\pm 2}$. Assume that $\mathbf{F}_p(\Omega) \neq \mathbf{F}$. Then we can find a non-trivial ρ which fixes $\mu^2 + \mu^{-2}$. Now $\mathbf{F}_p(\mu^2) = \mathbf{F}$, and so ρ does not fix μ^2. Therefore $(\mu^2)^\rho = \mu^{-2}$, and $|\rho| = 2$. Consequently $q - 1 \mid 2(q^{1/2} + 1)$, which forces $q^{1/2} - 1 \mid 2$. Therefore $q = 4$ or 9, and $\mathbf{F}_p(\Omega) = \mathbf{F}_2$ or $\mathbf{F}_3$, respectively. When $q = 4$, one easily checks that $\mathbf{F}_p(I) = \mathbf{F}_2$. But when $q = 9$, we see that $\mathbf{F}_p(I)$ contains $\mu + \mu^{-1} \notin \mathbf{F}_3$; therefore $\mathbf{F}_p(I) = \mathbf{F}$.

We now treat the remaining possibilities for Ω, and our goal is to show that $\mathbf{F}_p(\Omega) = \mathbf{F}$. It suffices to prove this when Ω is one of

$$\Omega_2^-(q),\ SU_3(q),\ SL_2(q),\ Sp_2(q),$$

for every one of the remaining classical groups contains one of these centralizing a subspace of codimension 2, 3, 2, 2, respectively. In the last two cases, the matrices

$\left(\begin{smallmatrix}\lambda & 1\\ -1 & 0\end{smallmatrix}\right)$ lie in Ω for all $\lambda \in \mathbf{F}$, so $\mathbf{F}_p(\Omega) = \mathbf{F}$, as required. Now consider $\Omega \approx SU_3(q)$, and let e_1, f_1, x be a standard basis as given in Proposition 2.3.2. Then Ω contains $\mathrm{diag}(\mu, \mu^{-q}, \mu^{q-1})$, with μ a generator for $\mathbf{F}_{q^2}^*$. Since $(\mu - \mu^q)(\mu - 1)(\mu^q - 1) \neq 0$, it follows that $\mu + \mu^{-q} + \mu^{q-1} \neq \mu^q + \mu^{-1} + \mu^{1-q}$, and hence the trace of this element of Ω does not lie in $\mathbf{F}_q$. However $\mathbf{F}_p(\Omega)$ certainly contains $\mathbf{F}_q$ in view of the argument given earlier for $SU_2(q)$. Hence $\mathbf{F}_p(\Omega) = \mathbf{F}_{q^2} = \mathbf{F}$, as required.

Finally, let $\Omega \approx \Omega_2^-(q)$. As described in §2.9, a generator for Ω has trace $\omega + \omega^{-1}$, where ω is a primitive $((q+1)/(2, q-1))^{\mathrm{th}}$ root of unity in $\mathbf{F}_{q^2}$. Arguing as in the $\Omega_2^+(q)$ case, we see that $\omega + \omega^{-1}$ lies in no proper subfield of $\mathbf{F}_q$, and hence $\mathbf{F}_p(\Omega) = \mathbf{F}$ here as well. This completes the proof of (i) and (ii).

(iii) We seek to find an element $g \in \Omega$ such that $\mathrm{tr}(g) \in \mathbf{F}_p^*$. Clearly we may assume that $p \mid n$, or else we can take $g = 1$. First assume that $n \geq 3$ and that q is odd. Then V contains a 2-space W such that $-1_W \in \Omega$ (choose W to be any 2-space in case **L**, a non-degenerate 2-space in cases **S** and **U**, and a non-degenerate 2-space of type $O_2^\epsilon(q)$ in case **O**, where $q \equiv \epsilon \pmod 4$). Thus we may take $g = -1_W$, since $\mathrm{tr}(-1_W) = n - 4$, which is not divisible by p. Next assume that $n \geq 3$ and q is even. Then there is a 2-space U in V and an element $g \in \Omega$ such that g induces an element of order 3 on U and centralizes V/U (in case **O** choose U to be a 2-space of type $O_2^\epsilon(q)$ where $q \equiv \epsilon \pmod 3$). Then $\mathrm{tr}(g) = n - 3$, which is not divisible by 2. The case $n = 2$ is easy, and is left to the reader. ∎

Corollary 2.10.10.

(i) *The group Ω can be realized over a proper subfield in $GL(V, \mathbf{F})$ if and only if $\Omega \approx SU_2(q)$, $\Omega_1(q)$ $(q > p)$, $\Omega_2^+(4)$ or $\Omega_2^+(9)$.*

(ii) *The group I can be realized over a proper subfield in $GL(V, \mathbf{F})$ if and only if $I \approx O_1(q)$ $(q > p)$ or $O_2^+(4)$.*

Remark. The exceptional cases in Corollary 2.10.10 are related in an obvious way to the isomorphisms $SU_2(q) \cong SL_2(q)$, $\Omega_2^+(4) \cong \Omega_2^-(2)$ and $\Omega_2^+(9) \cong SO_2^-(3)$

We now turn to some representation theoretic observations concerning groups acting on subspace decompositions of a vector space. By a *subspace decomposition* of V we mean a set of subspaces $V_1, \ldots, V_t$ of V with $t \geq 2$ such that

$$V = V_1 \oplus \cdots \oplus V_t.$$

Often we write $\mathcal{D} = \{V_1, \ldots, V_t\}$, the symbol $\mathcal{D}$ being a mnemonic for 'decomposition'. For $G \leq GL(V)$, the *stabilizer in G of $\mathcal{D}$* is the group $N_G\{V_1, \ldots, V_t\}$, which is the subgroup of G which permutes the spaces V_i amongst themselves. The *centralizer in G of $\mathcal{D}$* is the group $N_G(V_1, \ldots, V_t)$, which is the subgroup of G fixing each V_i. We write

$$\begin{aligned} G_{\mathcal{D}} &= N_G\{V_1, \ldots, V_t\} \\ G_{(\mathcal{D})} &= N_G(V_1, \ldots, V_t) \\ G^{\mathcal{D}} &= G_{\mathcal{D}}/G_{(\mathcal{D})}, \end{aligned} \tag{2.10.5}$$

so that $G^{\mathcal{D}}$ acts faithfully as permutation group on the t spaces in $\mathcal{D}$. We say that G *stabilizes* $\mathcal{D}$ if $G = G_{\mathcal{D}}$ and that G *centralizes* $\mathcal{D}$ if $G = G_{(\mathcal{D})}$.

Lemma 2.10.11. *Suppose that $G \leq GL(V)$ and that G centralizes the subspace decomposition $\mathcal{D} = \{V_1, \dots, V_t\}$. Also suppose that G acts irreducibly on each V_i and that the V_i are pairwise non-isomorphic as G-modules.*

(i) *The spaces V_i are the only irreducible G-submodules of V.*

(ii) *Any G-invariant subspace of V is a sum of some of the V_i.*

Proof. (i) We use induction on t, and we define $V_i^- = \bigoplus_{j \neq i} V_j$. Let W be an irreducible G-submodule of V. Assume for the moment that $W \not\leq V_i^-$ for some i. Then $V = V_i^- \oplus W = V_i^- \oplus V_i$, and hence $W \cong V_i$ as G-modules as they are both isomorphic to V/V_i^-. It follows that if $j \neq i$, then $W \leq V_j^-$, for otherwise the same reasoning would show that $W \cong V_j$, and this is impossible as $V_i \not\cong V_j$. We have therefore shown that $W \leq V_j^-$ for some j, and so by induction we conclude that $W = V_k$ for some k, as claimed.

(ii) Clear from (i). ■

Lemma 2.10.12. *Suppose that G centralizes $\mathcal{D}$ and that G acts irreducibly on each V_i. Suppose further that for a pair i, j there exists $g \in G$ such that $g^{V_i} \neq 1$ and $g^{V_j} = 1$. Then $V_i \not\cong V_j$ as G-modules.*

Proof. Clear. ■

Proposition 2.10.13. *Assume that G stabilizes $\mathcal{D}$ and that as $G_{(\mathcal{D})}$-modules, the spaces V_i are irreducible and pairwise non-isomorphic. Suppose further that $G^{\mathcal{D}}$ is transitive on $\mathcal{D}$.*

(i) *G is irreducible on V.*

(ii) *If, in addition, $G_{(\mathcal{D})}$ is absolutely irreducible on each V_i, then G is absolutely irreducible on V.*

Proof. Let U be a non-zero G-submodule of V. Applying Lemma 2.10.11 to the group $G_{(\mathcal{D})}$, we know that $V_i \leq U$ for some i. Hence $U = V$ by the transitivity of $G^{\mathcal{D}}$, and so (i) holds. To prove (ii), assume that $G_{(\mathcal{D})}^{V_i}$ is absolutely irreducible for each i. Take $c \in C_{GL(V)}(G)$. Clearly c must stabilize $\mathcal{D}$ by Lemma 2.10.11. Suppose that $V_i c = V_j$ for some i, j. Then c induces an isomorphism between V_i and V_j as $G_{(\mathcal{D})}$-modules, and so our assumption that V_i are pairwise non-isomorphic implies that $i = j$. Therefore c centralizes $\mathcal{D}$, and the assumption about absolute irreducibility shows that c acts as a scalar λ_i on V_i (see Lemma 2.10.1). Since $G^{\mathcal{D}}$ is transitive, we have $\lambda_i = \lambda_j$ for all i, j, which is to say c acts as a scalar on all of V. Therefore G is absolutely irreducible by Lemma 2.10.1. ■

We conclude this chapter with some relatively easy results concerning quasiequivalent representations, and the relationship between representations and non-degenerate forms. Assume that X is a group and that $\rho : X \to GL(V, \mathbf{F})$ is a representation of X. Observe that if $\theta \in Aut(X)$, then $\theta\rho$ is another such representation, given by $x(\theta\rho) = (x\theta)\rho$ $(x \in X)$. Two representations $\rho_1, \rho_2 : X \to GL(V, \mathbf{F})$ are said to be *quasiequivalent* if there is an automorphism $\theta \in Aut(X)$ such that ρ_1 is equivalent to $\theta\rho_2$. Evidently quasiequivalence is an equivalence relation on the set of representations of X to $GL(V, \mathbf{F})$.

Lemma 2.10.14. *Assume that ρ_1 and ρ_2 are faithful. Then $X\rho_1$ and $X\rho_2$ are conjugate in $GL(V, \mathbf{F})$ if and only if ρ_1 and ρ_2 are quasiequivalent.*

Proof. Assume that $(X\rho_1)^g = X\rho_2$ for some $g \in GL(V, \mathbf{F})$. Then there is a map $\theta : X \to X$ which satisfies $(x\rho_1)^g = (x\theta)\rho_2$. Clearly θ is a well-defined automorphism of X since the ρ_i are faithful. Thus ρ_1 is equivalent to $\theta\rho_2$, so ρ_1 and ρ_2 are quasiequivalent. The converse is just as easy, and the details are left to the reader. ∎

Given a group X and an absolutely irreducible representation ρ of X to $GL(V, \mathbf{F})$, it is important to know whether X fixes a non-degenerate symplectic form, symmetric bilinear form, a unitary form, or no non-degenerate form at all. Fix a basis β of V and consider all elments of $GL(V, \mathbf{F})$ as matrices relative to β. Fix $g \in GL(V, \mathbf{F})$ and observe that if $(x\rho)g(x\rho)^{\mathrm{t}} = g$ for all $x \in X$, then X fixes a non-degenerate bilinear symmetric form (respectively, symplectic form) provided $g^{\mathrm{t}} = g$ (respectively, $g^{\mathrm{t}} = -g$). The form is of course $(v, w) = vgw^{\mathrm{t}}$. And if $\mathbf{F}$ has an involutory automorphism ψ, and if $(x\rho)g(x\rho)^{\psi\mathrm{t}} = g$ for all $x \in X$, then X fixes a non-degenerate unitary form provided $g = g^{\psi\mathrm{t}}$. The following easy Lemma relates non-degenerate forms to conditions on the representation ρ of X.

Lemma 2.10.15. *Assume that $\rho : X \to GL(V, \mathbf{F})$ is an absolutely irreducible representation.*

(i) *$X\rho$ fixes a non-degenerate symplectic or symmetric bilinear form in $GL(V, \mathbf{F})$ if and only if ρ is equivalent to the dual ρ^*.*
(ii) *$X\rho$ fixes a non-degenerate unitary form in $GL(V, \mathbf{F})$ if and only if $|\mathbf{F} : \mathbf{F}_p|$ is even and ρ^ψ is equivalent to ρ^* (where ψ is the involutory automorphism of $\mathbf{F}$).*

Proof. The 'only if' assertions are clear from Lemma 2.1.8. We now prove the two 'if' assertions simultaneously. Assume that ρ^ψ is equivalent to ρ^*, where $\psi \in Aut(\mathbf{F})$ and $|\psi| \le 2$. Then there exists $g \in GL(V, \mathbf{F})$ such that $g^{-1}(x\rho)g = (x\rho)^{-1\mathrm{t}\psi}$, whence

$$(x\rho)g(x\rho)^{\mathrm{t}\psi} = g. \tag{2.10.6}$$

Taking transpose, multiplying on the left by $g^{-1\mathrm{t}}$, on the right by $(x\rho)^{-1\mathrm{t}}$ and applying ψ yields

$$g^{-1\mathrm{t}\psi}(x\rho)g^{\mathrm{t}\psi} = (x\rho)^{-1\mathrm{t}\psi}. \tag{2.10.7}$$

These two equations show that $g^{\mathrm{t}\psi}g^{-1} \in C_{GL(V,\mathbf{F})}(X\rho)$, and hence

$$g^{\mathrm{t}\psi} = \lambda g$$

for some $\lambda \in \mathbf{F}^*$ as $X\rho$ is absolutely irreducible (see Lemma 2.10.1). Evidently $g = (g^{t\psi})^{t\psi}$, and hence $\lambda\lambda^{\psi} = 1$. Thus if $\psi = 1$, then $\lambda = \pm 1$, and so it follows from (2.10.6) that $X\rho$ fixes a non-degenerate symplectic or symmetric bilinear form. If $|\psi| = 2$, then we may write $\mathbf{F} = \mathbf{F}_{q^2}$ and $\lambda^{q+1} = 1$. Thus there exists $\mu \in \mathbf{F}^*$ such that $\mu^{q-1} = \lambda^q$. We now see from (2.10.6) that $(x\rho)\mu g (x\rho)^{t\psi} = \mu g$ and $(\mu g)^t = \mu g$. Thus $X\rho$ fixes a non-degenerate unitary form, as desired. ■

Chapter 3
THE STATEMENT OF THE MAIN THEOREM

3.1 Introduction

Let $(V, \mathbf{F}, \kappa)$ be a classical geometry of dimension n as described in §2.1. Thus κ is either the zero form, or a non-degenerate symplectic, unitary or quadratic form on V. Let $X = X(V, \mathbf{F}, \kappa)$ where X ranges over the symbols in (2.1.15), and write $\overline{}$ for reduction modulo scalars. Assume that n is at least 2, 3, 4 or 7 in cases **L**, **U**, **S** or **O**, respectively. Thus by Theorem 2.1.3, the groups $\overline{\Omega}$ comprise the finite classical simple groups, apart from $L_2(2)$, $L_2(3)$, $U_3(2)$ and $Sp_4(2)$. Also, by Theorem 2.1.4, we know that $\overline{A} = Aut(\overline{\Omega})$, except when $\Omega \approx \Omega_8^+(q)$ or $Sp_4(2^f)$.

Now let G be a group satisfying $\overline{\Omega} \trianglelefteq G \leq \overline{A}$. Recall that $\mathcal{C}(G)$ is the collection of 'natural' subgroups of G appearing in [As$_1$], which we discussed briefly in Chapter 1. The members of $\mathcal{C}(G)$ are, by and large, stabilizers of various objects associated with the classical geometry $(V, \mathbf{F}, \kappa)$ corresponding to G, such as subspaces, direct sum decompositions, tensor decompositions, and so on. The detailed descriptions of these subgroups will appear in Chapter 4 (see also the definitions given in (3.1.3)-(3.1.5) below). Also recall the collection $\mathcal{S}$ of almost simple irreducible groups, defined in §1.2. We now state the Main Theorem of this book, which gives comprehensive information about the groups in $\mathcal{C}(G)$.

MAIN THEOREM. *Let $(V, \mathbf{F}, \kappa)$ and G be as above. The members of $\mathcal{C}(G)$ are listed in Tables 3.5.A-F.*

(A) *The group-theoretic structure of the members of $\mathcal{C}(G)$ can be read off from Tables 3.5.A-F, as described in §3.3.*

(B) *The conjugacy amongst the members of $\mathcal{C}(G)$ is given by Tables 3.5.A-F, along with the Action Table 3.5.G, as described in §3.2.*

(C) *Assume that $n \geq 13$. For a member $H \in \mathcal{C}(G)$, the precise conditions under which H is maximal in G are determined by Tables 3.5.A-I, as described in §3.4. Moreover, these tables also determine the set of overgroups of H lying in $\mathcal{C}(G) \cup \mathcal{S}$.*

Parts (A) and (B) of the Main Theorem cover all groups G whose socle is a simple classical group, apart from those G inducing a triality automorphism when $\mathrm{soc}(G) = P\Omega_8^+(q)$ and those G inducing a graph automorphism when $\mathrm{soc}(G) = Sp_4(2^f)$.

In order to describe how Tables 3.5.A-I achieve the goals of the Main Theorem, we must first introduce some terminology and notation. Define

$$\mathcal{C} = \bigcup \mathcal{C}(X),$$

where X ranges over the groups satisfying either $\Omega \leq X \leq A$ or $\overline{\Omega} \leq X \leq \overline{A}$. As we mentioned in Chapter 1, the collection $\mathcal{C}(X)$ is a union of families $\mathcal{C}_i(X)$, $i = 1, \ldots, 8$, and for each such i we put

$$\mathcal{C}_i = \bigcup \mathcal{C}_i(X),$$

where X ranges between Ω and A, and between $\overline{\Omega}$ and $\overline{A}$, as before.

Warning. Our definition of the collections $\mathcal{C}_i$ differs slightly from Aschbacher's original definition in [As$_1$]. One difference is that we enlarge $\mathcal{C}_1$ so that it includes Aschbacher's $\mathcal{C}_1'$ in case **L**. This is merely a cosmetic change which allows us to avoid some cumbersome notation in a few circumstances. The other differences are more substantial. Essentially, we delete certain families of groups which are members of Aschbacher's $\mathcal{C}$, so that our $\mathcal{C}$ is generally smaller. We justify these modifications by showing that each group which we delete is in fact contained in another member of $\mathcal{C}$ — thus Aschbacher's Theorem 1.2.1 still holds when his $\mathcal{C}$ is replaced by ours. In Table 3.5.J we summarize all such changes, and in the relevant sections of Chapter 4 it is proved that the groups which we delete are in fact contained in other members of $\mathcal{C}$. The reason for excluding these 'small' members of Aschbacher's $\mathcal{C}$ is that they have many overgroups, and so become somewhat unmanageable in the analysis of overgroups in Chapters 7 and 8.

Each collection $\mathcal{C}_i$ breaks up into several subcollections, which we call *types*. If **T** is a type in $\mathcal{C}_i$ and H is a member of **T**, then we usually say that H *is of type* **T**. The collections $\mathcal{C}_i$ appear in Column I of Tables 3.5.A-F, and the types contained in $\mathcal{C}_i$ appear in Column III. In most cases, the type provides the approximate group-theoretic structure of the subgroups H in $\mathcal{C}$ which are of that type; moreover, the type is written in such a way as to indicate which kind of object H stabilizes. For example, in Table 3.5.F with $\Omega \approx \Omega_n^-(q)$, a subgroup of type $O_m^\xi(q) \perp O_{n-m}^{-\xi}(q)$ is the stabilizer in G of a direct sum decomposition $V = W \perp W^\perp$, where W is a non-degenerate m-space which inherits an $O_m^\xi(q)$-geometry from $(V, \mathbf{F}, \kappa)$. And in Table 3.5.B with $\Omega \approx SU_n(q)$, a subgroup of type $GU_{n_1}(q) \otimes GU_{n_2}(q)$ is the stabilizer in G of a tensor decomposition $V = V_1 \otimes V_2$, where V_i is a unitary space of dimension n_i. There is one exception to this sort of notation — namely, we write P_i for the type corresponding to the stabilizers of the totally singular i-spaces in $(V, \mathbf{F}, \kappa)$. We do so because these stabilizers have a structure somewhat more difficult to describe than the other members of $\mathcal{C}$, and any attempt at succintly expressing this structure would fail to indicate what object the group is stabilizing. The notation P_i is apt because the stabilizer of a totally singular i-space is a *parabolic subgroup*, and hence our symbol has a mnemonic advantage. In case **L** there is also an occurrence of type $P_{n-i,i}$, which is the stabilizer of a suitable pair of subspaces — this type arises because of the inverse transpose automorphism mentioned after Theorem 1.2.1 (see (2.2.4)), and is explained in detail in §4.1. More generally, all the types will be described explicitly throughout Chapter 4. With little doubt, the best way for the reader to become familiar with our conventions concerning types is by way of examples.

Example 1. In Table 3.5.A, the collection $\mathcal{C}_2$ falls into $\tau(n) - 1$ types (where $\tau(n)$ is the number of divisors of n), indexed by m as m runs through the divisors of n satisfying $1 \le m < n$. In Table 3.5.E, we write the types falling under $\mathcal{C}_7$ in two rows — namely, the bottom two rows of Table 3.5.E. In the upper row corresponding to C_7, the number of types depends on the parity of q, as follows. When q is even, the number of types is just the number of ways of writing n in the form $n = m^t$ with $t \geq 2$ and $(m,q) \neq (2,2)$. And when q is odd, the number of types is the number of ways of writing n in the form $n = m^t$ with t even and $(m,q) \neq (2,3)$. As for the bottom row, these types occur only when q is odd, and the number of types depends on the sign $\epsilon = \pm$. When $\epsilon = -$, the number of types is the number of ways of writing n in the form $n = m^t$ with $m \geq 4$. When $\epsilon = +$, the number of types is the number of ways of writing n in the form $n = m^t$ with $m \geq 6$. To illustrate, take $\Omega \approx \Omega_{64}^{+}(q)$. If q is even and $q \geq 4$, then C_7 falls into the following types: $Sp_8(q) \wr S_2$, $Sp_4(q) \wr S_3$, $Sp_2(q) \wr S_6$. If $q = 2$, the only types are $Sp_8(2) \wr S_2$ and $Sp_4(2) \wr S_3$ (due to the condition $(q,m) \neq (2,2)$). If q is odd and $q \geq 5$, then the types occurring are $Sp_8(q) \wr S_2$, $Sp_2(q) \wr S_6$, $O_8^+(q) \wr S_2$, $O_8^-(q) \wr S_2$ and $O_4^-(q) \wr S_3$. When $q = 3$, we have the same types, except that $Sp_2(3) \wr S_6$ is to be ommitted.

Since Γ acts naturally on the geometry $(V, \mathbf{F}, \kappa)$, it is plain to see that if $H \in \mathcal{C}$ then any Γ-conjugate or $\overline{\Gamma}$-conjugate of H is of the same type as H. For example, if $\Omega \approx Sp_n(q)$ with $n = mt$ (m even and $t \geq 2$), and if $H \in \mathcal{C}_2(\Omega)$ is of type $Sp_m(q) \wr S_t$, then H is the stabilizer in Ω of a decomposition $V = V_1 \perp \cdots \perp V_t$, where each V_i is a non-degenerate subspace of dimension m. And if $x \in \Gamma$, the H^x is the stabilizer of the decomposition $V = V_1 x \perp \cdots \perp V_t x$, and so H^x is again of type $Sp_m(q) \wr S_t$. Now A is equal to Γ except in case **L** with $n \geq 3$, and so in this exceptional case some further remarks are required. Here, the group A acts on the set of groups of a given type, except those of type P_i. Elements of $A \backslash \Gamma$ interchange i-spaces with $(n-i)$-spaces, and hence interchange groups of type P_i with groups of type P_{n-i}. For the sake of having A act on the set of groups of given type, we regard the two types P_i and P_{n-i} as one and the same in case **L**. Thus we may assert

$$A \text{ acts on the set of groups of a given type } \mathbf{T}. \tag{3.1.1}$$

A fundamental starting point for our results on conjugacy is Theorem BΓ in [As$_1$], which in our terminology can be stated as follows:

Theorem 3.1.1 [As$_1$, Theorem BΓ].

(i) *The group A acts transitively on the groups in $\mathcal{C}(\Gamma)$ of a given type* $\mathbf{T}$.

(ii) *The group Γ also acts transitively, except for the fact that in case* **L** *with $n \geq 3$, Γ has two orbits on groups of type P_i when $i \neq n-i$.*

We have already discussed Columns I and III of Tables 3.5.A-F, and we now describe briefly the other columns. Sections 3.2-3.4 contain more detailed explanations. Assume that $\mathbf{T}$ is a type contained in $\mathcal{C}_i$ so that $\mathbf{T}$ appears in Column III and $\mathcal{C}_i$ appears in Column I. Column IV gives the conditions under which this type occurs. Column V gives information about how A acts on the set of subgroups of each type; these actions of A are encapsulated in the Action Table 3.5.G, which is described in more detail below.

Hence one can deduce how G acts and thus determine the conjugacy of members of $\mathcal{C}(G)$ — this process is described in §3.2. Column VI gives preliminary information about the overgroups of members of $\mathcal{C}(G)$, giving the conditions under which members of $\mathcal{C}(\overline{\Omega})$ are non-maximal in $\overline{\Omega}$. The complete information can be obtained using Columns V and VI, and Tables 3.5.G-I, as described in §3.4. Finally, Column II indicates where in Chapter 4 the relevant groups are discussed — indeed, the appropriate group is discussed in Proposition 4.i.j, where $H \in \mathcal{C}_i$ and j is given in Column II.

Assume here that X satisfies

$$\Omega \leq X \leq A \text{ or } \overline{\Omega} \leq X \leq \overline{A}. \tag{3.1.2}$$

We now formally define the collection $\mathcal{C}(X)$ in terms of the the collection $\mathcal{C}(\Gamma)$; the members of $\mathcal{C}(\Gamma)$ are described explicitly in Chapter 4. First, when $A \neq \Gamma$ (i.e., in case **L** with $n \geq 3$), set

$$\mathcal{C}(A) = \{N_A(H) \mid H \in \mathcal{C}(\Gamma)\}.$$

Now for $\Omega \leq X \leq A$, define

$$\mathcal{C}(X) = \begin{cases} \{H \cap X \mid H \in \mathcal{C}(\Gamma)\} & \text{if } X \leq \Gamma \\ \{H \cap X \mid H \in \mathcal{C}(A)\} & \text{if } X \not\leq \Gamma, \end{cases} \tag{3.1.3}$$

and

$$\mathcal{C}(\overline{X}) = \{\overline{H} \mid H \in \mathcal{C}(X)\}. \tag{3.1.4}$$

Thus for any group $H \in \mathcal{C}(\Gamma)$, there are corresponding groups $H_X \in \mathcal{C}(X)$ for all groups X as in (3.1.2). If both X and Y satisfy (3.1.2), then H_X will sometimes be referred to as the X*-associate* of H_Y. It is a fact that all members of $\mathcal{C}(\Gamma)$ contain the full group of scalars $\mathbf{F}^*$. Consequently $\overline{H \cap X} = \overline{H} \cap \overline{X}$ whenever $H \in \mathcal{C}(\Gamma)$ and $X \leq \Gamma$. The same holds with A replacing Γ, and thus

$$\mathcal{C}(\overline{X}) = \begin{cases} \{\overline{H} \cap \overline{X} \mid H \in \mathcal{C}(\Gamma)\} & \text{if } X \leq \Gamma \\ \{\overline{H} \cap \overline{X} \mid H \in \mathcal{C}(A)\} & \text{if } X \not\leq \Gamma. \end{cases} \tag{3.1.5}$$

For the rest of this chapter, assume that $H_\Gamma \in \mathcal{C}(\Gamma)$, and write H_X for the corresponding X-associate for all X satisfying (3.1.2). We set $H = H_G$ (where as before $\overline{\Omega} \leq G \leq \overline{A}$), so that

$$H_{\overline{\Omega}} \leq H \leq H_{\overline{A}}. \tag{3.1.6}$$

Note that by Theorem 3.1.1, the members of $\mathcal{C}(\overline{\Omega})$ which are of the same type as $H_{\overline{\Omega}}$ are precisely the $\overline{A}$-conjugates of $H_{\overline{\Omega}}$. In view of (3.1.3)-(3.1.5), we have

$$H_G = \begin{cases} H_{\overline{\Gamma}} \cap G & \text{if } G \leq \overline{\Gamma} \\ N_G(H_{\overline{\Gamma}}) & \text{if } G \not\leq \overline{\Gamma}. \end{cases} \tag{3.1.7}$$

In (3.1.6), $H_{\overline{\Omega}} \trianglelefteq H_{\overline{A}}$, and so $H \leq N_G(H_{\overline{\Omega}})$. However, it is not in general true that equality holds — that is, H need not be the full normalizer of $H_{\overline{\Omega}}$. To make precise statements about the conjugacy and structure of the members of $\mathcal{C}$, we shall need to know when equality fails. The essential information is recorded in the following Theorem.

Theorem 3.1.2.

(i) *For all* X *satisfying* $\overline{\Omega} \leq X \leq \overline{A}$, *at least one of the following holds:*

(a) $H_X = N_X(H_{\overline{\Omega}})$;

(b) $H_X\overline{\Omega} = X$.

(ii) *In case* **L** *we have* $H_{\overline{\Omega}} = N_{\overline{\Omega}}(H_{\overline{\Omega}})$, *except when* $H_{\overline{\Omega}}$ *is of type* $GL_1(q) \wr S_n$ *in* $\mathcal{C}_2$ *and either*

(c) $q = 2$ *and* n *is even, or*

(d) $(q, n) \in \{(5,2), (3,4), (4,3)\}$.

The proof of Theorem 3.1.2 is spread throughout Chapter 4, as we consider in detail the subgroups in the various families $\mathcal{C}_i$. For example, when $H \in \mathcal{C}_1$, then Theorem 3.1.2 is treated in Propositions 4.1.2, 4.1.8 and 4.1.16. As for the exceptional cases in Theorem 3.1.2.ii, $N_{\overline{\Omega}}(H_{\overline{\Omega}})$ is isomorphic to $2 \times H_{\overline{\Omega}} \cong 2 \times S_n$, $H_{\overline{\Omega}}.3 \cong A_4$, $H_{\overline{\Omega}}.2^2 \cong 3^2.Q_8$ or $H_{\overline{\Omega}}.S_3$ when $H_{\overline{\Omega}}$ is of type $GL_1(2) \wr S_n$ (n even), $GL_1(5) \wr S_2$, $GL_1(4) \wr S_3$ or $GL_1(3) \wr S_4$, respectively.

Proposition 3.1.3. *Assume that* $\overline{\Omega} \leq X \leq \overline{A}$. *Then* $H_{\overline{\Omega}} = H_X \cap \overline{\Omega}$, *except when* $X \not\leq \overline{\Gamma}$ *and one of the exceptional cases in Theorem 3.1.2.ii arises.*

Proof. If $X \leq \overline{\Gamma}$, then the fact that $H_X \cap \overline{\Omega} = H_{\overline{\Omega}}$ follows directly from (3.1.3)-(3.1.5). Assume now that $X \not\leq \overline{\Gamma}$, so that case **L** holds with $n \geq 3$. Here $H_X = N_X(H_{\overline{\Gamma}})$, and so by (3.1.7), $H_X \cap \overline{\Omega} = N_{\overline{\Omega}}(H_{\overline{\Gamma}}) \leq N_{\overline{\Omega}}(H_{\overline{\Omega}})$. So provided none of the exceptional cases in Theorem 3.1.2.ii occurs, it is clear that $H_X \cap \overline{\Omega} = H_{\overline{\Omega}}$, as desired. ∎

§3.2 How to determine the conjugacy amongst members of $\mathcal{C}$

Our strategy for describing the conjugacy amongst the members of $\mathcal{C}$ runs as follows. We focus on an $\overline{A}$-conjugacy class of groups in $\mathcal{C}$ and calculate the number of classes into which this splits under the action of $\overline{\Omega}$. This number will usually be called c, and thus we obtain a homomorphism from $\overline{A}$ to S_c, the symmetric group of degree c. Obviously $\overline{\Omega}$ lies in the kernel, and so we can regard this as a homomorphism, usually called π, from $\ddot{A}$ to S_c. Given enough information about π, one can read off how the group G (here $\overline{\Omega} \leq G \leq \overline{A}$) acts on these c classes, simply by restricting π to $\ddot{G}$. Now to describe a transitive permutation representation of a group, it suffices to identify the stabilizer of a point, as the action is then isomorphic to the action on the cosets of the point stabilizer. Thus our goal is to describe the stabilizer in $\ddot{A}$ of one of these c $\overline{\Omega}$-classes. In Chapter 2 we have given precise information about the structure of $\ddot{A}$ in terms of explicit generators and relations; in Chapter 4 we shall indicate precisely what the point stabilizer is, again explicitly in terms of these generators. In this way, we derive complete information about the action of $\ddot{A}$ on the c classes, and hence complete information about the action of $\ddot{G}$.

In the rest of this section we describe this strategy in more technical detail, and provide some examples so that the reader can see this strategy in action. We continue with the notation H_X introduced in §3.1.

For any groups X, Y satisfying $Y \leq \overline{\Omega} \leq X \leq \overline{A}$, define $[Y]_X$ as the set of X-conjugates of Y — that is, $[Y]_X = \{Y^x \mid x \in X\}$. Furthermore, write $[Y] = [Y]_{\overline{\Omega}}$ and define $[Y]^X$ to be the set of $\overline{\Omega}$-classes contained in the X-class $[Y]_X$. Thus $[Y]^X = \{[Y^x] \mid x \in X\}$ and obviously X acts on $[Y]^X$ via $[Y^x]^{x'} = [Y^{xx'}]$ $(x, x' \in X)$. It is an elementary exercise to show that $\left| [Y]^X \right| = |X : N_X(Y)\overline{\Omega}|$. Now define

$$c = \left| [H_{\overline{\Omega}}]^{\overline{A}} \right|, \tag{3.2.1}$$

and write

$$[H_{\overline{\Omega}}]^{\overline{A}} = \{[H_1], \ldots, [H_c]\} \tag{3.2.2}$$

for suitable $H_i \in \mathcal{C}(\overline{\Omega})$. Thus c is the number of $\overline{\Omega}$-classes into which the $\overline{A}$-class $[H_{\overline{\Omega}}]_{\overline{A}}$ splits. We arrange the indices so that $H_1 = H_{\overline{\Omega}}$. As we mentioned earlier, $\ddot{A} = \overline{A}/\overline{\Omega}$ acts transitively on the c $\overline{\Omega}$-classes $[H_i]$, and it is clear that the stabilizer in $\ddot{A}$ of $[H_i]$ is $N_{\overline{A}}(H_i)\overline{\Omega}/\overline{\Omega}$. In particular,

$$c = |\overline{A} : N_{\overline{A}}(H_i)\overline{\Omega}|. \tag{3.2.3}$$

More generally, for any group X as above, define $X_{[H_i]} = N_X(H_i)\overline{\Omega}$. Thus $\ddot{X}_{[H_i]}$ is the stabilizer in $\ddot{X}$ of $[H_i]$. Clearly

$$\ddot{X}_{[H_i]} = N_X(H_i)\overline{\Omega}/\overline{\Omega}. \tag{3.2.4}$$

When the context is clear, we write X_i instead of $X_{[H_i]}$. The action of $\ddot{A}$ on $[H_{\overline{\Omega}}]^{\overline{A}}$ induces a homomorphism $\pi = \pi_{H_{\overline{\Omega}}}$ from $\ddot{A}$ to the symmetric group $Sym([H_{\overline{\Omega}}]^{\overline{A}}) \cong S_c$. The actions π which occur are given in Column V of Tables 3.5.A-F, which uses symbols appearing in the left-hand column of the Action Table 3.5.G (if $c = 1$ then π is trivial and we just put the symbol 1 in Column V). As remarked in the first paragraph in this section, to specify π it suffices to describe $\ddot{A}_1 = \ddot{A}_{[H_1]}$, the stabilizer in $\ddot{A}$ of the element $[H_1] \in [H_{\overline{\Omega}}]^{\overline{A}}$. This stabilizer is given for each action in the right-hand column 6 of Table 3.5.G. The symbols $\ddot{\delta}$, $\ddot{\phi}$, $\ker(\ddot{\gamma})$, etc., are all described in the relevant sections of Chapter 2. Of course $c = |\ddot{A} : \ddot{A}_1|$ and $\mathrm{im}(\pi) \cong \ddot{A}/\ker(\pi)$; these are given in columns 3 and 5. Finally, column 2 gives the conditions under which the relevant action occurs. We illustrate these ideas with the following example.

Example 2. Suppose that $\Omega \approx SL_{14}(729)$ and that $H_{\overline{\Omega}}$ is a subfield subgroup in $\mathcal{C}_5$ of type $GL_{14}(9)$ (see Table 3.5.A). Then according to Column V of Table 3.5.A, $\pi = \pi_1(c_5)$. Upon referring to the information displayed below Table 3.5.A — specifically, the value of c_5 — we see that $c = 728/[8, 52] = 7$. Now by Proposition 2.2.3, $\ddot{A} = \langle \ddot{\delta}, \ddot{\phi}, \ddot{\iota} \rangle$, with the relations $|\ddot{\delta}| = 14$, $|\ddot{\phi}| = 6$, $|\ddot{\iota}| = 2$, $\ddot{\delta}^{\ddot{\phi}} = \ddot{\delta}^3$, $\ddot{\delta}^{\ddot{\iota}} = \ddot{\delta}^{-1}$ and $[\ddot{\phi}, \ddot{\iota}] = 1$. Thus $\ddot{\delta}^7$ and $\ddot{\phi}^3\ddot{\iota}$ are central involutions and $\langle \ddot{\delta}^2, \ddot{\phi} \rangle$ is a Frobenius group of order 42, so we may write

$$\ddot{A} = \langle \ddot{\delta}^7 \rangle \times \langle \ddot{\phi}^3 \ddot{\iota} \rangle \times \langle \ddot{\delta}^2, \ddot{\phi} \rangle \cong \mathbf{Z}_2 \times \mathbf{Z}_2 \times (\mathbf{Z}_7{:}\mathbf{Z}_6).$$

Then looking at the row in Table 3.5.G corresponding to $\pi_1(c)$, we see that $\ddot{A}_1 = \langle \ddot{\delta}^7, \ddot{\phi}, \ddot{\imath} \rangle$. Hence $\ker(\pi) = \langle \ddot{\delta}^7, \ddot{\phi}^3 \ddot{\imath} \rangle$ and $\mathrm{im}(\pi) \cong \mathbf{Z}_7{:}\mathbf{Z}_6$. The group $\mathbf{Z}_7{:}\mathbf{Z}_6$ is of course the Sylow 7-normalizer in $Sym([H_{\overline{\Omega}}]^{\overline{A}}) \cong S_7$.

We now describe how to determine the conjugacy amongst the members of $\mathcal{C}(G)$. For each $i = 1, \ldots, c$, write $H_{G,i}$ for the G-associate of H_i (this is defined after (3.1.4)). The following observation is our starting point for determining conjugacy.

Lemma 3.2.1.

(i) *If* $c = 1$, *then* $\ddot{H}_{G,i} = \ddot{G}$.

(ii) *If* $c \geq 2$, *then* $H_i = H_{G,i} \cap \overline{\Omega}$ *and* $H_{G,i} = N_G(H_i)$.

Proof. In this proof we may take $i = 1$, so that $H_i = H_{\overline{\Omega}}$ and $H_{G,i} = H$. Suppose first that $c = 1$, so that $N_{\overline{A}}(H_{\overline{\Omega}})\overline{\Omega} = \overline{A}$ by (3.2.3). Consequently $N_G(H_{\overline{\Omega}})\overline{\Omega} = G$. So in either case (a) or (b) of Theorem 3.1.2.i, we have $H\overline{\Omega} = G$. Thus $\ddot{H} = \ddot{G}$, as required. Now take $c \geq 2$. Then it follows from (3.2.3) that $N_{\overline{A}}(H_{\overline{\Omega}})\overline{\Omega} \neq \overline{A}$, and hence $H_{\overline{A}}\overline{\Omega} \neq \overline{A}$. Thus Theorem 3.1.2.i implies that $H = N_G(H_{\overline{\Omega}})$ and that $H_{\overline{\Omega}} = N_{\overline{\Omega}}(H_{\overline{\Omega}})$. The result is now clear. ∎

We now come to the fundamental result linking conjugacy in $\mathcal{C}(G)$ with the homomorphism π.

Lemma 3.2.2.

(i) *Every member of* $\mathcal{C}(G)$ *of the same type as* H *is* G*-conjugate to some* $H_{G,i}$.

(ii) $\ddot{H}_{G,i} = \ddot{G}_i$.

(iii) $H_{G,i}$ *and* $H_{G,j}$ *are* G*-conjugate if and only if* $[H_i]$ *and* $[H_j]$ *are in the same* $\pi(\ddot{G})$*-orbit.*

Proof. Part (i) is immediate from Theorem 3.1.1. Assertion (ii) follows directly from (3.2.4) provided $H_{G,i} = N_G(H_i)$. And if $H_{G,i} \neq N_G(H_i)$, then Lemma 3.2.1 implies that $c = 1$ and $\ddot{H}_{G,i} = \ddot{G} = \ddot{G}_i$, as required. For (iii), there is nothing to prove if $c = 1$, so assume that $c \geq 2$. Suppose first that $H_{G,j} = H_{G,i}^g$ for some $g \in G$. Now according to Lemma 3.2.1.ii, $H_k = H_{G,k} \cap \overline{\Omega}$ for all k, whence $H_j = H_i^g$. Therefore $[H_j] = [H_i^g] = [H_i]^g$, which means $[H_i]$ and $[H_j]$ lie in the same $\pi(\ddot{G})$-orbit. Conversely, suppose that $[H_j] = [H_i]^g$ for some $g \in G$. Then H_j and H_i^g are $\overline{\Omega}$-conjugate, and so H_j and H_i are G-conjugate. Lemma 3.2.1.ii implies $H_{G,k} = N_G(H_k)$ for all k, and thus it follows that $N_G(H_i) = H_{G,i}$ and $N_G(H_j) = H_{G,j}$ are G-conjugate, as desired. ∎

The task of determining conjugacy in $\mathcal{C}(G)$ has now been reduced to some easy calculations within $\ddot{A}$, as the following example demonstrates. The reader is also referred to Example 6, below, for similar sorts of calculations.

Example 3. Continuing with Example 2, we see that the number of orbits of $\pi(\ddot{G})$ on the seven elements in $[H_{\overline{\Omega}}]^{\overline{A}}$ is given by

$$\begin{cases} 1 & \text{if } 7 \mid |\pi(\ddot{G})| \\ 1 + \frac{6}{m} & \text{if } |\pi(\ddot{G})| = m \text{ and } m \mid 6. \end{cases} \tag{3.2.5}$$

The number in (3.2.5) is thus the number of G-classes of members of $\mathcal{C}_5(G)$ of type $GL_{14}(9)$. Thus if G is for instance the (unique) group which satisfies $\ddot{G} = \langle \ddot{\phi}^2, \ddot{\delta}^7 \rangle$, then $\ddot{G}$ fixes $[H_1]$ and has two orbits of size three on the remaining six classes.

§3.3 How to determine the structure of members of $\mathcal{C}$

We retain the notation developed so far in this chapter, so that the H_i are as in (3.2.2) and the $H_{G,i}$ are the corresponding G-associates. It is clear that

$$H_{G,i} \cong (H_{G,i} \cap \overline{\Omega}).\ddot{H}_{G,i},$$

and thus the structure of H_i may be determined provided we know that of $H_{G,i} \cap \overline{\Omega}$ and that of $\ddot{H}_{G,i}$. Now according to Lemma 3.2.2.ii, $\ddot{H}_{G,i} = \ddot{G}_i$, and hence

$$\ddot{H}_{G,i} = \ddot{G}_i \cong \ker_{\ddot{G}}(\pi).\pi(\ddot{G}_i). \tag{3.3.1}$$

Since the action of $\ddot{A}$ on $[H_{\overline{\Omega}}]^{\overline{A}}$ has been specified, the group $\ddot{A}_i$ is known, and hence $\ddot{G}_i$ is also known. The structure of $H_{G,i} \cap \overline{\Omega}$ may be determined as follows. Proposition 3.1.3 implies that $H_{G,i} \cap \overline{\Omega} = H_i$, unless $H \not\leq \overline{\Gamma}$ and H is of type $GL_1(2) \wr S_n$ (with n even), $GL_1(3) \wr S_4$ or $GL_1(4) \wr S_3$ in $\mathcal{C}_2$. The structure of $H_{\overline{\Omega}}$ (and hence of H_i) is given explicitly in Proposition 4.j.k, where $H \in \mathcal{C}_j$ and k appears in Column II of Tables 3.5.A-F. When $H \not\leq \overline{\Gamma}$ and H is of type $GL_1(2) \wr S_n$ with n even, we have $\overline{\Omega} = \overline{\Gamma}$ and $H \cap \overline{\Omega} = N_{\overline{\Omega}}(H_{\overline{\Omega}}) \cong S_n \times 2$. Here $H \cong S_n \times [4]$. We make no further remarks concerning groups of type $GL_1(3) \wr S_4$ or $GL_1(4) \wr S_3$ — complete information about the subgroup structure of the automorphism groups of $L_3(4)$ and $L_4(3)$ can be found in [At,Kl$_2$], for example.

Example 4. We return to our previous example with $\Omega \approx SL_{14}(729)$ and $H_{\overline{\Omega}} \in \mathcal{C}_5(\overline{\Omega})$ of type $GL_{14}(9)$. First of all, we saw in Example 2 that $\ker(\pi) = \langle \ddot{\delta}^7, \ddot{\phi}^3\ddot{\iota} \rangle$, and hence $\ker_{\ddot{G}}(\pi) = \ddot{G} \cap \langle \ddot{\delta}^7, \ddot{\phi}^3\ddot{\iota} \rangle$. So in view of (3.3.1), it remains to find $\pi(\ddot{G}_i)$. Now if $\pi(\ddot{G}) \cong \mathbf{Z}_7{:}\mathbf{Z}_m$ with $m \mid 6$, then $\pi(\ddot{G})$ is transitive and $\pi(\ddot{G}_i) \cong \mathbf{Z}_m$ for each i. So assume that 7 does not divide $|\pi(\ddot{G})|$, which means $\pi(\ddot{G})$ is intransitive. Write $|\pi(\ddot{G})| = m$, with $m \mid 6$. Then by (3.2.5) $\ddot{G}$ has $1 + \frac{6}{m}$ orbits on $[H_{\overline{\Omega}}]^{\overline{A}}$, with one orbit of size 1 and the others of size m. Thus if $[H_j]$ is in the orbit of size 1, we have

$$\ddot{H}_{G,i} \cong \begin{cases} \ker_{\ddot{G}}(\pi).\mathbf{Z}_m & \text{if } i = j \\ \ker_{\ddot{G}}(\pi) & \text{if } i \neq j. \end{cases}$$

§3.4 How to determine the overgroups of members of $\mathcal{C}$

Our results concerning maximality apply only when $n = \dim(V)$ is at least 13, so we adopt the hypothesis $n \geq 13$ in this section. The other notation carries over from before. Now there are precisely two ways in which H can fail to be maximal in G. Namely, (at least) one of the following holds:
(∗) $H < H\overline{\Omega} < G$;
(†) $H < K < G$ for some K not containing $\overline{\Omega}$.
We define a subgroup of G to be *unfaithful* if it contains $\overline{\Omega}$, and *faithful* otherwise; thus a subgroup is faithful if and only if the action of G on its cosets is faithful. (This terminology was introduced in [Wi5].) Thus the overgroup appearing in (∗) is unfaithful, while that appearing in (†) is faithful. We now explain how both unfaithful and faithful overgroups can be read off from Tables 3.5.A-I.

Before continuing, we warn the reader that by an *overgroup* of H in G we mean a group lying *strictly* between H and G.

Evidently the unfaithful overgroups of H correspond to the groups lying between $\ddot{H}$ and $\ddot{G}$. Recalling that $\ddot{H} = \ddot{H}_{G,1}$ (due to our convention $H_{\overline{\Omega}} = H_1$) and recalling Lemma 3.2.2.ii, we deduce that the unfaithful overgroups of H are in one-to-one correspondence with the groups X satisfying

$$\ddot{G}_1 \leq X < \ddot{G}. \tag{3.4.1}$$

We now turn to the faithful overgroups, and we suppose that K is a faithful overgroup as given in (†). Observe that by Aschbacher's Theorem 1.2.1, either K is contained in a member of $\mathcal{C}(G)$ or $K \in \mathcal{S}$. Determining *all* faithful overgroups of H in G can be quite an unpleasant undertaking in certain circumstances, so we content ourselves with the restriction $K \in \mathcal{C}(G) \cup \mathcal{S}$. We define the *overgroup sets* as follows:

$$\begin{aligned} \mathcal{G}^G_{\mathcal{C}}(H) &= \{K \mid H < K < G,\ K \in \mathcal{C}(G)\} \\ \mathcal{G}^G_{\mathcal{S}}(H) &= \{K \mid H < K < G,\ K \in \mathcal{S}\} \\ \mathcal{G}^G(H) &= \mathcal{G}^G_{\mathcal{C}}(H) \cup \mathcal{G}^G_{\mathcal{S}}(H) \end{aligned} \tag{3.4.2}$$

When the superscript ^{G} is omitted it is to be understood that $G = \overline{\Omega}$. We now demonstrate how all groups in $\mathcal{G}^G(H)$ can be identified using the Tables in §3.5.

Lemma 3.4.1. *If $K \in \mathcal{G}^G(H)$, then $H_{\overline{\Omega}} < K \cap \overline{\Omega} < \overline{\Omega}$.*

Proof. Assume for a contradiction that $H_{\overline{\Omega}} = K \cap \overline{\Omega}$. Then $H \neq N_G(H_{\overline{\Omega}})$, and consequently $H\overline{\Omega} = G$ by Theorem 3.1.2. But then $K\overline{\Omega} = G$, and it now follows that $|K| = |G : \overline{\Omega}||K \cap \overline{\Omega}| = |H|$, a contradiction. ∎

Now suppose that $K \in \mathcal{G}^G(H)$, so that by Lemma 3.4.1 we have $H_{\overline{\Omega}} < K \cap \overline{\Omega} < \overline{\Omega}$. In particular, $H_{\overline{\Omega}}$ is non-maximal in $\overline{\Omega}$; all such instances are indicated in Column VI of Tables 3.5.A-F. First suppose that $K \in \mathcal{G}^G_{\mathcal{C}}(H)$, and write $K_{\overline{\Omega}}$ for the $\overline{\Omega}$-associate of K. If $K_{\overline{\Omega}} = K \cap \overline{\Omega}$, then complete information about the possibilities for K can be gleaned from Table 3.5.H. If $K \cap \overline{\Omega} \neq K_{\overline{\Omega}}$, then K is one of the exceptional types appearing in Theorem 3.1.2.ii. If K is of type $GL_1(2) \wr S_n$ in $\mathcal{C}_2(GL_n(2))$ with n even, then $K \cap \overline{\Omega} = K_{\overline{\Omega}} \times 2 \cong S_n \times 2$. It is an easy exercise (once the reader is familiar with the members of $\mathcal{C}$) to see that no member of $\mathcal{C}$ can be contained in $K \cap \overline{\Omega}$, apart from groups of the same type as K. But if H and K are of the same type, then H *equals* K, and this runs contrary to our assumption that K is a proper overgroup. Therefore no example arises here. As usual, we omit any discussion of the case where K is of type $GL_1(3) \wr S_4$ or $GL_1(4) \wr S_3$. If $K \in \mathcal{G}_{\mathcal{S}}^{G}(H)$, then $K \cap \overline{\Omega} \in \mathcal{S}$ and so one may appeal to Table 3.5.I for the classification of the possibilities for K. We now explain how Tables 3.5.H and 3.5.I are to be read.

Reading Table 3.5.H

Table 3.5.H classifies all triples $(H_{\overline{\Omega}}, K_{\overline{\Omega}}, \overline{\Omega})$ where $H_{\overline{\Omega}} < K_{\overline{\Omega}} < \overline{\Omega}$ and $H_{\overline{\Omega}}, K_{\overline{\Omega}} \in \mathcal{C}(\overline{\Omega})$. Column 1 gives the group Ω, and columns 3 and 5 give the types of $H_{\overline{\Omega}}$ and $K_{\overline{\Omega}}$. In columns 2 and 4 we give $\mathcal{C}_i$ and $\mathcal{C}_j$, where $H_{\overline{\Omega}} \in \mathcal{C}_i(\overline{\Omega})$ and $K_{\overline{\Omega}} \in \mathcal{C}_j(\overline{\Omega})$. Conditions under which the triple occurs appear in Column 6. Each row in the table corresponds to a unique triple $(H_{\overline{\Omega}}, K_{\overline{\Omega}}, \overline{\Omega})$ up to conjugacy in $\overline{\Omega}$, except when the term '(two)' or '(four)' appears. When '(two)' appears in Column 5, there are precisely two groups $K_{\overline{\Omega}}$ containing $H_{\overline{\Omega}}$, giving exactly two non-conjugate triples. The last row of the table is more complicated and is explained in Example 6, below. Column 7 indicates when *G-novelties* arise. Here, if $X < Y < \overline{\Omega} \trianglelefteq G$, then $N_G(X)$ is said to be a *G-novelty with respect to* Y if $N_G(X) \not\leq N_G(Y)$. The symbol 'N' appears in Column 7 if $N_G(H_{\overline{\Omega}})$ is a G-novelty with respect to $N_G(K_{\overline{\Omega}})$ for some G. Notice that, except in the last row, whenever 'N' appears we have $|[H_{\overline{\Omega}}]^{\overline{A}}| = 1$ and $|[K_{\overline{\Omega}})]^{\overline{A}}| = 2$. It follows that in these cases, there is a G-novelty if and only if $\pi_{K_{\overline{\Omega}}}(\ddot{G}) \neq 1$. The term 'novelty' first appeared in [Wi$_5$] and [At]. All inclusions $H_{\overline{\Omega}} < K_{\overline{\Omega}}$ with $H_{\overline{\Omega}}, K_{\overline{\Omega}} \in \mathcal{C}(\overline{\Omega})$ are discussed in Chapter 6.

Example 5. Consider the first four rows of Table 3.5.H with $\Omega \approx SL_n(2)$, $H_{\overline{\Omega}} \in \mathcal{C}_2(\overline{\Omega})$ and $H_{\overline{\Omega}}$ of type $GL_1(2) \wr S_n$, so that $H_{\overline{\Omega}}$ is the stabilizer in $\overline{\Omega}$ of a decomposition $V = V_1 \oplus \cdots \oplus V_n$ ($\dim(V_i) = 1$). Then when n is odd, $\mathcal{G}_{\mathcal{C}}(H_{\overline{\Omega}})$ is a set of size three, consisting of the stabilizer of a 1-space, the stabilizer of an $(n-1)$-space, and the stabilizer of the 1-space and the $(n-1)$-space (a group of type $GL_1(2) \oplus GL_{n-1}(2)$). When n is even, $\mathcal{G}_{\mathcal{C}}(H_{\overline{\Omega}})$ has size four, consisting of the three stabilizers mentioned before (although the stabilizer of the pair is of type $P_{1,n-1}$ in this case), along with a member of $\mathcal{C}_8(\overline{\Omega})$, which is the stabilizer of a non-degenerate symplectic form on V. The symbol 'N' appears in Column 7 in the rows corresponding to P_1, P_{n-1}, because when $G = \overline{A}$ the group $N_G(H_{\overline{\Omega}})$ is a G-novelty with respect to $N_G(K_{\overline{\Omega}})$.

Example 6. Here we explain the last row of Table 3.5.H, which is by far the most complicated situation of all. We have $\Omega \approx \Omega_n^+(q)$, with q odd and $n = 4m$ with m even. Put $-\epsilon = (-)^{(q-1)n/16}$, so that according to Table 3.5.E, $\overline{\Omega} \approx P\Omega_n^+(q)$ has just two conjugacy classes of tensor product subgroups (which are members of $\mathcal{C}_4(\overline{\Omega})$) of type

$O_4^+(q) \otimes O_m^\epsilon(q)$. We let H_1, H_2 be representatives of these classes, and write $H_i = H_{G,i} \cap \overline{\Omega}$ with $H_{G,i} \in \mathcal{C}_4(G)$. Now Table 3.5.E implies that $\overline{\Omega}$ has just four classes of tensor product subgroups of type $Sp_2(q) \otimes Sp_{n/2}(q)$ in $\mathcal{C}_4(\overline{\Omega})$. Furthermore, the last row of Table 3.5.H asserts that each such group of type $Sp_2(q) \otimes Sp_{n/2}(q)$ contains a group of type $O_4^+(q) \otimes O_m^\epsilon(q)$. Proposition 6.3.4 in Chapter 6 implies that each H_i is contained in just two groups K_{2i-1}, K_{2i} of type $Sp_2(q) \otimes Sp_{n/2}(q)$ in $\mathcal{C}_4(\overline{\Omega})$. Thus we may write

Note that any group in $\mathcal{C}_4(\overline{\Omega})$ of type $Sp_2(q) \otimes Sp_{n/2}(q)$ contains a group of type $O_4^+(q) \otimes O_m^\epsilon(q)$, and by conjugating in $\overline{\Omega}$ we can assume it contains H_1 or H_2. But the four groups K_j are the only groups of type $Sp_2(q) \otimes Sp_{n/2}(q)$ in $\mathcal{C}_4(\overline{\Omega})$ containing the H_i. It now follows that K_1, K_2, K_3, K_4 are representatives for the four classes. In fact, K_1 is the unique member of $[K_1]$ which contains H_1; a similar statement holds for the other K_j. For $1 \leq i \leq 4$, let $K_i = K_{G,i} \cap \overline{\Omega}$ with $K_{G,i} \in \mathcal{C}_4(G)$.

Now $\ddot{A}$ acts on $[H_i]^{\overline{A}} = \{[H_1], [H_2]\}$ via the homomorphism $\pi_{H_1} = \pi_{H_2}$, and with a slight abuse of notation we refer to this as π_H. Similarly, $\ddot{A}$ acts on $[K_j]^{\overline{A}} = \{[K_1], [K_2], [K_3], [K_4]\}$ via the homomorphism $\pi_K = \pi_{K_j}$ $(1 \leq j \leq 4)$. As described in §3.2, using Column V of Table 3.5.E and the Action Table 3.5.G, we can read off what π_H and π_K are exactly. Recall that $\ddot{A}$ is given explicitly in Proposition 2.7.3, and $\ddot{A} = B \times \langle \ddot{\phi} \rangle$, where $B = \langle \ddot{r}_\square, \ddot{\delta} \rangle \cong D_8$. According to Table 3.5.E, $\pi_H = \pi_2$ and $\pi_K = \pi_5$. So by appealing to Table 3.5.G, we deduce

$$\ker(\pi_H) = \langle \ddot{r}_\square \ddot{r}_\boxtimes, \ddot{\delta}, \ddot{\phi} \rangle,$$

and

$$\ker(\pi_K) = \langle \ddot{\phi} \rangle.$$

In particular, $\ddot{\phi}$ lies in the kernel of both actions. Consequently, the action of $\ddot{G}$ on these classes is determined by $\ddot{G}_B$, the projection of $\ddot{G}$ in B. In particular, the action of $\ddot{A}$ is determined by the action of $\ddot{\delta}$ and $\ddot{r}_\square$. Since $\ddot{\delta} \in \ker(\pi_H)$ and $\ddot{r}_\square \notin \ker(\pi_H)$, we know that $\ddot{\delta}$ fixes both $[H_1]$ and $[H_2]$, and $\ddot{r}_\square$ interchanges them. Now $\pi_5(\ddot{\delta})$ moves two points and fixes two. And since $\ddot{\delta}$ fixes $[H_i]$, $\ddot{\delta}$ must fix $\{[K_{2i-1}], [K_{2i}]\}$. Therefore, with no loss we can assume that $\ddot{\delta}$ interchanges $[K_1]$ and $[K_2]$, and fixes $[K_3]$ and $[K_4]$. Thus as permutations, we may write

$$\begin{array}{lcc} \ddot{\delta} & 1 & ([K_1],[K_2]) \\ \ddot{r}_\square & ([H_1],[H_2]) & ([K_1],[K_3])([K_2],[K_4]). \end{array} \qquad (3.4.3)$$

Evidently B contains exactly eight conjugacy classes of subgroups with representatives 1, $\langle \ddot{r}_\square \rangle$, $\langle \ddot{\delta} \rangle$, $\langle \ddot{r}_\square \ddot{r}_\boxtimes \rangle$, $\langle \ddot{r}_\square, \ddot{r}_\boxtimes \rangle$, $\langle \ddot{r}_\square \ddot{r}_\boxtimes, \ddot{\delta} \rangle$, $\langle \ddot{r}_\square \ddot{\delta} \rangle$ and B.

First let us discuss the possible unfaithul overgroups of $H_{G,i}$ (see (*) above). Here we appeal to the second paragraph of this section. Now if $\ddot{G}_B \leq \ker(\pi_H)$, then $\ddot{G}_{[H_1]} = \ddot{G}_{[H_2]} = \ddot{G}$, and so there are no groups X satisfying (3.4.1), which is to say there are no unfaithful overgroups. On the other hand, if $\ddot{G}_B \not\leq \ker(\pi_H)$, then $\ddot{G}_{[H_1]} = \ddot{G}_{[H_2]} = \ker_{\ddot{G}}(\pi_H)$, a subgroup of index 2 in $\ddot{G}$, and so there is precisely one unfaithful overgroup, namely $H_{G,i}\overline{\Omega} = \ker_G(\pi_H)$ — here $\ker_G(\pi_H)$ denotes the preimage in G of $\ker_{\ddot{G}}(\pi_H)$, and so $\ker_G(\pi_H)$ is a subgroup of index 2 in G. Thus $H_{G,1}$ has an unfaithful overgroup if and only if $\ddot{G}_B$ is (up to conjugacy) one of $\langle \ddot{r}_\square \rangle$, $\langle \ddot{r}_\square, \ddot{r}_\boxtimes \rangle$, $\langle \ddot{r}_\square \ddot{\delta} \rangle$ and B.

We now consider the possible faithful overgroups. The principle which we use is:

$$\text{if } H_i \leq K_j \text{ and } \ddot{H}_{G,i} \leq \ddot{G}_{[K_j]}, \text{ then } H_{G,i} \leq K_{G,j}. \tag{3.4.4}$$

For we have already remarked that K_j is the unique member of the $\overline{\Omega}$-class $[K_j]$ which contains H_i, and so if $H_{G,i}$ fixes the class $[K_j]$ it must normalize K_j. Thus (3.4.4) follows as $K_{G,j} = N_G(K_j)$ by Lemma 3.2.1. For example, suppose initially that $\ddot{G}_B = 1$, that is, $\ddot{G} \leq \langle \ddot{\phi} \rangle$. Then $\ddot{G}$ fixes all six conjugacy classes $[H_i]$ $(i = 1, 2)$, $[K_j]$ $(j = 1, 2, 3, 4)$. In particular, $H_{G,1}$ fixes $[K_1]$, and hence $H_{G,1} \leq K_{G,1}$ by (3.4.4). Similarly $H_{G,i} < K_{G,j}$ for $(i, j) = (1, 2)$, $(2, 3)$ and $(2, 4)$. Thus $\mathcal{G}_{\mathcal{C}}^{G}(H_{G,i}) = \{K_{G,2i-1}, K_{G,2i}\}$ for $i = 1, 2$. Next suppose that $\ddot{G}_B = \langle \ddot{r}_\square \rangle$. Then $\ddot{G}$ interchanges $[H_1]$ and $[H_2]$, and so $\ddot{G}_{[H_1]} = \ddot{G}_{[H_2]} = \ker_{\ddot{G}}(\pi_H)$. Now $\ddot{G} \leq \langle \ddot{\phi}, \ddot{r}_\square \rangle$ (by our assumption $\ddot{G}_B \leq \langle \ddot{r}_\square \rangle$) and $\ker(\pi_H)$ is displayed above. Thus one easily sees that $\ker_{\ddot{G}}(\pi_H) = \ddot{G} \cap \ker(\pi_H) \leq \langle \ddot{\phi} \rangle \leq \ker(\pi_K)$. Putting this information together with Lemma 3.2.2, we deduce

$$\ddot{H}_{G,i} = \ddot{G}_{[H_i]} = \ker_{\ddot{G}}(\pi_H) \leq \ker(\pi_K).$$

Thus each $H_{G,i}$ fixes each $[K_j]$, and so as in the previous case, $H_{G,i} < K_{G,j}$ for $(i, j) = (1, 1)$, $(1, 2)$, $(2, 3)$ and $(2, 4)$. Now take the case where $\ddot{G}_B = \langle \ddot{\delta} \rangle$. Since $\ddot{\delta}$ fixes $[H_1]$ and $[H_2]$ (see (3.4.3)), we have by Lemma 3.2.2 $\ddot{H}_{G,i} = \ddot{G}_{[H_i]} = \ddot{G}$. Moreover, $\ddot{\delta}$ interchanges $[K_1]$ and $[K_2]$, and so $\ddot{K}_{G,i} = \ddot{G}_{[K_j]} < \ddot{G}$ for $i = 1, 2$. Therefore $\ddot{H}_{G,1} \not\leq \ddot{K}_{G,i}$ for $i = 1, 2$, and in particular $H_{G,1} \not\leq K_{G,i}$. Therefore $H_{G,1}$ is a G-novelty with respect to K_1 and K_2. However, $\ddot{\delta}$ fixes $[K_3]$, and hence so does $\ddot{G}$. In particular, $\ddot{H}_2 \leq \ddot{G}_{[K_3]}$, and so $H_{G,2} < K_{G,3}$ by (3.4.4); similarly $H_{G,2} < K_{G,4}$. The other possibilities for $\ddot{G}_B$ are treated similarly, and we record the conclusions in the Table 3.4.A.

We remark that π_2 is identical with the homomorphism $\ddot{\gamma}$ defined in §2.5. Thus in Table 3.4.A, one may substitute $\ker_{\ddot{G}}(\ddot{\gamma})$ for $\ker_G(\pi)$. Recall from §2.5 that $\ddot{G}$ lies in the kernel of $\ddot{\gamma}$ if and only if G preserves the two families of totally singular $\frac{n}{2}$-spaces in the $O_n^+(q)$-geometry. Also observe that in all but the first two rows of the table, G-novelties will arise. This concludes Example 6.

We conclude our discussion of Table 3.5.H with the observation that the G-novelties which occur are always maximal in G, except in the following cases:

(a) $\Omega \approx SL_n(2)$, $H_{\overline{\Omega}}$ of type $GL_1(2) \wr S_n$ with n even (here $N_G(H_{\overline{\Omega}}) < N_G(Sp_n(2))$;

(b) $\Omega \approx \Omega_n^+(2)$, $H_{\overline{\Omega}}$ of type $O_2^+(2) \wr S_{n/2}$ (here $N_G(H_{\overline{\Omega}}) < N_G(P_{n/2-1})$);

Table 3.4.A

$\ddot{G}_B$	unfaithful overgroups	(i,j) with $H_{G,i} < K_{G,j}$
1	$-$	$(1,1),(1,2),(2,3),(2,4)$
$\langle \ddot{r}_\square \rangle$	$\mathrm{ker}_G(\pi_H)$	$(1,1),(1,2),(2,3),(2,4)$
$\langle \ddot{\delta} \rangle$	$-$	$(2,3),(2,4)$
$\langle \ddot{r}_\square \ddot{r}_\boxtimes \rangle$	$-$	$-$
$\langle \ddot{r}_\square \ddot{r}_\boxtimes, \ddot{\delta} \rangle$	$-$	$-$
$\langle \ddot{r}_\square, \ddot{r}_\boxtimes \rangle$	$\mathrm{ker}_G(\pi_H)$	$-$
$\langle \ddot{r}_\square \ddot{\delta} \rangle$	$\mathrm{ker}_G(\pi_H)$	$-$
$\langle \ddot{r}_\square, \ddot{\delta} \rangle$	$\mathrm{ker}_G(\pi_H)$	$-$

(c) $(H_{\overline{\Omega}}, K_{\overline{\Omega}}, \overline{\Omega})$ as in the last row of Table 3.5.H and $\ddot{G}_B$ as in the last three rows of Table 3.4.A (here $N_G(H_{\overline{\Omega}})$ is contained in the unfaithful subgroup $\mathrm{ker}_G(\pi_H)$).

Reading Table 3.5.I

Table 3.5.I classifies all triples $(H_{\overline{\Omega}}, K_{\overline{\Omega}}, \overline{\Omega})$ with $H_{\overline{\Omega}} < K_{\overline{\Omega}} < \overline{\Omega}$ and $K_{\overline{\Omega}} \in \mathcal{S}$. The type of $H_{\overline{\Omega}}$ appears in the first column, with the assertion that in all cases $H_{\overline{\Omega}} \in \mathcal{C}_7(\overline{\Omega})$. The isomorphism type of $K_{\overline{\Omega}}$ is given in the second column, and it is to be understood that $\mathcal{G}_\mathcal{S}(H_{\overline{\Omega}})$ contains precisely one group for each isomorphism type appearing in the second column, subject to the conditions given in the fourth column. For example, suppose that $H \in \mathcal{C}_7(G)$ with $\Omega \approx \Omega^+_{2^t}(q)$, and that H is of type $Sp_2(q) \wr S_t$ with q even, $q \geq 4$ and $t \geq 4$. Then $\mathcal{G}_\mathcal{S}(H_{\overline{\Omega}})$ is a set of size one if t is even, and a set of size two if t is odd. If t is even, the unique member has structure $Sp_{2t}(q)$; and if t is odd, the two members have structures $Sp_{2t}(q)$ and $\Omega^+_{2t+2}(q)$. Moreover, in this latter case with t odd, we have the chain

$$Sp_2(q) \wr S_t \cong H_{\overline{\Omega}} < Sp_{2t}(q) < \Omega^+_{2t+2}(q) < \Omega \approx \Omega^+_{2^t}(q),$$

where the middle two groups are the two members of $\mathcal{G}_\mathcal{S}(H_{\overline{\Omega}})$. Recall that $H_{\overline{\Omega}} = H \cap \overline{\Omega}$ with $H \in \mathcal{C}(G)$, and hence H acts on $\mathcal{G}_\mathcal{S}(H_{\overline{\Omega}})$. Since there is a unique group of a given isomorphism type in $\mathcal{G}_\mathcal{S}(H_{\overline{\Omega}})$, it follows that $H < N_G(K_{\overline{\Omega}}) \in \mathcal{S}$. Thus the members of $\mathcal{G}^G_\mathcal{S}(H)$ are in one-to-one correspondence with the members of $\mathcal{G}_\mathcal{S}(H_{\overline{\Omega}})$. Finally, the phrase 'two classes' in the right hand column appears because each triple given in Table 3.5.I actually represents two conjugacy classes in $\overline{\Omega}$. For example, there are precisely two conjugacy classes of groups of type $Sp_4(q) \wr S_t$ in $\mathcal{C}_7(\Omega^+_{4^t}(q))$, there are precisely two conjugacy classes of groups $Sp_{4t}(q)$ in $\mathcal{S}$, and each $Sp_{4t}(q)$ contains a group of type $Sp_4(q) \wr S_t$. The action of $\overline{\Gamma}$ on these two classes is that of π_2.

§3.5 The tables

In this section we present Tables 3.5.A-3.5.I, as described in the preceding sections of this chapter. When Ω is an orthogonal group with $n = \dim(V)$ even, q odd, and with associated quadratic form Q, we set $D = D(Q) \in \{\square, \boxtimes\}$. In addition, we write $D_n^\epsilon(q)$ for the discriminant of a non-degenerate quadratic form on an n-dimensional vector space over $\mathbf{F}_q$ of type ϵ; thus according to Proposition 2.5.13.ii,

$$D_n^\epsilon(q) = \begin{cases} \square & \text{if } \epsilon = + \text{ and } (q-1)n/4 \text{ is even} \\ \square & \text{if } \epsilon = - \text{ and } (q-1)n/4 \text{ is odd} \\ \boxtimes & \text{otherwise.} \end{cases}$$

Table 3.5.A
$\Omega \approx SL_n(q)$

I	II	III	IV	V	VI
$\mathcal{C}_i$	j	type	conditions	π	$H_{\bar{\Omega}}$ non-max.
$\mathcal{C}_1$	17	P_m	$1 \le m \le n-1$	π_3 if $m \neq n/2$ 1 if $m = n/2$	–
	4	$GL_m(q) \oplus GL_{n-m}(q)$	$1 \le m < n/2$	1	always
	22	$P_{m,n-m}$	$1 \le m < n/2$	1	always
$\mathcal{C}_2$	9	$GL_m(q) \wr S_t$	$n = mt,\ t \ge 2$	1	$m = 1,\ q \le 4$; or $m = q = 2$
$\mathcal{C}_3$	6	$GL_m(q^r)$	$n = mr$, r prime	1	–
$\mathcal{C}_4$	10	$GL_{n_1}(q) \otimes GL_{n_2}(q)$	$n = n_1 n_2,\ 2 \le n_1 < \sqrt{n}$	$\pi_1((q-1, n_1, n_2))$	$n_1 = q = 2$
$\mathcal{C}_5$	3	$GL_n(q_o)$	$q = q_o^r$, r prime	$\pi_1(c_5)$ *	–
$\mathcal{C}_6$	5-7	$r^{2m}.Sp_{2m}(r)$	$n = r^m$, r prime, $r \neq p$, f odd, f minimal subject to $p^f \equiv 1(r_1)$, $r_1 = r(2,r)$	$\pi_1(c_6)$ *	–
$\mathcal{C}_7$	3	$GL_m(q) \wr S_t$	$n = m^t,\ m \ge 3$	$\pi_1(c_7)$ *	–
$\mathcal{C}_8$	3	$Sp_n(q)$	n even	$\pi_1((q-1, n/2))$	–
	4	$O_n(q)$	qn odd	$\pi_1((q-1, n))$	–
	4	$O_n^\epsilon(q)$	$\epsilon = \pm$, q odd, n even	$\pi_2((q-1,n)/2)$ if $\epsilon = -$ & $D_n^-(q) = \boxtimes$; $\pi_1((q-1,n)/2)$ otherwise	– –
	5	$U_n(q^{1/2})$	q a square	$\pi_1(c_8)$ *	–

* Here we display the values of c_i for $i = 5, 6, 7, 8$.

$$c_5 = \frac{q-1}{[q_o - 1, (q-1)/(q-1,n)]}$$

$$c_6 = \begin{cases} 1 & \text{if } n = 2 \text{ and } q \equiv \pm 3 \pmod 8 \\ 1 & \text{if } n = 3 \text{ and } q \equiv 4 \text{ or } 7 \pmod 9 \\ 2 & \text{if } n = 4 \text{ and } q \equiv 5 \pmod 8 \\ (q-1, n) & \text{otherwise.} \end{cases}$$

$$c_7 = \begin{cases} \frac{1}{2}(q-1, m) & \text{if } t = 2,\ m \equiv 2 \pmod 4 \text{ and } q \equiv 3 \pmod 4 \\ (q-1, \frac{n}{m}) & \text{otherwise} \end{cases}$$

$$c_8 = \frac{q-1}{[q^{1/2}+1, (q-1)/(q-1,n)]}$$

Table 3.5.B
$\Omega \approx SU_n(q),\ n \geq 3$

I	II	III	IV	V	VI
$\mathcal{C}_i$	j	type	conditions	π	$H_{\overline{\Omega}}$ non-max.
$\mathcal{C}_1$	18	P_m	$1 \leq m \leq [n/2]$	1	–
	4	$GU_m(q) \perp GU_{n-m}(q)$	$1 \leq m < n/2$	1	–
$\mathcal{C}_2$	9	$GU_m(q) \wr S_t$	$n = mt,\ t \geq 2$	1	$m = q = 2$
	4	$GL_{n/2}(q^2).2$	n even	1	–
$\mathcal{C}_3$	6	$GU_m(q^r)$	$n = mr,\ r$ prime, $r \geq 3$	1	–
$\mathcal{C}_4$	10	$GU_{n_1}(q) \otimes GU_{n_2}(q)$	$n = n_1 n_2,\ 2 \leq n_1 < \sqrt{n}$	$\pi_1((q+1, n_1, n_2))$	$n_1 = q = 2$
$\mathcal{C}_5$	3	$GU_n(q_o)$	$q = q_o^r,\ r$ prime, $r \geq 3$	$\pi_1(c_5)$ *	–
	5	$O_n(q)$	qn odd	$\pi_1((q+1,n))$	–
	5	$O_n^{\epsilon}(q)$	q odd, n even	$\pi_1((q+1,n)/2)$ if $D_n^{\epsilon}(q) = \square$ $\pi_2((q+1,n)/2)$ otherwise	–
	6	$Sp_n(q)$	n even	$\pi_1((q+1,n/2))$	–
$\mathcal{C}_6$	5, 6	$r^{2m}.Sp_{2m}(r)$	$n = r^m,\ r$ prime, $r \neq p$, f even, f minimal subject to $p^f \equiv 1\,(r_1),\ r_1 = r(2,r)$	$\pi_1(c_6)$ *	–
$\mathcal{C}_7$	3	$GU_m(q) \wr S_t$	$n = m^t,\ m \geq 3,\ (q,m) \neq (2,3)$	$\pi_1(c_7)$ *	–

* Here we display the values of c_i for $i = 5, 6, 7$.

$$c_5 = \frac{q+1}{[q_o+1, (q+1)/(q+1,n)]}$$

$$c_6 = \begin{cases} 1 & \text{if } n = 3 \text{ and } q \equiv 2 \text{ or } 5 \pmod 9 \\ 2 & \text{if } n = 4 \text{ and } q \equiv 3 \pmod 8 \\ (q+1,n) & \text{otherwise} \end{cases}$$

$$c_7 = \begin{cases} \frac{1}{2}(q+1,m) & \text{if } t = 2,\ m \equiv 2 \pmod 4 \text{ and } q \equiv 1 \pmod 4 \\ (q+1,n/m) & \text{otherwise.} \end{cases}$$

Table 3.5.C

$\Omega \approx Sp_n(q)$, n even, $n \geq 4$

I	II	III	IV	V	VI
$\mathcal{C}_i$	j	type	conditions	π	$H_{\overline{\Omega}}$ non-max.
$\mathcal{C}_1$	19 3	P_m $Sp_m(q) \perp Sp_{n-m}(q)$	$1 \leq m \leq n/2$ m even, $2 \leq m < n/2$	1 1	– –
$\mathcal{C}_2$	10 5	$Sp_m(q) \wr S_t$ $GL_{n/2}(q).2$	$n = mt$, m even, $t \geq 2$ q odd	1 1	$m = q = 2$
$\mathcal{C}_3$	10 7	$Sp_m(q^r)$ $GU_{n/2}(q)$	$n = rm$, r prime, m even q odd	1 1	– –
$\mathcal{C}_4$	11	$Sp_m(q) \otimes O^{\epsilon}_{n/m}(q)$	$m \mid n$, m even, $n/m \geq 3$, q odd, $\epsilon \in \{+, -, \circ\}$	1	$q = n/m = 3$
$\mathcal{C}_5$	4	$Sp_n(q_o)$	$q = q_o^r$, r prime	$c = (2, r, q-1)$, π_1 if $c = 2$	–
$\mathcal{C}_6$	9	$2^{1+2m}.O^-_{2m}(2)$	$n = 2^m$, $q = p \geq 3$	$c \leq 2$, $c = 2$ iff $p \equiv \pm 1(8)$, π_1 if $c = 2$	–
$\mathcal{C}_7$	4	$PSp_m(q) \wr S_t$	$n = m^t$, qt odd, $t \geq 3$, m even, $(q, m) \neq (3, 2)$	1	–
$\mathcal{C}_8$	6	$O^{\pm}_n(q)$	q even	1	–

Table 3.5.D

$\Omega \approx \Omega_n(q)$, nq odd

I	II	III	IV	V	VI
$\mathcal{C}_i$	j	type	conditions	π	$H_{\overline{\Omega}}$ non-max.
$\mathcal{C}_1$	20 6	P_m $O_m(q) \perp O^{\epsilon}_{n-m}(q)$	$1 \leq m \leq (n-1)/2$ $1 \leq m < n$, m odd, $\epsilon = \pm$	1 1	– $(q, m, \epsilon) = (3, n-2, +)$
$\mathcal{C}_2$	14 15	$O_m(q) \wr S_t$ $O_1(q) \wr S_n$	$n = mt$, $t \geq 2$, $m > 1$ $q = p$	1 1 if $q \equiv \pm 3\,(8)$ π_1 if $q \equiv \pm 1(8)$	$q = m = 3$ –
$\mathcal{C}_3$	17	$O_{n/r}(q^r)$	$r \mid n$, r prime, $r \neq n$	1	–
$\mathcal{C}_4$	18	$O_m(q) \otimes O_{n/m}(q)$	$m \mid n$, $m < \sqrt{n}$	1	$q = m = 3$
$\mathcal{C}_5$	8	$O_n(q_o)$	$q = q_o^r$, r prime	$(2, r)$; π_1 if $r = 2$	–
$\mathcal{C}_7$	8	$O_m(q) \wr S_t$	$n = m^t$, $(q, m) \neq (3, 3)$	1	–

Table 3.5.E
$\Omega \approx \Omega_n^+(q),\ n \geq 8$

I	II	III	IV	V	VI
$\mathcal{C}_i$	j	type	conditions	π	$H_{\overline{\Omega}}$ non-max.
$\mathcal{C}_1$	20	P_m	$1 \leq m \leq \frac{n}{2}$	1 if $m < n/2$; π_2 if $m = n/2$	$m = \frac{n}{2} - 1$
	6	$O_m^\epsilon(q) \perp O_{n-m}^\epsilon(q)$	$1 \leq m < n/2$, $\epsilon \in \{+,-,\circ\}$, q odd if m odd	1 if m even; π_1 if m odd	$(\epsilon, m) = (+,2),\ q \leq 3$
	7	$Sp_{n-2}(q)$	q even	1	–
$\mathcal{C}_2$	11, 14	$O_m^\epsilon(q) \wr S_t$	$n = mt,\ t \geq 2$, $\epsilon^{n/m} = +$ if m even; q odd & $D = \square$ if m odd	1 if m even; π_1 if m odd	$O_m^\epsilon(q) \approx O_2^+(2)$, $O_2^\pm(3)$, $O_2^+(4)$, $O_2^+(5)$, $O_3(3)$, $O_4^+(2)$
	15	$O_1(q) \wr S_n$	$q = p \geq 3,\ D = \square$	π_1 if $q \equiv \pm 3\,(8)$; π_4 if $q \equiv \pm 1\,(8)$	–
	7	$GL_{n/2}(q).2$		1 if $\frac{n}{2}$ odd; π_2 if $\frac{n}{2}$ even	$\frac{n}{2}$ odd
	16	$O_{n/2}(q)^2$	$\frac{1}{2}qn$ odd, $q \equiv 3(4)$	1	–
$\mathcal{C}_3$	18	$GU_{n/2}(q)$	$n \equiv 0(4)$	π_2	
	14	$O_{n/r}^+(q^r)$	$r \mid n$, r prime, $\frac{n}{r}$ even, $\frac{n}{r} \geq 4$	1 if r odd; π_2 if $r = 2$	–
	20	$O_{n/2}(q^2)$	$qn/2$ odd	1 if $q \equiv 3\,(4)$; π_3 if $q \equiv 1\,(4)$	
$\mathcal{C}_4$	12	$Sp_{n_1}(q) \otimes Sp_{n_2}(q)$	$n = n_1 n_2,\ n_i$ even	π_2 if q even; π_2 if $n \equiv 4(8)$; π_5 otherwise	$q = n_1 = 2$
	14-17	$O_{n_1}^{\epsilon_1}(q) \otimes O_{n_2}^{\epsilon_2}(q)$	$n = n_1 n_2,\ n_i \geq 3$, n_1 even, q odd, $\Omega_{n_1}^{\epsilon_1}(q) \not\cong \Omega_{n_2}^{\epsilon_2}(q)$, $\epsilon_1 = +$ if n_2 odd	1 if n_2 odd; π_5 if $q \equiv 1\,(4),\ n \equiv 0\,(8)$, $\epsilon_1 = \epsilon_2 = +$; π_5 if $q \equiv 3\,(4)$, $n_1 \equiv n_2 \equiv 0\,(4),\ \epsilon_1 = \epsilon_2 = +$; π_5 if $q \equiv 3\,(4),\ n_1 \equiv 0\,(4)$, $n_2 \equiv 2\,(4),\ \epsilon_1 = + = -\epsilon_2$; π_2 otherwise	$q = n_2 = 3$; $n_1 = 4,\ \epsilon_1 = +$, n_2 odd or $D_{n_2}^{\epsilon_2}(q) = \boxtimes$
$\mathcal{C}_5$	10	$O_n^+(q_o)$	$q = q_o^r$, r prime	1 if r odd or q even; π_1 if $r = 2,\ n \equiv 2(4),\ q_o \equiv 1(4)$; π_4 otherwise	–
	10	$O_n^-(q^{1/2})$	q a square	1 if q even; π_4 if $n \equiv 2(4),\ \sqrt{q} \equiv 1(4)$; π_1 otherwise	
$\mathcal{C}_6$	8	$2_+^{1+2m}.O_{2m}^+(2)$	$n = 2^m,\ q = p \geq 3$	π_5 if $q \equiv \pm 3(8)$; π_6 if $q \equiv \pm 1(8)$	–
$\mathcal{C}_7$	5	$Sp_m(q) \wr S_t$	$n = m^t$, m & qt even, $(q,m) \notin \{(2,2),(3,2)\}$	1 if $t = 2$ and $m \equiv 2(4)$; π_2 if q even, $t \geq 3$; π_2 if q even, $m \equiv 0\,(4)$; π_5 otherwise	$m \leq 4$, q even
	6, 7	$O_m^\epsilon(q) \wr S_t$	$n = m^t$, q odd, $m \geq 4$ if $\epsilon = -$, $m \geq 6$ if $\epsilon = +$	1 if $t = 2,\ m \equiv 2(4)$; π_2 if $t = 3,\ m \equiv 2(4),\ q \equiv -\epsilon(4)$; π_2 if $t = 2,\ m \equiv 0(4),\ \epsilon = -$; π_5 otherwise	–

Table 3.5.F
$\Omega \approx \Omega_n^-(q),\ n \geq 8$

I	II	III	IV	V	VI
$\mathcal{C}_i$	j	type	conditions	c	$H_{\overline{\Omega}}$ non-max.
$\mathcal{C}_1$	20	P_m	$1 \leq m \leq \frac{n}{2}-1$,	1	–
	6	$O_m^{\xi}(q) \perp O_{n-m}^{-\xi}(q)$	$1 \leq m \leq n/2,\ \xi \in \{+,-,\circ\}$ $(m,\xi) \neq (\frac{n}{2},\circ)$, m odd $\Rightarrow q$ odd	π_1 if qm odd 1 otherwise	$(\xi, m) = (+,2),\ q \leq 3$
	7	$Sp_{n-2}(q)$	q even	1	–
$\mathcal{C}_2$	11, 14	$O_m^{\xi}(q) \wr S_{n/m}$	$m \mid n,\ 2 \leq m \leq \frac{n}{2},\ \xi \in \{-,\circ\}$ m odd $\Rightarrow q$ odd and $D = \square$ m even $\Rightarrow \frac{n}{m}$ odd and $\xi = -$	π_1 if qm odd 1 otherwise	$(q,m) \in \{(3,2),(3,3)\}$
	15	$O_1(q) \wr S_n$	$q = p \geq 3,\ D = \square$	π_1 if $q \equiv 3\,(8)$ π_2 if $q \equiv 7\,(8)$	–
	16	$O_{n/2}(q)^2$	$\frac{n}{2}$ odd, $q \equiv 1(4)$	1	–
$\mathcal{C}_3$	18	$GU_{n/2}(q)$	$\frac{n}{2}$ odd	1	–
	16	$O_{n/r}^-(q^r)$	$r \mid n,\ r$ prime, $n/r \geq 3$	1	–
	20	$O_{n/2}(q^2)$	$qn/2$ odd	1 if $q \equiv 1\,(4)$ π_3 if $q \equiv 3\,(4)$	– –
$\mathcal{C}_4$	17	$O_m(q) \otimes O_{n/m}^-(q)$	$m \mid n,\ qm$ odd, $n/m \geq 4$	1	$q = m = 3$
$\mathcal{C}_5$	10	$O_n^-(q_o)$	$q = q_o^r$, r an odd prime	1	–

Action Table 3.5.G
Non-trivial Actions of $\ddot{A}$

case; π	conditions	c	$\ker(\pi)$	$\mathrm{im}(\pi)$	stabilizer
L; $\pi_1(c)$		c	$\ddot{A}_1(c)$ *	$\ddot{A}/\ddot{A}_1(c)$	$\langle \ddot{\delta}^c, \ddot{\phi}, \ddot{\iota} \rangle$
L; $\pi_2(c)$	q odd, n even, $n \geq 4$	$\frac{1}{2}(q-1, n)$	$\ddot{A}_2(c)$ *	$\ddot{A}/\ddot{A}_2(c)$	$\langle \ddot{\delta}^c, \ddot{\phi}\ddot{\delta}^{(p-1)/2}, \ddot{\iota}\ddot{\delta}^{-1} \rangle$
L; π_3	$n \geq 3$	2	$\ddot{\Gamma}$	$\mathbf{Z}_2$	$\ddot{\Gamma}$
U; $\pi_1(c)$		c	$\ddot{A}_1(c)$	$\ddot{A}/\ddot{A}_1(c)$ *	$\langle \ddot{\delta}^c, \ddot{\phi} \rangle$
U; $\pi_2(c)$	q odd, n even	c	$\ddot{A}_2(c)$ *	$\ddot{A}/\ddot{A}_2(c)$	$\langle \ddot{\delta}^c, \ddot{\phi}\ddot{\delta}^{(p-1)/2} \rangle$
S; π_1	q odd	2	$\langle \ddot{\phi} \rangle$	$\mathbf{Z}_2$	$\langle \ddot{\phi} \rangle$
O$^\circ$; π_1		2	$\langle \ddot{\phi} \rangle$	$\mathbf{Z}_2$	$\langle \ddot{\phi} \rangle$
O$^+$; π_1	q odd	2	$\ker_{\ddot{\Gamma}}(\ddot{\tau})$	$\mathbf{Z}_2$	$\ker_{\ddot{\Gamma}}(\ddot{\tau})$
O$^+$; π_2		2	$\ker_{\ddot{\Gamma}}(\ddot{\gamma})$	$\mathbf{Z}_2$	$\ker_{\ddot{\Gamma}}(\ddot{\gamma})$
O$^+$; π_3	q odd, $D = \square$, $\frac{1}{4}n(p-1)$ even	2	$\langle \ddot{r}_\square\ddot{\delta}, \ddot{r}_\boxtimes\ddot{\delta}, \ddot{\phi} \rangle$	$\mathbf{Z}_2$	$\langle \ddot{r}_\square\ddot{\delta}, \ddot{r}_\boxtimes\ddot{\delta}, \ddot{\phi} \rangle$
	q odd, $D = \square$, $\frac{1}{4}n(p-1)$ odd	2	$\langle \ddot{r}_\square\ddot{\delta}, \ddot{r}_\boxtimes\ddot{\delta}, \ddot{r}_\square\ddot{\phi} \rangle$	$\mathbf{Z}_2$	$\langle \ddot{r}_\square\ddot{\delta}, \ddot{r}_\boxtimes\ddot{\delta}, \ddot{r}_\square\ddot{\phi} \rangle$
O$^+$; π_4	q odd, $D = \square$, $\frac{1}{4}n(p-1)$ even	4	$\langle \ddot{\phi} \rangle$	D_8	$\langle \ddot{\phi}, \ddot{r}_\square \rangle$
	q odd, $D = \square$, $\frac{1}{4}n(p-1)$ odd	4	$\langle \ddot{r}_\square\ddot{\phi} \rangle$	D_8	$\langle \ddot{\phi}, \ddot{r}_\square \rangle$
O$^+$; π_5	q odd, $n/2$ even	4	$\langle \ddot{\phi} \rangle$	D_8	$\langle \ddot{\phi}, \ddot{\delta} \rangle$
O$^+$; π_6	$q = p \geq 3$	8	1	D_8	1
O$^+$; π_7	$q \equiv 3 \pmod 4$, $n/2$ odd	4	$\langle \ddot{\phi} \rangle$	$\mathbf{Z}_2 \times \mathbf{Z}_2$	$\langle \ddot{\phi} \rangle$
O$^-$; π_1	q odd	2	$\ker_{\ddot{\Gamma}}(\ddot{\tau})$	$\mathbf{Z}_2$	$\ker_{\ddot{\Gamma}}(\ddot{\tau})$
O$^-$; π_2	$q = p \equiv 7 \pmod 8$, $n/2$ odd	4	1	D_8	$\langle \ddot{r}_\square \rangle$
O$^-$; π_3	q odd, $D = \square$	2	$\langle \ddot{r}_\square\ddot{r}_\boxtimes, \ddot{\delta}, \ddot{\phi} \rangle$	$\mathbf{Z}_2$	$\langle \ddot{r}_\square\ddot{r}_\boxtimes, \ddot{\delta}, \ddot{\phi} \rangle$

* Actions described further on following pages.

Description of π in case $\mathbf{L}^{\pm}$

For a triple x, y, z of integers, define

$$B(x,y,z) = \{m \in \mathbf{Z} \mid m > 0,\ x^m \equiv y \pmod{z}\},$$

and if $B(x,y,z)$ is non-empty, let $b(x,y,z)$ be its minimal member. Now let $x = b(p,1,c)$ (this exists because p does not divide c), and when p is odd, let $y = b(p,1,2c)$.

First consider $\pi = \pi_1(c)$ in case $\mathbf{L}$. Here $\ker(\pi) = \ddot{A}_1(c)$ is defined as

$$\ddot{A}_1(c) = \begin{cases} \langle \ddot{\delta}^c, \ddot{\phi}^x \rangle & \text{if } B(p,-1,c) \text{ is empty} \\ \langle \ddot{\delta}^c, \ddot{\phi}^x, \ddot{\iota}\ddot{\phi}^{\,b(p,-1,c)} \rangle & \text{otherwise.} \end{cases}$$

Recall from §2.2 that $|\ddot{\delta}| = (q-1,n)$, $|\ddot{\phi}| = f$ and $|\ddot{\iota}| = 2$, subject to the relations given in Proposition 2.2.3. The best way to think of the action of $\ddot{A}$ on $[H_{\overline{\Omega}}]^{\overline{A}}$ is to identify $[H_{\overline{\Omega}}]^{\overline{A}}$ with the set $\mathbf{Z}_c$ (the cyclic group of order c). We may then regard π as a homomorphism from $\ddot{A}$ to $Sym(\mathbf{Z}_c)$, as follows. Writing $\mathbf{Z}_c$ additively, we may view π in the following way:

$$\begin{aligned} \pi(\ddot{\delta}) &: i \mapsto i+1 \\ \pi(\ddot{\phi}) &: i \mapsto ip \\ \pi(\ddot{\iota}) &: i \mapsto -i, \end{aligned} \tag{3.5.1}$$

where $i \in \mathbf{Z}_c$. (Note that when $n = 2$, we have $\iota = 1$ by definition, and also $c \leq 2$; this is compatible with the fact that $i \equiv -i \pmod 2$ in (3.5.1).) Then

$$\begin{aligned} |\pi(\ddot{\delta})| &= c \\ |\pi(\ddot{\phi})| &= x \\ |\pi(\ddot{\iota})| &= \begin{cases} 2 & \text{if } c \geq 3 \\ 1 & \text{if } c = 2. \end{cases} \end{aligned} \tag{3.5.2}$$

Furthermore

$$|\ddot{A}/\ddot{A}_1(c)| = |\mathrm{im}(\pi)| = \begin{cases} cx & \text{if } B(p,-1,c) \text{ is non-empty} \\ 2cx & \text{otherwise.} \end{cases}$$

Next take $\pi = \pi_2(c)$ in case $\mathbf{L}$. Here q is odd, $c = \frac{1}{2}(q-1,n)$ and $\ker(\pi) = \ddot{A}_2(c)$ is defined as

$$\ddot{A}_2(c) = \begin{cases} \langle \ddot{\delta}^c, \ddot{\phi}^x \ddot{\delta}^{(p^x-1)/2}, \ddot{\iota}\ddot{\phi}^{\,b(p,-1,2c)} \rangle & \text{if } B(p,-1,2c) \text{ is non-empty} \\ \langle \ddot{\delta}^c, \ddot{\phi}^x \ddot{\delta}^{(p^x-1)/2}, \ddot{\iota}\ddot{\phi}^{b(p,c-1,2c)}\ddot{\delta}^{c/2} \rangle & \text{if } B(p,c-1,2c) \text{ is non-empty} \\ \langle \ddot{\delta}^c, \ddot{\phi}^x \ddot{\delta}^{(p^x-1)/2} \rangle & \text{otherwise.} \end{cases}$$

As before, identify $[H_{\overline{\Omega}}]^{\overline{A}}$ with the set $\mathbf{Z}_c$, so that π may be written as:

$$\begin{aligned} \pi(\ddot{\delta}) &: i \mapsto i+1 \\ \pi(\ddot{\phi}) &: i \mapsto ip - \tfrac{1}{2}(p-1) \\ \pi(\ddot{\iota}) &: i \mapsto 1-i, \end{aligned} \tag{3.5.3}$$

where $i \in \mathbf{Z}_c$. Then

$$\begin{aligned} |\pi(\ddot{\delta})| &= c \\ |\pi(\ddot{\phi})| &= y \\ |\pi(\ddot{\iota})| &= 2. \end{aligned} \tag{3.5.4}$$

Furthermore

$$|\ddot{A}/\ddot{A}_2(c)| = |\mathrm{im}(\pi)| = \begin{cases} 2cx & \text{if } B(p,-1,c) \text{ is empty} \\ cx & \text{otherwise.} \end{cases}$$

As for $\pi = \pi_1(c)$ in case **U**, $\ker(\pi) = \ddot{A}_1(c)$ is defined as

$$\ddot{A}_1(c) = \langle \ddot{\delta}^c, \ddot{\phi}^x \rangle. \tag{3.5.5}$$

Moreover

$$\begin{aligned} |\pi(\ddot{\delta})| &= c \\ |\pi(\ddot{\phi})| &= x \end{aligned} \tag{3.5.6}$$

and we may write

$$\begin{aligned} \pi(\ddot{\delta}) &: i \mapsto i+1 \\ \pi(\ddot{\phi}) &: i \mapsto ip. \end{aligned} \tag{3.5.7}$$

Furthermore

$$|\ddot{A}/\ddot{A}_1(c)| = |\mathrm{im}(\pi)| = cx. \tag{3.5.8}$$

Finally, take $\pi = \pi_2(c)$ in case **U**. Here

$$\ddot{A}_2(c) = \langle \ddot{\delta}^c, \ddot{\phi}^x \ddot{\delta}^{(p^x-1)/2} \rangle.$$

Moreover (3.5.6) holds with y replacing x, and we may write

$$\begin{aligned} \pi(\ddot{\delta}) &: i \mapsto i+1 \\ \pi(\ddot{\phi}) &: i \mapsto ip - \tfrac{1}{2}(p-1). \end{aligned} \tag{3.5.9}$$

In addition, $|\ddot{A} : \ddot{A}_2(c)| = |\mathrm{im}(\pi)| = cx$.

Table 3.5.H

Triples $(H_\Omega, K_\Omega, \Omega)$ with $H_{\overline{\Omega}} < K_{\overline{\Omega}} < \overline{\Omega}$ and $H_{\overline{\Omega}}, K_{\overline{\Omega}} \in \mathcal{C}(\overline{\Omega})$, $n \geq 13$

Ω	$\mathcal{C}_i$	type of H_Ω	$\mathcal{C}_j$	type of K_Ω	conditions	novelties
$SL_n(2)$	$\mathcal{C}_2$	$GL_1(2) \wr S_n$	$\mathcal{C}_1$	P_1, P_{n-1}		N
			$\mathcal{C}_1$	$P_{1,n-1}$	n even	
			$\mathcal{C}_1$	$GL_1(2) \oplus GL_{n-1}(2)$	n odd	
			$\mathcal{C}_8$	$Sp_n(2)$	n even	
$SL_n(2)$	$\mathcal{C}_2$	$GL_2(2) \wr S_{n/2}$	$\mathcal{C}_8$	$Sp_n(2)$	n even	
$SL_n(2)$	$\mathcal{C}_4$	$GL_2(2) \otimes GL_{n/2}(2)$	$\mathcal{C}_3$	$GL_{n/2}(4)$	n even	
$SL_n(3)$	$\mathcal{C}_2$	$GL_1(3) \wr S_n$	$\mathcal{C}_8$	$O_n(3)$	n odd	
			$\mathcal{C}_8$	$O_n^\epsilon(3)$	n even, $\epsilon = (-)^{n/2}$	
$SL_n(4)$	$\mathcal{C}_2$	$GL_1(4) \wr S_n$	$\mathcal{C}_8$	$U_n(2)$		
$SL_n(q)$	$\mathcal{C}_1$	$GL_m(q) \oplus GL_{n-m}(q)$	$\mathcal{C}_1$	P_m, P_{n-m}	$1 \leq m < n/2$	N
$SL_n(q)$	$\mathcal{C}_1$	$P_{m,n-m}$	$\mathcal{C}_1$	P_m, P_{n-m}	$1 \leq m < n/2$	N
$SU_n(2)$	$\mathcal{C}_2$	$GU_2(2) \wr S_{n/2}$	$\mathcal{C}_2$	$GU_1(2) \wr S_n$	n even	
$SU_n(2)$	$\mathcal{C}_4$	$GU_2(2) \otimes GU_{n/2}(2)$	$\mathcal{C}_2$	$GU_{n/2}(2) \wr 2$	n even	
$Sp_n(2)$	$\mathcal{C}_2$	$Sp_2(2) \wr S_{n/2}$	$\mathcal{C}_8$	$O_n^\epsilon(2)$	$\epsilon = (-)^{n/2}$	
$Sp_n(3)$	$\mathcal{C}_4$	$O_3(3) \otimes Sp_{n/3}(3)$	$\mathcal{C}_2$	$Sp_{n/3}(3) \wr S_3$	$3 \mid n$	
$\Omega_n^+(q)$	$\mathcal{C}_1$	$P_{n/2-1}$	$\mathcal{C}_1$	$P_{n/2}$ (two)		N
$\Omega_n^\pm(2)$	$\mathcal{C}_1$	$O_2^+(2) \perp O_{n-2}^\pm(2)$	$\mathcal{C}_1$	$Sp_n(2)$	n even	
$\Omega_n^\epsilon(3)$	$\mathcal{C}_1$	$O_2^+(3) \perp O_{n-2}^\epsilon(3)$	$\mathcal{C}_1$	$O_1(3) \perp O_{n-1}(3)$ (two)	n even	N
				$O_1(3) \perp O_{n-1}^\pm(3)$	n odd	
$\Omega_n^+(2)$	$\mathcal{C}_2$	$O_2^+(2) \wr S_{n/2}$	$\mathcal{C}_1$	P_1	$n/2$ even	
			$\mathcal{C}_1$	$Sp_{n-2}(2)$	$n/2$ odd	
			$\mathcal{C}_1$	$P_{n/2-1}$		
			$\mathcal{C}_1$	$P_{n/2}$ (two)		N
			$\mathcal{C}_2$	$O_2^-(2) \wr S_{n/2}$	$n/2$ even	
$\Omega_n^+(3)$	$\mathcal{C}_2$	$O_2^+(3) \wr S_{n/2}$	$\mathcal{C}_2$	$O_{n/2}^\epsilon(3) \wr 2$	$\epsilon = \pm$, $\frac{n}{2}$ even, $\epsilon = (-)^{n/4}$	
			$\mathcal{C}_2$	$O_{n/2}(3) \times O_{n/2}(3)$	$\frac{n}{2}$ odd	
$\Omega_n^+(4)$	$\mathcal{C}_2$	$O_2^+(4) \wr S_{n/2}$	$\mathcal{C}_5$	$O_n^\epsilon(2)$	$\epsilon = (-)^{n/2}$	
$\Omega_n^+(5)$	$\mathcal{C}_2$	$O_2^+(5) \wr S_{n/2}$	$\mathcal{C}_2$	$O_1(5) \wr S_n$ (two)		N
$\Omega_n^\epsilon(3)$	$\mathcal{C}_2$	$O_2^-(3) \wr S_{n/2}$	$\mathcal{C}_2$	$O_1(3) \wr S_n$ (two)	$\epsilon = (-)^{n/2}$	N
$\Omega_n^\epsilon(3)$	$\mathcal{C}_2$	$O_3(3) \wr S_{n/3}$	$\mathcal{C}_2$	$O_1(3) \wr S_n$	$D = \square$	
$\Omega_n^+(2)$	$\mathcal{C}_2$	$O_4^+(2) \wr S_{n/4}$	$\mathcal{C}_2$	$O_2^-(2) \wr S_{n/2}$	$4 \mid n$	
$\Omega_n^+(q)$	$\mathcal{C}_2$	$GL_{n/2}(q).2$	$\mathcal{C}_1$	$P_{n/2}$ (two)	$n/2$ odd	N
$\Omega_n^+(2)$	$\mathcal{C}_4$	$Sp_2(2) \otimes Sp_{n/2}(2)$	$\mathcal{C}_3$	$GU_{n/2}(2)$	n even	
$\Omega_n^\epsilon(3)$	$\mathcal{C}_4$	$O_3(3) \otimes O_{n/3}^\epsilon(3)$	$\mathcal{C}_2$	$O_{n/3}^\epsilon(3) \wr S_3$	$3 \mid n$	
$\Omega_n^+(q)$	$\mathcal{C}_4$	$O_4^+(q) \otimes O_{n/4}(q)$	$\mathcal{C}_4$	$Sp_2(q) \otimes Sp_{n/2}(q)$ (two)	q and $\frac{n}{4}$ odd	N
	$\mathcal{C}_4$	$O_4^+(q) \otimes O_{n/4}^\epsilon(q)$ (two)	$\mathcal{C}_4$	$Sp_2(q) \otimes Sp_{n/2}(q)$ (four)	q odd, $\frac{n}{4}$ even, $-\epsilon = (-)^{(q-1)n/16}$	N

Table 3.5.I
Triples $(H_{\overline{\Omega}}, K_{\overline{\Omega}}, \overline{\Omega})$ with $H_{\overline{\Omega}} < K_{\overline{\Omega}} < \overline{\Omega}$ and $H_{\overline{\Omega}} \in \mathcal{C}$, $K_{\overline{\Omega}} \in \mathcal{S}$, $n \geq 13$

type of $H_{\overline{\Omega}}$	$K_{\overline{\Omega}}$	Ω	conditions	
$Sp_2(q) \wr S_t \in \mathcal{C}_7$	$Sp_{2t}(q)$	$\Omega^+_{2^t}(q)$	q even, $q \geq 4$, $t \geq 4$	two classes
$Sp_2(q) \wr S_t \in \mathcal{C}_7$	$\Omega^+_{2t+2}(q)$	$\Omega^+_{2^t}(q)$	q even, $q \geq 4$, t odd, $t \geq 5$	two classes
$Sp_4(q) \wr S_t \in \mathcal{C}_7$	$Sp_{4t}(q)$	$\Omega^+_{4^t}(q)$	q even	two classes

Table 3.5.J

case	$\mathcal{C}_i$	our extra conditions
L	$\mathcal{C}_1$	include $\mathcal{C}_1'$ in $\mathcal{C}_1$
L$^{\pm}$	$\mathcal{C}_7$	$m \geq 3$ when H is of type $GL^{\pm}_m(q) \wr S_t$
L	$\mathcal{C}_8$	$n \geq 3$
U	$\mathcal{C}_5$	q odd when H is of type $O^{\epsilon}_n(q)$
S	$\mathcal{C}_2$	q odd when H is of type $GL_{n/2}(q).2$
S	$\mathcal{C}_3$	q odd when H is of type $GU_{n/2}(q)$
S	$\mathcal{C}_4$	$m_2 \geq 3$ when H is of type $Sp_{m_1}(q) \otimes O^{\epsilon}_{m_2}(q)$
O	$\mathcal{C}_2$	$q = p$ when H is of type $O_1(q) \wr S_n$
O	$\mathcal{C}_3$	$m \geq 3$ when H is of type $O^{\epsilon}_m(q^r)$
O	$\mathcal{C}_4$	$m_i \geq 3$ when H is of type $O^{\epsilon_1}_{m_1}(q) \otimes O^{\epsilon_2}_{m_2}(q)$

Chapter 4
THE STRUCTURE AND CONJUGACY OF THE MEMBERS OF $\mathcal{C}$

§4.0 Introduction

In this chapter we describe the natural collections $\mathcal{C}_i$ of subgroups of classical groups, and in addition give complete information about their structure and conjugacy. The results of this chapter yield a proof of parts (A) and (B) of the Main Theorem as stated in §3.1.

Let $(V, \mathbf{F}, \kappa)$ be a classical geometry as described in Chapter 2, and let $X = X(V, \mathbf{F}, \kappa)$ as X ranges over the symbols Ω, S, I, Δ, Γ and A (see (2.1.15)). Recall from §3.1 that the members of $\mathcal{C}$ are defined in terms of the members of $\mathcal{C}(\Gamma)$ (see (3.1.3) and (3.1.4)). In §4.i ($1 \leq \mathrm{i} \leq 8$) we give an explicit description of the collections $\mathcal{C}_i(\Gamma)$.

We now fix the following notation. Let H_Γ be a member of $\mathcal{C}_i(\Gamma)$ for some i, and let H_X be the corresponding X-associate for all groups X satisfying (3.1.2) (see the discussion between (3.1.4) and (3.1.5)). As a matter of convenience we define

$$H = H_\Delta \in \mathcal{C}(\Delta) \tag{4.0.1}$$

Recall from §3.2 the definitions of $[H_{\overline{\Omega}}]^{\overline{A}}$, c and π. Thus c is the size of the set $[H_{\overline{\Omega}}]^{\overline{A}}$ and π is the natural homomorphism from $\ddot{A}$ to $Sym([H_{\overline{\Omega}}]^{\overline{A}}) \cong S_c$. Our goals in this chapter are to

(I) determine c and the homomorphism π from $\ddot{A}$ to $Sym([H_{\overline{\Omega}}]^{\overline{A}})$;
(II) determine the structure of $H_{\overline{\Omega}}$;
(III) determine when $H_{\overline{\Omega}}$ is local, and determine $\mathrm{soc}(H_{\overline{\Omega}})$ when $H_{\overline{\Omega}}$ is non-local.

As explained in §§3.2,3.3, once goals (I) and (II) are achieved, parts (A) and (B) of the Main Theorem will be established. Once goal (III) is achieved, Corollary 1.2.4 will follow as soon as we determine (in Chapters 7 and 8) which members of $\mathcal{C}$ are maximal. The case where $H \in \mathcal{C}_i$ is handled in §4.i and the relevant Propositions 4.i.j are divided into parts (I), (II) and (III), accordingly. The action occurring in (I) is described in these Propositions in terms of the homomorphism π appearing in the Action Table 3.5.G. We assume in all of these Propositions that

$$\begin{gathered} n \geq 2,3,4,7 \text{ in cases } \mathbf{L},\ \mathbf{U},\ \mathbf{S},\ \mathbf{O}, \text{ respectively,} \\ (q,n) \neq (2,2),\ (3,2) \text{ in case } \mathbf{L}, \\ (q,n) \neq (2,3) \text{ in case } \mathbf{U}. \end{gathered} \tag{4.0.2}$$

However, sometimes in various Lemmas used to prove the Propositions condition (4.0.2) is relaxed. Moreover some of the Propositions contain parts (O) or (IV) which provide certain extra pieces of information.

We have seen in Chapter 3 that questions concerning structure and conjugacy amongst members of $\mathcal{C}$ depend on Theorem 3.1.2. We mentioned after the statement of Theorem 3.1.2 that is proof is spread throughout Chapter 4. Here we provide a means of establishing this Theorem.

Proposition 4.0.1.

(i) *If either $H_\Gamma\Omega = \Gamma$ or $N_\Gamma(H_\Omega) \leq H_\Gamma$, then Theorem 3.1.2.i holds.*

(ii) *If $N_\Gamma(H_\Omega) \leq H_\Gamma$, then $H_{\overline{\Omega}} = N_{\overline{\Omega}}(H_{\overline{\Omega}})$.*

Proof. Suppose that $N_\Gamma(H_\Omega) \leq H_\Gamma$. Since $H_\Omega = H_\Gamma \cap \Omega$ (by definition), $H_\Omega \trianglelefteq H_\Gamma$, and hence $H_\Gamma = N_\Gamma(H_\Omega)$. Since H_Γ contains the full group of scalars $\mathbf{F}^*$, it follows that $H_{\overline{\Gamma}} = N_{\overline{\Gamma}}(H_{\overline{\Omega}})$, Therefore (ii) holds. Moreover, if $\overline{\Omega} \leq X \leq \overline{\Gamma}$, then $H_X = H_{\overline{\Gamma}} \cap X$ by (3.1.5), whence $H_X = N_X(H_{\overline{\Omega}})$, which shows part (a) of Theorem 3.1.2.i holds. If $\overline{\Omega} \leq X \leq \overline{A}$ and $X \not\leq \overline{\Gamma}$, then by (3.1.7), $H_X = N_X(H_{\overline{\Gamma}})$. Furthermore, $N_X(H_{\overline{\Omega}}) \leq N_X(N_{\overline{\Gamma}}(H_{\overline{\Omega}})) = N_X(H_{\overline{\Gamma}}) = H_X$, and as $H_{\overline{\Omega}} \trianglelefteq H_X$, we deduce $H_X = N_X(H_{\overline{\Omega}})$, which yields part (a) of Theorem 3.1.2.i again.

To complete the proof, assume that $H_\Gamma\Omega = \Gamma$. Then $H_{\overline{\Gamma}}\overline{\Omega} = \overline{\Gamma}$, and hence part (b) of Theorem 3.1.2.i holds provided $X \leq \overline{\Gamma}$. Suppose then that $\overline{\Omega} \leq X \leq \overline{A}$ and $X \not\leq \overline{\Gamma}$. Here $H_X = H_{\overline{A}} \cap X$. If H_Γ is of type P_i for some i (recall from §3.1 that a group in C is said to be of type P_i if it is the stabilizer of a totally singular i-space in the geometry $(V, \mathbf{F}, \kappa)$), then according to Proposition 4.1.16 below, $H_\Gamma = N_\Gamma(H_\Omega)$, whence part (a) of Theorem 3.1.2.i holds, by the argument in the preceding paragraph. Thus we can assume that H_Γ is not of type P_i. Therefore, it follows from [As₁, 13.2(1)] that $H_{\overline{A}}\overline{\Delta} = \overline{A}$. This fact, together with $H_{\overline{\Gamma}}\overline{\Omega} = \overline{\Gamma}$, yields $H_{\overline{A}}\overline{\Omega} = \overline{A}$. Thus $H_X\overline{\Omega} = X$, which yields part (b) of Theorem 3.1.2.i. ■

The remaining results in this section will serve as our fundamental tools for achieving goals (I), (II) and (III).

Proposition 4.0.2 (Aschbacher [As₁, Theorem BΔ]). *Precisely one of the following holds:*

(i) *$\ddot{\Delta}$ is transitive on $[H_{\overline{\Omega}}]^{\overline{A}}$*

(ii) *case* **L** *holds, $H \in C_1(\Delta)$, and H is of type P_i with $i \neq \frac{n}{2}$.*

In part (ii) of Proposition 4.0.2, notice that as elements in $A \backslash \Gamma$ interchange i-spaces with $(n-i)$-spaces, $[H_{\overline{\Omega}}]^{\overline{A}}$ consists of stabilizers of i-spaces and of $(n-i)$-spaces, and so Δ has two orbits.

Lemma 4.0.3. *Assume that part* (i) *of Proposition 4.0.2 occurs. Then the following hold.*

(i) *$c \leq |\ddot{\Gamma} : \ddot{H}_\Gamma|$ and $c \leq |\ddot{\Delta} : \ddot{H}|$.*

(ii) *$|\overline{H} : H_{\overline{\Omega}}| \leq |\ddot{\Delta}|/c$.*

(iii) *If G satisfies $\Omega \leq G \leq \Delta$ and $HG = \Delta$, then $c \leq |\ddot{G} : \ddot{H}_G|$.*

(iv) *If G is as in* (iii), *then $|H_{\overline{G}} : H_{\overline{\Omega}}| \leq |\ddot{G}|/c$.*

(v) *In case* $\mathbf{L}^\pm$ *we have $c \leq (q \mp 1)/|\det(H_I)|$.*

(vi) *In case* **S** *we have $c \leq (q-1)/|\tau(H)|$.*

(vii) *In case* $\mathbf{O}^\circ$ *we have $c \leq |\ddot{S} : \ddot{H}_S|$.*

(viii) *In case* $\mathbf{O}^\pm$ *with q even we have $c \leq |\ddot{I} : \ddot{H}_I|$.*

(ix) *If* $H_\Gamma = N_\Gamma(H_\Omega)$, *then equality holds in* (i)-(viii).

Proof. We first prove (iii). Since Proposition 4.0.2.i holds, $c = |\ddot{\Delta} : \ddot{\Delta}_{[H_{\overline{\Omega}}]}|$, where (recall (3.2.4)) $\ddot{\Delta}_{[H_{\overline{\Omega}}]} = N_{\overline{\Delta}}(H_{\overline{\Omega}})\overline{\Omega}/\overline{\Omega}$. Clearly $\overline{H} \leq N_{\overline{\Delta}}(H_{\overline{\Omega}})$, and so $\overline{G}N_{\overline{\Delta}}(H_{\overline{\Omega}}) = \overline{\Delta}$. Now an easy counting argument shows $c = |\ddot{G} : \ddot{G}_{[H_{\overline{\Omega}}]}|$. Since $H_\Omega \trianglelefteq H_G$, we have $\ddot{H}_G \leq \ddot{G}_{[H_{\overline{\Omega}}]}$, and this proves (iii). The same reasoning shows that (iii) holds with Δ replaced by Γ, and hence it is clear that (i) holds. Assertion (ii) follows directly from (i). Also (iv) is immediate from (iii). To prove (v), note first that $\Delta = I$ in case $\mathbf{L}^+$ and that $\Delta = I\mathbf{F}^*$ in case $\mathbf{L}^-$. So as $\mathbf{F}^* \leq H$, it follows from (iii) that $c \leq |\ddot{I} : \ddot{H}_I| = |I : H_I\Omega|$. But $|H_I\Omega : \Omega| = |\det(H_I)|$, and so $|I : H_I\Omega| = (q \mp 1)/|\det(H_I)|$. The proof of (vi) is entirely similar, the only difference being $|H\Omega : \Omega| = |\tau(H)|$. As for (vii), note that $\Delta = S \times \mathbf{F}^*$, which means $HS = \Delta$. The result now follows from (iii). Part (viii) is similar, for $\Delta = I \times \mathbf{F}^*$ in this case. To prove (ix), assume that $H_\Gamma = N_\Gamma(H_\Omega)$. It then follows that $H_{\overline{\Gamma}} = N_{\overline{\Gamma}}(H_{\overline{\Omega}})$, whence the reasoning in the beginning of this proof yields $c = |\ddot{\Gamma} : \ddot{H}_\Gamma|$. Thus equality holds in the first part of (i). Similarly, running through the argument used in (i)-(viii) shows that equality holds throughout. ■

We also record the following easy facts:

$$|H_{\overline{I}} : H_{\overline{\Omega}}| = |\ddot{H}_I|, \qquad |\overline{H} : H_{\overline{\Omega}}| = |\ddot{H}|. \tag{4.0.3}$$

We use the next two results to identify $\mathrm{soc}(H_{\overline{\Omega}})$ when $H_{\overline{\Omega}}$ is non-local. Recall that a group is said to be *semisimple* if it is a direct product of non-abelian simple groups. Thus $H_{\overline{\Omega}}$ is non-local if and only if $\mathrm{soc}(H_{\overline{\Omega}})$ is semisimple.

Lemma 4.0.4. *Assume that* $X \trianglelefteq Y$, *that* X *is semisimple, and that* $C_Y(X) = 1$. *Then* $X = \mathrm{soc}(Y)$.

Proof. Let N be a minimal normal subgroup of Y. Then $[X, N] \trianglelefteq Y$ and $[X, N] \leq N \cap X$. Therefore by minimality, either $[X, N] = 1$ or $[X, N] = N$. Hence the condition $C_Y(X) = 1$ ensures that $N = [X, N] \leq X$, which proves $\mathrm{soc}(Y) \leq X$. Now write $X = X_1 \times \cdots \times X_k$, where each X_i is non-abelian and simple. Now the only simple normal subgroups of X are its subgroups X_i, and so by relabelling if necessary, we have $X_1^Y = X_1 \times \cdots \times X_j$ for some j. It follows that X_1^Y is a minimal normal subgroup of Y, whence $X \leq \mathrm{soc}(Y)$. The result is now proved. ■

Lemma 4.0.5. *Assume that* K *is a perfect subgroup of* Ω.

(i) $\overline{C_{GL(V)}(K)} = C_{PGL(V)}(\overline{K})$.

(ii) *If* K *is absolutely irreducible on* V, *then* $C_{PGL(V)}(\overline{K}) = 1$.

(iii) *If* K *is irreducible and* $C_{PGL(V)}(\overline{K}) = 1$, *then* K *is absolutely irreducible.*

(iv) *Assume that* $K \trianglelefteq H_\Omega$ *and that* $V = V_1 \oplus \cdots \oplus V_t$. *Moreover, assume that* K *fixes each space* V_i, $1 \leq i \leq t$. *Then* $\overline{K} = \mathrm{soc}(H_{\overline{\Omega}})$ *provided all of the following hold:*

(a) $\overline{K}$ *is semisimple;*

(b) K *is absolutely irreducible on each* V_i;

(c) $V_i \not\cong V_j$ *as* K*-modules for* $i \neq j$;

Proof. Let C be the full preimage of $C_{PGL(V)}(\overline{K})$ in $GL(V)$. Thus $[K,C] = [C,K] \leq \mathbf{F}^*$, and hence $[K,C,K] = [C,K,K] = 1$. So by the Three-Subgroup Lemma, $[K,K,C] = 1$, and thus $[K,C] = 1$ as K is perfect. Therefore $C \leq C_{GL(V)}(K)$. The reverse inclusion is obvious, and so $C = C_{GL(V)}(K)$. Part (i) now follows. Assertion (ii) is immediate from (i) and Schur's Lemma. As for (iii), observe that (i) implies that $C_{GL(V)}(K) = \mathbf{F}^*$, and so K must be absolutely irreducible by Lemma 2.10.1. We now prove (iv). First note that condition (c) along with Lemma 2.10.11 implies that $C_{GL(V)}(K)$ fixes each V_i, and condition (b) guarantees that C acts as scalars on each V_i. Therefore (d) and part (i) show that $C_{H_{\overline{\Omega}}}(\overline{K}) = 1$, and hence the result follows from Lemma 4.0.4. ∎

In the rest of this chapter, the notation introduced in Chapters 2 and 3 will remain in effect. To avoid excessive length in Chapter 4, we usually leave the proofs of the Propositions for cases **S** and **O°** to the reader. These proofs are in general rather easier than those in the other cases.

§4.1 The reducible subgroups $\mathcal{C}_1$

We begin this section with the definition of the members of $\mathcal{C}_1(\Gamma)$.

Definition. The members of $\mathcal{C}_1(\Gamma)$ are the groups $N_\Gamma(W)$, or $N_\Gamma(W,U)$, where $\dim(W) = m$ with $1 \leq m \leq \frac{n}{2}$ and $\dim(U) = n - m$, as appearing in Table 4.1.A. The abbreviations 't.s.', 'n.d.' and 'n.s.' stand for totally singular, non-degenerate and non-singular, respectively. (Recall these terms are defined in §2.1.)

Remarks on the conditions. The condition $m \leq \frac{n}{2}$ in the first row of Table 4.1.A applies to case **L** because we are regarding the types P_m and P_{n-m} to be one and the same (since elements in $A\backslash\Gamma$ interchange stabilizers of m-spaces with stabilizers of $(n-m)$-spaces). It also applies to cases **U**, **S** and **O** in view of Corollary 2.1.7. The condition $m < \frac{n}{2}$ in rows 3-5 and the condition $(m,\epsilon_1) \neq (n-m,\epsilon_2)$ in row 7 occur in order to avoid overlap with members of $\mathcal{C}_2$. For if $V = V_1 \perp V_2$ and (V_1,κ) is similar to (V_2,κ),

Table 4.1.A

case	type	description	conditions
all	P_m	$N_\Gamma(W)$, W t.s.	$m \leq \frac{n}{2}$
L	$P_{m,n-m}$	$N_\Gamma(W,U)$, $W < U$	$m < \frac{n}{2}$
L	$GL_m(q) \oplus GL_{n-m}(q)$	$N_\Gamma(W,U)$, $W \cap U = 0$	$m < \frac{n}{2}$
U	$GU_m(q) \perp GU_{n-m}(q)$	$N_\Gamma(W)$, W n.d.	$m < \frac{n}{2}$
S	$Sp_m(q) \perp Sp_{n-m}(q)$	$N_\Gamma(W)$, W n.d.	$m < \frac{n}{2}$
O	$O^{\epsilon_1}_m(q) \perp O^{\epsilon_2}_{n-m}(q)$	$N_\Gamma(W)$, W n.d.	$\epsilon_1 = \mathrm{sgn}(W)$, $\epsilon_2 = \mathrm{sgn}(W^\perp)$ $(m,\epsilon_1) \neq (n-m,\epsilon_2)$
O$^\pm$	$Sp_{n-2}(q)$	$N_\Gamma(W)$, W n.s., $m = 1$	q and n even

(d) *the subgroup of K acting as scalars on each V_i acts as scalars on all of V.*

then $N_\Gamma(V_1, V_2) < N_\Gamma\{V_1, V_2\}$, as there is an element in Δ which interchanges V_1 and V_2. Therefore such decompositions are included in $\mathcal{C}_2$ (see §4.2).

For the rest of this section, we have $H_\Gamma \in \mathcal{C}_1(\Gamma)$, and thus either $H_\Gamma = N_\Gamma(W)$ or $H_\Gamma = N_\Gamma(U, W)$ as described in Table 4.1.A. Therefore $H = N_\Delta(W)$ or $H = N_\Delta(U, W)$.

We begin our study of the reducible subgroups by considering the case where W is non-degenerate in cases **U**, **S** and **O**, and where H is of type $GL_m(q) \oplus GL_{n-m}(q)$ in case **L**. As a convenience we set $U = W^\perp$ in cases **U**, **S** and **O**, so that in all four cases we know that H fixes the decomposition $V = W \oplus U$. Here, for $X = \Omega$, S or I we make the identification

$$X(W) = \{g \oplus 1_U \mid g \in X(W, \kappa)\} \leq I, \tag{4.1.1}$$

so that $X(W)$ acts naturally on W while centralizing U. This notation was first introduced in the discussion after Lemma 2.1.5 in §2.1. We define $X(U)$ analogously. The following Lemma is our starting point.

Lemma 4.1.1. *Assume that* $V = W \perp U$ *in cases* **S**, **O** *or* **U**, *or that* $V = W \oplus U$ *in case* **L**.

(i) $H_I = I(W) \times I(U)$.
(ii) $H_\Omega \geq \Omega(W) \times \Omega(U)$.
(iii) $X(W) \cap \Omega = \Omega(W)$ *and* $X(U) \cap \Omega = \Omega(U)$ *where* X *is one of* Ω, S *and* I.
(iv) $H_\Omega^W = I(W)$.
(v) $H_\Omega^U = I(U)$, *unless case* **O** *holds and* $\dim(W) = 1$.

Proof. The first statement is clear, and (ii) follows from (iii). Here we concentrate on (iii), and it suffices to prove that $\Omega(W) = \Omega \cap X(W)$, for the proof for U is identical. There is no work to be done in case **S**, for here $\Omega = I$. And as for case $\mathbf{L}^\pm$, note that if $g \in X(W)$, then $g \in \Omega$ if and only if $\det(g) = 1$, which holds if and only if $\det_W(g_W) = 1$. It remains to prove (iii) in case **O**, and to do so we fix $g \in X(W) \cap \Omega$. Suppose first that q is odd. Then Proposition 2.5.6 ensures that $g = r_{w_1} \dots r_{w_k}$, where each w_i is a non-singular vector in W. Since $\det(g) = 1$, we know that k must be even. Moreover, by Description 1 in §2.5 we see that $\prod_{i=1}^k (w_i, w_i)$ is a square, and hence $g \in \Omega(W)$. Thus $I(W) \cap \Omega \leq \Omega(W)$, and the reverse inclusion is proved in the same way. Thus $I(W) \cap \Omega = \Omega(W)$, and so $X(W) \cap \Omega = \Omega(W)$ for $X = S$ and Ω, also. Assume now that q is even and that $I(W) \not\approx O_4^+(2)$. Then again, g is a product of k reflections in non-singular vectors in W, and since a product of an odd number of reflections never lies in Ω, we know that k is even. Therefore $g \in \Omega(W)$, and we deduce $I(W) \cap \Omega = \Omega(W)$, as desired. Finally, assume that $I(W) \approx O_4^+(2)$. Note that $I(W)$ contains reflections, and so $I(W) \not\leq \Omega$. Therefore $I(W) \cap \Omega$ has index 2 in $I(W)$, and our goal is to show that this intersection is $\Omega(W)$ (see Proposition 2.5.9). Thus it suffices to show that

$$\Omega(W) \leq \Omega. \tag{4.1.2}$$

Take $h \in \Omega(W)$. If h is a product of reflections in non-singular vectors in W, then as before we see that h is a product of an even number of reflections, and so lies in Ω, as desired. Thus it may be assumed that h is not a product of reflections. We now prove (4.1.2) by induction on $\dim(W^\perp)$. Consider first the case in which $\dim(W^\perp) = 2$. If $\mathrm{sgn}(W^\perp) = +$, then h fixes a totally singular 1-space Y in $W^\perp$. Now let X be a totally singular 2-space in W. By the definition of $\Omega_4^+(2)$ in Description 4 in §2.5, we know that $\dim(X) - \dim(X \cap Xh)$ is even. Therefore $\dim(X \perp Y) - \dim((X \perp Y) \cap (X \perp Y)h)$ is also even, and since $X \perp Y$ is a maximal totally singular subspace of V, we know that $h \in \Omega$ by Description 4. Assume now that $\mathrm{sgn}(W^\perp) = -$, and let e_1, e_2, f_1, f_2, x, y be a standard basis as given in Proposition 2.5.3.ii, with $W = \langle e_1, e_2, f_1, f_2 \rangle$. Let h' be the involution in $I(W)$ which acts as $e_1 \leftrightarrow e_2$, $f_1 \leftrightarrow f_2$. We leave it as an exercise to the reader to prove that $h' \in \Omega(W)$ yet h' is not a product of reflections in $I(W)$. Since neither h nor h' is the product of reflections in $I(W)$, it follows from Proposition 2.5.9.i that hh' is a product of reflections in $I(W)$. And since h and h' both lie in $\Omega(W)$, we know that $hh' \in \Omega(W)$. Since a product of an odd number of reflections can never lie in $\Omega(W)$, it follows that hh' is the product of an even number of reflections in $I(W)$, and so it suffices to show that h' is the product of an even number of reflections in I. One checks that $h' = r_{f_1+f_2+x} r_{e_1+e_2+x} r_{e_1+e_2+f_1+f_2+x} r_x$, and so the result is proved. Completing the argument is now easy — for assume that $\dim(W^\perp) \geq 4$. Then $W^\perp$ contains a non-degenerate 2-space W_o, and we have $\Omega(W) \leq \Omega(W \perp W_o)$ by the above. We demonstrated in a previous case that $\Omega(W \perp W_o) \leq \Omega$, and (iii) now follows.

Using (iii), we have $|I(U)\Omega : \Omega| = |I(U) : \Omega(U)|$, and according to Table 2.1.C $|I(U) : \Omega(U)| = |I : \Omega|$ (this uses the fact that $\dim(U) \neq 1$, which holds in view of (4.0.2) and the fact that $\dim(U) \geq \frac{n}{2}$). Consequently $I(U)\Omega = I$. Thus for all $g \in I(W)$, there exists $g' \in I(U)$ such that $gg' \in \Omega$, whence $H_\Omega^W = I(W)$, proving (iv). The proof of of (v) is analogous. ■

We now begin the proof of Theorem 3.1.2 with the following.

Proposition 4.1.2. *If W is non-degenerate in case* **U**, **S** *or* **O**, *or if H is of type $GL_m(q) \oplus GL_{n-m}(q)$ in case* **L**, *then $H_\Gamma = N_\Gamma(H_\Omega)$ and Theorem 3.1.2 holds.*

Proof. First consider case **L** with H of type $GL_m(q) \oplus GL_{n-m}(q)$. Here $H_\Omega^W \geq SL(W)$, and so H_Ω is irreducible on W. Similarly H_Ω is irreducible on U, and as W and U have different dimensions, W is the unique H_Ω-invariant m-space and U is the unique H_Ω-invariant $(n-m)$-space. Therefore $N_\Gamma(H_\Omega) \leq N_\Gamma(W, U) = H_\Gamma$. It is obvious that $H_\Gamma \leq N_\Gamma(H_\Omega)$, as $H_\Omega = H_\Gamma \cap \Omega$. Therefore $H_\Gamma = N_\Gamma(H_\Omega)$, and hence Theorem 3.1.2 holds in view of Proposition 4.0.1. Next suppose that κ is non-degenerate and that $m < \frac{n}{2}$. Then $\dim(U) > \frac{n}{2}$, and so by Proposition 2.10.6.i and (4.0.2), $\Omega(U)$ is irreducible on U. Therefore H_Ω is irreducible on U by Lemma 4.1.1, and so U is the unique H_Ω-invariant $(n-m)$-space. Consequently $N_\Gamma(H_\Omega) \leq N_\Gamma(U) \leq N_\Gamma(U^\perp) = N_\Gamma(W) = H_\Gamma$. The desired results follow as before. Finally, assume that κ is non-degenerate and that $m = \frac{n}{2}$. Then case $\mathbf{O}^-$ holds, $\mathrm{sgn}(W) = \pm$ and $\mathrm{sgn}(U) = \mp$. Moreover, $L := \Omega(W) \times \Omega(U)$ acts irreducibly on W and U by Proposition 2.10.6, and they are non-isomorphic L-modules (see Lemma 2.10.12). Therefore Lemma 2.10.11.i ensures that W and U are the only

H_{Ω}-invariant m-subspaces. As $\mathrm{sgn}(W) = -\mathrm{sgn}(U)$, we deduce that Γ cannot interchange W and U, whence $N_{\Gamma}(H_{\Omega}) \leq H_{\Gamma}$, as before. ∎

Proposition 4.1.3. *Assume that case* **S** *holds and that H is of type $Sp_m(q) \perp Sp_{n-m}(q)$, so that W is non-degenerate.*

(I) $c = 1$.

(II) $H_{\overline{\Omega}} \cong (2, q-1).(PSp_m(q) \times PSp_{n-m}(q))$.

(III) $H_{\overline{\Omega}}$ *is local if and only if either q is odd or $q = m = 2$. When $H_{\overline{\Omega}}$ is non-local,* $\mathrm{soc}(H_{\overline{\Omega}}) \cong Sp_m(q)' \times Sp_{n-m}(q)'$.

Remark. We take the derived group in part (III) to account for the fact that $Sp_4(2)$ is almost simple, with a simple subgroup of index 2 (see Proposition 2.9.2.ii).

Proof. It follows from Witt's Lemma 2.1.6, and the fact that all non-degenerate m-spaces are isometric, that Ω is transitive on non-degenerate m-spaces. Therefore $c = 1$. Furthermore, Lemma 4.1.1 yields $H_I = I(W) \times I(U) \cong Sp_m(q) \times Sp_{n-m}(q)$. Since $I = \Omega$ we see now that (II) holds, upon factoring out the scalar group $\langle -1 \rangle$. If q is odd then $H_{\overline{\Omega}}$ centralizes $-\overline{1}_W$, and if $q = m = 2$, then $H_{\overline{\Omega}}$ normalizes $I(W)' \cong \mathbf{Z}_3$. Conversely, if these two conditions fail, then $H_{\overline{\Omega}} = H_I \cong Sp_m(q) \times Sp_{n-m}(q)$ with both $Sp_m(q)$ and $Sp_{n-m}(q)$ almost simple, by Proposition 2.9.2. Therefore $H_{\overline{\Omega}}$ is non-local, and $\mathrm{soc}(H_{\overline{\Omega}}) \cong \mathrm{soc}(Sp_m(q)) \times \mathrm{soc}(Sp_{n-m}(q)) = Sp_m(q)' \times Sp_{n-m}(q)'$. This proves (III). ∎

We now consider the unitary groups. It turns out that the structure of the stabilizers of non-degenerate subspaces in the unitary geometry closely resembles the structure of members of $\mathcal{C}_1$ of type $GL_m(q) \oplus GL_{n-m}(q)$. Therefore we treat these two cases together. It is convenient to make the following definition. For $\epsilon = \pm$ and for integers $n_1, \ldots, n_k$ define

$$\begin{aligned} a^{\epsilon}_{n_1,\ldots,n_k} &= \left| \{(\lambda_1, \ldots, \lambda_k) \in \mathbf{F}^*_{q^2} \times \cdots \times \mathbf{F}^*_{q^2} \mid \lambda_i^{q-\epsilon} = 1, \prod_{i=1}^{k} \lambda_i^{n_i} = 1\} \right| \\ b^{\epsilon}_{n_1,\ldots,n_k} &= \frac{(q-\epsilon)^{k-1}}{a^{\epsilon}_{n_1,\ldots,n_k}} \prod_{i=1}^{k} (q-\epsilon, n_i) \end{aligned} \tag{4.1.3}$$

Proposition 4.1.4. *Suppose that case* $\mathbf{L}^{\epsilon}$ *holds. If $\epsilon = +$ assume that H is of type $GL_m(q) \oplus GL_{n-m}(q)$, so that $V = U \oplus W$. If $\epsilon = -$ assume that H is of type $GU_m(q) \perp GU_{n-m}(q)$, so that W is non-degenerate.*

(I) $c = 1$.

(II) $H_{\overline{\Omega}} \cong \left[a^{\epsilon}_{m,n-m}/(q-\epsilon, n)\right].(L^{\epsilon}_m(q) \times L^{\epsilon}_{n-m}(q)).[b^{\epsilon}_{m,n-m}]$.

(III) $H_{\overline{\Omega}}$ *is local if and only if*

(a) $a^{\epsilon}_{m,n-m} > (q-\epsilon, n)$; *or*

(b) $m = q = 2$; *or*

(c) $(m, n, q, \epsilon) = (1, 3, 2, +)$.

If $H_{\overline{\Omega}}$ is non-local then $\mathrm{soc}(H_{\overline{\Omega}}) \cong L^{\epsilon}_m(q) \times L^{\epsilon}_{n-m}(q)$.

Proof. By Lemma 4.1.1 we have $H_I = I(W) \times I(U) \cong GL^{\epsilon}_m(q) \times GL^{\epsilon}_{n-m}(q)$. In particular, we have $|\det(H_I)| = |\det_W(I(W))| = q - \epsilon$. Therefore (I) holds in view of Lemma 4.0.3.v. To determine H_Ω, note that $|H_I : H_\Omega| = q - \epsilon$, and $H_\Omega \cong (SL^{\epsilon}_m(q) \times SL^{\epsilon}_{n-m}(q)).[q-\epsilon]$. Moreover if C is the subgroup of H_Ω inducing scalars on W and U, then $C \trianglelefteq H_\Omega$ and it is clear that

$$H_\Omega \cong C.(L^{\epsilon}_m(q) \times L^{\epsilon}_{n-m}(q)).[b], \tag{4.1.4}$$

where $b = (q-\epsilon, m)(q-\epsilon, n-m)(q-\epsilon)/|C|$. However $|C| = a^{\epsilon}_{m,n-m}$ and $\mathbf{F}^* \cap \Omega \leq H_\Omega$, and hence (II) follows from (4.1.4) upon factoring out the scalars. Part (III) follows from (II) and Lemma 4.0.5.iv. ■

We now consider the orthogonal groups, so that $\kappa = Q$ is a non-degenerate quadratic form. Here the situation is a bit more difficult. For this next result, recall the homomorphism γ introduced in §§2.5,2.8. Further, if X is one of the symbols Ω, S, I and if Y is a non-degenerate subspace of V, we adopt the notation $\ddot{X}(Y)$ to denote the image of $X(Y)$ in $\ddot{I}$. That is, $\ddot{X}(Y) = \overline{X(Y)\Omega/\Omega}$.

Lemma 4.1.5. *Assume that case* **O** *holds and let Y be a non-degenerate subspace in V.*

(i) $\ddot{\Omega}(Y) = 1$

(ii) *We have*

$$\ddot{S}(Y) = \begin{cases} \ddot{S} & \text{if } \dim(Y) \geq 2 \\ 1 & \text{if } \dim(Y) = 1. \end{cases}$$

(iii) *We have*

$$\ddot{I}(Y) = \begin{cases} \ddot{I} & \text{if } \dim(Y) \geq 2 \\ \langle \ddot{r}_y \rangle & \text{if } Y = \langle y \rangle. \end{cases}$$

(iv) *If* $\dim(V)$ *and* $\dim(Y)$ *are even, then* $\ker_{\ddot{I}(Y)}(\ddot{\gamma}) = \ker_{\ddot{I}}(\ddot{\gamma})$.

Proof. Assertion (i) is stated in Lemma 4.1.1, and (iv) follows from (iii). So it suffices to prove (ii) and (iii). Let y be a non-singular vector in Y. If $Y = \langle y \rangle$, then $S(Y) = 1$ and $I(Y) = \langle r_y \rangle$, whence the result is clear. And if q is even, then the result is again clear as $S = I = \Omega\langle r_y \rangle$ (see (2.5.12)). So we can assume that q is odd and $\dim(Y) \geq 2$, and we choose $g \in S(Y)\backslash\Omega(Y)$. Thus the spinor norm of g is a non-square in the (Y,Q)-geometry, and hence it is also a non-square in the (V,Q)-geometry. Therefore $g \in S\backslash\Omega$, which means $\ddot{S}(Y) = \ddot{S}$. Furthermore, $r_y \in I\backslash S$ (since $\det(r_y) = -1$) and so $\ddot{I}(Y) = \langle \ddot{r}_y \rangle \ddot{S}(Y) = \langle \ddot{r}_y \rangle \ddot{S} = \ddot{I}$, as desired. ■

We are now in a position to prove the first results for members of $\mathcal{C}_1$ in case **O**. We remind the reader that the definition of π_1 appears in Table 3.5.G (also see §3.2).

Proposition 4.1.6. *Assume that case* $\mathbf{O}^{\epsilon}$ *holds and that W is non-degenerate. Put* $W_1 = W$, $W_2 = U$, $m_i = \dim(W_i)$, $\epsilon_i = \operatorname{sgn}(W_i)$, $D_i = D(W_i)$ $(i = 1,2)$, *so that H is of type* $O^{\epsilon_1}_{m_1}(q) \perp O^{\epsilon_2}_{m_2}(q)$. *Then the following hold.*

(O) *$\epsilon = \pm$ if $\epsilon_1 = \epsilon_2 = \circ$, and $\epsilon = \epsilon_1\epsilon_2$ otherwise.*

(I) *$c = (qm_1m_2 - 1, 2)$, and when $c = 2$, we have $\pi = \pi_1$.*

(II)

$$H_{\overline{\Omega}} \cong \begin{cases} \Omega_{n-1}(q) & m_1 = 1,\ q \text{ odd},\ n \text{ even},\ D = \square \\ (\Omega^{\epsilon_1}_{m_1}(q) \times \Omega^{\epsilon_2}_{m_2}(q)).2 & q \text{ even; or} \\ & m_1 = 1 \text{ and } m_2 \text{ even; or} \\ & m_1 = 1,\ m_2 \text{ odd},\ D = \boxtimes\text{; or} \\ & m_1, m_2 \text{ odd},\ m_1 \geq 3,\ D = \square\text{; or} \\ & m_1, m_2 \text{ even},\ D_1 = D_2 = \boxtimes \\ (\Omega^{\epsilon_1}_{m_1}(q) \times \Omega^{\epsilon_2}_{m_2}(q)).[4] & m_1, m_2 \text{ odd},\ m_1 \geq 3,\ D = \boxtimes\text{; or} \\ & n \text{ odd},\ m_1 \geq 2\text{; or} \\ & m_1, m_2 \text{ even},\ D = \boxtimes \\ 2.(P\Omega^{\epsilon_1}_{m_1}(q) \times P\Omega^{\epsilon_2}_{m_2}(q)).[4] & m_1, m_2 \text{ even},\ q \text{ odd},\ D_1 = D_2 = \square. \end{cases}$$

(III) *$H_{\overline{\Omega}}$ is local if and only if either*

(a) *$\Omega^{\epsilon_i}_{m_i}(q)$ is local for some i (see Proposition 2.9.3); or*

(b) *$I(W_1, Q) \approx O_2^+(3)$ and n is odd.*

If $H_{\overline{\Omega}}$ is non-local then $\mathrm{soc}(H_{\overline{\Omega}}) \cong \Omega^{\epsilon_1}_{m_1}(q) \times \Omega^{\epsilon_2}_{m_2}(q)$.

Remark. Whenever m_i or n is odd, then it is to be understood that q is odd, in view of Lemma 2.5.1. There is one subtle point here, in that there are no non-degenerate 1-spaces in even characteristic, even though some 1-spaces are non-singular. Also, whenever the condition $D = \boxtimes$ or $D_i = \boxtimes$ appears in part (II) then again q is taken to be odd, as there are no non-squares in even characteristic. Thus q is odd throughout part (II), except in the second line.

Proof. Throughout this proof we condense the notation by writing $X_i = X(W_i)$, where X ranges over the symbols Ω, S and I.

(O). These conditions on $\epsilon, \epsilon_1, \epsilon_2$ are obvious, except when both ϵ_i lie in $\{+, -\}$. To prove $\epsilon = \epsilon_1\epsilon_2$ use Proposition 2.5.11.ii.

(I). First of all, by Lemma 4.1.1, $\Omega_1 \times \Omega_2 \leq H_\Omega$, and recall that we are assuming $m_1 = m \leq \frac{n}{2}$. Moreover, according to Proposition 4.1.2 we have $N_\Gamma(H_\Omega) = H_\Gamma$. Therefore by Lemma 4.0.3.i,ix we know that $c = |\ddot{\Delta} : \ddot{H}|$. Now by Lemma 4.1.1,

$$H_I = I_1 \times I_2 \cong O^{\epsilon_1}_{m_1}(q) \times O^{\epsilon_2}_{m_2}(q), \tag{4.1.5}$$

and thus by Lemma 4.1.5.iii we have $\ddot{H}_I = \ddot{I}$ (since either m or $n - m$ is greater than 1). Therefore $c \leq |\ddot{\Delta} : \ddot{I}| \leq 2$. Now when q is even or when qn is odd we know that $\ddot{\Delta} = \ddot{I}$ (see (2.6.1), (2.6.2), (2.7.2), (2.8.3)), and so in these cases $c = 1$. Thus we can assume that q is odd and n is even. If m_1 and m_2 are even, then there exists $g_i \in \Delta(W_i, \kappa)$ satisfying $\tau_{W_i}(g_i) = \mu$ (recall from (2.1.23) that $\langle\mu\rangle = \mathbf{F}^*$). It follows that $g = g_1g_2 \in H$ and $\tau(g) = \mu$. Therefore $\ddot{g} \in \ddot{\Delta}\backslash\ddot{I}$, and hence $\ddot{H} = \ddot{\Delta}$, proving $c = 1$. Now suppose that m_1 and m_2 are odd, and take $g \in H_\Gamma$. Then by Lemma 2.1.9.i, $g_W \in \Gamma(W, Q)$ and $\tau(g) = \tau_W(g_W)$. Thus by Lemma 2.6.4 we have $\ddot{g} \in \ker(\ddot{\tau})$, and hence $\ddot{H}_\Gamma \leq \ker(\ddot{\tau})$. We have $c = |\ddot{\Gamma} : \ddot{H}_\Gamma|$ by Propositions 4.0.3.i,ix and 4.1.2, and therefore $c \geq |\ddot{\Gamma} : \ker(\ddot{\tau})| = 2$. We have already seen that $c \leq 2$, and hence $c = 2$ and $\ddot{H}_\Gamma = \ker(\ddot{\tau})$. Thus $\pi = \pi_1$ (see Table 3.5.G), and this establishes (I).

(II). We determine the structure of $H_{\overline{\Omega}}$ in a few cases here, leaving the rest as exercises for the reader. Let w_1 be a non-singular vector in W_1. Since $m_2 \geq [\frac{n}{2}] \geq 4$ (recall (4.0.2)), it follows easily from Proposition 2.5.3 that W_2 contains a non-singular vector w_2 with the same norm as w_1. Now set $r_i = r_{w_i}$ for $i = 1, 2$, so that $r_1 r_2 \in H_\Omega$.

Case q even. Here $I_i = \Omega_i\langle r_i\rangle$, and therefore $H_\Omega = (\Omega_1\langle r_1\rangle \times \Omega_2\langle r_2\rangle) \cap \Omega = (\Omega_1 \times \Omega_2)\langle r_1 r_2\rangle \cong (\Omega^{\epsilon_1}_{m_1}(q) \times \Omega^{\epsilon_2}_{m_2}(q)).2$.

Case $m_1 = 1$. Here q is odd by Lemma 2.5.1. Take $h \in H_\Omega$, and write $h = h_1 h_2$ with $h_i \in I_i$. Since $I_1 = \langle r_1\rangle$, we know that h_1 is either 1 or r_1. Now if $h_1 = 1$, then $h = h_2 \in I_2 \cap \Omega = \Omega_2$, by Lemma 4.1.1.iii. And if $h_1 = r_1$, then the same reasoning shows that $r_1 r_2 h \in \Omega_2$, and we have now proved that $H_\Omega = \Omega_2\langle r_1 r_2\rangle \cong \Omega^{\epsilon_2}_{n-1}(q).2$. When n is odd or when n is even and $D = \boxtimes$, we have $H_{\overline{\Omega}} = H_\Omega$, which shows $H_{\overline{\Omega}}$ is as given in (II). When n is even and $D = \square$, we actually have $H_\Omega = \Omega_2 \times \langle -1\rangle$, and so upon factoring out $\langle -1\rangle$ we obtain $H_{\overline{\Omega}} \cong \Omega_{n-1}(q)$.

Case q is odd and $m_1 \geq 2$. Let s_i be an element of $S_i \backslash \Omega_i$, so that $s_i \in S \backslash \Omega$. As in the previous case, write $h \in H_\Omega$ as $h = h_1 h_2$, with $h_i \in I_i$. Now $I_i = \langle \Omega_i, r_i, s_i\rangle$, and $\langle s_1 s_2, r_1 r_2\rangle \leq H_\Omega$. Thus we may multiply h by some element of $\langle s_1 s_2, r_1 r_2\rangle$ to ensure that $h_1 \in \Omega_1 \leq \Omega$. Then $h_2 \in I_2 \cap \Omega = \Omega_2$, and so we have proved $H_\Omega = (\Omega_1 \times \Omega_2)\langle s_1 s_2, r_1 r_2\rangle \cong (\Omega^{\epsilon_1}_{m_1}(q) \times \Omega^{\epsilon_2}_{m_2}(q)).2^2$. If n is odd or if n is even and $D = \boxtimes$, we again have $H_{\overline{\Omega}} = H_\Omega$. So it remains to consider n even and $D = \square$. If for example m_i is even and $D(W_i) = \boxtimes$ for both i, then we can choose $s_i = -1_{W_i}$ (see Proposition 2.5.13). It is clear that $H_\Omega = (\Omega_1 \times \Omega_2)\langle r_1 r_2\rangle \times \langle -1\rangle$, and so $H_{\overline{\Omega}}$ is as in (II). Other cases are treated similarly.

(III). Since $\Omega_i \trianglelefteq H_\Omega$ and since $\Omega_i \cong \overline{\Omega}_i$, we know that $H_{\overline{\Omega}}$ is local if Ω_i is local for some i. Assume now that (b) holds. Then $W_1 = \langle w_\square\rangle \perp \langle w_\boxtimes\rangle$, where $\langle w_\square\rangle$ (respectively, $\langle w_\boxtimes\rangle$) is the unique $\square$1-space (respectively, $\boxtimes$1-space) in W_1 (if e_1, f_1 is a standard basis for (W_1, Q), take $w_\square = e_1 - f_1$, $w_\boxtimes = e_1 + f_1$). Thus $I_1 = \langle r_{w_\square}, r_{w_\boxtimes}\rangle \cong 2^2$. And since n is odd, either $r_{w_\square} \oplus -1_{W_2}$ or $r_{w_\boxtimes} \oplus -1_{W_2}$ is a non-scalar in $Z(H_\Omega)$, and so $H_{\overline{\Omega}}$ is local. Conversely, suppose that (a) and (b) fail. Then $K := \Omega_1 \times \Omega_2$ is semisimple (see §2.9). Furthermore, K is absolutely irreducible on W_2. Now if K fails to be absolutely irreducible on W_1, then Proposition 2.10.6 forces $m_1 = 2$; thus Ω_1 is cyclic, and so $\Omega_1 = 1$. Therefore, either K is absolutely irreducible on W_1 or $I_1 \approx O^+_2(q)$ with $q \leq 3$ (see Proposition 2.9.2.i). Suppose first that K is absolutely irreducible on W_1. Since (a) fails, neither -1_{W_1} nor -1_{W_2} is contained in Ω when q is odd; therefore the subgroup of Ω inducing scalars on W_1 and W_2 is just $\langle -1\rangle$, and so $H_{\overline{\Omega}}$ is non-local and $\mathrm{soc}(H_{\overline{\Omega}}) = K$ by Lemma 4.0.5.iv. The remaining case with $I_1 \approx O^+_2(q)$, $q \leq 3$, is left to the reader. ■

Proposition 4.1.7. *Assume that case* $\mathbf{O}^\pm$ *holds and that* H *is of type* $Sp_{n-2}(q)$, *so that* q *is even and* W *is a non-singular 1-space.*

(I) $c = 1$.

(II) $H_{\overline{\Omega}} \cong Sp_{n-2}(q)$.

Proof. Write $W = \langle w \rangle$, with w a non-singular vector. Then $r_w \in H_I \backslash \Omega$ and consequently $\ddot{H}_I \geq \langle \ddot{r}_w \rangle = \ddot{I}$, by (2.5.12). Thus $\ddot{H}_I = \ddot{I}$, and so $c = 1$ by Lemma 4.0.3.viii, proving (I). Note that the bilinear form $\mathbf{f}$ induces a non-degenerate form $\widetilde{\mathbf{f}}$ on $W^\perp/W$ given by $\widetilde{\mathbf{f}}(x + W, y + W) = \mathbf{f}(x, y)$ for all $x, y \in W^\perp$. Thus each element in H_Ω induces a map in $I(W^\perp/W, \widetilde{\mathbf{f}}) \cong Sp_{n-2}(q)$, and hence there is a homomorphism from H_Ω to $Sp_{n-2}(q)$ with kernel $C = C_\Omega(W^\perp/W)$. We first show that this homomorphism is surjective. Take $h \in I(W^\perp/W, \widetilde{\mathbf{f}})$ and for each $x + W \in W^\perp/W$ pick $x_1 \in W^\perp$ such that $(x + W)h = x_1 + W$. Then for $\lambda \in \mathbf{F}$ we have $Q(x_1 + \lambda w) = Q(x_1) + \lambda^2 Q(w)$, and so there is a unique choice of λ for which $Q(x_1 + \lambda w) = Q(x)$. Thus there is a well-defined map g on $W^\perp$ as follows: xg is the unique element of $(x + W)h$ for which $Q(xg) = Q(x)$. Observe that if $x, y \in W^\perp$, $\mathbf{f}(xg, yg) = \widetilde{\mathbf{f}}(xg + W, yg + W) = \widetilde{\mathbf{f}}((x+W)h, (y+W)h) = \widetilde{\mathbf{f}}(x+W, y+W) = \mathbf{f}(x, y)$. Therefore $Q((x+y)g) = Q(xg+yg)$. And since $xg+yg \in (x+W)h+(y+W)h = \big((x+y)+W\big)h$, it follows from the definition of g that $xg + yg = (x + y)g$. A similar argument shows that $(\lambda x)g = \lambda(xg)$, and hence $g \in GL(W^\perp)$. Furthermore, g preserves Q on $W^\perp$, and hence by Witt's Lemma 2.1.6, g extends to an isometry of (V, Q), which we also call g. Evidently $g \in H_I$ and the map which g induces on $W^\perp/W$ is precisely h. We may multiply g by r_w if necessary in order to ensure that $g \in H_\Omega$. Since r_w centralizes $W^\perp/W$, the (possibly new) element g still induces h on $W^\perp/W$, and so the homomorphism is indeed surjective. Thus to complete the proof it suffices to show that $C = 1$. So take $k \in C$ and observe that for any $x \in W^\perp$ we have $xk = x + \lambda w$ for some $\lambda \in \mathbf{F}$. Therefore $Q(x) = Q(x+\lambda w) = Q(x)+\lambda^2 Q(w)$, and so $\lambda = 0$. Therefore $k \in C_\Omega(W^\perp)$. Now pick any $v \in V \backslash W^\perp$ and observe that $Y = \langle w, v \rangle$ is a non-degenerate 2-space. Thus $V = Y \perp Y^\perp$, and k centralizes $Y^\perp$. Thus k acts on Y, and with respect to the basis w, v, we know that $k = \left(\begin{smallmatrix} 1 & 0 \\ \lambda & 1 \end{smallmatrix}\right)$ for some λ. Therefore k_Y has order 1 or 2. But the elements of $I(Y)$ of order 2 lie in $I(Y) \backslash \Omega(Y)$, and hence lie in $I \backslash \Omega$. Therefore $k_Y = 1$, and hence $k = 1$, completing the proof. ∎

We now prove the second special case of Theorem 3.1.2, the first being Proposition 4.1.2.

Proposition 4.1.8. *Assume that case* $\mathbf{O}^\pm$ *holds and that* H *is of type* $Sp_{n-2}(q)$*, so that* q *is even and* W *is a non-singular 1-space. Then* $H_\Gamma = N_\Gamma(H_\Omega)$*, and hence Theorem 3.1.2 holds.*

Proof. It suffices to prove that H_Ω fixes a unique non-singular 1-space. For then $N_\Gamma(H_\Omega) \leq N_\Gamma(W) = H_\Gamma$, and all assertions will follows as in the proof of Proposition 4.1.2. So suppose for a contradiction that H_Ω also fixes the non-singular 1-space $Y \neq W$. Then H_Ω acts faithfully on the $(n-2)$-space $Z = W^\perp \cap Y^\perp$. However H_Ω acts irreducibly as $Sp_{n-2}(q)$ on $W^\perp/W$, whence H_Ω acts irreducibly as $Sp_{n-2}(q)$ on Z. Therefore either Z is totally singular or Q_Z is a non-degenerate quadratic form on Z.

But $n-2 > \frac{n}{2}$ (see (4.0.2)) so Z is not totally singular; and $|Sp_{n-2}(q)|$ does not divide $|O^{\pm}_{n-2}(q)|$, so Q_Y cannot be non-degenerate. This contradiction completes the proof. ■

We now turn to the case in which W is totally singular. The stabilizers of the totally singular subspaces of V are in fact *parabolic* subgroups of the classical groups. These are very well understood via the Lie theory of the Chevalley groups (see [Ca$_1$, Ch. 8], for example). In the following discussion we describe how to determine their structure with an elementary approach using the associated classical geometry. Recall the integer u and the field automorphism α defined in (2.1.17) and (2.1.18).

Lemma 4.1.9. *Assume that $n = 2k$ is even and that case* **U**, **S** *or* **O**$^+$ *holds. Let $\beta = \{e_1, \ldots, f_k\}$ be a standard basis as described in Propositions 2.3.2, 2.4.1 and 2.5.3.i. Put $W_1 = \langle e_1, \ldots, e_k\rangle$, $W_2 = \langle f_1, \ldots, f_k\rangle$, and $K = N_I(W_1, W_2)$.*

(i) *$K \cong GL_k(q^u)$ and K acts naturally on W_1.*

(ii) *As K-modules we have $W_2 \cong W_1^{\alpha *}$.*

(iii) *$K \cap \Omega = \{k \in K \mid \det_{W_1}(k) \in (\mathbf{F}^*)^x\}$, where $x = q+1$, 1, 2 in cases* **U**, **S** *and* **O**$^+$*, respectively. In particular, $SL(W_1) \leq K^{W_1}$.*

Proof. Recall $\kappa = \mathbf{f}$ in cases **U** and **S**, and $\mathbf{f} = \mathbf{f}_Q$ in case **O**$^+$ with $\kappa = Q$. Therefore

$$\mathbf{f}_\beta = \begin{pmatrix} 0 & \mathbf{I}_k \\ \pm\mathbf{I}_k & 0 \end{pmatrix}$$

where the sign is $+$ in cases **O** and **U** and is $-$ in case **S**. It follows directly from Lemma 2.1.8 that

$$K_\beta = \left\{ \begin{pmatrix} \mathbf{A} & 0 \\ 0 & \mathbf{A}^{-1\alpha t} \end{pmatrix} \,\middle|\, \mathbf{A} \in GL_k(q^u) \right\}. \tag{4.1.6}$$

Assertions (i) and (ii) are now clear, and so it remains to determine the value of x in (iii). Obviously we may take $x = 1$ in case **S**, for here $I = \Omega$. As for case **U**, we see that $k \in K \cap \Omega$ if and only if $\det_{W_1}(k)^{1-q} = 1$, and so $x = q+1$. This leaves case **O**$^+$. Note that when q is even, K fixes W_1, and so $K \leq \Omega$ by Description 4 in §2.5. Thus the result holds here, as $(\mathbf{F}^*)^2 = \mathbf{F}^*$. It remains to consider q odd. Now it is a standard fact that $SL_n(q)$ is generated by elements of order p, and hence $O^p(SL(W_1)) = SL(W_1)$. And since $GL(W_1)/SL(W_1) \cong \mathbf{Z}_{q-1}$, it follows that K has a unique subgroup of index 2, denoted $\frac{1}{2}K$. Evidently $\frac{1}{2}K = \{k \in K \mid \det_{W_1}(k) \in (\mathbf{F}^*)^2\}$. Obviously $K \leq S$, and hence by Proposition 2.5.7 we have $\frac{1}{2}K \leq \Omega$. Thus it suffices to show that $K \not\leq \Omega$. But clearly $k = r_{e_1+f_1} r_{e_1+\mu f_1} \in K \backslash \Omega$, since the spinor norm of k is μ (see Description 1 in §2.5). Thus (iii) is proved. ■

Corollary 4.1.10. *Assume that W is totally singular.*

(i) *$SL(W) \leq H_\Omega^W$.*

(ii) *If $\dim(W) \geq 2$, then no Sylow p-subgroup of Ω can centralize W.*

Proof. (i). This is straightforward in case **L**, so let us assume that κ is non-degenerate. Using Witt's Lemma 2.1.6, we can assume that W has a basis $e_1, \ldots, e_m$ which is part of the standard basis β given in Propositions 2.3.2, 2.4.1 and 2.5.3 in the respective cases **U**, **S** and **O**. Now let $Y = \langle f_1, \ldots, f_m \rangle$, where the f_i are also taken from the standard basis β. Then $W \oplus Y$ is a non-degenerate space of dimension $2m$, and by Lemma 4.1.9.iii we know that $SL(W) \leq N_{\Omega(W \oplus Y)}(W)^W$. The result now follows, for $\Omega(W \oplus Y) \leq \Omega$ by Lemma 4.1.1.ii.

(ii). This is immediate from (i) and the fact that $SL(W)$ has non-trivial Sylow p-subgroups when $\dim(W) \geq 2$. ■

In the following discussion we will make use of the concept of a maximal flag. A *maximal flag* in $(V, \mathbf{F}, \kappa)$ is a chain

$$0 = V_o < V_1 < V_2 < \cdots < V_r, \tag{4.1.7}$$

where each V_i is a totally singular i-space and V_r is a maximal totally singular subspace. Thus $r = n$, $[\frac{n}{2}]$, $\frac{n}{2}$, $\frac{n-1}{2}$, $\frac{n}{2}$ and $\frac{n}{2} - 1$ in cases **L**, **U**, **S**, **O°**, **O⁺** and **O⁻**, respectively. It is easy to see that I is transitive on maximal flags by Witt's Lemma 2.1.6. We will usually write $\mathcal{F}$ for a maximal flag, and r is often called the *rank* of $\mathcal{F}$. Note that in the geometries of type $O_1(q)$, $GU_1(q)$ and $O_2^-(q)$, we have $r = 0$. Thus these geometries have a unique maximal flag — namely, the zero flag V_o. As for the $O_2^+(q)$-geometry, one checks that there are precisely two singular 1-spaces, given by $\langle e_1 \rangle$ and $\langle f_1 \rangle$, where e_1, f_1 appear in Proposition 2.5.3.i. Thus this geometry has precisely two maximal flags, interchanged by the elements in $O_2^+(q) \backslash SO_2^+(q)$ (respectively, $O_2^+(q) \backslash \Omega_2^+(q)$) with q odd (respectively, q even). We denote by $N_I(\mathcal{F})$ the subgroup of I which fixes each space V_i in $\mathcal{F}$.

Lemma 4.1.11. *Assume that g is a p-element in $N_I(\mathcal{F})$. Then $g \in \Omega$ if and only if g centralizes $V_r^{\perp}/V_r$.*

Proof. Define $X = V_r^{\perp}/V_r$. There is nothing to prove in cases **L**$^{\pm}$, **S** and **O°**, for here $|I : \Omega|$ is relatively prime to p and $\dim(X) \leq 1$. Now assume that case **O⁺** holds. Here $X = 0$, and so we must show that $g \in \Omega$. When q is odd, this is obvious since $|I : \Omega|$ divides 4. And when q is even, we again see that $g \in \Omega$ in view of Description 4 in §2.5 and the fact that g fixes $V_{n/2}$. It remains to consider case **O⁻**. Write Q_X for the induced form on X, as described in (2.1.20). If p is odd, then $g \in \Omega$ as $|I : \Omega|$ divides 4. Moreover $g^X \in I(X, Q_X) \cong D_{2(q+1)}$, which is a p'-group; therefore $g^X = 1$, which shows that the Lemma holds with q odd. So assume that q is even. Since I is transitive on maximal flags, we can assume that $V_i = \langle e_1, \ldots, e_i \rangle$ for $1 \leq i \leq \frac{n}{2} - 1$, and that $V_{n/2-1}^{\perp} = V_{n/2-1} \oplus \langle x, y \rangle$, where e_i, x and y are as given in Proposition 2.5.3.ii. Write $xg = v + x'$ and $yg = w + y'$, with $v, w \in \langle e_1, \ldots, e_{n/2-1} \rangle$ and $x', y' \in \langle x, y \rangle$. Since g fixes Q, we know that $Q(x') = (x', y') = 1$ and $Q(y') = Q(y) = \zeta$, as in Lemma 2.5.2. So defining $g_1 = r_{w+x'} r_{x'} r_{\zeta v + y'} r_{y'}$, we see that gg_1 fixes $\langle x, y \rangle$. Thus $gg_1 = g_2 g_3$, where $g_2 \in I(\langle x, y \rangle)$ and $g_3 \in I(\langle x, y \rangle^{\perp})$. Now g_1 is a product of reflections in vectors in $V_{n/2-1}^{\perp}$, and hence g_1 centralizes $V_{n/2-1}$. Furthermore, g_2 centralizes $V_{n/2-1}$ and g fixes $V_{n/2-1}$. Consequently g_3 fixes $V_{n/2-1}$, which is a maximal totally singular space in $\langle x, y \rangle^{\perp}$. Thus using Description 4 in §2.5, we know that $g_3 \in \Omega(\langle x, y \rangle^{\perp})$, and hence $g_3 \in \Omega$ by

Lemma 4.1.1.ii. Also $g_1 \in \Omega$ by Description 2. Consequently $g \in \Omega$ if and only if $g_2 \in \Omega$. Now observe that g_1 and g_3 centralize X, and hence $g^X = g_2^X$. Thus if g centralizes X, then $g_2 = 1$ and hence $g \in \Omega$, as desired. On the other hand, if g does not centralize X, then g_2 is a non-trivial 2-element in $I(\langle x, y\rangle)$. But all non-trivial 2-elements in $O_2^-(q)$ with q even are reflections, and hence g_2 is a reflection. Therefore $g_2 \notin \Omega$, and hence $g \notin \Omega$, completing the proof. ∎

Lemma 4.1.12. *Assume that κ is non-degenerate and that W is totally singular. Then there exist spaces X, Y such that $H_I = C{:}L$, where $C = C_I(W, W^\perp/W, V/W^\perp)$ and $L = N_I(W, Y, X)$, and the following hold.*

(i) *$V = (W \oplus Y) \perp X$, with Y a totally singular m-space, $W \oplus Y$ non-degenerate and $X = (W \oplus Y)^\perp$.*

(ii) *$L = K \times I(X)$, where $GL_m(q^u) \cong K \leq I(W \oplus Y)$ and K acts naturally on W; and as K-modules we have $Y \cong W^{\alpha *}$.*

(iii) *C is a p-group and $C \leq \Omega$.*

Proof. Using Witt's Lemma 2.1.6, we can assume that W has a basis $e_1, \ldots, e_m$, where the vectors e_i are taken from a standard basis given in Proposition 2.5.3. Then let $Y = \langle f_1, \ldots, f_m\rangle$, as in the proof of Corollary 4.1.10, and put $X = (W \oplus Y)^\perp$. Evidently $Y \oplus W$ is a non-degenerate $2m$-space in V and hence X is a non-degenerate $(n-2m)$-space.

Now put $L = N_I(W, Y, X)$. Clearly $I(X) \leq L$ and hence $L = K \times I(X)$, where $K = N_{I(W \oplus Y)}(W, Y)$. We now prove that $H_I = C{:}L$. Take $h \in H_I$ and note that as $K^W = GL(W)$ we may multiply h by an element in H_I in order to assume that $h \in C_I(W)$. But now an easy calculation, similar to the one used for (4.1.6), shows that as H_I-modules, $V/W^\perp \cong W^{\alpha *}$, and hence h also centralizes $V/W^\perp$. Now consider $h^{W^\perp/W}$. Each element in $W^\perp/W$ can be written uniquely in the form $x + W$ for some $x \in X$. Thus we obtain a linear transformation $h_1 : X \to X$ satisfying $xh_1 + W = xh + W$. Moreover, W is totally singular, $X \leq W^\perp$, and h preserves κ, whence it is easy to see that h_1 preserves κ. In other words, $h_1 \in I(X)$. Therefore $hh_1^{-1} \in C$, which proves $H_I = CL$. It is clear that $C \cap L = 1$, and so we have in fact $H_I = C{:}L$, as desired.

It remains to demonstrate that $C \leq \Omega$. In case **S** this is trivial since $I = \Omega$, and in case **U** it is clear for $\det(C) = 1$. As for case **O**, let $0 = X_o < X_1 < \cdots < X_r$ be a maximal flag in X. Then putting $W_i = \langle e_1, \ldots, e_i\rangle \leq W$, we see that $0 < W_1 < \cdots < W_m = W < W \oplus X_1 < \cdots < W \oplus X_r$ is a maximal flag in V, and C centralizes each quotient space in the flag. Thus $C \leq \Omega$ by Lemma 4.1.11, proving (iii). ∎

In Lemma 4.1.12 we assumed that κ is non-degenerate. In Lemma 4.1.13 we state the corresponding result when κ is identically 0, that is, for case **L**. The proof is easier than that of Lemma 4.1.12, so we leave it to the reader.

Lemma 4.1.13. *Assume that case **L** holds. Then there exists a complement Y to W in V such that $H_I = C{:}L$, where $C = C_I(W, V/W)$ and $L = N_I(W, Y)$. Moreover, C is a p-group and $C \leq \Omega$.*

In Lemmas 4.1.12 and 4.1.13, the structure of the p-group C and the action of L on C can be calculated without too much difficulty. The easiest example is case **L**, where C is elementary abelian of order $q^{m(n-m)}$ and is isomorphic to $W \otimes Y$ as an L-module. In Lie theoretic terminology, C is called the ***unipotent radical*** of the parabolic subgroup H_I, and L is the *Levi factor*. For more information on unipotent radicals and Levi factors, see [Ca, Ch.8] and [G-L, §4].

Proposition 4.1.14. *Assume that $P \in Syl_p(\Omega)$.*

(i) *Provided case $\mathbf{O}^+$ does not hold, P fixes a unique maximal flag $V_1 < \cdots < V_r$, and hence V_i is the unique P-invariant totally singular i space for $i = 1, \ldots, r$.*

(ii) *If case $\mathbf{O}^+$ holds, then P fixes precisely two maximal flags $V_1 < \cdots < V_{n/2-1} < V_{n/2}^j$ with $j = 1, 2$. Thus V_i is the unique P-invariant totally singular i-space for $1 \leq i \leq \frac{n}{2} - 1$, and $V_{n/2}^j$ $(j = 1, 2)$ are the only two P-invariant totally singular $\frac{n}{2}$-spaces. These two maximal flags are interchanged by an element of $N_I(P) \backslash \ker(\gamma)$.*

Proof. First suppose that case **L** holds, and let $\{v_1, \ldots, v_n\}$ be a basis for V. It is clear that the set of lower triangular matrices with entries 1 along the diagonal is a Sylow p-subgroup of $\Omega = SL(V)$, and so P does indeed fix a maximal flag. To prove uniqueness, define $g \in \Omega = SL(V)$ by $v_1 g = v_1$, $v_i g = v_{i-1} + v_i$ for $2 \leq i \leq n$. It is not hard to verify that $\langle v_1, \ldots, v_i \rangle$ is the only g-invariant i-space, and uniqueness follows.

For the rest of this proof we assume that κ is non-degenerate. We proceed by induction on n. The result is trivial when $n = 1$, so we take $n \geq 2$. Because there are $q^{un} - 1$ non-zero vectors in V, we know that P must fix one, v say, and we put $V_1 = \langle v \rangle$. If V_1 is non-degenerate, then $P \leq I(V_1) \times I(V_1^\perp)$. It follows that case **U** does not hold, for $GU_1(q) \times GU_{n-1}(q)$ does not contain a Sylow p-subgroup of $SU_n(q)$. Therefore case **O** holds, and $O_1(q) \times O_{n-1}^\xi(q)$ contains a Sylow p-subgroup of $\Omega_n^\epsilon(q)$. Using Table 2.1.C, it follows that $n = 2$. In this case, $\Omega \approx \Omega_2^\epsilon(q) \cong \mathbf{Z}_{(q-\epsilon)/(2,q-1)}$, and hence $P = 1$. Our remarks concerning maximal flags in $O_2^\pm(q)$ preceding Lemma 4.1.11 indicate that assertions (i) and (ii) are in fact true. Next assume that V_1 is degenerate but non-singular, so that case $\mathbf{O}^\pm$ holds with n and q even. Here $N_\Omega(V_1) \cong Sp_{n-2}(q)$, which does not contain a Sylow 2-subgroup of Ω unless $n = 2$. This case we have already treated.

We may assume for the rest of this proof that all P-invariant 1-spaces are totally singular. In particular, V_1 is totally singular. Let $Y = V_1^\perp / V_1$ and let κ_Y be the induced form on Y (see (2.1.20)). We now claim that $P^Y \leq \Omega(Y, \kappa_Y)$. This is obvious in cases **U** and **S**, and also in case **O** with q odd. So assume that case **O** holds in even characteristic. According to Lemma 4.1.12, $N_I(V_1) = C{:}L$, where $C = C_I(V_1, V_1^\perp/V_1, V/V_1^\perp)$ and $L = N_I(V_1, W_1, X)$ for suitable spaces W_1 and X. By Lemma 4.1.12, $C \leq O_p(N_\Omega(V_1)) \leq P$, and so $P = C{:}P_1$, where $P_1 = P \cap L$. Clearly P_1 centralizes V_1 and W_1, which means $P_1 \leq I(X)$. Since $P_1 \leq \Omega$, it follows from Lemma 4.1.1.iii that $P_1 \leq \Omega(X)$, and so by induction P_1 fixes precisely k flags in X, where $k = 2$ if X is an $O_{n-2}^+(q)$-space, and $k = 1$

otherwise. We write these flags as $0 = X_o < X_1 < \cdots < X_{r-2} < X^i_{r-1}$ $(1 \le i \le k)$. Evidently P fixes the flags $\mathcal{F}_i$ $(1 \le i \le k)$ given by $0 = V_o < V_1 < \cdots < V_{r-1} < V^i_r$, where $V_{j+1} = V_1 \oplus X_j$ for $j = 1, \ldots, r-1$.

To complete the proof we must show that every P-invariant maximal flag is equal to some $\mathcal{F}_i$. Assume therefore that $\mathcal{F}'$ is a P-invariant flag given by $0 = V'_o < V'_1 < \cdots < V'_r$. Assume first that $V_1 = V'_1$. Then $V'_i = V_1 \oplus X'_i$, where $X'_i = X \cap V'_i$. We obtain the desired result using induction in the (X, κ)-geometry. We are left with the case where $V_1 \ne V'_1$. Here P centralizes $W = V_1 \oplus V'_1$. We have already seen that all P-invariant 1-spaces must be totally singular, and hence W is totally singular. But this contradicts Corollary 4.1.10.ii. The proof is complete. ■

Proposition 4.1.14 and the fact that I is transitive on maximal flags give the following result.

Corollary 4.1.15. *Assume that H is of type P_m, so that W is totally singular.*
(i) *Every maximal flag of (V, κ) is stabilized by a Sylow p-subgroup of Ω.*
(ii) *H contains a Sylow p-subgroup of Ω.*
(iii) *$|C|_p|L \cap \Omega|_p = |\Omega|_p$, where C and L are as in Lemma 4.1.12.*

We also obtain the proof of Theorem 3.1.2 in the case where H is of type P_m.

Proposition 4.1.16. *Assume that H is of type P_m, so that W is totally singular. Then $H_\Gamma = N_\Gamma(H_\Omega)$, and hence Theorem 3.1.2 holds.*

Proof. As in the proofs of Propositions 4.1.2 and 4.1.8, Theorem 3.1.2 will follow from Proposition 4.0.1 and the fact that $H_\Gamma = N_\Gamma(H_\Omega)$. Take $g \in N_\Gamma(H_\Omega)$. By the Frattini argument, we can assume that g normalizes $P \in Syl_p(H_\Omega)$. Note that $P \in Syl_p(\Omega)$ by Corollary 4.1.15.ii. Therefore Proposition 4.1.14 applies. Clearly, if W is the unique P-invariant totally singular m-space, then g must fix W, which means $g \in H_\Gamma$. If this fails, then Proposition 4.1.14.ii implies that W is a totally singular $\frac{n}{2}$-space in case $\mathbf{O}^+$, and P fixes precisely two totally singular $\frac{n}{2}$-spaces W and W'. Thus $g \in N_\Gamma\{W, W'\}$. Now if g interchanges W and W', then H_Ω fixes the $(\frac{n}{2} - 1)$-space $W \cap W'$. However, this is impossible, for by Corollary 4.1.10.i, $SL(W) \le H_\Omega^W$, which means that H_Ω acts irreducibly on W. This contradiction ensures that g fixes W, which is to say $h \in H_\Gamma$. The proof is finished. ■

We are now in a position to determine the structure of these parabolic subgroups of the simple group $\overline{\Omega}$. This first result uses the notation appearing in (4.1.3).

Proposition 4.1.17. *Assume that case* **L** *holds and that H is of type P_m, so that $H = N_\Delta(W)$.*
(I)
$$c = \begin{cases} 1 & \text{if } m = \frac{n}{2} \\ 2 & \text{otherwise,} \end{cases}$$
and if $c = 2$ then $\pi = \pi_3$ (see Table 3.5.G).

(II) $H_{\overline{\Omega}} \cong [q^{m(n-m)}].[a^+_{m,n-m}/(q-1,n)].(L_m(q) \times L_{n-m}(q)).[b^+_{m,n-m}]$.
(III) $H_{\overline{\Omega}}$ *is p-local.*

Proof. First of all, $\Omega = SL(V)$ is transitive on $\mathcal{U}_k$ for all k, and since Γ acts on $\mathcal{U}_k$, we deduce that $\ddot{H}_\Gamma = \ddot{\Gamma}$, and hence

$$\ddot{\Gamma} \le \ddot{A}_{[H_{\overline{\Omega}}]}. \tag{4.1.8}$$

Evidently (4.1.8) implies that $c = |\ddot{A} : \ddot{A}_{[H_{\overline{\Omega}}]}| \le |\ddot{A} : \ddot{\Gamma}| = 2$. Now when $n = 2$, we have $A = \Gamma$, and hence $c = 1$. We assume therefore that $n \ge 3$. Now the inverse transpose automorphism $\iota \in A$ described in §2.2 takes stabilizers of m-spaces to stabilizers of $(n-m)$-spaces. Thus if $m = \frac{n}{2}$, then H_Ω^ι is Ω-conjugate to H_Ω, which means $\ddot{\iota} \in \ddot{A}_{[H_{\overline{\Omega}}]}$. Therefore from (4.1.8) we deduce $\ddot{A}_{[H_{\overline{\Omega}}]} = \ddot{A}$, and so $c = 1$. Thus we can assume that $m < \frac{n}{2}$. Suppose for a contradiction that $c = 1$. Then $h\iota \in N_A(H_\Omega)$ for some $h \in \Gamma$, and hence H_Ω fixes some $(n-m)$-space U. According to Corollary 4.1.15.ii, H_Ω contains a Sylow p-subgroup P of Ω, and by Proposition 4.1.14, P fixes a unique $(n-m)$-space. In addition, this P-invariant $(n-m)$-space contains W, and hence $W < U$. However, Lemma 4.1.13 shows that $H_\Omega^{V/W} \ge SL(V/W)$, and so H_Ω must be irreducible on V/W. This contradicts the fact that H_Ω fixes U/W. Therefore, $c = 2$, and it follows that $\ddot{A}_{[H_{\overline{\Omega}}]} = \ddot{\Gamma}$. Thus from Table 3.5.G we see that $\pi = \pi_3$, proving (I). To prove (II), write $H_I = C{:}L$ as described in Lemma 4.1.13. Then $H_{\overline{\Omega}} \cong C{:}\overline{L \cap \Omega}$, and the structure of $\overline{L \cap \Omega}$ has been determined already in Proposition 4.1.4. The order of C is easily seen to be $q^{m(n-m)}$, and thus (II) holds. Part (III) is immediate from (II). ■

Remark. The previous proof shows that if $n \ge 3$, then the stabilizer of a 1-space in $SL_n(q)$ does not fix an $(n-1)$-space. This implies that the natural representation of $SL_n(q)$ is neither self-dual nor unitary. For in a self-dual or unitary representation, the stabilizer of a 1-space also fixes a hyperplane (namely, the 'perp' of the 1-space).

Proposition 4.1.18. *Assume that case* **U** *holds and that H is of type P_m, so that W is totally singular.*
(I) $c = 1$.
(II) $H_{\overline{\Omega}} \cong [q^{m(2n-3m)}]{:}[a/(q+1,n)].(L_m(q^2) \times U_{n-2m}(q)).[b]$, *where*

$$a = \left|\{(\lambda_1, \lambda_2) \mid \lambda_i \in \mathbf{F}_{q^2},\ \lambda_2^{q+1} = 1,\ \lambda_1^{m(q-1)}\lambda_2^{n-2m} = 1\}\right|$$

and

$$b = (q^2-1)(q^2-1,m)(q+1,n-2m)/a.$$

(III) $H_{\overline{\Omega}}$ *is p-local.*

Proof. Write $H_I = C{:}L$ and $L = K \times I(X)$ as in Lemma 4.1.12. Since $K^W = GL(W)$ we can choose $g \in K$ satisfying $\det_W(g) = \mu$, where $\langle \mu \rangle = \mathbf{F}^*$. Then $\det(g) = \mu^{1-q}$, which has order $q+1$. Consequently $|\det(K)| = q+1$, and so (I) holds by Lemma 4.0.3.v. As in the proof of Proposition 4.1.17, we have $H_{\overline{\Omega}} \cong C.\overline{L \cap \Omega}$. Now $|L : L \cap \Omega| = q+1$ (since $\det(K) = q+1$) and so $L \cap \Omega = \frac{1}{q+1}(GL_m(q^2) \times GU_{n-2m}(q)) \cong (SL_m(q^2) \times$

$SU_{n-2m}(q)).[q^2-1]$. Moreover, the subgroup acting as scalars on W, Y and X (where Y and X are as in Lemma 4.1.12) has order a, and so $L \cap \Omega \cong a.(L_m(q^2) \times U_{n-2m}(q)).b$. The order of C can now be determined using Corollary 4.1.15.iii and Table 2.1.C, and so (II) holds upon factoring out the scalars in $\mathbf{F}^*$ (a group of order $(q+1,n)$). Assertion (III) is clear. ■

Proposition 4.1.19. *Assume that case* **S** *holds and that* W *is totally singular.*

(I) $c = 1$.

(II) $H_{\overline{\Omega}} \cong [q^a].(q-1).(PGL_m(q)) \times PSp_{n-2m}(q))$, *where* $a = \frac{m}{2} - \frac{3m^2}{2} + mn$.

(III) $H_{\overline{\Omega}}$ *is p-local.*

The proof is left to the reader.

We now come to the orthogonal groups. Before we state the main structure result, it is convenient to make the following definition. When q is odd and $\frac{1}{2}(q-1)m$ is even, define the group J as follows. As we mentioned in the proof of Lemma 4.1.9, when q is odd $GL_m(q)$ has a unique subgroup of index 2, denoted $\frac{1}{2}GL_m(q)$. Moreover, when $\frac{1}{2}(q-1)m$ is even, $\frac{1}{2}GL_m(q)$ has a unique central involution z say, and J is defined as $\frac{1}{2}GL_m(q)/\langle z \rangle$. An easy check shows that J has the following structure:

$$J \cong \begin{cases} (\frac{1}{2}(q-1)).L_m(q).(\frac{1}{2}(q-1,m)) & m \text{ even} \\ (\frac{1}{4}(q-1)).L_m(q).(q-1,m) & m \text{ odd.} \end{cases}$$

We are now ready for the main result on the structure of the parabolics in the orthogonal groups.

Proposition 4.1.20. *Assume that case* $\mathbf{O}^\epsilon$ *holds and that* H *is of type* P_m, *so that* W *is totally singular.*

(I)

$$c = \begin{cases} 2 & \text{in case } \mathbf{O}^+ \text{ with } m = \frac{n}{2} \\ 1 & \text{otherwise,} \end{cases}$$

and if $c = 2$, *then* $\pi = \pi_2$.

(II)

$$H_{\overline{\Omega}} \cong \begin{cases} [q^a]{:}(GL_m(q) \times \Omega^\epsilon_{n-2m}(q)) & q \text{ even} \\ [q^a]{:}\frac{1}{2}GL_m(q) & qn \text{ odd and } n = 2m+1 \\ [q^a]{:}\frac{1}{2}GL_m(q) & q \text{ odd, } n = 2m,\ \frac{1}{2}m(q-1) \text{ odd} \\ [q^a]{:}J & q \text{ odd, } n = 2m,\ \frac{1}{2}m(q-1) \text{ even} \\ [q^a]{:}(\frac{1}{2}GL_m(q) \times \Omega^\epsilon_{n-2m}(q)) & q \text{ odd, } -1 \in \Omega,\ \frac{1}{2}m(q-1) \text{ odd} \\ [q^a]{:}2.(J \times P\Omega^\epsilon_{n-2m}(q)).2 & q \text{ odd, } -1 \in \Omega,\ \frac{1}{2}m(q-1) \text{ even, } \frac{n}{2} > m, \\ [q^a]{:}(\frac{1}{2}GL_m(q) \times \Omega^\epsilon_{n-2m}(q)).2 & q \text{ odd, } -1 \notin \Omega,\ n-2m \geq 2 \end{cases}$$

where $a = mn - \frac{m}{2}(3m+1)$.

(III) $H_{\overline{\Omega}}$ *is p-local.*

Proof. Let C, L, K, Y and X be as in Lemma 4.1.12, so that $H_I = C{:}L$ and $H_\Omega = C{:}(L \cap \Omega)$. Now I is transitive on totally singular m-spaces by Witt's Lemma 2.1.6, and so $HI = \Delta$. Therefore $c = |\ddot{I} : \ddot{H}_I|$ by Lemma 4.0.3.iii,ix and Proposition 4.1.16. Now X is a non-degenerate $(n-2m)$-space, and so according to Lemma 4.1.5 we have

$$\ddot{I}(X) = \begin{cases} \ddot{I} & \text{if } m < \frac{1}{2}(n-1) \\ \langle \ddot{r}_x \rangle & \text{if } nq \text{ is odd, } m = \frac{1}{2}(n-1) \text{ and } X = \langle x \rangle \\ 1 & \text{if } \epsilon = + \text{ and } m = \frac{1}{2}n. \end{cases} \tag{4.1.9}$$

We now determine $\ddot{K}$. Now when q is even, we have $K \le \Omega(W \oplus Y) \le \Omega$, and so $\ddot{K} = 1$. When q is odd, it follows from Lemma 4.1.9.iii that K contains elements with spinor norm a non-square, and so $\ddot{K} = \ddot{S}$. Summarizing,

$$\ddot{K} = \ker_{\ddot{I}}(\ddot{\gamma}) = \begin{cases} \ddot{S} & \text{if } q \text{ is odd} \\ 1 & \text{if } q \text{ is even.} \end{cases} \tag{4.1.10}$$

So combining (4.1.9) and (4.1.10) we deduce

$$\ddot{H}_I = \ddot{L} = \begin{cases} \ddot{I} & \text{if } m < \frac{n}{2} \\ \ker_{\ddot{I}}(\ddot{\gamma}) & \text{if } \epsilon = + \text{ and } m = \frac{n}{2}. \end{cases} \tag{4.1.11}$$

Thus when $m < \frac{n}{2}$ we know that $c = 1$, and when $m = \frac{n}{2}$ we have $c = 2$. Moreover, when $c = 2$ it follows from the definition of γ that $\ddot{H}_\Gamma \le \ker(\ddot{\gamma})$. On the other hand, Proposition 4.1.16 and Lemma 4.0.3.i,ix ensure that $\ddot{H}_\Gamma = \ddot{\Gamma}_{[H_{\overline{\Omega}}]} = \frac{1}{2}\ddot{\Gamma}$, and so all these groups must be equal to $\ker(\ddot{\gamma})$. Therefore $\pi = \pi_2$, as claimed.

We now determine the structure of $H_{\overline{\Omega}}$ in a few cases, leaving the others as exercises. As before, we have

$$H_{\overline{\Omega}} \cong C{:}(\overline{L \cap \Omega}). \tag{4.1.12}$$

As we mentioned in the previous paragraph, when q is even, $K \le \Omega$, and so $L \cap \Omega = K \times (I(X) \cap \Omega) = K \times \Omega(X) \cong GL_m(q) \times \Omega^\epsilon_{n-2m}(q)$. Therefore (II) holds in view of (4.1.12) and Corollary 4.1.15.iii. Assume for the rest of this paragraph that q is odd. We have $K \le S$, and so $L \cap S = K \times S(X) \cong GL_m(q) \times SO^\epsilon_{n-2m}(q)$. Assume first that $n - 2m \le 1$. Then $S(X) = 1$, and hence $L \cap \Omega = K \cap \Omega$, and $K \cap \Omega = \frac{1}{2}K$, since $K \not\le \Omega$. Thus when $-1 \notin \Omega$, we have $H_{\overline{\Omega}} = H_\Omega = C{:}(\frac{1}{2}K) \cong [q^a].\frac{1}{2}GL_m(q)$. And when $-1 \in \Omega$, we have $H_{\overline{\Omega}} \cong C{:}J$. For the rest of this proof we assume that $n - 2m \ge 2$, so that both K and $S(X)$ contain elements with spinor norm a non-square. The product of these two elements has spinor norm a square, hence lies in Ω, and therefore $L \cap \Omega = ((K \cap \Omega) \times \Omega^\epsilon_{n-2m}(q)).2 \cong (\frac{1}{2}GL_m(q) \times \Omega^\epsilon_{n-2m}(q)).2$. If $-1 \notin \Omega$, then (II) follows as $H_{\overline{\Omega}} = H_\Omega$. The cases where $-1 \in \Omega$ are left as exercises. Observe that (III) is immediate from (II). ■

At this point it is convenient to prove a technical Lemma, which will come in useful at later stages in this chapter.

Lemma 4.1.21. *Assume that case* **O** *holds,* W *is totally singular,* $g \in H_I = N_I(W)$, g *centralizes* $W^\perp/W$ *and* $\langle \det_W(g) \rangle = \mathbf{F}^*$. *Then* $\langle \ddot{g} \rangle = \ker_{\bar{I}}(\ddot{\gamma})$.

Proof. Using the notation of Lemma 4.1.12 we see that $g \in CK$, since g centralizes $W^\perp/W$. Therefore $\ddot{g} \in \ddot{C}\ddot{K} = \ddot{K}$ (note $\ddot{C} = 1$ by Lemma 4.1.12.iii). Therefore $\ddot{g} \in \ker_{\bar{I}}(\ddot{\gamma})$ by (4.1.10). When q is even there is nothing to prove, as $\ker_{\bar{I}}(\ddot{\gamma}) = 1$. When q is odd, it follows from Lemma 4.1.9.iii and the fact that $\det_W(g)$ is a non-square that $g \notin \Omega$, and hence $\Omega\langle g \rangle = S$. We may now deduce $\langle \ddot{g} \rangle = \ddot{S} = \ker_{\bar{I}}(\ddot{\gamma})$, as desired. ■

In general, the stabilizers of totally singular spaces in the classical geometries are maximal parabolic subgroups. However, there are some exceptions. For example, we have seen that in case $\mathbf{O}^+$, each group of type $P_{n/2-1}$ in $\mathcal{C}_1(\Omega)$ is in fact contained in two groups of type $P_{n/2}$ (corresponding to the two families of totally singular $n/2$-spaces). Thus groups of type $P_{n/2-1}$ are sometimes non-maximal. Yet if G contains an element interchanging these two families (that is, G does not lie in $\ker(\gamma)$) then groups of type $P_{n/2-1}$ in $\mathcal{C}_1(G)$ are indeed maximal. This is an instance of the 'novelty' phenomenon described in §3.4. A similar situation arises in case **L**. Here we consider groups of type $P_{m,n-m}$, which are stabilizers of pairs (W, U), with $W < U$, and $\dim(W) = m = n - \dim(U)$. Evidently, if G lies in Γ or $\overline{\Gamma}$, then groups of type $P_{m,n-m}$ in $\mathcal{C}_1(G)$ are non-maximal, for they are contained in groups of type P_m (and of type P_{n-m}). Yet when G contains an element in the coset of the inverse transpose automorphism, G contains elements which interchange the stabilizers of m-spaces with the stabilizers of $(n-m)$-spaces, and so the groups of type $P_{m,n-m}$ emerge as maximal subgroups (novelties, once again). At this stage we state the structure result for the groups of type $P_{m,n-m}$ in $\mathcal{C}_1$ when case **L** holds. We use the notation introduced in (4.1.3).

Proposition 4.1.22. *Assume that case* **L** *holds and that* H *is of type* $P_{m,n-m}$, *so that* $H = N_\Delta(U, W)$ *with* $W < U$.

(I) $c = 1$.

(II) $H_{\overline{\Omega}} \cong \left[q^{2mn-3m^2}\right] : \left[a^+_{m,m,n-2m}/(q-1,n)\right] . (L_m(q)^2 \times L_{n-2m}(q)).[b^+_{m,m,n-2m}]$.

(III) $H_{\overline{\Omega}}$ *is p-local.*

Proof. Left to the reader. ■

We now treat the portion of Theorem 3.1.2 dealing with groups of type $P_{m,n-m}$.

Proposition 4.1.23. *Assume that* H *is of type* $P_{m,n-m}$, *so that* $H = N_\Delta(W, U)$ *with* $W < U$. *Then* $H_\Gamma = N_\Gamma(H_\Omega)$, *and hence Theorem 3.1.2 holds.*

Proof. As in the proof of Proposition 4.1.16, we see that W and U are respectively the unique H_Ω-invariant m-space and $n-m$-space. Therefore $N_\Gamma(H_\Omega) \leq N_\Gamma(W, U) = H_\Gamma$, as required. In addition, Theorem 3.1.2 holds in view of Proposition 4.0.1. ■

§4.2 The imprimitive subgroups $\mathcal{C}_2$

The members of $\mathcal{C}_2$ are the stabilizers of suitable subspace decompositions of V. By a *subspace decomposition of* V we mean a set of non-zero subspaces $V_1, \ldots, V_t$ of V with

$t \geq 2$ such that

$$V = V_1 \oplus \cdots \oplus V_t. \tag{4.2.1}$$

Often we write $\mathcal{D} = \{V_1, \ldots, V_t\}$. The stabilizer in G of such a subspace decomposition is the group $N_G\{V_1, \ldots, V_t\}$, which is the subgroup of G which permutes the spaces V_i amongst themselves. When the symbol $\mathcal{D}$ is in use, recall the notation introduced in (2.10.5). If the spaces V_i in the subspace decomposition $\mathcal{D}$ all have the same dimension m, say, then $\mathcal{D}$ is called an *m-space decomposition* or simply an *m-decomposition.* If the spaces V_i are all non-degenerate and V_i is orthogonal to V_j for all $i \neq j$, then $\mathcal{D}$ is said to be *non-degenerate.* If the spaces V_i are all totally singular, then $\mathcal{D}$ will be called totally singular. If the spaces V_i are all isometric, then $\mathcal{D}$ will be called isometric. Note for example that if $\mathcal{D}$ is a non-degenerate m-space decomposition in case **S**, then $\mathcal{D}$ is automatically isometric in view of the uniqueness statement in Proposition 2.4.1.

Definition. The members of $\mathcal{C}_2(\Gamma)$ are the stabilizers in Γ of m-space decompositions $\mathcal{D}$ of V as given in (4.2.1) such that the conditions of Table 4.2.A hold.

Table 4.2.A

case	type	description of $\mathcal{D}$	conditions
L	$GL_m(q) \wr S_t$	m-decomposition	
U	$GU_m(q) \wr S_t$	non-degenerate	
S	$Sp_m(q) \wr S_t$	non-degenerate	
O	$O_m^{\xi}(q) \wr S_t$	non-degenerate, $\xi = \mathrm{sgn}(V_i)$	if $m = 1$, then $q = p$
U, S, O$^+$	$GL_{n/2}(q^u).2$	totally singular, $m = n/2$	q odd in case **S**
O$^{\pm}$	$O_{n/2}(q)^2$	non-degenerate, non-isometric, $\frac{1}{2}qn$ odd	

Remarks on the conditions. Our definition of $\mathcal{C}_2$ differs slightly from that in [As$_1$] in that we impose the conditions appearing in the right hand column. We now justify these restrictions. Suppose first that $\mathcal{D}$ is a non-degenerate, isometric 1-decomposition in case **O** with q odd, and assume further that $q = p^f$ with $f > 1$. We may choose $v_i \in V_i$ such that $(v_i, v_i) = \lambda$, for some λ independent of i. Replacing $\mathcal{D}$ by some Δ-conjugate if n is even, or replacing Q by $\lambda^{-1}Q$ if n is odd, we may assume that $\lambda = 1$, so that $\beta = \{v_1, \ldots, v_n\}$ is orthonormal. Let $r_i = r_{v_i}$, and for a permutation $\rho \in S_n$ let $g_\rho \in GL(V, \mathbf{F})$ satisfy $v_i g_\rho = v_{i\rho}$. Then it is easy to see that $I_{\mathcal{D}} = \langle r_i, g_\rho \mid 1 \leq i \leq n, \rho \in S_n \rangle \cong \mathbf{Z}_2 \wr S_n$. Moreover $\Delta_{\mathcal{D}} = I_{\mathcal{D}} \mathbf{F}^*$; to see this, take $g \in \Delta_{\mathcal{D}}$. Since $I_{\mathcal{D}}$ acts as the full symmetric group S_n on $\mathcal{D}$, we can assume that $g \in \Delta_{(\mathcal{D})}$, which means $g = \mathrm{diag}_\beta(\lambda_1, \ldots, \lambda_n)$ with $\lambda_i \in \mathbf{F}^*$. Then $\tau(g) = (v_i g, v_i g) = \lambda_i^2$ for all i, so $\lambda_i = \pm\lambda_j$ for all i, j. So multiplying g by some element of $I_{(\mathcal{D})}$, we can assume that the λ_i are all equal, which is to say $g \in \mathbf{F}^*$. This proves $\Delta_{\mathcal{D}} = I_{\mathcal{D}} \mathbf{F}^*$, as claimed. Now observe that $I = \{g \in GL(V, \mathbf{F}) \mid g_\beta g_\beta^{t} = \mathbf{I}_n\}$, and $X = \{g \in I \mid (g_\beta)_{ij} \in \mathbf{F}_p \text{ for all } i, j \leq n\}$ is a subfield subgroup in $\mathcal{C}_5$ (see §4.5) isomorphic to $O_n^{\xi}(p)$ for some ξ. And since $q > p$, we have $I_{\mathcal{D}} < X < I$. Now let $\langle \nu \rangle = Aut(\mathbf{F})$ and put $\phi = \phi_{\beta}(\nu)$, so that $\phi \in \Gamma$ and $\sigma(\phi) = \nu$. Then $\Gamma_{\mathcal{D}} = \Delta_{\mathcal{D}}\langle \phi \rangle = I_{\mathcal{D}}\mathbf{F}^*\langle \phi \rangle < Y \leq \Gamma$, where $Y = \langle X, \mathbf{F}^*, \phi \rangle$. But clearly ϕ normalizes X, and so $Y^{\infty} \cong \Omega_n^{\xi}(p)$, while $\Gamma^{\infty} \cong \Omega_n^{\epsilon}(q)$. Therefore $Y \neq \Gamma$ and so $G_{\mathcal{D}} < Y \cap G < G$ for all G satisfying $\Omega \leq G \leq \Gamma$. Hence we have the restriction $q = p$ when $m = 1$ in case **O**.

The condition q odd when $\mathcal{D}$ is totally singular in case **S** occurs for the following reason. Suppose that q is even and that case **S** holds with $n = 2m$ and $\kappa = \mathbf{f}$. Let $e_1, \ldots, e_m, f_1, \ldots, f_m$ be a symplectic basis and put $U = \langle e_1, \ldots, e_m \rangle$ and $W = \langle f_1, \ldots, f_m \rangle$. Also let Q be the quadratic form on V defined by $Q(\sum_{i=1}^m \alpha_i e_i + \beta_i f_i) = \sum_{i=1}^m \alpha_i \beta_i$. Then a direct calculation shows that $\mathbf{f}_Q = \mathbf{f}$ and that $N_{\Gamma}\{U, W\} < N_{\Gamma}(Q) < \Gamma$. The subgroup $N_{\Gamma}(Q)$ lies in $\mathcal{C}_8$ (see §4.8).

For the rest of the section we assume that $H \in \mathcal{C}_2(\Delta)$. Thus $H = \Delta_{\mathcal{D}}$ for an m-space decomposition $\mathcal{D}$ as given in (4.2.1) which satisfies the conditions in Table 4.2.A. We begin with a useful Lemma.

Lemma 4.2.1. *Assume that H is of type $GL_m^{\pm}(q) \wr S_t$, $Sp_m(q) \wr S_t$ or $O_m^{\xi}(q) \wr S_t$. Then there is a subgroup $J \leq H_I$ which satisfies:*
(i) *$J \cong S_t$ and J acts naturally on $\mathcal{D}$;*
(ii) *$H = \Delta_{(\mathcal{D})} J$.*

Proof. The spaces V_i are all isometric, and so there are bases $\beta_i = \{v_{1,i}, \ldots, v_{m,i}\}$ of V_i such that the map $v_{r,i} \mapsto v_{r,j}$ induces an isometry from V_i to V_j. The existence of J is now clear. ■

Corollary 4.2.2.
(i) *$H^{\mathcal{D}} \cong S_t$.*
(ii) *Provided H is not of type $O_{n/2}(q)^2$, we have $H_I^{\mathcal{D}} \cong S_t$.*
(iii) *If H is not of type $O_{n/2}(q)^2$ and if $I_{(\mathcal{D})}\Omega = I$, then $H_{\Omega} \cong \Omega_{(\mathcal{D})}.S_t$.*

Proof. Evidently (iii) follows directly from (ii), so we focus on (i) and (ii). In view of Lemma 4.2.1, we need consider only the case where H is of type $GL_{n/2}(q^u).2$ or $O_{n/2}(q)^2$. Suppose first that H is of type $GL_{n/2}(q^u).2$, so that $\mathcal{D}$ is totally singular and κ is non-degenerate. Then we can take $V_1 = \langle e_1, \ldots, e_{n/2} \rangle$ and $V_2 = \langle f_1, \ldots, f_{n/2} \rangle$, where the e_i and f_i form a standard basis of V (see Propositions 2.3.2, 2.4.1, 2.5.3.i). Then in cases **U** and $\mathbf{O}^+$, the map which interchanges e_i and f_i for all i lies in H_I. And in case **S**, the map $e_i \mapsto f_i$, $f_i \mapsto -e_i$ lies in H_I. Now assume that H is of type $O_{n/2}(q)^2$ with q odd, so that $\mathcal{D}$ is non-isometric. Then there are bases $\{v_1, \ldots, v_{n/2}\}$ of V_1 and $\{w_1, \ldots, w_{n/2}\}$ of V_2 such that $(v_i, v_j) = \delta_{ij}$ and $(w_i, w_j) = \mu\delta_{ij}$, where $\langle \mu \rangle = \mathbf{F}^*$. Then the map g defined by $v_i g = w_i$, $w_i g = \mu v_i$ lies in Δ and $\tau(g) = \mu$. The proof is now complete. ■

The next few results take care of the case in which κ is non-degenerate and $\mathcal{D}$ is totally singular. We begin with with a result which is essentially a restatement of Lemma 4.1.9.

Lemma 4.2.3. *Assume that κ is non-degenerate and that $\mathcal{D}$ is a totally singular $\frac{n}{2}$-decomposition. Then $I_{(\mathcal{D})} \cong GL_{n/2}(q^u)$, and $V_1 \cong V_2^{*\alpha}$ as $I_{(\mathcal{D})}$-modules.*

We now prove the structure results for the groups in $\mathcal{C}_2$, with κ non-degenerate and $\mathcal{D}$ totally singular. We begin with the unitary groups.

Proposition 4.2.4. *Assume that case* **U** *holds and H is of type $GL_{n/2}(q^2).2$, so that $\mathcal{D}$ is a totally singular $\frac{n}{2}$-decomposition.*

(I) $c = 1$.

(II) $H_{\overline{\Omega}} \cong \left[\frac{(q-1)(q+1,\frac{n}{2})}{(q+1,n)}\right].L_{n/2}(q^2).\left[\frac{(q^2-1,\frac{n}{2})}{(q+1,\frac{n}{2})}\right].2.$

(III) *$H_{\overline{\Omega}}$ is non-local if and only if*

(a) $q = 2$, or

(b) $q = 3$ *and* $n \not\equiv 0 \pmod 8$.

In these two cases $\mathrm{soc}(H_{\overline{\Omega}}) \cong L_{n/2}(q^2)$.

Proof. Take $g \in I_{(\mathcal{D})}$ and observe that $\det_{V_2}(g) = \det_{V_1}(g)^{-q}$ by Lemma 4.2.3. Consequently $|\det(I_{(\mathcal{D})})| = |(\mathbf{F}^*)^{q-1}| = q+1$, and hence $c = 1$ by Lemma 4.0.3.v. Moreover $I_{(\mathcal{D})}\Omega = I$, and hence by Corollary 4.2.2.iii we have $H_{\overline{\Omega}} = \overline{\Omega}_{(\mathcal{D})}.2$. Clearly $\Omega_{(\mathcal{D})}$ is isomorphic to the subgroup of $GL_{n/2}(q^2)$ consisting of elements whose determinant is a $(q-1)^{\text{th}}$ root of unity. Thus

$$\Omega_{(\mathcal{D})} \cong \left[(q-1)(q+1,\frac{n}{2})\right].L_{n/2}(q^2).\left[(q^2-1,\frac{n}{2})/(q+1,\frac{n}{2})\right],$$

and so (II) now follows upon factoring out the scalars. To prove (III), assume that $H_{\overline{\Omega}}$ is non-local. Then $(q+1,\frac{n}{2})(q-1) = (q+1,n)$, which forces (a) or (b) to hold. Conversely, if (a) or (b) holds, then $\mathrm{soc}(H_{\overline{\Omega}}) = L_{n/2}(q^2)$ by Lemma 4.0.5.iv. ■

Next we come to the symplectic groups, and as usual we leave the proof to the reader.

Proposition 4.2.5. *Assume that case* **S** *holds and that H is of type $GL_{n/2}(q).2$, so that $\mathcal{D}$ is a totally singular $\frac{n}{2}$-decomposition with q odd.*

(I) $c = 1$.

(II) $H_{\overline{\Omega}} \cong \frac{(q-1)}{2}.PGL_{n/2}(q).2$.

(III) *$H_{\overline{\Omega}}$ is non-local if and only if $q = 3$ and $n \geq 6$, in which case* $\mathrm{soc}(H_{\overline{\Omega}}) \cong L_{n/2}(3)$.

Before continuing with the structure results, we contribute to the proof of Theorem 3.1.2.

Proposition 4.2.6. *Assume that κ is non-degenerate and that $\mathcal{D}$ is totally singular, so that H is of type $GL_{n/2}(q^u).2$. Then Theorem 3.1.2 holds.*

Proof. Assume first that $SL_{n/2}(q^u)$ is perfect. Then according to Lemma 4.2.3, $H_{\Omega}^{\infty} \cong SL_{n/2}(q^u)$. Moreover, it follows from the Remark after Proposition 4.1.17 and Lemmas 2.10.15 and 4.2.3 that V_1 and V_2 are non-isomorphic as H_{Ω}^{∞}-modules. Therefore by Lemma 2.10.11 we have $N_\Gamma(H_\Omega) \leq N_\Gamma(H_\Omega^\infty) \leq N_\Gamma\{V_1, V_2\} = H_\Gamma$. Now use Proposition 4.0.1.i. If $SL_{n/2}(q^u)$ is not perfect, then $\Omega \approx Sp_4(3)$ and $SL_{n/2}(q) \cong SL_2(3)$. The map g which negates V_1 and centralizes V_2 lies in H and satisfies $\tau(g) = -1$. Therefore $H\Omega = \Delta = \Gamma$, and so once again we can appeal to Proposition 4.0.1.i. ■

Proposition 4.2.7. *Assume that case* $\mathbf{O}^+$ *holds and that* H *is of type* $GL_{n/2}(q).2$, *so that* $\mathcal{D}$ *is a totally singular* $\frac{n}{2}$*-decomposition.*
(I) $c = (\frac{n}{2}, 2)$, *and if* $c = 2$ *then* $\pi = \pi_2$.
(II)

$$H_{\overline{\Omega}} \cong \begin{cases} GL_{n/2}(q).(\frac{n}{2}, 2) & q \text{ even} \\ ((q-1)/2).L_{n/2}(q).(q-1, \frac{n}{2}) & qn/2 \text{ odd, } (q-1)/2 \text{ odd} \\ ((q-1)/4).L_{n/2}(q).(q-1, \frac{n}{2}) & qn/2 \text{ odd, } (q-1)/2 \text{ even} \\ ((q-1)/2).L_{n/2}(q).(\frac{1}{2}(q-1), \frac{n}{4}).2 & q \text{ odd, } n/2 \text{ even} \end{cases}$$

(III) $H_{\overline{\Omega}}$ *is non-local if and only if either*
(a) $q \leq 3$ *or*
(b) $q = 5$ *and* $\frac{n}{2}$ *is odd.*
In these cases $\mathrm{soc}(H_{\overline{\Omega}}) \cong L_{n/2}(q)$.

Proof. It follows from Proposition 4.2.6 and Lemma 4.0.3.i,ix that

$$c = |\ddot{\Gamma} : \ddot{\Gamma}_{\mathcal{D}}|. \tag{4.2.3}$$

We may choose $e_i \in V_1$ and $f_j \in V_2$ such that $\beta = \{e_1, \ldots, e_m, f_1, \ldots, f_m\}$ is a standard basis of V. Define $\phi = \phi_\beta(\nu)$ and $\delta = \delta_{Q,\beta}(\mu)$ as in §2.7, where (recall) $\langle\nu\rangle = Aut(\mathbf{F})$ and $\langle\mu\rangle = \mathbf{F}^*$. Also choose $x \in I_{(\mathcal{D})}$ satisfying $\det_{V_1}(x) = \mu$ and let $y \in H_I$ be the element which interchanges e_i and f_i for all i, so that y interchanges V_1 and V_2. Thus if $K = I_{(\mathcal{D})}^\infty$ (a group which acts as $SL(V_1)$ on V_1), then $\Gamma_{\mathcal{D}} = \langle K, x, y, \delta, \phi\rangle$, and so

$$\ddot{\Gamma}_{\mathcal{D}} = \langle \ddot{x}, \ddot{y}, \ddot{\delta}, \ddot{\phi}\rangle. \tag{4.2.4}$$

We now argue that

$$\ddot{\Gamma}_{(\mathcal{D})} = \ker_{\ddot{\Gamma}}(\ddot{\gamma}), \tag{4.2.5}$$

where $\ddot{\gamma}$ is the homomorphism described in §2.7. Certainly $\ddot{\Gamma}_{(\mathcal{D})} \leq \ker_{\ddot{\Gamma}}(\ddot{\gamma})$, and the reverse inclusion is essentially a direct consequence of Proposition 2.7.4.iii. One need only observe that if q is odd and D is a square, then $\ddot{x} = \ddot{r}_\square \ddot{r}_\boxtimes$ (for both x and $r_\square r_\boxtimes$ are elements of $S\backslash\Omega$ with spinor norm a non-square). We are now in a position to prove (I). For when $\frac{n}{2}$ is even, $y \in \ker(\gamma)$, and so by (4.2.5) $\ddot{\Gamma}_{\mathcal{D}} = \ker_{\ddot{\Gamma}}(\ddot{\gamma})$, and hence by (4.2.3) we have $c = 2$ and $\pi = \pi_2$. But when $\frac{n}{2}$ is odd, $y \notin \ker(\gamma)$, which means $\ddot{\Gamma}_{\mathcal{D}} = \ddot{\Gamma}$, and thus $c = 1$.

Observe that if $\frac{n}{2}$ is even, then $y \in \Omega$. On the other hand, if $\frac{n}{2}$ is odd then no element in Ω can interchange V_1 and V_2, and hence

$$H_\Omega = \begin{cases} \Omega_{(\mathcal{D})}.2, & n/2 \text{ even} \\ \Omega_{(\mathcal{D})}, & n/2 \text{ odd.} \end{cases} \tag{4.2.6}$$

And according to Lemma 4.1.9.iii, we have

$$\Omega_{(\mathcal{D})} = \{g \in I_{(\mathcal{D})} \mid \det_{V_1}(g) \in (\mathbf{F}^*)^2\} \cong \tfrac{1}{(2,q-1)} GL_m(q). \tag{4.2.7}$$

Therefore (II) follows from (4.2.6), (4.2.7) and Corollary 2.5.13.ii upon factoring out scalars. Assertion (III) follows from (II) and Lemma 4.0.5.iv. ∎

For the rest of this section we assume that either case **L** holds, or κ and $\mathcal{D}$ are non-degenerate. Thus H is of type $GL_m^{\pm}(q) \wr S_t$, $Sp_m(q) \wr S_t$, $O_m^{\xi}(q) \wr S_t$, or (in the case where $\mathcal{D}$ is non-isometric) $O_{n/2}(q)^2$. We use notation similar to that introduced in §4.1. Namely, we write

$$X_i = X(V_i, \kappa), \tag{4.2.8}$$

where X ranges over the symbols Ω, S, I, Δ, and it is to be understood that X_i acts naturally on V_i, while centralizing $\bigoplus_{j \neq i} V_j$. Furthermore, we define $\tau_i = \tau_{(V_i,\kappa)}$ for all i. Here are some useful observations.

Lemma 4.2.8. *Suppose that $\mathcal{D}$ is non-degenerate.*
(i) $I_{(\mathcal{D})} = I_1 \times \cdots \times I_t$.
(ii) *If $\tau_i(\Delta_i) = \tau(\Delta)$ for all i then*
  (a) $HI = \Delta$
  (b) $H_I^{\mathcal{D}} \cong H^{\mathcal{D}} \cong S_t$.
(iii) *If $\mathcal{D}$ is isometric, then $H_I = I_{(\mathcal{D})} J \cong I_1 \wr S_t$, with J as in Lemma 4.2.1.*

Proof. Assertion (i) is obvious and (iii) is immediate from (i) and Lemma 4.2.1. To prove (ii), write $\tau(\Delta) = \langle \lambda \rangle$ (for some $\lambda \in \mathbf{F}^*$) and take $g_i \in \tau_i^{-1}(\lambda)$ for $1 \leq i \leq t$. Then $g = g_1 \ldots g_t \in \Delta_{(\mathcal{D})}$ and $\tau(g) = \lambda$. Thus $\langle g \rangle I = \Delta$, and it follows that $HI = \Delta$ and $H_I \Delta_{(\mathcal{D})} = H$. Therefore (a) and (b) are now clear. ∎

Proposition 4.2.9. *Assume that case $\mathbf{L}^{\epsilon}$ holds, and that H is of type $GL_m^{\epsilon}(q) \wr S_t$.*
(I) $c = 1$.
(II) $H_{\overline{\Omega}} \cong \left[\dfrac{(q-\epsilon)^{t-1}(q-\epsilon,m)}{(q-\epsilon,n)}\right].L_m^{\epsilon}(q)^t.[(q-\epsilon,m)^{t-1}].S_t$.
(III) *$H_{\overline{\Omega}}$ is non-local if and only if $\epsilon = +$ and one of the following holds:*
  (a) *$q = 2$, $m = 1$ and $n \geq 5$, in which case $\mathrm{soc}(H_{\overline{\Omega}}) \cong A_n$;*
  (b) *$q = 2$ and $m \geq 3$, in which case $\mathrm{soc}(H_{\overline{\Omega}}) \cong L_m(2)^t$;*
  (c) *$q = 3$, m is odd and $t = 2$, in which case $\mathrm{soc}(H_{\overline{\Omega}}) \cong L_{n/2}(3)^2$.*

Proof. First of all, $|\det(I_{(\mathcal{D})})| = q - \epsilon$, and so $I_{(\mathcal{D})}\Omega = I$. Therefore (I) holds by Lemma 4.0.3.v, and also $H_{\Omega}^{\mathcal{D}} = H_I^{\mathcal{D}} \cong S_t$. Thus

$$H_{\Omega} \cong \Omega_{(\mathcal{D})}.S_t. \tag{4.2.9}$$

Now define C as the subgroup of $I_{(\mathcal{D})}$ acting as scalars on each V_i. Then $I_{(\mathcal{D})}/C \cong PGL_m^{\epsilon}(q)^t$ and it is clear that $L_m^{\epsilon}(q)^t \trianglelefteq \Omega_{(\mathcal{D})}/(C \cap \Omega) \trianglelefteq PGL_m^{\epsilon}(q)^t$. Consequently

$$\Omega_{(\mathcal{D})} \cong (C \cap \Omega).L_m^{\epsilon}(q)^t.[a], \tag{4.2.10}$$

for some divisor a of $(q-\epsilon,m)^t$. Evidently $|C| = (q-\epsilon)^t$ and $|\det(C)| = (q-\epsilon)/(q-\epsilon,m)$, which means $|C \cap \Omega| = (q-\epsilon)^{t-1}(q-\epsilon,m)$. Since $|C \cap \Omega|$ is known and since $|\Omega_{(\mathcal{D})}| = |I_{(\mathcal{D})}|/(q-\epsilon) = |GL^{\epsilon}_m(q)^t|/(q-\epsilon)$, we may use (4.2.10) to solve for a, and we obtain $a = (q-\epsilon,m)^{t-1}$. Thus (II) now follows from (4.2.9) and (4.2.10) upon factoring out the scalars.

To prove (III) we assume that $H_{\overline{\Omega}}$ is non-local. Then $(q-\epsilon)^{t-1}(q-\epsilon,m) = (q-\epsilon,n)$, and hence $\epsilon = +$ and $q \leq 3$. First take $q = 2$; then we cannot have $m = 2$, as $L_2(2)$ is local. Thus either $m = 1$ or $m \geq 3$. If $m = 1$ then $H_{\overline{\Omega}} \cong S_n$, and so $n \geq 5$ and $\mathrm{soc}(H_{\overline{\Omega}}) \cong A_n$, giving (a). If $m \geq 3$, then $K = SL(V_1) \times \cdots \times SL(V_t) \cong L_m(2)^t$ and K satisfies the conditions of Lemma 4.0.5.iv. Thus $\mathrm{soc}(H_{\overline{\Omega}}) = \overline{K} = K$, giving (b). Now take $q = 3$. Then $2^{t-1}(2,m) = (2,n)$, which forces $t = 2$ and m odd. This time $K = SL(V_1) \times SL(V_2) \cong L_m(3)^2$ and this time we obtain (c). ∎

Proposition 4.2.10. *Assume that case* **S** *holds and that H is of type $Sp_m(q) \wr S_t$.*

(I) $c = 1$.

(II) $H_{\overline{\Omega}} \cong (2,q-1)^{t-1}.(PSp_m(q) \wr S_t)$.

(III) *If $H_{\overline{\Omega}}$ is non-local if and only if q is even and $(q,m) \neq (2,2)$, and if this occurs then $\mathrm{soc}(H_{\overline{\Omega}}) \cong (Sp_m(q)')^t$.*

We are now left with the orthogonal groups, and to analyse them we write $\mathrm{sgn}(Q_{V_i}) = \xi$ for all i, so that

$$I_i = I(V_i, Q) \approx O^{\xi}_m(q).$$

Furthermore, put $D_i = D(Q_{V_i})$, and note that Lemma 4.1.5 yields

$$\ddot{I}_{(\mathcal{D})} = \begin{cases} \ddot{I} & \text{if } m \geq 2 \\ \langle \ddot{r}_{v_1} \rangle & \text{if } m = 1 \text{ and } V_1 = \langle v_1 \rangle. \end{cases} \tag{4.2.11}$$

Proposition 4.2.11. *Assume that case* $\mathbf{O}^{\epsilon}$ *holds, H is of type $O^{\xi}_m(q) \wr S_t$, and that m is even, so that $\epsilon, \xi \in \{+,-\}$.*

(I) $c = 1$.

(II) *We have $\epsilon = \xi^t$ and*

$$H_{\overline{\Omega}} \cong \begin{cases} \Omega^{\xi}_m(q)^t.2^{t-1}.S_t & q \text{ even} \\ 2^{t-1}.P\Omega^{\xi}_m(q)^t.2^{2(t-1)}.S_t & q \text{ odd, } D_i = \square \\ (2^{t-1} \times \Omega^{\xi}_m(q)^t.2^{t-1}).S_t & q \text{ odd, } D = \boxtimes \\ (2^{t-2} \times \Omega^{\xi}_m(q)^t.2^{t-1}).S_t & \text{otherwise.} \end{cases}$$

(III) *$H_{\overline{\Omega}}$ is local if and only if one of the following holds:*
   (a) *$\Omega^{\xi}_m(q)$ is local (see Proposition 2.9.3);*
   (b) *$O^{\xi}_m(q) \approx O^{+}_2(q)$ with $q \leq 3$;*
   (c) *q is odd and $t \geq 3$.*

If $H_{\overline{\Omega}}$ is non-local then $\mathrm{soc}(H_{\overline{\Omega}}) \cong \Omega^{\xi}_m(q)^t$.

Proof. (I). Since each τ_i is surjective (see the definition of δ just before (2.7.2) and (2.8.2)) we know that $HI = \Delta$, by Lemma 4.2.8. Moreover $H_I\Omega = I$ by (4.2.11), and hence $H\Omega = \Delta$. Thus $\ddot{H} = \ddot{\Delta}$, and so (I) holds by Lemma 4.0.3.i.

(II). First notice that $\epsilon = \xi^m$ by Proposition 2.5.11.ii. We now determine the structure of $H_{\overline{\Omega}}$. Evidently $H_I^{\mathcal{D}} \cong S_t$ by Corollary 4.2.2.ii. The fact that $I_{(\mathcal{D})}\Omega = I$ implies that $H_{\Omega}^{\mathcal{D}} = H_I^{\mathcal{D}}$, and hence $H_{\Omega} = \Omega_{(\mathcal{D})}.S_t$. Thus we have

$$H_{\overline{\Omega}} \cong \overline{\Omega}_{(\mathcal{D})}.S_t. \tag{4.2.12}$$

Hence it remains to determine the structure of $\overline{\Omega}_{(\mathcal{D})}$. Observe that $I_{(\mathcal{D})}/\Omega_{(\mathcal{D})} \cong I/\Omega \cong 2^d$, where $d = (2, q-1)$. And since $I_{(\mathcal{D})} \cong O^{\xi}_m(q)^t \cong \Omega^{\xi}_m(q)^t.2^{dt}$, we now know that $\Omega_{(\mathcal{D})} \cong \Omega^{\xi}_m(q)^t.2^{d(t-1)}$. Thus when q is even it is clear from (4.2.12) that (II) holds. We may therefore assume for the rest of this paragraph that q is odd. If $D(Q_{V_i}) = \square$, then $-1_{V_i} \in \Omega_i$, and hence $\Omega_{(\mathcal{D})} \cong 2^t.P\Omega^{\xi}_m(q)^t.2^{2(t-1)}$ and -1 lies in the normal 2^t. Therefore $H_{\overline{\Omega}}$ is as given in (II). The other cases are left to the reader.

(III). It is easy to check that if (a), (b) or (c) holds then $H_{\overline{\Omega}}$ is local. Conversely, suppose that (a), (b) and (c) fail. If q is odd, then we must have $t = 2$, and so D is a square. And since $\Omega^{\xi}_m(q)$ is non-local, D_i is a non-square. Therefore $-1_{V_i} \notin \Omega$, and so Lemma 4.0.5.iv applies in view of Propositions 2.9.2 and 2.10.6. If q is even, then Ω_i is not $\Omega^{\pm}_2(q)$ or $\Omega^{+}_4(2)$, and hence Ω_i is semisimple by Proposition 2.9.2. Therefore Lemma 4.0.5.iv applies once again. ■

At this stage it is convenient to prove the relevant part of Theorem 3.1.2.

Proposition 4.2.12. *Assume that H is of type $GL^{\pm}_m(q) \wr S_t$, $Sp_m(q) \wr S_t$, $O^{\xi}_m(q) \wr S_t$ or $O_{n/2}(q)^2$.*
(ii) *Theorem 3.1.2 holds.*
(ii) *In case **O** with m odd we have $H_{\Gamma} = N_{\Gamma}(H_{\Omega})$.*

The proof of Proposition 4.2.12 requires a preliminary Lemma. For a vector $v = v_1 + \cdots + v_t$, with $v_i \in V_i$, we define the *$\mathcal{D}$-length of v* to be the number of non-zero vectors v_i. For a subgroup K of Δ, write $\mathrm{orb}_K(v)$ for the orbit of K on V containing v.

Lemma 4.2.13. *Assume that H is of type $GL^{\pm}_m(q) \wr S_t$, $Sp_m(q) \wr S_t$ or $O^{\xi}_m(q) \wr S_t$. Let $v \in V$ have $\mathcal{D}$-length k, and write a and b for the sizes of smallest orbits of I_1 and Ω_1 on $V_1 \backslash 0$, respectively. Then*

$$|\mathrm{orb}_{H_{\Omega}}(v)| \geq \max\left\{ \binom{t}{k} b^k,\ \binom{t}{k} a^{k-1} \right\}.$$

Proof. Since $\Omega_1 \times \cdots \times \Omega_t \leq \Omega_{(\mathcal{D})}$ by Lemma 4.1.1, it is clear that $|\mathrm{orb}_{\Omega_{(\mathcal{D})}}(v)| \geq b^k$. As the groups I_i are all conjugate in I, it follows that $I_i\Omega = I_j\Omega$ for all i, j. Thus for all elements $g_1 \ldots g_{t-1} \in I_1 \ldots I_{t-1}$ (with $g_i \in I_i$), there exists $g_t \in I_t$ such that $g_1 \ldots g_t \in \Omega$. Consequently $|\mathrm{orb}_{\Omega_{(\mathcal{D})}}(v)| \geq a^{k-1}$. Since $Alt(\mathcal{D}) \leq H_{\Omega}^{\mathcal{D}}$, we know that H_{Ω} is transitive on the set of k-subsets of $\mathcal{D}$, which means $|\mathrm{orb}_{H_\Omega}(v)| \geq \binom{t}{k}|\mathrm{orb}_{\Omega_{(\mathcal{D})}}(v)|$. The Lemma now follows. ∎

We are now ready for the

Proof of Proposition 4.2.12. First assume that case **U** holds. Let β be an orthonormal basis of V which is the union of orthonormal bases for each V_i. Then $\phi_\beta(\nu) \in \Gamma_{(\mathcal{D})}$ and $\Delta\langle\phi\rangle = \Gamma$. We saw in the proof of Proposition 4.2.9 that $H_I\Omega = I$, and since $\Delta = I\mathbf{F}^*$ (see (2.3.3)), it follows that $H\Omega = \Delta$. We now conclude $H_\Gamma\Omega = \Gamma$. Next assume that case **O** holds with m even. In the proof of Proposition 4.2.11 we saw that $H\Omega = \Delta$. And if $\phi_i \in \Gamma(V_i, \kappa)$ satisfies $\sigma_{(V_i,\kappa)}(\phi_i) = \nu$, then $\phi = \phi_1 \dots \phi_t \in \Gamma_{(\mathcal{D})}$ and $\sigma(\phi) = \nu$. Once again we have $H_\Gamma\Omega = \Gamma$. Similarly, one proves that $H_\Gamma\Omega = \Gamma$ in case **S**. Therefore, in view of Proposition 4.0.1.i, Theorem 3.1.2 holds in cases **U** and **S**, and in case **O** with m even.

For the rest of this proof we assume that either case **L** holds, or that case **O** holds with m odd.

Suppose first that Ω_i is insoluble. Then by Proposition 2.9.2, $K := \Omega_1' \times \cdots \times \Omega_t'$ is a central product of quasisimple groups. Moreover $K = \Omega_{(\mathcal{D})}^\infty$, and so $K \trianglelefteq H_\Omega$. Therefore K is generated by components of H_Ω. Suppose for a contradiction that H_Ω contains a component not contained in K. Then as $\Omega_{(\mathcal{D})}/K$ is soluble, this component must induce non-trivial permutations on $\mathcal{D}$, and hence on the groups Ω_i. But this violates [As8, 31.5], and hence K is the group generated by the components of H_Ω. In particular, K is characteristic in H_Ω, whence $N_\Gamma(H_\Omega) \leq N_\Gamma(K)$. Now $V_1, \dots, V_t$ are mutually non-isomorphic irreducible K-submodules of V, and hence by Lemma 2.10.11.i, $N_\Gamma(K) \leq \Gamma_{\mathcal{D}} = H_\Gamma$. The result follows by Proposition 4.0.1.

This leaves us with the case where Ω_i is soluble. First consider case **O** (with m odd). By Proposition 2.9.2.i, $\Omega_i \approx \Omega_1(p)$ or $\Omega_3(3)$. And as $n \geq 7$ (recall (4.0.2)), H is of type $O_1(p) \wr S_n$ or $O_3(3) \wr S_{n/3}$.

Let $x \in N_\Gamma(H_\Omega)$ and let $W_i = V_i x$, so that $H_\Omega = \Omega_{\mathcal{D}'}$, where $\mathcal{D}' = \{W_1, \dots, W_t\}$. The remainder of the proof in case **O** is devoted to showing that $\mathcal{D}' = \mathcal{D}$. It will then follow that $x \in \Gamma_{\mathcal{D}} = H_\Gamma$, as required.

Let a and b be as defined in Lemma 4.2.13 for $\mathcal{D}$. Thus $(a, b) = (2, 1)$ or $(6, 4)$ according to whether H is of type $O_1(p) \wr S_n$ or $O_3(3) \wr S_{n/3}$ (see Lemma 2.10.5). Note that a and b are also the appropriate parameters for $\mathcal{D}'$. Choose v lying in the smallest orbit of I_1 on $V_1 \backslash 0$. Then clearly $|\mathrm{orb}_{H_\Omega}(v)| \leq ta$, and so if k is the $\mathcal{D}'$-length of v, Lemma 4.2.13 yields

$$ta \geq |\mathrm{orb}_{H_\Omega}(v)| \geq \max\left\{\binom{t}{k}b^k,\ \binom{t}{k}a^{k-1}\right\}.$$

Thus if H is of type $O_1(p) \wr S_n$, we have $2n \geq \binom{n}{k}2^{k-1}$, which forces $k = 1$. Consequently $V_1 \in \mathcal{D}'$, and similarly $V_i \in \mathcal{D}'$ for all i. Thus $\mathcal{D} = \mathcal{D}'$, as desired. If H is of type $O_3(3) \wr S_{n/3}$, then $6t \geq \binom{t}{k}4^k$, which forces $k = 1$ again. Relabelling the W_i if necessary, we may write $v \in (V_1 \cap W_1)\backslash 0$. It follows that $N := N_H(v)$ acts reducibly on V_1 and

W_1. Since $\Omega_2 \times \cdots \times \Omega_t \leq N$, all irreducible constituents of N on V/V_1 have dimension at least 3. However N acts reducibly on the 3-space W_1, hence $W_1 = V_1$. Similarly, each V_i is equal to some W_j, and we may now conclude that $\mathcal{D} = \mathcal{D}'$, finishing the proof in case **O**.

We now handle the case where Ω_i is soluble and case **L** holds. Thus either $m = 1$ or $\Omega_i \approx SL_2(q)$ with $q \leq 3$. If $\Omega_i \approx SL_2(2)$, then $H_\Omega \cong SL_2(2) \wr S_t \cong S_3 \wr S_t$, and so $O_3(H_\Omega) \cong 3^t$. As in the first paragraph of this proof, the spaces V_i are mutually non-isomorphic irreducible $O_3(H_\Omega)$-submodules of V, and hence $N_\Gamma(H_\Omega) \leq H_\Gamma$. Now use Proposition 4.0.1. When $\Omega_i \approx SL_2(3)$, we have $O_2(H_\Omega) \cong Q_8^t$, with each factor Q_8 acting non-trivially on exactly one of the spaces V_i. Thus the same reasoning holds here. For the remainder of the proof we take $m = 1$, so that $H_\Omega \cong (q-1)^{n-1}.S_n$. Assume for the moment that $q = 2$, so that $H_\Omega \cong S_n$ acting naturally on the vectors $v_i \in V_i \backslash 0$. Here $\Gamma = \Omega = GL_n(2)$, and we suppose that there exists $x \in N_\Omega(H_\Omega) \backslash H_\Omega$. We reintroduce the notation W_i appearing earlier in the proof, and let $W_i = \langle w_i \rangle$. Then w_i has $\mathcal{D}$-length k for some $k \geq 2$. Since $\mathrm{orb}_{H_\Omega}(w_i) = \binom{n}{k}$, it follows that $k = n-1$. Therefore $w_i = v_j^*$ for some j, where $v_j^* = \sum_{i \neq j} v_i$. Adjusting by a suitable element of H_Ω we can take x to be the map $v_i \mapsto v_i^*$ for all i. It follows that n must be even (otherwise x is singular) and $N_\Omega(H_\Omega) = H_\Omega \times \langle x \rangle \cong H_\Omega \times 2$. This yields part (c) of Theorem 3.1.2.ii. Finally, we take $q \geq 3$. As $\Omega \not\approx SL_2(3)$, we know that the V_i are mutually non-isomorphic modules for $\Omega_{(\mathcal{D})} \cong (q-1)^{n-1}$. Therefore the argument employed earlier in the proof goes through provided $\Omega_{(\mathcal{D})}$ is characteristic in H_Ω. This is certainly the case when $n \geq 5$, for here $\Omega_{(\mathcal{D})}$ is the soluble radical of H_Ω. Easy calculations also show that $\Omega_{(\mathcal{D})}$ is characteristic when $(q, n) \notin \{(5,2), (4,3), (3,4)\}$. These three exceptional cases lead to part (d) of Theorem 3.1.2.ii. This completes the proof. ■

We resume the structure results.

Proposition 4.2.14. *Assume that case* $\mathbf{O}^\epsilon$ *holds and* H *is of type* $O_m(q) \wr S_t$, *with* qm *odd and* $m > 1$.

(O) *If* t *is even, then* $\epsilon = (-)^{(q-1)n/4}$.

(I) $c = (2, t)$ *and if* $c = 2$ *then* $\pi = \pi_1$.

(II) *We have*

$$H_{\bar{\Omega}} \cong \begin{cases} (2^{t-2} \times \Omega_m(q)^t.2^{t-1}).S_t & t \text{ even} \\ (2^{t-1} \times \Omega_m(q)^t.2^{t-1}).S_t & t \text{ odd.} \end{cases}$$

(III) $H_{\bar{\Omega}}$ *is non-local if and only if* $t = 2$, *and in this case* $\mathrm{soc}(H_{\bar{\Omega}}) \cong \Omega_{n/2}(q)^2$.

Proof. First observe that part (O) follows directly from Propositions 2.5.11 and 2.5.13.

We now prove (I), (II) and (III). Here $\tau_i(\Delta_i) = (\mathbf{F}^*)^2$ by (2.6.3), and hence $\tau(\Delta_{(\mathcal{D})}) = (\mathbf{F}^*)^2$. Consequently by Lemma 4.2.1.ii, $\tau(H) = (\mathbf{F}^*)^2$, and so $H = H_I\mathbf{F}^*$, which means $\ddot{H} = \ddot{H}_I$. However

$$\ddot{H}_I = \ddot{I} \tag{4.2.13}$$

by Lemma 4.1.5, and it follows from Table 2.1.D that $|\ddot{\Delta} : \ddot{I}| = (2, n) = (2, t)$. Combining this information with Proposition 4.2.12 and Lemma 4.0.3.i now yields

$$c = |\ddot{\Delta} : \ddot{H}| = |\ddot{\Delta} : \ddot{I}| = (2, n) = (2, t). \tag{4.2.14}$$

If $c = 2$, then as in the proof of Proposition 4.1.6 we see that $\pi = \pi_1$, and so we have established (I).

The argument of the proof of Proposition 4.2.11(II) shows $H_\Omega \cong \overline{\Omega}_{(\mathcal{D})}.S_t$ and $\Omega_{(\mathcal{D})} \cong \Omega_m(q)^t.2^{2(t-1)}$. Since $I_1 = S_1 \times \langle -1_{V_1} \rangle$, we see that $\Omega_{(\mathcal{D})} \cong 2^{t-1} \times \Omega_m(q).2^{t-1}$, where the central 2^{t-1} corresponds to the group $\langle -1_{V_i \perp V_j} \mid 1 \le i, j \le t \rangle$. Therefore we obtain (II) upon factoring out -1 when t is even.

If $t \ge 3$ then it is clear from (II) that $H_{\overline{\Omega}}$ is 2-local. Conversely, if $t = 2$, then $m \ge 5$ (see (4.0.2)) and so Ω_i is simple by Proposition 2.9.2. Therefore (III) holds in view of Lemma 4.0.5.iv and Proposition 2.10.6. ■

In the next Proposition, it will be helpful to recall our extra condition $q = p$ appearing in the right hand column of Table 4.2.A.

Proposition 4.2.15. *Assume that case* $\mathbf{O}^\epsilon$ *holds and* H *is of type* $O_1(q) \wr S_n$, *so that* $q = p \ge 3$.

(O) *We have*

$$\epsilon = \begin{cases} \circ & \text{if } n \text{ is odd} \\ - & \text{if } n \equiv 2 \pmod 4 \text{ and } q \equiv 3 \pmod 4 \\ + & \text{otherwise.} \end{cases}$$

(I)

$$c = \begin{cases} 2(n,2) & \text{if } q \equiv \pm 1 \pmod 8 \\ (n,2) & \text{if } q \equiv \pm 3 \pmod 8 \end{cases}$$

and

$$\pi = \begin{cases} \pi_1 & \text{if } n \text{ is odd and } q \equiv \pm 1 \pmod 8 \\ \pi_1 & \text{if } n \text{ is even and } q \equiv \pm 3 \pmod 8 \\ \pi_4 & \text{if } q \equiv \pm 1 \pmod 8 \text{ and } \epsilon = + \\ \pi_2 & \text{if } q \equiv 7 \pmod 8 \text{ and } \epsilon = - \end{cases}$$

(II) *We have*

$$H_{\overline{\Omega}} \cong \begin{cases} 2^{n-1}.A_n & \text{if } n \text{ is odd and } q \equiv \pm 3 \pmod 8 \\ 2^{n-2}.A_n & \text{if } n \text{ is even and } q \equiv \pm 3 \pmod 8 \\ 2^{n-1}.S_n & \text{if } n \text{ is odd and } q \equiv \pm 1 \pmod 8 \\ 2^{n-2}.S_n & \text{if } n \text{ is even and } q \equiv \pm 1 \pmod 8 \end{cases}$$

(III) $H_{\overline{\Omega}}$ *is 2-local.*

Proof. Reasoning as we did just after Table 4.2.A, we may assume that there exist $v_i \in V_i$ such that $\{v_1, \dots, v_n\}$ is an orthonormal basis. In particular, D is a square when n is even, and so (O) is immediate from Proposition 2.5.10. Putting $r_i = r_{v_i}$, we have

$$\begin{aligned} I_{(\mathcal{D})} &= \langle r_i \mid 1 \le i \le n \rangle \cong 2^n, \\ \Omega_{(\mathcal{D})} &= \langle r_i r_j \mid 1 \le i, j \le n \rangle \cong 2^{n-1}. \end{aligned} \tag{4.2.15}$$

Therefore

$$\overline{\Omega}_{(\mathcal{D})} \cong 2^{n-(2,n)}. \tag{4.2.16}$$

We now argue that

$$H_I = H_\Omega \langle r_1, r_{v_1-v_2} \rangle. \tag{4.2.17}$$

For take $h \in H_I$. Multiplying h by $r_{v_1-v_2}$ if necessary, we can assume that h induces an even permutation on $\mathcal{D}$ (observe that $r_{v_1-v_2}$ interchanges V_1 and V_2). Furthermore $A_n \cong J' \leq I' \leq \Omega$, and so multiplying h by an element of J', we can ensure that $h \in I_{(\mathcal{D})}$. But then either h or hr_1 lies in Ω, and so (4.2.17) is proved.

Reasoning as in the proof of Proposition 4.2.14, but using (4.2.17) instead of (4.2.13), we obtain

$$c = |\ddot{\Delta} : \ddot{H}| = |\ddot{\Delta} : \ddot{H}_I| = |\ddot{\Delta} : \langle \ddot{r}_1, \ddot{r}_{v_1-v_2} \rangle|. \tag{4.2.18}$$

Since the spinor norm of $r_1 r_{v_1-v_2}$ is 2, we know that that $r_1 r_{v_1-v_2}$ lies in Ω if and only if $2 \in (\mathbf{F}^*)^2$. Therefore (4.2.17) implies

$$\ddot{H} = \begin{cases} \ddot{I} & \text{if } p \equiv \pm 3 \pmod 8) \\ \langle \ddot{r}_1 \rangle = \langle \ddot{r}_\Box \rangle & \text{if } p \equiv \pm 1 \pmod 8). \end{cases} \tag{4.2.19}$$

We now consider several cases separately.

Case $p \equiv \pm 3 \pmod 8$. Here (4.2.19) and Table 2.1.D imply $c = |\ddot{\Delta} : \ddot{I}| = (n, 2)$. Moreover, if n is even, then as in the proof of Proposition 4.1.6 we see that $\pi = \pi_1$, and hence (I) holds. Furthermore, since $r_{v_1-v_2} \notin \Omega \langle r_1 \rangle$, it follows that $H_{\overline{\Omega}}$ cannot induce the transposition (V_1, V_2) on $\mathcal{D}$, and hence $H_{\overline{\Omega}}^{\mathcal{D}} \cong A_n$. Consequently (II) holds in view of (4.2.16).

Case $p \equiv \pm 1 \pmod 8$ *and* n *odd.* Here $r_1 r_{v_1-v_2}$ has spinor norm 2, which is a square, and so $r_1 r_{v_1-v_2} \in H_\Omega$, which shows $H_{\overline{\Omega}}^{\mathcal{D}} \cong S_n$. Therefore (II) holds by (4.2.16). Now $-r_1 = r_2 \ldots r_n \in \Omega$, and so $\ddot{r}_1 = 1$. Consequently $\ddot{H} = 1$, and so $c = |\ddot{\Delta}| = 2$. Obviously $\pi = \pi_1$ since $|\ddot{\Gamma}| = 2$.

Case $p \equiv \pm 1 \pmod 8$ *and* n *even.* As in the previous case we see that (II) holds. This time, however, $\ddot{\Delta} \cong D_8$ as $D = \Box$, and by (4.2.19) we have $\ddot{H} = \langle \ddot{r}_\Box \rangle \cong \mathbf{Z}_2$. Thus $c = 4$, and it is clear from Table 3.5.G that $\pi = \pi_4$ if $\epsilon = +$ and $\pi = \pi_2$ if $\epsilon = -$. ∎

Proposition 4.2.16. *Assume that case* $\mathbf{O}^\epsilon$ *holds and that* H *is of type* $O_{n/2}(q)^2$, *so that* $\mathcal{D}$ *is non-degenerate and non-isometric,* qm *is odd and* $m = \frac{n}{2} \geq 5$.

(O) *We have* $\epsilon = (-)^{(q+1)/2}$.

(I) $c = 1$.

(II) $H_{\overline{\Omega}} \cong SO_m(q) \times SO_m(q)$.

(III) $H_{\overline{\Omega}}$ *is non-local and* $\mathrm{soc}(H_{\overline{\Omega}}) \cong \Omega_m(q)^2$.

Proof. Here V_1 is non-isometric to V_2, and hence D is a non-square. Thus (O) follows from Proposition 2.5.10. Since J interchanges V_1 and V_2, we know that $\tau(J) \nleq (\mathbf{F}^*)^2$, and hence $HI = \Delta$, which means $\ddot{H}\ddot{I} = \ddot{\Delta}$. And because $m \geq 5$ we have $\ddot{I}_{(\mathcal{D})} = \ddot{I}$ by Lemma 4.1.5. Consequently $\ddot{H} = \ddot{\Delta}$, and so (I) holds by Lemma 4.0.3.i. Moreover

$H_I = I_1 \times I_2 \cong \Omega_m(q).2^2 \times \Omega_m(q).2^2$. And since $|H_I : H_\Omega| = |I : \Omega| = 4$, we deduce $H_\Omega \cong (\Omega_m(q) \times \Omega_m(q)).[4]$. Take elements $s_i \in S_i \backslash \Omega_i$ for $i = 1, 2$. Then $s_i \in S \backslash \Omega$, and since $-1 \in S \backslash \Omega$ (see Proposition 2.5.13), we have $-s_i \in \Omega$. Therefore $H_{\overline{\Omega}} = H_\Omega = \Omega_1\langle -s_1\rangle \times \Omega_2\langle -s_2\rangle \cong SO_m(q) \times SO_m(q)$, and so (II) holds. Assertion (III) is immediate from (II) and Lemma 4.0.5.iv. ■

We conclude this section by recording the following fact, which will be put to use in the proof of Proposition 7.2.5. The Proposition may be verified by extracting the appropriate portions of the proofs of the results throughout this chapter.

Proposition 4.2.17. *Assume that Ω_i is insoluble, and set $L = \Omega_1' \times \cdots \times \Omega_t'$, a central product of quasisimple groups. Then H_Ω is transitive on the components of L, except when H is of type $O_{n/2}(q)^2$, as in Proposition 4.2.16.*

§4.3 The field extension subgroups $\mathcal{C}_3$

Let $\mathbf{F}_\sharp$ be a field extension of $\mathbf{F}$ of degree $r > 1$, where $r \mid n$. Then V acquires the structure of an $\mathbf{F}_\sharp$-vector space in a natural way; namely, there is an $\mathbf{F}$-vector space isomorphism between V and an m-dimensional vector space over $\mathbf{F}_\sharp$, where $m = \frac{n}{r}$, and so $\mathbf{F}_\sharp$ acts on V via this isomorphism. Thus $\mathbf{F}_\sharp$ embeds in $\mathrm{End}_\mathbf{F}(V)$ and in this situation we usually write $V_\sharp$ or $(V_\sharp, \mathbf{F}_\sharp)$ or $(V, \mathbf{F}_\sharp)$ for V regarded as a vector space over $\mathbf{F}_\sharp$. Hence in case **L** we obtain an embedding

$$GL_m(q^r) \cong GL(V_\sharp, \mathbf{F}_\sharp) \leq GL(V, \mathbf{F}) \cong GL_n(q). \tag{4.3.1}$$

Note that the group $GL(V_\sharp, \mathbf{F}_\sharp)$ is irreducible since it is transitive on non-zero vectors, but it is not absolutely irreducible as $C_{GL(V,\mathbf{F})}(GL(V, \mathbf{F}_\sharp)) = \mathbf{F}_\sharp^*$ (see Lemma 2.10.1).

Suppose now that case **S** holds, so that κ is symplectic. Furthermore, suppose that $\kappa_\sharp$ is a non-degenerate symplectic form on $(V_\sharp, \mathbf{F}_\sharp)$. Writing $\mathrm{T} = \mathrm{T}_\mathbf{F}^{\mathbf{F}_\sharp}$ (the trace map from $\mathbf{F}_\sharp$ to $\mathbf{F}$), we see that $\mathrm{T}\kappa_\sharp$ is a non-degenerate symplectic form on $(V, \mathbf{F})$. To check this, note that $\mathrm{T}(\kappa_\sharp(v, w)) = \mathrm{T}(-\kappa_\sharp(w, v)) = -\mathrm{T}(\kappa_\sharp(w, v))$, which shows that $\mathrm{T}\kappa_\sharp$ is skew-symmetric. One may also verify bilinearity and non-degeneracy quite easily. Consequently

$$Sp_m(q^r) \cong I(V_\sharp, \mathbf{F}_\sharp, \kappa_\sharp) \leq I(V, \mathbf{F}, T\kappa_\sharp) \cong Sp_n(q). \tag{4.3.2}$$

Recall, however, that all symplectic geometries in dimension n over $\mathbf{F}$ are isometric (see Proposition 2.4.1). Thus without loss of generality we may take $\kappa = \mathrm{T}\kappa_\sharp$, and hence we obtain an embedding $I_\sharp = I(V_\sharp, \mathbf{F}_\sharp, \kappa_\sharp) \leq I$.

In a similar fashion we obtain various other embeddings, listed in Table 4.3.A, of an isometry group $I_\sharp = I(V_\sharp, \mathbf{F}_\sharp, \kappa_\sharp)$ in the isometry group $I = I(V, \mathbf{F}, \kappa)$. These embeddings will be discussed explicitly in the rest of this section.

If $I_\sharp \leq I$ is an embedding given in Table 4.3.A, write $X_\sharp = X(V_\sharp, \mathbf{F}_\sharp, \kappa_\sharp)$ for all symbols X in (2.1.15), and put $\tau_\sharp = \tau_{(V_\sharp, \mathbf{F}_\sharp, \kappa_\sharp)}$ and $\sigma_\sharp = \sigma_{(V_\sharp, \mathbf{F}_\sharp, \kappa_\sharp)}$. Furthermore, define

$$\Gamma_{\sharp,\mathbf{F}} = \{g \in \Gamma_\sharp \mid \tau_\sharp(g) \in \mathbf{F}\}. \tag{4.3.3}$$

Then one checks that $\Gamma_{\sharp,\mathbf{F}} \leq \Gamma$.

Definition. The members of $\mathcal{C}_3(\Gamma)$ are the groups $\Gamma_{\sharp,\mathbf{F}}$, with $r = |\mathbf{F}_\sharp : \mathbf{F}|$ prime, which arise with κ and $\kappa_\sharp$ as in Table 4.3.A. In the Table, $m = \frac{n}{r}$.

Table 4.3.A

case	type	description of $\kappa_\sharp$	conditions
L	$GL_m(q^r)$	$\kappa = \kappa_\sharp = 0$	
U	$GU_m(q^r)$	$\kappa_\sharp$ unitary, $\kappa = \mathrm{T}\kappa_\sharp$, r odd	
S	$Sp_m(q^r)$	$\kappa_\sharp$ symplectic, $\kappa = \mathrm{T}\kappa_\sharp$	
S	$GU_{n/2}(q)$	$\kappa_\sharp$ unitary, $\kappa = \mathrm{T}(\lambda\kappa_\sharp)$ $\mathrm{T}(\lambda) = 0$, $\lambda \in \mathbf{F}_\sharp^*$, $r = 2$	q odd
O$^\epsilon$	$O_m^\epsilon(q^r)$	$\kappa_\sharp$ quadratic, $\kappa = \mathrm{T}\kappa_\sharp$, $\mathrm{sgn}(\kappa_\sharp) = \mathrm{sgn}(\kappa)$	$m \geq 3$
O$^\epsilon$	$O_{n/2}(q^2)$	$\kappa_\sharp$ quadratic, $\kappa = \mathrm{T}\kappa_\sharp$, qm odd, $r = 2$	
O$^\epsilon$	$GU_{n/2}(q)$	$\kappa_\sharp$ unitary, $\kappa(v) = \kappa_\sharp(v,v)$ $r = 2$, $\epsilon = (-)^m$	

Remarks on the conditions. Our definition here differs from that given in [As$_1$]. Aschbacher defines $\mathcal{C}_3$ as the set of subgroups $N_\Gamma(\mathbf{F}_\sharp)$, where $\mathbf{F} \subseteq \mathbf{F}_\sharp \subseteq \mathrm{End}_{\mathbf{F}}(V)$ with $\mathbf{F}_\sharp$ a field extension of prime degree r over $\mathbf{F}$, and $C_I(\mathbf{F}_\sharp)$ is irreducible on $(V, \mathbf{F})$. By Proposition 4.3.3.i below, our collection $\mathcal{C}_3(\Gamma)$ is a subset of Aschbacher's. The groups occurring in Aschbacher's collection, but not in ours, are groups of type $O_2^\pm(q^r)$ in $O_{2r}^\pm(q)$ and groups of type $GU_{n/2}(q)$ in $Sp_n(q)$ with q even; these are excluded by the conditions appearing in the right hand column of Table 4.3.A. The exclusion of these groups is justified in the discussions before Proposition 4.3.7 and Lemma 4.3.12, below.

For the remainder of this section we have $H_\Gamma = \Gamma_{\sharp,\mathbf{F}}$ for some embedding in Table 4.3.A. The notation $X_\sharp$, $\tau_\sharp$, $\sigma_\sharp$ carries over from the paragraph just before the Definition, above. We write

$$m = \frac{n}{r} = \dim_{\mathbf{F}_\sharp}(V_\sharp),$$

where r is the prime $[\mathbf{F}_\sharp : \mathbf{F}]$. Also set $\mathrm{N} = \mathrm{N}_{\mathbf{F}}^{\mathbf{F}_\sharp}$, the norm map from $\mathbf{F}_\sharp$ to $\mathbf{F}$. Note that $\Omega_\sharp$ is either the trivial group (in cases $\mathbf{L}^\pm$ with $n = r$) or is a central product of quasisimple groups (see Proposition 2.9.2). Thus in all cases, $\Omega_\sharp$ is perfect which means

$$\Omega_\sharp \leq \Gamma^\infty = \Omega. \tag{4.3.4}$$

Now $\Omega_\sharp$ is absolutely irreducible on $(V_\sharp, \mathbf{F}_\sharp)$ by Proposition 2.10.6.i, but this does not imply that $\Omega_\sharp$ is irreducible on $(V, \mathbf{F})$. Indeed, take case $\mathbf{L}^\pm$ with $n = r$; here $\Omega_\sharp = 1$, and hence cannot be irreducible on $(V, \mathbf{F})$. Lemma 4.3.2, however, shows that $\Omega_\sharp$ is usually irreducible on $(V, \mathbf{F})$. The proof of Lemma 4.3.2 relies on the following Lemma, which uses the existence of primitive prime divisors. For integers x and k, with $x, k \geq 2$, a *primitive prime divisor* of $x^k - 1$ is a prime divisor of $x^k - 1$ which does not divide $x^{k'} - 1$ for all $1 \leq k' < k$. By a theorem of Zsigmondy [Zs], stated as Theorem 5.2.14

in Chapter 5, there always exists at least one primitive prime divisor of $x^k - 1$, unless $(x, k) = (2, 6)$ or $k = 2$ and $x + 1$ is a power of 2. We denote such a prime by x_k. Note that if $(x, k) \neq (2, 3)$, then $x^k + 1$ is divisible by a primitive prime divisor x_{2k} of $x^{2k} - 1$. As a convenience, we sometimes write x_{-k} for a primitive prime divisor of $x^{2k} - 1$ (when $(x, k) \neq (2, 3)$). More information about primitive prime divisors appears at the end of §5.2.

Lemma 4.3.1. *Assume that $m \geq 2$. Then $\Omega_\sharp$ is a central product of quasisimple groups and one of the following holds:*

(i) *H is of type $GL_2(8)$ or $GL_3(4)$ in $GL_6(2)$; or H is of type $GU_2(8)$ in $GU_6(2)$; or H is of type $Sp_2(8)$ in $Sp_6(2)$; or H is of type $GU_4(2)$ in $O_8^+(2)$; or H is of type $O_4^+(8)$ in $O_{12}^+(2)$; or H is of type $GU_2(q)$ in $Sp_4(q)$.*

(ii) *the order of each non-abelian composition factor of $\Omega_\sharp$ is divisible by p_{fk}, where*

$$k = \begin{cases} n & \text{in case } \mathbf{L} \\ 2n & \text{in case } \mathbf{U} \text{ with } n \text{ odd} \\ 2(n-r) & \text{in case } \mathbf{U} \text{ with } n \text{ even} \\ n & \text{if } H \text{ is of type } Sp_{n/r}(q^r) \\ n & \text{if } H \text{ is of type } GU_{n/2}(q) \text{ with } n/2 \text{ odd} \\ n-2 & \text{if } H \text{ is of type } GU_{n/2}(q) \text{ with } n/2 \text{ even}, \frac{n}{2} \geq 3 \\ n & \text{if } H \text{ is of type } O^-_{n/r}(q^r) \\ n-2r & \text{if } H \text{ is of type } O^+_{n/r}(q^r) \\ n-2 & \text{if } H \text{ is of type } O_{n/2}(q^2) \text{ with } n/2 \text{ odd} \\ n-r & \text{if } H \text{ is of type } O_{n/r}(q^r) \text{ with } n \text{ odd.} \end{cases}$$

Proof. The first assertion follows from Proposition 2.9.2 (note that $\Omega_\sharp$ is in fact quasisimple, except when H is of type $O_4^+(q^{n/4})$ in $O_n^+(q)$, in which case $\Omega_\sharp \cong SL_2(q^{n/4}) \circ SL_2(q^{n/4})$). The second is a direct consequence of Table 2.1.C. For example, suppose that H is of type $O^+_{n/r}(q^r)$ in $O_n^+(q)$. Then $|\Omega_\sharp|$ is divisible by $q^{r(n/r-2)} - 1$, and so q_{n-2r} divides $|\Omega_\sharp|$, unless $n - 2r = 6$ and $q = 2$. This occurs only when $n = 12$ and $r = 3$, so that H is of type $O_4^+(2^3)$ in $O_{12}^+(2)$. Thus either (i) or (ii) holds. All other cases are handled in the same way. ■

Lemma 4.3.2. *If $m \geq 2$, then either $\Omega_\sharp$ is irreducible on $(V, \mathbf{F})$ or H is of type $GU_2(q)$ in $Sp_4(q)$.*

Proof. We assume throughout this proof that $m \geq 2$ and that H is not of type $GU_2(q)$ with $\Omega \approx Sp_4(q)$. If $\Omega_\sharp \approx SL_m(q^r)$ or $Sp_m(q^r)$, then $\Omega_\sharp$ is transitive on non-zero vectors in V, and hence it is irreducible. Thus we may assume that $\Omega_\sharp \approx SU_m(q^r)$, $\Omega_m^{\epsilon_\sharp}(q^r)$ or $SU_{n/2}(q)$ (with $m = n/2 \geq 3$). According to Proposition 2.10.6 and the condition $m \geq 3$ given in Table 4.3.A, we know that $\Omega_\sharp$ is irreducible on $(V_\sharp, \mathbf{F}_\sharp)$. In other words, $\Omega_\sharp \mathbf{F}_\sharp^*$ is irreducible on $(V, \mathbf{F})$. Therefore by Clifford's Theorem, V is completely reducible as an $\mathbf{F}\Omega_\sharp$-module. In addition, since $\mathbf{F}_\sharp$ centralizes $\Omega_\sharp$, the proof of Clifford's Theorem actually shows that $\Omega_\sharp$ acts homogeneously on V. So let W be an irreducible $\mathbf{F}\Omega_\sharp$-submodule W

of V and set $\ell = \dim_{\mathbf{F}}(W)$. Then $\Omega_\sharp$ acts faithfully on W, and if $J = N_{\Omega_\sharp \mathbf{F}_\sharp^*}(W)$, then

$$|\Omega_\sharp \mathbf{F}_\sharp^* : J| = \frac{n}{\ell}. \tag{4.3.5}$$

We aim to show that $W = V$. Suppose that $g \in J \backslash \Omega_\sharp \mathbf{F}^*$. Multiplying g by some element in $\Omega_\sharp$, we can assume that $g \in \mathbf{F}_\sharp^* \backslash \mathbf{F}^*$. Since $|\mathbf{F}_\sharp : \mathbf{F}|$ is prime, it follows that the field generated by $\mathbf{F}$ and g is $\mathbf{F}_\sharp$, and hence W is in fact $\mathbf{F}_\sharp$-invariant. Therefore W is $\Omega_\sharp \mathbf{F}_\sharp^*$-invariant, which means $W = V$, as required. Thus we can assume that $J = \Omega_\sharp \mathbf{F}^*$. Also we can suppose that $W \neq V$, so that $\ell \leq [\frac{n}{2}]$. Therefore, as $\Omega_\sharp$ acts faithfully on W, we have embeddings $\Omega_\sharp \preceq GL_\ell(q^u) \preceq GL_{[n/2]}(q^u)$. We now apply Lemma 4.3.1, recalling that H is not of type $GU_2(q)$ in $Sp_4(q)$, by assumption. Clearly Lemma 4.3.1.i cannot occur. For example, $\Omega_4^+(8) \cong L_2(8) \times L_2(8)$ does not embed in $GL_6(2)$, so H cannot be of type $O_4^+(8)$ in $O_{12}^+(2)$. Therefore Lemma 4.3.1.ii occurs, and since $\Omega_\sharp$ is a central product of quasisimple groups, it follows that $q_k \mid |GL_\ell(q^u)|$, where k is as in Lemma 4.3.1.ii. Consequently $k \leq u\ell \leq \frac{un}{2}$. If $u = 2$, then case **U** holds and since $k \neq 2n$, we must have $2(n-r) = k \leq n$, forcing H to be of type $GU_2(q^r)$ in $GU_{2r}(q)$. If $u = 1$, then $k \leq \frac{n}{2}$, and so H must be of type $O_4^+(q^r)$ in $O_{4r}^+(q)$. In either case, the order of each non-abelian composition factor of $\Omega_\sharp$ is divisible by $q_{un/2}$, and so we must have $\ell = \frac{n}{2}$. But then (4.3.5) and the fact that $J = \Omega_\sharp \mathbf{F}^*$ yields $|\mathbf{F}_\sharp^*||\Omega_\sharp \cap \mathbf{F}^*|/|\mathbf{F}^*||\Omega_\sharp \cap \mathbf{F}_\sharp^*| = 2$. However $\Omega_\sharp \cap \mathbf{F}_\sharp^*$ is just the group of scalars in $\Omega_\sharp$, namely $\langle -1 \rangle$ (of order $(2, q-1)$). Therefore $|\Omega_\sharp \cap \mathbf{F}_\sharp^*| = |\Omega_\sharp \cap \mathbf{F}^*| = (2, q-1)$, and this yields $(q^{ur}-1)/(q^u-1) \leq 2$, which is absurd. ■

In the next Proposition we prove that our definition of $\mathcal{C}_3$ is essentially the same as that of Aschbacher's corresponding collection, except that our collection is slightly smaller, owing to the restrictions occurring in the right-hand column of Table 4.3.A. We gave Aschbacher's definition in the Remark after Table 4.3.A, and it may also be found in [As$_1$, p.472]. Here we introduce Aschbacher's notation $\Gamma_{\mathbf{F}}(V_\sharp, \mathbf{F}_\sharp, \kappa_\sharp)$, which is defined to be the same group as $\Gamma_{\sharp,\mathbf{F}}$ given in (4.3.3). This has the advantage of explicitly expressing dependence on $\kappa_\sharp$. Thus if $\kappa_\sharp'$ is some other form on $(V_\sharp, \mathbf{F}_\sharp)$, we have $\Gamma_{\mathbf{F}}(V_\sharp, \mathbf{F}_\sharp, \kappa_\sharp') = \{g \in \Gamma(V_\sharp, \mathbf{F}_\sharp, \kappa_\sharp') \mid \tau_{\kappa_\sharp'}(g) \in \mathbf{F}\}$.

Proposition 4.3.3.

(i) *$I_\sharp$ is irreducible on $(V, \mathbf{F})$.*

(ii) *$H_\Gamma = N_\Gamma(\mathbf{F}_\sharp)$.*

(iii) *Our collection $\mathcal{C}_3$ is the same as Aschbacher's, except for the groups excluded by the conditions in the right hand column of Table 4.3.A.*

Proof. (i) In view of Lemma 4.3.2, we may take either $m = 1$ or H to be of type $GU_2(q)$ in $Sp_4(q)$. If $m = 1$, then $\mathbf{Z}_{q^r - \epsilon} \cong GL_1^\epsilon(q^r) \approx I_\sharp < I \approx GL_r^\epsilon(q)$. When $\epsilon = +$, the group $I_\sharp$ is transitive on the non-zero vectors of V, so it is obviously irreducible. If $\epsilon = -$, then $|I_\sharp|$ is divisible by a primitive prime divisor q_{2r} of $q^{2r} - 1$ (note that $I \not\approx GU_3(2)$ by (4.0.2)).

Since q_{2r} does not divide $|GL_{r-1}(q^2)|$, it follows that a group of order q_{2r} in $I_\sharp$ must be irreducible in $GL(V, \mathbf{F})$, and so the result is proved. Next assume that H is of type $GU_2(q)$ in $Sp_4(q)$. Here $I_\sharp \approx GU_2(q)$ and $L := I_\sharp \cap \mathbf{F}_\sharp^* \cong \mathbf{Z}_{q+1}$. We may write $V = W_1 \oplus W_2$, where the W_i are 2-dimensional irreducible $\mathbf{F}L$-modules. Thus if $I_\sharp$ is reducible, it must fix each W_i. But in this case $I_\sharp \preceq C_{GL(W_1,\mathbf{F})}(L^{W_1}) \times C_{GL(W_2,\mathbf{F})}(L^{W_2}) \cong \mathbf{Z}_{q^2-1} \times \mathbf{Z}_{q^2-1}$, which is impossible. Therefore $I_\sharp$ is in fact irreducible.

(ii). Clearly $I_\sharp \leq C_I(\mathbf{F}_\sharp)$, and so (i) shows that $C_I(\mathbf{F}_\sharp)$ is irreducible on $(V, \mathbf{F})$. Therefore, $N_\Gamma(\mathbf{F}_\sharp)$ is a member of Aschbacher's collection $\mathcal{C}_3(\Gamma)$ as he defines it in [As$_1$, p.472]. Now quoting the complete statement of [As$_1$, Theorem A(1)] we find that $N_\Gamma(\mathbf{F}_\sharp) = \Gamma_\mathbf{F}(V_\sharp, \mathbf{F}_\sharp, \kappa_\sharp')$ for some form $\kappa_\sharp'$. Thus as H_Γ normalizes $\mathbf{F}_\sharp$, we have

$$\Gamma_\mathbf{F}(V_\sharp, \mathbf{F}_\sharp, \kappa_\sharp) = H_\Gamma \leq N_\Gamma(\mathbf{F}_\sharp) = \Gamma_\mathbf{F}(V_\sharp, \mathbf{F}_\sharp, \kappa_\sharp'). \tag{4.3.6}$$

If case **L** holds, then $\kappa_\sharp$ and $\kappa_\sharp'$ are the zero form, and so it is clear that equality holds in (4.3.6). Assume therefore that case **U**, **S** or **O** holds. Here both κ and $\kappa_\sharp$ are non-degenerate forms. If $m = 1$, then case **U** holds and both κ and $\kappa_\sharp$ are unitary. Up to scalar mulitiples there is a unique non-degenerate unitary form on the 1-space $(V_\sharp, \mathbf{F}_\sharp)$, so once again equality holds in (4.3.6). Next assume that $m \geq 2$. Then $\Omega_\sharp$ is a central product of quasisimple groups and hence

$$\Omega_\sharp \leq \Gamma_\mathbf{F}(V_\sharp, \mathbf{F}_\sharp, \kappa_\sharp')^\infty \leq \Omega(V_\sharp, \mathbf{F}_\sharp, \kappa_\sharp'). \tag{4.3.7}$$

Suppose first that $\kappa_\sharp$ is symplectic. Then Table 4.3.A shows that case **S** holds, and hence according to [As$_1$, Theorem A(1)] $\kappa_\sharp'$ is either symplectic or unitary. If $\kappa_\sharp'$ is symplectic, then equality holds in (4.3.7) and hence in (4.3.6). If however $\kappa_\sharp'$ is unitary, then $r = 2$ and we obtain an embedding $Sp_m(q^2) \leq SU_m(q)$, which is impossible by Lagrange's Theorem. The same proof works if $\kappa_\sharp$ is quadratic. Assume therefore that $\kappa_\sharp$ is unitary. As before, if $\kappa_\sharp'$ is also unitary, then equality holds in (4.3.7) and hence in (4.3.6). So we are left to consider the case in which $\kappa_\sharp'$ is either symplectic or quadratic. In this situation, $r = 2$, $\Omega_\sharp \approx SU_m(q)$ and $\Omega(V_\sharp, \mathbf{F}_\sharp, \kappa_\sharp') \approx Sp_m(q^2)$ or $\Omega_m^\delta(q^2)$. Let $\mu_\sharp$ be a generator for $\mathbf{F}_\sharp^*$. Then by (2.3.2) we have $\tau_{\kappa_\sharp}(\mu_\sharp) = \mu_\sharp^{q+1} \in \mathbf{F}^*$, which means $\mu_\sharp \in \Gamma_\mathbf{F}(V_\sharp, \mathbf{F}_\sharp, \kappa_\sharp) \leq \Gamma_\mathbf{F}(V_\sharp, \mathbf{F}_\sharp, \kappa_\sharp')$. But $\tau_{\kappa_\sharp'}(\mu_\sharp) = \mu_\sharp^2 \notin \mathbf{F}^*$, a contradiction. Part (ii) is proved.

(iii). According to (i) and (ii), H_Γ is a member of Aschbacher's collection $\mathcal{C}_3(\Gamma)$. On the other hand, Theorem A(1) of [As$_1$] shows that every member of Aschbacher's collection (apart from groups of type $O_2^\pm(q^r)$ in $O_{2r}^\pm(q)$ and $GU_m(q)$ in $Sp_{2m}(q)$ with q even) appears in Table 4.3.A. Therefore (iii) holds. ■

We now investigate the relationship between the $(V_\sharp, \mathbf{F}_\sharp, \kappa_\sharp)$-geometry and the $(V, \mathbf{F}, \kappa)$-geometry. We begin with an elementary observation.

Lemma 4.3.4. *If $W_\sharp$ is a subspace of $V_\sharp$, then* $\dim_\mathbf{F}(W_\sharp) = r \dim_{\mathbf{F}_\sharp}(W_\sharp)$.

Lemma 4.3.5. *If $h \in H_\Gamma$, then*

(i) $\tau(h) = \tau_\sharp(h)$
(ii) $\sigma_\sharp(h)$ *and* $\sigma(h)$ *agree on* $\mathbf{F}$.

Proof. In case $\mathbf{L}$ we have $\tau(h) = \tau_\sharp(h) = 1$ (see the remarks after (2.1.9)); moreover (ii) is clear from the definitions of σ and $\sigma_\sharp$. So assume that κ is non-degenerate. Excluding the last row of Table 4.3.A, we have $\kappa = \mathrm{T}(\lambda\kappa_\sharp)$, where $\lambda \in \mathbf{F}_\sharp^*$ and either $\lambda = 1$ or λ satisfies $\mathrm{T}(\lambda) = 0$. Thus using the notation in the proof of Lemma 2.1.2, for all $\mathbf{v} \in V^\ell$ we have

$$\begin{aligned}\tau(h)\kappa(\mathbf{v})^{\sigma(h)} &= \kappa(\mathbf{v}h) = \mathrm{T}(\lambda\kappa_\sharp(\mathbf{v}h)) \\ &= \mathrm{T}\big(\tau_\sharp(h)\lambda\kappa_\sharp(\mathbf{v})^{\sigma_\sharp(h)}\big) = \tau_\sharp(h)\kappa(\mathbf{v})^{\sigma_\sharp(h)}.\end{aligned}$$

(Note that this equation uses the fact that $\tau_\sharp(h) \in \mathbf{F}$.) Thus (i) and (ii) hold by the uniqueness assertions in Lemma 2.1.2.i,iii. The embedding in the last row is entirely similar, for we have $\tau(h)\kappa(v)^{\sigma(h)} = \kappa(vh) = \kappa_\sharp(vh, vh) = \tau_\sharp(h)\kappa_\sharp(v, v)^{\sigma_\sharp(h)}$. ■

By Chapter 2 (see (2.3.4), (2.4.5), (2.6.4), (2.7.3) and (2.8.4)) we can choose $\phi_\sharp \in \Gamma_\sharp$ such that $\langle\sigma_\sharp(\phi_\sharp)\rangle = Aut(\mathbf{F}_\sharp)$ and $\tau_\sharp(\phi_\sharp) = 1$. Thus $\phi_\sharp \in H_\Gamma$ and by Lemma 4.3.5, $\tau(\phi_\sharp) = 1$ and $\langle\sigma(\phi_\sharp)\rangle = Aut(\mathbf{F})$. In addition, $\Gamma_\sharp = \Delta_\sharp\langle\phi_\sharp\rangle$, and consequently

$$H_\Gamma = \Gamma_{\sharp,\mathbf{F}} = (\Gamma_{\sharp,\mathbf{F}} \cap \Delta)\langle\phi_\sharp\rangle = \Delta_{\sharp,\mathbf{F}}\langle\phi_\sharp\rangle, \tag{4.3.8}$$

where

$$\Delta_{\sharp,\mathbf{F}} = \{g \in \Delta_\sharp \mid \tau_\sharp(g) \in \mathbf{F}^*\}.$$

Putting $\psi = \phi_\sharp^{uf}$, it is clear that $\psi \in GL(V, \mathbf{F})$ and hence

$$\langle\psi\rangle = \langle\phi_\sharp\rangle \cap \Delta = \langle\phi_\sharp\rangle \cap I. \tag{4.3.9}$$

We now obtain from (4.3.8) and (4.3.9),

$$H = \Delta_{\sharp,\mathbf{F}}\langle\psi\rangle,$$

and

$$H_I = (\Delta_{\sharp,\mathbf{F}} \cap I)\langle\psi\rangle. \tag{4.3.10}$$

However τ and $\tau_\sharp$ agree on $\Delta_{\sharp,\mathbf{F}}$ by Lemma 4.3.5.i, and hence $\Delta_{\sharp,\mathbf{F}} \cap I = I_\sharp$. Thus (4.3.10) yields

$$H_I = I_\sharp\langle\psi\rangle \cong I_\sharp.\mathbf{Z}_r. \tag{4.3.11}$$

Lemma 4.3.5.i also implies that

$$\mathbf{F}_\sharp^* \cap I = \mathbf{F}_\sharp^* \cap I_\sharp. \tag{4.3.12}$$

First we consider the case where H is of type $GL_m^\pm(q^r)$. For the unitary case note that r must be odd, for if $r = 2$, then $\mathrm{T}(\kappa_\sharp(v, w)) = \mathrm{T}(\kappa_\sharp(w, v))$, and so $\mathrm{T}\kappa_\sharp$ cannot be unitary.

Proposition 4.3.6. *Assume that case* $\mathbf{L}^\epsilon$ *holds, so that* H *is of type* $GL^\epsilon_m(q^r)$.
(I) $c = 1$.
(II) $H_{\overline{\Omega}} \cong \mathbf{Z}_a.L^\epsilon_m(q^r).\mathbf{Z}_b.\mathbf{Z}_r$, *where* $a = \dfrac{(q-\epsilon,m)(q^r-\epsilon)}{(q-\epsilon)(q-\epsilon,n)}$ *and* $b = \dfrac{(q^r-\epsilon,m)}{(q-\epsilon,m)}$.
(III) $H_{\overline{\Omega}}$ *is local.*

Proof. Here κ and $\kappa_\sharp$ are both trivial when $\epsilon = +$ and are both unitary when $\epsilon = -$, and r is odd when $\epsilon = -$. Furthermore $I_\sharp \approx GL^\epsilon_m(q^r)$ and $\Omega_\sharp \approx SL^\epsilon_m(q^r)$. It is a consequence of Lemma 2.10.2 that for $x \in I_\sharp$,

$$\det_{\mathbf{F}}(x) = \mathrm{N}(\det_{\mathbf{F}_\sharp}(x)), \qquad (4.3.13)$$

and since N is surjective, $|\det(I_\sharp)| = q - \epsilon$. Consequently (I) holds by Lemma 4.0.3.v and moreover

$$I_\sharp\Omega = I. \qquad (4.3.14)$$

Therefore

$$H_\Omega \cong (I_\sharp \cap \Omega).\mathbf{Z}_r. \qquad (4.3.15)$$

Clearly (4.3.14) implies that

$$I_\sharp \cap \Omega = \tfrac{1}{q-\epsilon} I_\sharp = \Omega_\sharp.\left(\frac{q^r-\epsilon}{q-\epsilon}\right). \qquad (4.3.16)$$

On the other hand, $I_\sharp \cap \Omega = \Omega_\sharp(\mathbf{F}^*_\sharp \cap \Omega).\mathbf{Z}_b$ for some b, and so it follows from (4.3.16) that

$$b = \frac{(q^r-\epsilon)|\Omega_\sharp|}{(q-\epsilon)|\Omega_\sharp(\mathbf{F}^*_\sharp \cap \Omega)|} = \frac{(q^r-\epsilon)|\mathbf{F}^*_\sharp \cap \Omega_\sharp|}{(q-\epsilon)|\mathbf{F}^*_\sharp \cap \Omega|}.$$

Obviously $|\mathbf{F}^*_\sharp \cap \Omega_\sharp| = (q^r - \epsilon, m)$. Moreover $\det(\mathbf{F}^*_\sharp) = (\mathbf{F}^*)^m$, and so

$$|\mathbf{F}^*_\sharp \cap \Omega| = \frac{|\mathbf{F}^*_\sharp|}{|\det(\mathbf{F}^*_\sharp)|} = \frac{(q-\epsilon,m)(q^r-\epsilon)}{(q-\epsilon)}. \qquad (4.3.17)$$

Thus $b = (q^r - \epsilon, m)/(q - \epsilon, m)$ and so from (4.3.15) and (4.3.17) we deduce

$$H_\Omega \cong (\mathbf{F}^*_\sharp \cap \Omega).P\Omega_\sharp.\mathbf{Z}_b.\mathbf{Z}_r \cong \left(\frac{(q-\epsilon,m)(q^r-\epsilon)}{(q-\epsilon)}\right).L^\epsilon_m(q^r).\mathbf{Z}_b.\mathbf{Z}_r. \qquad (4.3.18)$$

We obtain $H_{\overline{\Omega}}$ by factoring out the group of scalars of order $(q - \epsilon, n)$ from the normal cyclic subgroup of order $(q - \epsilon, m)(q^r - \epsilon)/(q - \epsilon)$. This establishes (II). Note that $H_{\overline{\Omega}}$ is always local, as $(q - \epsilon, m)(q^r - \epsilon)/(q - \epsilon) > (q - \epsilon, n)$, proving (III). ■

We now come to groups of type $GU_{n/2}(q)$ in $Sp_n(q)$. Here $\kappa_\sharp$ is unitary, $r = 2$ and $\lambda \in \mathbf{F}^*_\sharp$ satisfies $\mathrm{T}(\lambda) = 0$. Then for $v, w \in V$ we have

$$\begin{aligned}\mathrm{T}(\lambda\kappa_\sharp(v,w)) &= \lambda\kappa_\sharp(v,w) + \lambda^q\kappa_\sharp(v,w)^q \\ &= -\lambda^q\kappa_\sharp(w,v)^q - \lambda\kappa_\sharp(w,v) = -\mathrm{T}(\lambda\kappa_\sharp(w,v)).\end{aligned}$$

Thus $T(\lambda\kappa_\sharp)$ is skew-symmetric, $T(\lambda\kappa_\sharp(v,v)) = 0$, and similarly one checks bilinearity and non-degeneracy. Thus putting $\kappa = T(\lambda\kappa_\sharp)$ we obtain $GU_m(q) \approx I_\sharp \le I \approx Sp_{2m}(q)$. If we choose another field element $\lambda' \in \mathbf{F}_\sharp^*$ satisfying $T(\lambda') = 0$, then $T(\lambda'\kappa_\sharp) = \lambda_o\kappa$ for some $\lambda_o \in \mathbf{F}^*$, and so we obtain the same embedding. Observe that if $\mu_\sharp$ is a generator for $\mathbf{F}_\sharp$ satisfying $\mu_\sharp^{q+1} = \mu$, then $\kappa(\mu_\sharp v, \mu_\sharp w) = T(\lambda\kappa_\sharp(\mu_\sharp v, \mu_\sharp w)) = T(\lambda\mu_\sharp^{q+1}\kappa_\sharp(v,w)) = \mu T(\lambda\kappa_\sharp(v,w)) = \mu\kappa(v,w)$ for all $v, w \in V$. Thus $\mu_\sharp \le \Delta_{\sharp,\mathbf{F}}$ and $\tau(\mu_\sharp) = \mu$. This shows $\Delta_\sharp \le \Delta$ and that $\Delta_\sharp\Omega = \Delta$. Since $\phi_\sharp \in H_\Gamma$ and $\langle\sigma(\phi_\sharp)\rangle = Aut(\mathbf{F})$ (see (4.3.8)), it now follows that $H_\Gamma\Omega = \Gamma$, and hence Theorem 3.1.2 holds for these groups by Proposition 4.0.1. When q is even, we may take $\lambda = 1$ and we define $Q(v) = \kappa_\sharp(v,v)$ for all $v \in V$. Then Q is a non-degenerate quadratic form such that $\mathbf{f}_Q = T\kappa_\sharp$. Moreover $\Gamma_{\sharp,\mathbf{F}} < \Gamma(V,\mathbf{F},Q) < \Gamma$. This group $\Gamma(V,\mathbf{F},Q)$ is actually a member of $\mathcal{C}_8(\Gamma)$ (see §4.8), and for this reason we impose the restriction q odd in Table 4.3.A.

Proposition 4.3.7. *Assume that case* **S** *holds and H is of type $GU_{n/2}(q)$, so that $\kappa_\sharp$ is unitary with q odd.*

(I) $c = 1$.

(II) $H_{\overline{\Omega}} \cong \frac{(q+1)}{2}.PGU_m(q).2$.

(III) $H_{\overline{\Omega}}$ *is local.*

Now we prove Theorem 3.1.2 for the groups in $\mathcal{C}_3$.

Proposition 4.3.8.

(i) *Theorem 3.1.2 holds for groups in $\mathcal{C}_3$.*

(ii) $H_\Gamma = N_\Gamma(H_\Omega)$ *in case* **O**.

Proof. First assume that $\Omega_\sharp$ is irreducible on V. Here $\mathbf{E} = \mathrm{End}_{\mathbf{F}\Omega_\sharp}(V)$ is a field extension of $\mathbf{F}$ by Schur's Lemma. Clearly $\mathbf{F} \subseteq \mathbf{F}_\sharp \subseteq \mathbf{E}$, and since $\mathbf{E}$ is a field it follows that $\mathbf{E}$ commutes with both $\mathbf{F}_\sharp$ and $\Omega_\sharp$. Therefore $\mathbf{E} \subseteq \mathrm{End}_{\mathbf{F}_\sharp\Omega_\sharp}(V)$. It is obvious that any element of $\mathrm{End}_\mathbf{F}(V)$ that commutes with $\mathbf{F}_\sharp\Omega_\sharp$ also commutes with $\mathbf{F}\Omega_\sharp$, and hence $\mathrm{End}_{\mathbf{F}_\sharp\Omega_\sharp}(V) \subseteq \mathbf{E}$. We conclude $\mathbf{E} = \mathrm{End}_{\mathbf{F}_\sharp\Omega_\sharp}(V)$. In view of the stipulation $m \ge 3$ when H is of type $O_m^\epsilon(q^r)$, it follows from Proposition 2.10.6 that $\Omega_\sharp$ is absolutely irreducible on $V_\sharp$. Hence $\mathrm{End}_{\mathbf{F}_\sharp\Omega_\sharp}(V) = \mathbf{F}_\sharp$ by Lemma 2.10.1, whence $\mathbf{E} = \mathbf{F}_\sharp$. Moreover $\Omega_\sharp = H_\Omega^\infty$, and so $N_\Gamma(H_\Omega) \le N_\Gamma(\Omega_\sharp) \le N_\Gamma(\mathbf{F}_\sharp)$. By Proposition 4.3.3.ii, $N_\Gamma(\mathbf{F}_\sharp) = H_\Gamma$, so the result follows by Proposition 4.0.1.

Now assume that $\Omega_\sharp$ is reducible on V, so that by Lemma 4.3.2, either $m = 1$ or H is of type $GU_2(q)$ in $Sp_4(q)$. In our description of the embedding of $GU_2(q)$ in $Sp_4(q)$ before Propositon 4.3.7, we proved that $H_\Gamma\Omega = \Gamma$, and hence that Theorem 3.1.2 holds (using Proposition 4.0.1.i). Thus we may assume that $m = 1$, and here H is of type $GL_1^\pm(q^n)$ in $SL_n^\pm(q)$, with n prime. We have $I_\sharp = \mathbf{F}_\sharp^*$, and so according to (4.3.15) in the proof of Proposition 4.3.6, $H_\Omega \cong (\mathbf{F}_\sharp^* \cap \Omega).\mathbf{Z}_n \cong \mathbf{Z}_k.\mathbf{Z}_n$, where $k = (q^n - \epsilon)/(q - \epsilon)$. When $n = 2$, we have $\epsilon = +$ and $H_\Omega \cong \mathbf{Z}_{q+1}.\mathbf{Z}_2$; it is well known that these are groups are self-normalizing in Ω (see [Dic$_1$, Ch.XII] for example), and so Theorem 3.1.2 holds in this case. Thus we can take $n \ge 3$. Using Zsigmondy's Theorem (stated before Lemma 4.3.1), there is a primitive prime divisor q_{en} of k (note we can exclude the case $\Omega \approx SU_3(2)$, for $SU_3(2)$ is soluble). Evidently $q_{en} \neq n$ and so $O_{q_{en}}(H_\Omega) \leq \mathbf{F}_\sharp^* \cap \Omega$. It follows that $O_{q_{en}}(H_\Omega)$ generates $\mathbf{F}_\sharp$ over $\mathbf{F}$. Consequently $N_\Gamma(H_\Omega) \leq N_\Gamma(O_{q_{en}}(H_\Omega)) \leq N_\Gamma(\mathbf{F}_\sharp)$. As before, we complete the proof using Propositions 4.3.3.ii and 4.0.1. ■

In view of Propositions 4.3.6 and 4.3.7, we can assume for the rest of this section that case **S** or **O** holds, and moreover that H is not of type $GU_{n/2}(q)$ in case **S**. Thus $m \geq 2$ and $\Omega_\sharp$ is irreducible on $(V, \mathbf{F})$ by Lemma 4.3.2.

Lemma 4.3.9. *Assume that case* **S** *or* **O** *holds and that $\kappa_\sharp$ is symplectic or orthogonal, respectively. Then* $\mathrm{soc}(H_{\overline{\Omega}}) = P\Omega_\sharp$.

Proof. In case **S**, we have $\Omega_\sharp \cong Sp_m(q^r)$ and $\overline{\Omega}_\sharp \cong P\Omega_\sharp$, which is simple (see Proposition 2.9.2). Similarly, in case **O** we have $\overline{\Omega}_\sharp \cong P\Omega_\sharp$, which is semisimple due to the restriction $m \geq 3$. Moreover $\Omega_\sharp \trianglelefteq H_\Omega$, and so in view of Lemmas 4.0.4 and 4.0.5.iii it suffices to show that $C_\Omega(\Omega_\sharp) \leq \langle -1 \rangle$. Because $\Omega_\sharp$ is absolutely irreducible on $V_\sharp$, it follows that $C_{GL(V,\mathbf{F})}(\Omega_\sharp) = \mathbf{F}_\sharp^*$. However $\mathbf{F}_\sharp \cap \Omega \leq \langle -1 \rangle$ by (4.3.12), and so the result follows. ■

Proposition 4.3.10. *Assume that case* **S** *holds and that H is of type $Sp_m(q^r)$, so that $\kappa_\sharp$ is symplectic.*

(I) $c = 1$.

(II) $H_{\overline{\Omega}} \cong PSp_m(q^r).r$.

(III) $H_{\overline{\Omega}}$ *is non-local and* $\mathrm{soc}(H_{\overline{\Omega}}) \cong PSp_m(q^r)$.

We now turn to case **O**, and consider the last three embeddings in Table 4.3.A. Thus for the rest of this section $\kappa = Q$ is a non-degenerate quadratic form. In the event that $\kappa_\sharp$ is also a quadratic form we will sometimes write $\kappa_\sharp = Q_\sharp$. We begin with an easy technical Lemma concerning the O_2^--geometry. Recall the homomorphism γ defined in §§2.7, 2.8.

Lemma 4.3.11. *Let $I \approx O_2^-(q)$ and put* $K = \ker_I(\gamma)$.

(i) $K \cong \mathbf{Z}_{q+1}$.

(ii) $\mathrm{End}_{\mathbf{F}K}(V) = \mathbf{F}_\flat$ *is a field extension of degree 2 over* $\mathbf{F}$.

(iii) *There is a unique non-degenerate unitary form $\mathbf{f}_\flat$ on the 1-dimensional $\mathbf{F}_\flat$-space $V_\flat = (V, \mathbf{F}_\flat)$ such that $\mathbf{f}_\flat(v, v) = Q(v)$ for all $v \in V$.*

(iv) $I(V_\flat, \mathbf{F}_\flat, \mathbf{f}_\flat) = \ker_I(\gamma)$.

Proof. Assertion (i) is immediate from Propositions 2.8.3.ii and 2.9.1.iii. Obviously any cyclic subgroup of $GL(V, \mathbf{F})$ of order $q+1$ must be irreducible on V, and so (ii) now follows from Schur's Lemma. Take $v \in V \backslash 0$ and observe that any element of V can be written as λv for some $\lambda \in \mathbf{F}_\flat$ (since $\dim_{\mathbf{F}_\flat}(V_\flat) = 1$). So define $\mathbf{f}_\flat(\lambda_1 v, \lambda_2 v) = \lambda_1 \lambda_2^q Q(v)$ for all $\lambda_i \in \mathbf{F}_\flat$. It is easy to check that $\mathbf{f}_\flat$ satisfies the description in (iii). Now $I(V_\flat, \mathbf{F}_\flat, \mathbf{f}_\flat) \approx GU_1(q) \cong \mathbf{Z}_{q+1}$, and $I \approx O_2^-(q) \cong D_{2(q+1)}$. Since I has a unique cyclic subgroup of order $q+1$, part (iv) now follows from (i). ■

Now we justify the condition $m \geq 3$ in the right hand column of Table 4.3.A. If $m = 1$, then case $\mathbf{O}^o$ holds and n is prime, and we show that Aschbacher's collection $\mathcal{C}_3$ is void. For a contradiction, assume that $C_I(\mathbf{F}_\sharp)$ is irreducible on V. Now $C_{GL(V,\mathbf{F})}(\mathbf{F}_\sharp) \cong GL_1(q^n) \cong \mathbf{Z}_{q^n-1}$, and hence $C_I(\mathbf{F}_\sharp)$ is cyclic of order dividing $(q^n - 1, |I|)$. Now $|I|$ is given in Table 2.1.C and since $n \geq 7$ and n is prime, we find that $(q^n - 1, |I|) = (q-1)(q-1,n)$. Observe that every subgroup of $C_I(\mathbf{F}_\sharp)$ with order dividing $q-1$ must lie in $\mathbf{F}^*$. Thus $C_I(\mathbf{F}_\sharp)$ has a subgroup of index $(q-1,n)$ which acts as scalars on V. But the only scalars in I are ± 1, and so $|C_I(\mathbf{F}_\sharp)| \mid q-1$, forcing $C_I(\mathbf{F}_\sharp) \leq \mathbf{F}^*$, which is absurd.

Next assume that $m = 2$ and that $O_2^\epsilon(q^r) \approx I_\sharp \leq I \approx O_{2r}^\epsilon(q)$, so that $n = 2r$. If $\epsilon = +$, then $(V_\sharp, Q_\sharp)$ has precisely two totally singular 1-spaces $U_\sharp$ and $W_\sharp$, and it follows that $U_\sharp$ and $W_\sharp$ are totally singular r-spaces in (V,Q). Therefore $H_\Gamma < N_\Gamma\{U_\sharp, W_\sharp\} \in \mathcal{C}_2(\Gamma)$. Suppose now that $\epsilon = -$. Using Lemma 4.3.11, we may write $\ker_{I_\sharp}(\gamma_{Q_\sharp}) = I(V_\flat, \mathbf{F}_\flat, \mathbf{f}_\flat)$, where $\mathbf{F}_\flat = \mathbf{F}_{q^{2r}}$ and $\mathbf{f}_\flat$ is a non-degenerate unitary form on the 1-dimensional $\mathbf{F}_\flat$-vector space $V_\flat$. Now let $\mathbf{F}_\natural$ be the subfield of $\mathbf{F}_\flat$ of order q^2. Then, as in the second row of Table 4.3.A we obtain $GU_1(q^r) \approx I(V_\flat, \mathbf{F}_\flat, \mathbf{f}_\flat) < I(V_\natural, \mathbf{F}_\natural, \mathbf{f}_\natural) \approx GU_r(q)$, where $\mathbf{f}_\natural = \mathrm{T}^{\mathbf{F}_\flat}_{\mathbf{F}_\natural}\mathbf{f}_\flat$. But now observe using Lemma 4.3.11.iii that for all $v \in V$ we have

$$Q(v) = \mathrm{T}(Q_\sharp(v)) = \mathrm{T}(\mathbf{f}_\flat(v,v)) = \mathbf{f}_\natural(v,v),$$

and hence there is an embedding $I(V_\natural, \mathbf{F}_\natural, \mathbf{f}_\natural) < I(V, \mathbf{F}, Q)$ as in the last row of Table 4.3.A. We now conclude

$$H_\Gamma \leq N_\Gamma(\ker_{I_\sharp}(\gamma_{Q_\sharp})) = N_\Gamma(I(V_\flat, \mathbf{F}_\flat, \mathbf{f}_\flat)) < N_\Gamma(I(V_\natural, \mathbf{F}_\natural, \mathbf{f}_\natural)) < \Gamma.$$

Hence the restriction $m \geq 3$.

As a corollary to the preceding discussion we obtain

Lemma 4.3.12. *Assume that $J \leq I$, and that J is irreducible but not absolutely irreducible on $(V, \mathbf{F})$. Then $N_\Gamma(J)$ is contained in some member of $\mathcal{C}_2(\Gamma) \cup \mathcal{C}_3(\Gamma) \cup \mathcal{C}_8(\Gamma)$.*

Proof. Put $\mathbf{E} = \mathrm{End}_{\mathbf{F}J}(V)$, and let $\mathbf{E}_o$ be a field of prime index over $\mathbf{F}$ such that $\mathbf{F} \subseteq \mathbf{E}_o \subseteq \mathbf{E}$. Then $N_\Gamma(J) \leq N_\Gamma(\mathbf{E}) \leq N_\Gamma(\mathbf{E}_o)$, and so by the definition given in [As₁, p. 472], $N_\Gamma(\mathbf{E}_o)$ is a member of Aschbacher's collection $\mathcal{C}_3(\Gamma)$. Now if $N_\Gamma(\mathbf{E}_o)$ is of type $O_2^\pm(q^r)$ in $O_{2r}^\pm(q)$, then the discussion preceding this Lemma shows that $N_\Gamma(J)$ is contained in a member of $\mathcal{C}_2(\Gamma) \cup \mathcal{C}_3(\Gamma)$. And if $N_\Gamma(\mathbf{E}_o)$ is of type $GU_m(q)$ in $Sp_{2m}(q)$ with q even, then according to the paragraph before Proposition 4.3.7, $N_\Gamma(J)$ is contained in a group of type $O_{2m}^\pm(q)$ in $\mathcal{C}_8(\Gamma)$. If $N_\Gamma(\mathbf{E}_o)$ is neither of these types, then $N_\Gamma(\mathbf{E}_o)$ coincides with one of the members of $\mathcal{C}_3(\Gamma)$ by Proposition 4.3.3.iii. This finishes the proof. ∎

In the next Lemma we allude to the two equivalence classes in $\mathcal{U}_{n/2}$ in case $\mathbf{O}^+$, which we discussed in Description 4 in §2.5. Recall that Lemma 2.5.8.ii asserts that these equivalence classes are precisely the Ω-orbits on $\mathcal{U}_{n/2}$.

Lemma 4.3.13. *Assume that case* $\mathbf{O}^+$ *holds and* H *is of type* $O_m^+(q^r)$, *so that* $\kappa_\sharp = Q_\sharp$ *is quadratic and* m *is even.*

(i) *Totally singular* $\frac{m}{2}$-*spaces in* $(V_\sharp, \mathbf{F}_\sharp, Q_\sharp)$ *are totally singular* $\frac{n}{2}$-*spaces in* $(V, \mathbf{F}, Q)$.

(ii) *If* $r = 2$, *then every totally singular* $\frac{m}{2}$-*space in* $(V_\sharp, \mathbf{F}_\sharp, Q_\sharp)$ *lies in the same* Ω-*orbit of totally singular* $\frac{n}{2}$-*spaces of* $(V, \mathbf{F}, Q)$. *Hence* $H_\Gamma \le \ker(\gamma)$.

(iii) *If* r *is odd, then totally singular* $\frac{m}{2}$-*spaces in distinct* $\Omega_\sharp$-*orbits are in distinct* Ω-*orbits. Hence* $H_\Gamma \not\le \ker(\gamma)$.

Proof. Assertion (i) is clear from the fact that $Q = \mathrm{T}Q_\sharp$. Also (ii) and (iii) follow from Description 4 in §2.5. ∎

Proposition 4.3.14. *Suppose that case* $\mathbf{O}^+$ *holds and that* H *is of type* $O_m^+(q^r)$, *so that* m *is even and* $m \ge 4$.

(I) $c = (r, 2)$, *and if* $c = 2$ *then* $\pi = \pi_2$.

(II) $H_{\overline{\Omega}} \cong P\Omega_m^+(q^r).[cr]$.

(III) $H_{\overline{\Omega}}$ *is non-local and* $\mathrm{soc}(H_{\overline{\Omega}}) \cong P\Omega_m^+(q^r)$.

Proof. By Proposition 4.3.8.ii and Lemma 4.0.3.ix, equality holds throughout Lemma 4.0.3. Observe that $\tau_\sharp(\Delta_\sharp) = \mathbf{F}_\sharp^*$, and so $\tau(\Delta_{\sharp,\mathbf{F}}) = \mathbf{F}^*$. Therefore by Lemma 4.0.3.iii we have $c = |\ddot{I} : \ddot{H}_I|$. According to (2.7.3) we may take $\phi_\sharp$ to fix a totally singular $\frac{m}{2}$-subspace of $(V_\sharp, \kappa_\sharp)$ (see (4.3.8)). Thus $\phi_\sharp$, and hence $\psi = \phi_\sharp^f$, fixes a totally singular $\frac{n}{2}$-space in (V, κ), which means

$$\ddot{\psi} \in \ker_{\ddot{I}}(\ddot{\gamma}). \tag{4.3.19}$$

Now let W be a totally singular $\frac{m}{2}$-space in $V_\sharp$, and let $x \in N_{I_\sharp}(W)$ satisfy $\langle \det_{(W,\mathbf{F}_\sharp)}(x)\rangle = \mathbf{F}_\sharp^*$. According to (4.3.13), $\langle \det_{(W,\mathbf{F})}(x)\rangle = \mathbf{F}^*$, and so by Lemma 4.1.21, we have

$$\langle \ddot{x} \rangle = \ker_{\ddot{I}}(\ddot{\gamma}). \tag{4.3.20}$$

Next, take a non-singular vector $v \in V_\sharp$ and let $r_v \in I_\sharp$ be the corresponding reflection. Since r_v interchanges the two $\Omega_\sharp$-orbits of totally singular $\frac{m}{2}$-spaces in $(V_\sharp, \mathbf{F}_\sharp, \kappa_\sharp)$, Lemma 4.3.13 implies

$$\ddot{r}_v \in \begin{cases} \ddot{I} \backslash \ker_{\ddot{I}}(\ddot{\gamma}) & \text{if } r \text{ is odd} \\ \ker_{\ddot{I}}(\ddot{\gamma}) & \text{if } r \text{ is even.} \end{cases} \tag{4.3.21}$$

Now $I_\sharp = \Omega_\sharp\langle x, r_v\rangle$ and so $H_I = \Omega_\sharp\langle x, r_v, \psi\rangle$ by (4.3.11). So combining (4.3.19), (4.3.20) and (4.3.21) we have

$$\ddot{H}_I = \begin{cases} \ker_{\ddot{I}}(\ddot{\gamma}) & \text{if } r \text{ is even} \\ \ddot{I} & \text{if } r \text{ is odd.} \end{cases}$$

Thus $c = |\ddot{I} : \ddot{H}_I| = (2, r)$, as claimed. And if $c = 2$, then H_Γ fixes each Ω-orbit of totally singular $\frac{n}{2}$-spaces, which shows $H_\Gamma \le \ker(\gamma)$ and hence $\pi = \pi_2$. This establishes (I), and

(III) is immediate from Lemma 4.3.9. To prove (II), observe that $|I:\Omega| = 2(2,q-1) = 2(2,q^r-1) = |I_\sharp:\Omega_\sharp|$. Therefore, since $c = |I : H_I\Omega|$, we deduce

$$|H_\Omega| = \frac{c|H_I|}{|I:\Omega|} = \frac{cr|I_\sharp|}{|I_\sharp:\Omega_\sharp|} = rc|\Omega_\sharp|,$$

and thus (II) now follows easily. ■

We have seen in Lemma 4.3.1 that groups of type $O_m^-(q^r)$ have order divisible by primitive prime divisors of q^n-1. The next Lemma finds the Sylow q_n-normalizer in Ω, which in turn helps us to find the structure of groups of type $O_m^-(q^r)$ in Proposition 4.3.16.

Lemma 4.3.15. *Suppose that case* $\mathbf{O}^-$ *holds. Then a Sylow q_n-normalizer in Ω has order* $\dfrac{n(q^{n/2}+1)}{2(2,q-1)}$.

Proof. We begin the proof by setting up a situation similar to that in the discussion before Lemma 4.3.12. Write $\mathbf{F} \subseteq \mathbf{F}_\flat \subseteq \mathbf{F}_\natural \subseteq \mathrm{End}_{\mathbf{F}}(V)$, where $\mathbf{F}_\natural$ and $\mathbf{F}_\flat$ are fields of order q^n and $q^{n/2}$, respectively. Thus V is 1-dimensional vector space over $\mathbf{F}_\natural$ and a 2-dimensional space over $\mathbf{F}_\flat$. Now let $\mathbf{f}_\natural$ be a non-degenerate unitary form on $V_\natural = (V, \mathbf{F}_\natural)$. Then the map $v \mapsto \mathbf{f}_\natural(v,v) \in \mathbf{F}_\flat$ is a non-degenerate quadratic form on $V_\flat = (V,\mathbf{F}_\flat)$ of type O_2^-. Therefore $\mathrm{T}_{\mathbf{F}}^{\mathbf{F}_\flat}(\mathbf{f}_\natural(v,v))$ is a non-degenerate quadratic form on $(V,\mathbf{F})$ of type $O_n^-(q)$, and so we may identify this form with Q. Observe

$$\mathbf{Z}_{q^{n/2}+1} \cong GU_1(q^{n/2}) \approx I(V_\natural, \mathbf{F}_\natural, \mathbf{f}_\natural) \leq I.$$

Thus $I_\natural$ contains a subgroup $J \in Syl_{q_n}(\Omega)$. Since q_n does not divide $|GL_{n-1}(q)|$, the group J is irreducible on V. Thus Schur's Lemma implies that $\mathbf{E} = \mathrm{End}_{\mathbf{F}J}(V)$ is a field extension of $\mathbf{F}$. Moreover $|\mathbf{E}:\mathbf{F}|$ divides n and clearly $\mathbf{F}_\natural \subseteq \mathbf{E}$. Therefore $\mathbf{E} = \mathbf{F}_\natural$, whence

$$C_I(J) = \mathbf{F}_\natural^* \cap I.$$

Now take $\lambda \in \mathbf{F}_\natural^* \cap I$, so that $\mathbf{f}_\natural(\lambda v, \lambda v) = \lambda^{q^{n/2}+1}\mathbf{f}_\natural(v,v)$ for all $v \in V$. On the other hand, $\lambda \in I$, and so $Q(\lambda v) = Q(v)$ for all $v \in V$. Hence for all $v \in V$,

$$\mathrm{T}_{\mathbf{F}}^{\mathbf{F}_\flat}((\lambda^{q^{n/2}+1}\mathbf{f}_\natural(v,v)) = \mathrm{T}_{\mathbf{F}}^{\mathbf{F}_\flat}(\mathbf{f}_\natural(\lambda v, \lambda v)) = Q(\lambda v) = Q(v) = \mathrm{T}_{\mathbf{F}}^{\mathbf{F}_\flat}(\mathbf{f}_\natural(v,v)).$$

Since $\mathbf{f}_\natural(v,v)$ takes on all values in $\mathbf{F}_\flat$ as v ranges over V, it follows that for all $\eta \in \mathbf{F}_\flat$ we have $\mathrm{T}_{\mathbf{F}}^{\mathbf{F}_\flat}(\eta(\lambda^{q^{n/2}+1}-1)) = 0$. Consequently $\lambda^{q^{n/2}+1} = 1$, and it now follows that $C_I(J) = I(V_\natural, \mathbf{F}_\natural, \mathbf{f}_\natural) \cong \mathbf{Z}_{q^{n/2}+1}$.

Since $\mathbf{F}_\natural = \mathrm{End}_{\mathbf{F}J}(V)$, there is a natural homomorphism from $N_I(J)$ to $Gal(\mathbf{F}_\natural/\mathbf{F})$ with kernel $C_I(J)$. We show this homomorphism is surjective. For let $\phi_\natural \in \Gamma(V_\natural, \mathbf{F}_\natural, \mathbf{f}_\natural)$ satisfy $\tau_\natural(\phi_\natural) = 1$ and $\sigma_\natural(\phi_\natural) = \nu_\natural$, where $\nu_\natural$ is a generator for $Aut(\mathbf{F}_\natural) \cong \mathbf{Z}_{fn}$. Then as in the beginning of this section we have $\phi_\natural^f \in I$ and so $\phi_\natural^f \in N_I(J)$. Also $\nu_\natural^f$ induces an automorphism of order n on $\mathbf{F}_\natural$ and hence an automorphism of order n on J. It follows from our discussion thus far that $N_I(J) = I(V_\natural, \mathbf{F}_\natural, \mathbf{f}_\natural)\langle\phi_\natural^f\rangle \cong \mathbf{Z}_{q^{n/2}+1}.\mathbf{Z}_n$. The Frattini argument implies that $|N_I(J) : N_\Omega(J)| = |I:\Omega|$, and the Lemma now follows since $|I:\Omega| = 2(2,q-1)$. ■

Proposition 4.3.16. *Suppose that case* $\mathbf{O}^-$ *holds and* H *is of type* $O_m^-(q^r)$, *so that* m *is even and* $m \geq 4$.

(I) $c = 1$.
(II) $H_{\overline{\Omega}} \cong P\Omega_m^-(q^r).r$.
(III) $H_{\overline{\Omega}}$ *is non-local and* $\mathrm{soc}(H_{\overline{\Omega}}) \cong P\Omega_m^-(q^r)$.

Proof. Arguing as in the proof of Proposition 4.3.14, we deduce $c = |\ddot{I} : \ddot{H}_I|$. Clearly $H_\Omega \leq N_\Omega(\Omega_\sharp)$, and $\Omega_\sharp$ contains a subgroup $J \in Syl_{q_n}(\Omega)$. Thus the Frattini argument yields

$$H_\Omega \leq \Omega_\sharp N_\Omega(J). \tag{4.3.22}$$

Since q_n does not divide $|GL_{n-1}(q)|$, the group J is irreducible (but not absolutely irreducible) on V, and hence $\mathbf{E} = \mathrm{End}_{\mathbf{F}J}(V)$ is a field. Indeed $\mathbf{E}$ is a field extension of degree n over $\mathbf{F}$. Since $q_n > n > r$, it follows that J centralizes $\mathbf{F}_\sharp$, which is to say $\mathbf{F}_\sharp \subseteq \mathbf{E}$. Since there is a unique field of order q^r in $\mathbf{E}$, we deduce that $N_\Omega(J) \leq N_\Omega(\mathbf{E}) \leq N_\Omega(\mathbf{F}_\sharp) = H_\Omega$, and hence by (4.3.22) we obtain

$$H_\Omega = \Omega_\sharp N_\Omega(J).$$

Consequently $|H_\Omega : \Omega_\sharp| = |N_\Omega(J) : N_{\Omega_\sharp}(J)|$. And according to the Lemma 4.3.15, $|N_\Omega(J) : N_{\Omega_\sharp}(J)| = \frac{n}{2}/\frac{m}{2} = r$, and so $|H_\Omega : \Omega_\sharp| = r$, which shows $H_{\overline{\Omega}} \cong P\Omega_m^-(q^r).r$, proving (II). And of course (III) follows from Lemma 4.3.9. Furthermore $|H_I : I_\sharp| = r$ by (4.3.11), so $|H_I : I_\sharp| = |H_\Omega : \Omega_\sharp|$. Consequently $|H_I : H_\Omega| = |I_\sharp : \Omega_\sharp| = |I : \Omega|$, which implies $H_I\Omega = I$. Therefore $c = 1$, as required. ■

Proposition 4.3.17. *Assume that case* $\mathbf{O}^\circ$ *holds and* H *is of type* $O_m(q^r)$, *with* qn *odd and* $m \geq 1$.

(I) $c = 1$.
(II) $H_{\overline{\Omega}} \cong \Omega_m(q^r).r$.
(III) $H_{\overline{\Omega}}$ *is non-local and* $\mathrm{soc}(H_{\overline{\Omega}}) \cong \Omega_m(q^r)$.

Proposition 4.3.18. *Assume that case* $\mathbf{O}^\epsilon$ *holds and* H *is of type* $GU_{n/2}(q)$, *so that* $r = 2$.

(O) $\epsilon = (-)^m$.
(I) $c = (m, 2)$ *and if* $c = 2$ *then* $\pi = \pi_2$.
(II) $H_{\overline{\Omega}} \cong (q + 1/a).U_m(q).[b(q + 1, \frac{n}{2})]$, *where* $a = (q + 1, 3 - \epsilon)$ *and* $b = c$ *if* q *is even,* $b = 1$ *if* q *is odd.*
(III) $H_{\overline{\Omega}}$ *is non-local if and only if* $(q, \epsilon) = (3, -)$, *and when this occurs* $\mathrm{soc}(H_{\overline{\Omega}}) \cong U_{n/2}(3)$.

Proof. Since $\tau_\sharp(\Delta_\sharp) = \mathbf{F}^*$ by (2.3.2), it follows as before that $c = |\ddot{I} : \ddot{H}_I|$. By (2.3.4) we may take $\phi_\sharp$ to centralize an orthonormal basis $\beta = \{v_1, \ldots, v_{n/2}\}$ of $V_\sharp$. Put $V_i = v_i\mathbf{F}_\sharp$, so that $V_1 \perp \cdots \perp V_{n/2}$ is a non-degenerate 1-decomposition of $(V_\sharp, \mathbf{F}_\sharp, \kappa_\sharp)$. Since $Q(v) = \kappa_\sharp(v, v)$ for all $v \in V$, it is clear that V_i is a non-degenerate -2-space in $(V, \mathbf{F}, Q)$, and so the V_i also form a non-degenerate -2-space decomposition of $(V, \mathbf{F}, Q)$. Assertion (O) is now a consequence of Lemma 2.5.11.ii. Following the discussion in §2.3 (see (2.3.1)), let $\delta_\sharp = \text{diag}_\beta(\omega, 1, \ldots, 1) \in I_\sharp$, where ω is a primitive $(q+1)^{\text{th}}$ root of unity in $\mathbf{F}_\sharp$, so that $I_\sharp = \Omega_\sharp\langle\delta_\sharp\rangle$. Thus $\langle\delta_\sharp\rangle = I(V_1, \kappa_\sharp) = \ker_{I(V_1,Q)}(\gamma)$, by Lemma 4.3.11.iv. Hence by Lemma 4.1.5.iv

$$\ddot{I}_\sharp = \langle\ddot{\delta}_\sharp\rangle = \ker_{\ddot{I}(V_1,Q)}(\ddot{\gamma}) = \ker_{\ddot{I}}(\ddot{\gamma}). \tag{4.3.23}$$

Notice that ψ inverts $I(V_i, \kappa_\sharp)$ for each i (by conjugation), and therefore $\psi_{V_i} \in I(V_i, Q)\backslash\ker_{I(V_i,Q)}(\gamma)$ for all i. Thus

$$\ddot{\psi} \in \begin{cases} \ker_{\ddot{I}}(\ddot{\gamma}) & \text{if } \frac{n}{2} \text{ is even} \\ \ddot{I}\backslash\ker_{\ddot{I}}(\ddot{\gamma}) & \text{if } \frac{n}{2} \text{ is odd.} \end{cases} \tag{4.3.24}$$

Therefore (4.3.23) and (4.3.24) together yield

$$c = |\ddot{I} : \ddot{H}_I| = |\ddot{I} : \langle\ddot{\delta}_\sharp, \ddot{\psi}\rangle| = |\ddot{I} : \ker_{\ddot{I}}(\ddot{\gamma})\langle\ddot{\psi}\rangle| = (2, \tfrac{n}{2}).$$

Now suppose that $c = 2$, so that $\frac{n}{2}$ is even. Clearly H_Γ acts on the set of totally singular $\frac{n}{4}$-spaces in $(V_\sharp, \mathbf{F}_\sharp, \kappa_\sharp)$, and each such space is a totally singular $\frac{n}{2}$-space in $(V, \mathbf{F}, Q)$. And as in the proof of Lemma 4.3.13, any two such spaces lie in the same Ω-orbit. Consequently $H_\Gamma \leq \ker(\gamma)$, and so we see now that (I) holds.

To prove (II), first note that $|H_I| = 2|GU_m(q)|$, and hence

$$|H_\Omega| = \frac{c|GU_m(q)|}{(2, q-1)}. \tag{4.3.25}$$

Now write $\langle z\rangle = Z(I_\sharp)$, so that $z = \text{diag}_\beta(\omega, \ldots, \omega)$, where ω is as above. As in the proof of (4.3.23) we see that $z \in \ker_I(\gamma)$ and that $z \in \Omega$ if and only if $q\frac{n}{2}$ is even. It follows that $|\langle z\rangle \cap \Omega| = \dfrac{q+1}{(2, q\frac{n}{2} - 1)}$, and hence

$$H_\Omega \cong \left[\frac{q+1}{(2, q\frac{n}{2} - 1)}\right].U_{n/2}(q).\left[\frac{(q+1, \frac{n}{2})c}{((2, q\frac{n}{2}), q-1)}\right].$$

We now obtain (II) upon factoring out scalars, noting that $-1 \notin \Omega$ if and only if $\frac{n}{2}$ is odd and $q \equiv 1 \pmod 4$.

As for (III), observe that $3 - \epsilon = q + 1$ only when $(q, \epsilon) = (3, -)$, and in this case the result follows from Lemma 4.3.9. ■

In this next result we refer to the discriminant, which was defined in §2.5 (see (2.5.14)).

Lemma 4.3.19. *Assume that case* $\mathbf{O}^\pm$ *holds, with* $\kappa_\sharp$ *quadratic,* q *odd and* $r = 2$. *Let* $v_\sharp \in V_\sharp$ *satisfy* $\mathbf{f}_\sharp(v_\sharp, v_\sharp) = \lambda \in \mathbf{F}_\sharp^*$, *so that* $r_{v_\sharp} \in I_\sharp$.

(i) $\text{span}_{\mathbf{F}_\sharp}(v_\sharp)$ *is a non-degenerate 2-space in* V *with discriminant* $\mu\text{N}(\lambda)$.

(ii)

$$r_{v_\sharp} \in \begin{cases} \Omega & \text{if } \lambda \notin (\mathbf{F}_\sharp^*)^2 \\ S\backslash\Omega & \text{if } \lambda \in (\mathbf{F}_\sharp^*)^2. \end{cases}$$

Proof. Choose $\eta \in \mathbf{F}_\sharp \backslash \mathbf{F}$ such that $\eta^2 = \mu$, so that $\mathrm{T}(\eta) = 0$. Thus $W = \mathrm{span}_{\mathbf{F}_\sharp}(v_\sharp)$ has a basis $\beta = \{v_\sharp, \eta v_\sharp\}$ and $r_{v_\sharp} = -1_W$. Now write $\lambda = a + b\eta$, with $a, b \in \mathbf{F}$. Then on W,

$$\mathbf{f}_\beta = \begin{pmatrix} \mathrm{T}(\lambda) & \mathrm{T}(\eta\lambda) \\ \mathrm{T}(\eta\lambda) & \mathrm{T}(\mu\lambda) \end{pmatrix} = \begin{pmatrix} 2a & 2\mu b \\ 2\mu b & 2\mu a \end{pmatrix},$$

and so $D(W) \equiv \mu(a^2 - b^2\mu) \equiv \mu\mathrm{N}(\lambda) \pmod{(\mathbf{F}^*)^2}$. Assertion (i) is now clear. As for (ii), note that if $\lambda \in (\mathbf{F}_\sharp^*)^2$, then $\mathrm{N}(\lambda) \in (\mathbf{F}^*)^2$, and so $D(W)$ is a non-square; since $r_{v_\sharp}$ negates W, this implies that $r_{v_\sharp} \in S \backslash \Omega$. Use the same reasoning when $\lambda \notin (\mathbf{F}_\sharp^*)^2$. ∎

We may now analyse groups of type $O_{n/2}(q^2)$ in case $\mathbf{O}^\pm$.

Proposition 4.3.20. *Assume that case* $\mathbf{O}^\epsilon$ *holds and that* H *is of type* $O_{n/2}(q^2)$ *with* n *even and* $q^{\frac{n}{2}}$ *odd, so that* $r = 2$ *and* $n \geq 10$.

(I)

$$c = \begin{cases} 1 & \text{if } D = \boxtimes \\ 2 & \text{if } D = \square. \end{cases}$$

If $c = 2$, *then* $\pi = \pi_3$.

(II) $H_{\overline{\Omega}} \cong \Omega_{n/2}(q^2).2$.

(III) $H_{\overline{\Omega}}$ *is non-local and* $\mathrm{soc}(H_{\overline{\Omega}}) \cong \Omega_{n/2}(q^2)$.

Proof. Part (III) follows from Lemma 4.3.9. It remains to prove (I) and (II). Since by (2.6.3) we have $\tau_\sharp(\Delta_\sharp) = (\mathbf{F}_\sharp^*)^2 \supseteq \mathbf{F}^*$, we deduce as before that $HI = \Delta$, and hence $c = |\ddot{I} : \ddot{H}_I|$. According to (4.3.13), $\det_{\mathbf{F}}(I_\sharp) = 1$, and so $I_\sharp \leq S$. However, by Lemma 4.3.19.ii, a reflection in a vector of $\mathbf{f}_\sharp$-norm 1 in $V_\sharp$ lies in $S \backslash \Omega$. Therefore

$$\ddot{I}_\sharp = \ddot{S}. \tag{4.3.26}$$

Thus to complete (I) we must determine whether ψ lies in S or in $I \backslash S$.

According to (2.6.4), we can take the element $\phi_\sharp$ to centralize a non-degenerate 1-space decomposition $\langle w_1 \rangle \perp \cdots \perp \langle w_{n/2} \rangle$ of $V_\sharp$ with $\mathbf{f}_\sharp(w_i, w_j) = \lambda\delta_{ij}$, where $\lambda = 1$ if $D_\sharp = D(Q_\sharp) = \square$ and $\langle \lambda \rangle = \mathbf{F}_\sharp^*$ if $D_\sharp = \boxtimes$. Moreover $w_i\phi_\sharp = \lambda^{(p-1)/2} w_i$, and so for all $\xi \in \mathbf{F}_\sharp$,

$$(\xi w_i)\psi = \xi^q \lambda^{(q-1)/2} w_i. \tag{4.3.27}$$

We put $W_i = \mathrm{span}_{\mathbf{F}_\sharp}(w_i)$, a non-degenerate 2-space in V with $D(W_i) = \mu\mathrm{N}(\lambda)$ (see Lemma 4.3.19.i). Indeed, $V = W_1 \perp \cdots \perp W_{n/2}$ is a non-degenerate 2-space decomposition of V. We now consider two cases separately.

Case $D = \boxtimes$. According to Proposition 2.5.11.i, $D(W_i) = \boxtimes$ for each i. Consequently $\mu\mathrm{N}(\lambda) \notin (\mathbf{F}^*)^2$, and hence $\mathrm{N}(\lambda) \in (\mathbf{F}^*)^2$. Therefore $\lambda \in (\mathbf{F}_\sharp^*)^2$, and so $\lambda = 1$. Now let η satisfy $\eta^2 = \mu$ as in the proof of Lemma 4.3.19, and let β_i be the basis $\{w_i, \eta w_i\}$ of W_i. Then by (4.3.27), $\psi_{\beta_i} = \mathrm{diag}(1, -1)$, and so $\det(\psi) = -1$. Therefore $\psi \in I \backslash S$, and so by (4.3.26) we obtain $\ddot{H}_I = \ddot{I}_\sharp\langle\ddot{\psi}\rangle = \ddot{I}$, which proves $c = 1$. Therefore as $-1 \notin \Omega$, we have $H_{\overline{\Omega}} = H_\Omega = \frac{1}{4}H_I$, and since $H_I = I_\sharp.2 = \Omega_\sharp.[8]$, we deduce $H_{\overline{\Omega}} \cong \Omega_m(q^2).2$, proving (I) and (II).

Case $D = \square$. Reasoning as in the previous case we deduce $D(W_i) = \square$ for all i, and hence $\langle\lambda\rangle = \mathbf{F}_\sharp^*$. This time we take a basis β_i of W_i given by $\{w_i, \lambda^{(q-1)/2}w_i\}$. According to (4.3.27),

$$\psi_{\beta_i} = \begin{pmatrix} 0 & 1 \\ -1 & 0 \end{pmatrix}.$$

In particular, $\det(\psi) = 1$, so $\psi \in S$, which proves $\ddot{H}_I = \ddot{S}$. Therefore $c = 2$, proving the first part of (I). Moreover $H_\Omega = \frac{1}{2}H_I = \Omega_\sharp.[4]$, and so (II) holds upon factoring out $\langle -1\rangle$.

It remains to determine π when $c = 2$. Thus we continue to assume that $D = \square$, so that $\langle\lambda\rangle = \mathbf{F}_\sharp^*$. Note that λ generates $\mathbf{F}_\sharp$ and μ generates $\mathbf{F}^*$, and so we may assume that $\lambda^{q+1} = \mu$. Put $\eta = \lambda^{(q+1)/2}$, and observe that $\tau_\sharp(\eta) = \eta^2 = \mu$, and hence $\mathbf{F}_\sharp^* \cap \Delta = \langle\eta\rangle$. Since $\Delta_\sharp = I_\sharp\mathbf{F}_\sharp^*$, it follows that $\Delta_{\sharp,\mathbf{F}} = I_\sharp\langle\eta\rangle$ (the definition of $\Delta_{\sharp,\mathbf{F}}$ is given after (4.3.8)). As we have already seen, $\ddot{I}_\sharp = \ddot{S}$ and $\ddot{\psi} \in \ddot{S}$, and so by the discussion between (4.3.8) and (4.3.11), we have

$$\ddot{H}_I = \ddot{S}, \quad \ddot{H} = \langle\ddot{S}, \ddot{\eta}\rangle, \quad \ddot{H}_\Gamma = \ker_{\ddot{\Gamma}}(\pi) = \ddot{S}\langle\ddot{\eta}, \ddot{\phi}_\sharp\rangle. \tag{4.3.28}$$

Case $\epsilon = -$. Since $D = \square$, we know that $q \equiv 3 \pmod 4$ by Proposition 2.5.10, and by Proposition 2.8.2, $\ddot{\Gamma} \cong D_8 \times \mathbf{Z}_f$, with f odd. Indeed, $\ddot{\Gamma}$ has just three subgroups of index 2 which contain $\ddot{\phi}$, and in the notation of Proposition 2.8.2 they are

(a) $\ker_{\ddot{\Gamma}}(\ddot{\tau}) = \langle\ddot{r}_\square\rangle \times \langle\ddot{r}_\boxtimes\rangle \times \langle\ddot{\phi}\rangle \cong \mathbf{Z}_2 \times \mathbf{Z}_2 \times \mathbf{Z}_f$
(b) $\ker_{\ddot{\Delta}}(\ddot{\gamma}) \times \langle\ddot{\phi}\rangle = \langle\ddot{r}_\square\ddot{\delta}\rangle \times \langle\ddot{\phi}\rangle \cong \mathbf{Z}_4 \times \mathbf{Z}_f$
(c) $\langle\ddot{r}_\square\ddot{r}_\boxtimes\rangle \times \langle\ddot{\delta}\rangle \times \langle\ddot{\phi}\rangle \cong \mathbf{Z}_2 \times \mathbf{Z}_2 \times \mathbf{Z}_f$.

However $\ddot{\eta} \notin \ker(\ddot{\tau})$ (since μ is a non-square in $\mathbf{F}$), and so $\ker(\pi)$ is either (b) or (c). Furthermore, $\eta^2 = \lambda^{(q+1)} = \mu \in \mathbf{F}^*$, which shows $|\ddot{\eta}| = 2$. Since $\ddot{S} = \langle\ddot{r}_\square\ddot{r}_\boxtimes\rangle$ is the centre of $\ddot{\Delta} \cong D_8$, it follows from (4.3.28) that $\ddot{H} \cong \mathbf{Z}_2 \times \mathbf{Z}_2$ (as opposed to $\mathbf{Z}_4$). This eliminates (b), and so $\ker_{\ddot{\Gamma}}(\pi)$ is given in (c). Thus $\pi = \pi_3$, as desired.

Case $\epsilon = +$. Here $q \equiv 1 \pmod 4$, and we appeal to Proposition 2.7.3, which shows that $\ddot{\Gamma} \cong D_8 \times \mathbf{Z}_f$. Further, $|\ddot{\delta}| = 4$ and $|\ddot{r}_\square\ddot{\delta}| = |\ddot{r}_\boxtimes\ddot{\delta}| = 2$. As in the previous paragraph, $\ddot{H} \cong \mathbf{Z}_2 \times \mathbf{Z}_2$, and so this time $\ddot{H} = \langle\ddot{r}_\square\ddot{\delta}, \ddot{r}_\boxtimes\ddot{\delta}\rangle$, for this is the unique Klein fours group in $\ddot{\Delta}$ not contained in $\ddot{I}$. Now $\ddot{\phi}_\sharp$ centralizes $\ddot{S}$ (indeed $\ddot{S} \le Z(\ddot{\Gamma})$) and $\eta^{\phi_\sharp} = \eta^p = \eta\mu^{(p-1)/2}$, which shows $\ddot{\phi}_\sharp$ also centralizes $\ddot{\eta}$. Consequently

$$\ddot{H}_\Gamma = C_{\ddot{\Gamma}}(\ddot{H}) = \begin{cases} \langle\ddot{r}_\square\ddot{\delta}, \ddot{r}_\boxtimes\ddot{\delta}, \ddot{\phi}\rangle & \text{if } p \equiv 1 \pmod 4 \\ \langle\ddot{r}_\square\ddot{\delta}, \ddot{r}_\boxtimes\ddot{\delta}, \ddot{r}_\square\ddot{\phi}\rangle & \text{if } p \equiv 3 \pmod 4, \end{cases}$$

by Proposition 2.7.3.iii. Thus $\pi = \pi_3$, and the proof is complete. ∎

§4.4 The tensor product subgroups $\mathcal{C}_4$

For $1 \le i \le t$, let V_i be a vector space over $\mathbf{F}$ of dimension n_i, so that $V_1 \otimes \cdots \otimes V_t$ is a vector space of dimension $\prod_{i=1}^t n_i$ over $\mathbf{F}$. Throughout this section we shall assume that $\prod_{i=1}^t n_i = n$, and so we may identify V with $V_1 \otimes \cdots \otimes V_t$. For $g_i \in GL(V_i)$ $(1 \le i \le t)$ define the element $g_1 \otimes \cdots \otimes g_t \in GL(V)$ by setting

$$(v_1 \otimes \cdots \otimes v_t)(g_1 \otimes \cdots \otimes g_t) = v_1g_1 \otimes \cdots \otimes v_tg_t \quad (v_i \in V_i),$$

and then extending linearly. Thus when case **L** holds we obtain the inclusion

$$\begin{aligned} GL_{n_1}(q) \circ \cdots \circ GL_{n_t}(q) &\cong GL(V_1, \mathbf{F}) \otimes \cdots \otimes GL(V_t, \mathbf{F}) \\ &\leq GL(V, \mathbf{F}) \cong GL_n(q). \end{aligned} \tag{4.4.1}$$

Suppose now that $\mathbf{F} = \mathbf{F}_q$, that q is odd, and that $\mathbf{f}_i$ is a non-degenerate bilinear form on V_i, so that $(V_i, \mathbf{F}, \mathbf{f}_i)$ is either a symplectic or orthogonal geometry of dimension n_i. Put $X_i = X(V_i, \mathbf{f}_i)$, where X ranges over the symbols in (2.1.15). We may now define a non-degenerate bilinear form $\mathbf{f} = \mathbf{f}_1 \otimes \cdots \otimes \mathbf{f}_t$ on $V_1 \otimes \cdots \otimes V_t$ by setting $\mathbf{f}(v_1 \otimes \cdots \otimes v_t, w_1 \otimes \cdots \otimes w_t) = \prod_{i=1}^t \mathbf{f}_i(v_i, w_i)$ (here $v_i, w_i \in V_i$) and then extending bilinearly. It is easy to check that $\mathrm{sym}(\mathbf{f}) = \prod_{i=1}^t \mathrm{sym}(\mathbf{f}_i)$. In fact, let us arrange the forms so that $\mathbf{f}_1, \ldots, \mathbf{f}_s$ are symplectic and $\mathbf{f}_{s+1}, \ldots, \mathbf{f}_t$ are symmetric. Then $\mathrm{sym}(\mathbf{f}) = (-1)^s$, and putting $\epsilon_i = \mathrm{sgn}(\mathbf{f}_i)$ for $s+1 \leq i \leq t$ we obtain an embedding

$$\begin{aligned} Sp_{n_1}(q) \circ \cdots \circ Sp_{n_s}(q) \circ O^{\epsilon_{s+1}}_{n_{s+1}}(q) \circ \cdots \circ O^{\epsilon_t}_{n_t}(q) &\cong I_1 \otimes \cdots \otimes I_t \leq \\ I(V, \mathbf{f}) &\cong \begin{cases} Sp_n(q) & \text{if } s \text{ is odd} \\ O^{\epsilon}_n(q) & \text{if } s \text{ is even,} \end{cases} \end{aligned} \tag{4.4.2}$$

where $\epsilon = \mathrm{sgn}(\mathbf{f})$ when s is even. The relationship between ϵ and the ϵ_i is described in Lemma 4.4.2, below.

Now consider the case where $\mathbf{F} = \mathbf{F}_q$ with q even, and each $\mathbf{f}_i$ is a non-degenerate bilinear form on V_i, so that $(V_i, \mathbf{F}, \mathbf{f}_i)$ is a symplectic geometry. Again set $X_i = X(V_i, \mathbf{f}_i)$. As in the previous paragraph, we obtain a non-degenerate symmetric bilinear form $\mathbf{f} = \mathbf{f}_1 \otimes \cdots \otimes \mathbf{f}_t$ on V. In addition, there is a unique non-degenerate quadratic form $Q = Q(\mathbf{f}_1, \ldots, \mathbf{f}_t)$ on V such that

(a) $Q(v_1 \otimes \cdots \otimes v_t) = 0$ for all $v_i \in V_i$, and

(b) $\mathbf{f}_Q = \mathbf{f}_1 \otimes \cdots \otimes \mathbf{f}_t$.

If W is a totally singular $\frac{n_1}{2}$-space in $(V_1, \mathbf{f}_1)$, then $W \otimes V_2 \otimes \cdots \otimes V_t$ is a totally singular $\frac{n}{2}$-space in (V, Q), and so $\mathrm{sgn}(Q) = +$. This yields

$$\begin{aligned} Sp_{n_1}(q) \otimes \cdots \otimes Sp_{n_t}(q) &\cong I_1 \otimes \cdots \otimes I_t \\ &\leq I(V, Q) \approx O^+_n(q). \end{aligned} \tag{4.4.3}$$

Finally, suppose that $\mathbf{F} = \mathbf{F}_{q^2}$ and that $(V_i, \mathbf{F}, \mathbf{f}_i)$ is a unitary geometry of dimension n_i, and as before put $X_i = X(V_i, \mathbf{f}_i)$. We may now define a non-degenerate unitary form $\mathbf{f} = \mathbf{f}_1 \otimes \cdots \otimes \mathbf{f}_t$ on $V_1 \otimes \cdots \otimes V_t$ by setting $\mathbf{f}(v_1 \otimes \cdots \otimes v_t, w_1 \otimes \cdots \otimes w_t) = \prod_{i=1}^t \mathbf{f}_i(v_i, w_i)$ and then extending sesquilinearly. This yields an embedding

$$\begin{aligned} GU_{n_1}(q) \circ \cdots \circ GU_{n_t}(q) &\cong I_1 \otimes \cdots \otimes I_t \\ &\leq I(V, \mathbf{f}) \approx GU_n(q). \end{aligned} \tag{4.4.4}$$

Assume now that one of the embeddings given in (4.4.1)-(4.4.4) occurs. For the sake of consistency, in the first paragraph let $\mathbf{f}_i$ be the zero form on V_i, so that $\mathbf{f} = \mathbf{f}_1 \otimes \cdots \otimes \mathbf{f}_t$

is the zero form on V. Moreover, we set $\kappa = \mathbf{f}_1 \otimes \cdots \otimes \mathbf{f}_t$ in (4.4.1), (4.4.2) and (4.4.4) and $\kappa = Q(\mathbf{f}_1, \ldots, \mathbf{f}_t)$ in (4.4.3). We may then write (in all four cases)

$$(V, \kappa) = (V_1, \mathbf{f}_1) \otimes \cdots \otimes (V_t, \mathbf{f}_t), \tag{4.4.5}$$

and we refer to this expression as a *tensor decomposition* of (V, κ). Sometimes we use the symbol $\mathcal{D}$ to denote such a tensor decomposition. In all four cases, set $X_i = X(V_i, \mathbf{F}, \mathbf{f}_i)$ and $X = X(V, \mathbf{F}, \kappa)$, where X ranges over the symbols in (2.1.15), and set $x_i = x_{(V_i, \mathbf{f}_i)}$ and $x = x_{(V,\kappa)}$, where x is σ or τ. One checks that if $g_i \in \Delta_i$, then $g = g_1 \otimes \cdots \otimes g_t \in \Delta$ and that

$$\tau(g) = \prod_{i=1}^{t} \tau_i(g_i). \tag{4.4.6}$$

Therefore $\Delta_1 \otimes \cdots \otimes \Delta_t \leq \Delta$. Furthermore, suppose that $\phi_i \in \Gamma_i$ satisfies $\sigma_i(\phi_i) = \nu$ (recall ν is a generator for $Aut(\mathbf{F})$). Then we may define $\phi_{\mathcal{D}} = \phi_1 \otimes \cdots \otimes \phi_t$ as the unique element of $\Gamma(V, \mathbf{F})$ which satisfies $\sigma(\phi_{\mathcal{D}}) = \nu$ and $(v_1 \otimes \cdots \otimes v_t)\phi_{\mathcal{D}} = v_1\phi_1 \otimes \cdots \otimes v_t\phi_t$. A direct calculation shows that $\phi_{\mathcal{D}} \in \Gamma$. Evidently $\phi_{\mathcal{D}}$ normalizes $\Delta_1 \otimes \cdots \otimes \Delta_t$, and we write

$$\Gamma_{(\mathcal{D})} = (\Delta_1 \otimes \cdots \otimes \Delta_t)\langle \phi_{\mathcal{D}} \rangle. \tag{4.4.7}$$

Observe that $\Gamma_{(\mathcal{D})}$ is independent of the choice of the ϕ_i. For any subgroup $G \leq \Gamma$ we write $G_{(\mathcal{D})} = G \cap \Gamma_{(\mathcal{D})}$. Thus for example

$$\Delta_{(\mathcal{D})} = \Delta_1 \otimes \cdots \otimes \Delta_t. \tag{4.4.8}$$

Definition. The members of $\mathcal{C}_4(\Gamma)$ are the groups $(\Delta_1 \otimes \Delta_2)\langle \phi_{\mathcal{D}} \rangle$ described in (4.4.7) (arising from embeddings (4.4.1)-(4.4.4) with $t = 2$) such that following conditions hold:
(a)

$$\begin{cases} (V, Q) \cong (V_1 \otimes V_2, Q(\mathbf{f}_1 \otimes \mathbf{f}_2)) & \text{in case } \mathbf{O}^+ \text{ with } q \text{ even} \\ (V, \mathbf{f}) \cong (V_1 \otimes V_2, \mathbf{f}_1 \otimes \mathbf{f}_2) & \text{otherwise,} \end{cases}$$

(b) $(V_1, \mathbf{f}_1)$ is not similar to $(V_2, \mathbf{f}_2)$, and
(c) the forms $\mathbf{f}_i$ are as described in Table 4.4.A, below.

Table 4.4.A

case	type	description of $\mathcal{D}$	conditions
L	$GL_{n_1}(q) \otimes GL_{n_2}(q)$	$n_1 < n_2$, $\mathbf{f}_i = 0$	
U	$GU_{n_1}(q) \otimes GU_{n_2}(q)$	$n_1 < n_2$, $\mathbf{f}_i$ unitary	
S	$Sp_{n_1}(q) \otimes O^{\epsilon}_{n_2}(q)$	$\mathbf{f}_1$ symplectic, $\mathbf{f}_2$ symmetric	q odd, $n_2 \geq 3$
$\mathbf{O}^+$	$Sp_{n_1}(q) \otimes Sp_{n_2}(q)$	$n_1 < n_2$, $\mathbf{f}_i$ symplectic	
O	$O^{\epsilon_1}_{n_1}(q) \otimes O^{\epsilon_2}_{n_2}(q)$	$\mathbf{f}_i$ symmetric, $(n_1, \epsilon_1) \neq (n_2, \epsilon_2)$	q odd, $n_i \geq 3$

Remarks on the conditions. The condition $n_i \geq 3$ appearing in rows 3 and 5 of Table 4.4.A does not appear in Aschbacher's description in [As₁]. We justify imposition of this condition in Proposition 4.4.4, below.

Before analysing the groups in $\mathcal{C}_4$, we first make some general remarks about properties of tensor decompositions. So assume that $\mathcal{D}$ is a tensor decomposition as given in (4.4.5) and take $g_i \in \Delta_i$. Then it is easy to see that

$$\det(g_1 \otimes \cdots \otimes g_t) = \prod_{i=1}^{t} \det(g_i)^{n/n_i}. \tag{4.4.9}$$

Now suppose that $\beta_i = \{v_{i,1}, \ldots, v_{i,n_i}\}$ is a basis of V_i. Then we may form the tensor product basis $\beta = \beta_1 \otimes \cdots \otimes \beta_t$, whose elements are the vectors $v_{1,i_1} \otimes \cdots \otimes v_{t,i_t}$ with $1 \leq i_j \leq n_j$, and taken with a lexicographical ordering, so that $v_{1,1} \otimes \cdots \otimes v_{t,1}$ is first, $v_{1,1} \otimes \cdots \otimes v_{t-1,1} \otimes v_{t,2}$ is second, and $v_{1,n_1} \otimes \cdots \otimes v_{t,n_t}$ is last. Note for example that in the second and fourth paragraphs of this section, $(\mathbf{f}_1)_{\beta_1} \otimes \cdots \otimes (\mathbf{f}_t)_{\beta_t} = \mathbf{f}_{\beta_1 \otimes \cdots \otimes \beta_t}$. Consequently,

Lemma 4.4.1. *Assume that q is odd and that $\mathbf{f}_i$ is a non-degenerate symmetric bilinear form for all i. Then $D(\mathbf{f}) = \prod_{i=1}^{t} D(\mathbf{f}_i)^{n/n_i}$.*

Lemma 4.4.2. *If q is odd and n is even, then ϵ given in (4.4.2) satisfies*

$$\epsilon = \begin{cases} - & \text{if } s = 0,\ \epsilon_j = - \text{ for some } j \\ & \text{and } n_i \text{ is odd for all } i \neq j \\ + & \text{otherwise.} \end{cases}$$

Proof. Clearly we may assume that n is even. Observe that if $(V_i, \mathbf{f}_i)$ contains a totally singular $n_i/2$-space W_i, then $V_1 \otimes \cdots \otimes V_{i-1} \otimes W_i \otimes V_{i+1} \otimes \cdots \otimes V_t$ is a totally singular $n/2$-space in $(V, \mathbf{f})$, and so $\epsilon = +$. Thus it may be assumed that $s = 0$ and $\epsilon_i \in \{-, \circ\}$ for all i. If $\epsilon_j = \epsilon_k = -$ for some j, k with $j \neq k$, then n/n_i is even for all i, and so Lemma 4.4.1 implies that $D(\mathbf{f})$ is a square; and since $n \equiv 0 \pmod 4$, we deduce from Proposition 2.5.10 that $\epsilon = +$. Therefore we can assume that $\epsilon_j = -$ for some j and n_i is odd for all $i \neq j$. Then Lemma 4.4.1 implies that $D(\mathbf{f}) = D(\mathbf{f}_j)$ and also $n \equiv n_j \pmod 4$. Therefore $\epsilon = -$ by Proposition 2.5.10. ■

For a subgroup M_i of $GL(V_i)$ we identify M_i with the subgroup $1 \otimes \cdots \otimes 1 \otimes M_i \otimes 1 \otimes \cdots \otimes 1$. Observe that V is *homogeneous* as a $GL(V_i)$-module; that is, V is a direct sum of isomorphic irreducible $GL(V_i)$-submodules. To see this, take $i = 1$ for convenience, and let $x_1, \ldots, x_{n/n_1}$ be a basis for $V_2 \otimes \cdots \otimes V_t$. Then $V = \bigoplus_i (V_1 \otimes x_i)$ is the desired direct sum decomposition.

Lemma 4.4.3. *Assume that $M_i \leq GL(V_i)$ for all i.*

(i) *If M_t is absolutely irreducible on V_t, then $C_{GL(V)}(M_t) = GL(V_1 \otimes \cdots \otimes V_{t-1})$.*

(ii) *If M_t is absolutely irreducible on V_t, then $N_{GL(V)}(M_t) = GL(V_1 \otimes \cdots \otimes V_{t-1}) \otimes N_{GL(V_t)}(M_t)$.*

(iii) *If each M_i is absolutely irreducible on V_i, then $\bigcap_i N_{GL(V)}(M_i) = N_{GL(V_1)}(M_1) \otimes \cdots \otimes N_{GL(V_t)}(M_t)$.*

(iv) *If M_t is absolutely irreducible on V_t, then the irreducible submodules for M_t are precisely the spaces $v \otimes V_t$ with $0 \neq v \in V_1 \otimes \cdots \otimes V_{t-1}$.*

(v) *If M_1 is irreducible on V_1 and M_i is absolutely irreducible on V_i for $i \geq 2$, then $M_1 \otimes \cdots \otimes M_t$ is irreducible on V.*

(vi) *If each M_i is absolutely irreducible on V_i, then $M_1 \otimes \cdots \otimes M_t$ is absolutely irreducible on V.*

Proof. We prove the result for $t = 2$; the general case then follows by induction.

Let $\beta_1 = \{x_1, \ldots, x_{n_1}\}$ be a basis for V_1, and $\beta_2 = \{y_1, \ldots, y_{n_2}\}$ a basis for V_2. Then $\beta = \{x_i \otimes y_j \mid 1 \leq i \leq n_1,\ 1 \leq j \leq n_2\}$ is a basis for $V = V_1 \otimes V_2$. Using the lexicographical ordering of β described above, for each $g \in M_2$ we have

$$g_\beta = \begin{pmatrix} g_{\beta_2} & & \\ & \ddots & \\ & & g_{\beta_2} \end{pmatrix}. \tag{4.4.10}$$

Now take $h \in N_{GL(V)}(M_2)$, and write

$$h_\beta = \begin{pmatrix} \mathbf{x}_{11} & \cdots & \mathbf{x}_{1n_1} \\ \vdots & & \vdots \\ \mathbf{x}_{n_1 1} & \cdots & \mathbf{x}_{n_1 n_1} \end{pmatrix}$$

with $\mathbf{x}_{ij} \in M_{n_2}(q)$, and observe that

$$g_{\beta_2}\mathbf{x}_{ij} = \mathbf{x}_{ij}(g^h)_{\beta_2} \tag{4.4.11}$$

for all i, j.

(i). Assume that M_2 is absolutely irreducible on V_2 and that h centralizes M_2. Then Schur's Lemma implies that $\mathbf{x}_{ij}$ is a scalar λ_{ij} times the identity matrix for all i, j. It follows that $h \in GL(V_1)$ and $(h_{\beta_1})_{ij} = \lambda_{ij}$.

(ii). Assume that M_2 is absolutely irreducible on V_2. Here, (4.4.11) and Schur's Lemma imply that $\mathbf{x}_{ij}$ is either 0 or non-singular. Thus for some pair i, j, we know that $\mathbf{x} = \mathbf{x}_{ij}$ is non-singular, and hence $g^h = g^k$, where $k \in GL(V)$ satisfies

$$k_\beta = \begin{pmatrix} \mathbf{x} & & \\ & \ddots & \\ & & \mathbf{x} \end{pmatrix}. \tag{4.4.12}$$

Therefore $k \in N_{GL(V_2)}(M_2)$, and $hk^{-1} \in C_{GL(V)}(M_2) = GL(V_1)$, by (i).

(iii). This is clear from (ii).

(iv). Assume that M_2 is absolutely irreducible and let W be an irreducible M_2-submodule of V. Denote by $\pi_i : W \mapsto x_i \otimes V_2$ the ith projection map. Then $\ker(\pi_i)$ is 0 or W, and hence there exists i such that $\ker(\pi_i) = 0$. With no loss $i = 1$, and thus $V = W \oplus \left(\bigoplus_{2 \leq i \leq n_1} x_i \otimes V_2\right)$. Since M_2 is homogeneous on V, we can choose a basis β' of V, contained in $W \cup \bigcup_{2 \leq i \leq n_1}(x_i \otimes V_2)$, such that for all $g \in M_2$, the matrix $g_{\beta'}$ is exactly the same matrix as given in (4.4.10). It is thus clear that there is an element

$z \in C_{GL(V)}(M_2)$ interchanging W and $x_2 \otimes V_2$. Using the fact that M_2 is absolutely irreducible on V_2, we know that $C_{GL(V)}(M_2) = GL(V_1)$ by (i). Consequently $z \in GL(V_1)$, and so $W = (x_2 \otimes V_2)z = v \otimes V_2$, where $v = x_2 z$.

(v). Assume that M_1 is irreducible on V_1 and that M_2 is absolutely irreducible on V_2. Let U be an irreducible $M_1 \otimes M_2$-submodule of V, and let W be an irreducible M_2-submodule of U. By (iii) we have $W = v \otimes V_2$ for some $v \in V_1$. Since M_1 is irreducible on V_1, we know that vM_1 spans V_1, which proves $U = V$, as desired.

(vi). Finally, assume that M_i is absolutely irreducible on V_i for $i = 1, 2$. Then $M_1 \otimes M_2$ is irreducible on V by (v), and so by Lemma 2.10.1 it suffices to show that $C_{GL(V)}(M_1 \otimes M_2) = \mathbf{F}^*$. Notice that (i) implies that $C_{GL(V)}(M_1 \otimes M_2) = GL(V_1) \cap GL(V_2) = \mathbf{F}^*$, and so the result is proved. ■

Remark. Observe that if $K \leq GL(V)$ and K acts homogeneously, then there is a basis of V such that all elements of K_β have the shape appearing in (4.4.12). This shows that $K \leq GL(V_1) \leq GL(V_1) \otimes GL(V_2)$ for some tensor decomposition $V = V_1 \otimes V_2$.

We now account for the conditions $n_i \geq 3$ in Table 4.4.A.

Proposition 4.4.4. *Assume that $\mathcal{D}$ is a tensor decompostion $(V, \kappa) = (V_1, \mathbf{f}_1) \otimes (V_2, \mathbf{f}_2)$ as described above, and assume that $I(V_1, \mathbf{f}_1) \approx O_2^\pm(q)$ with q odd. Then $\Gamma_{(\mathcal{D})}$ is contained in another member of $\mathcal{C}(\Gamma)$.*

Proof. Write $X_i = X(V_i, \mathbf{f}_i)$ as usual. Suppose first that $I_1 \approx O_2^+(q)$. Then $(V_1, \mathbf{f}_1)$ contains just two singular points U, W, say, and Γ_1 acts on these two points. Evidently $U \otimes V_2$ and $W \otimes V_2$ are totally singular $\frac{n}{2}$-spaces in (V, κ), and hence $\Gamma_{(\mathcal{D})} < \Gamma_{\mathcal{D}_1} \in \mathcal{C}_2(\Gamma)$, where $\mathcal{D}_1$ is the totally singular $\frac{n}{2}$-space decomposition $(U \otimes V_2) \oplus (W \otimes V_2)$. Next assume that $I_1 \approx O_2^-(q)$. Then $I_1' \cong \mathbf{Z}_{q+1}$ is irreducible (but not absolutely irreducible) on V_1. It follows from Lemma 4.4.3.v that $J = I_1' \otimes I_2$ is irreducible on V. However $I_1' \leq Z(J)$, and hence $C_{GL(V)}(J) \neq \mathbf{F}^*$. Therefore J is not absolutely irreducible on V by Lemma 2.10.1. Since $\Gamma_{(\mathcal{D})} \leq N_\Gamma(J)$, the result now follows from Lemma 4.3.12. ■

For the remainder of this section we shall assume that $H_\Gamma \in \mathcal{C}_4(\Gamma)$, so that $H_\Gamma = \Gamma_{(\mathcal{D})}$, where $\mathcal{D}$ is a tensor decomposition $(V, \kappa) = (V_1, \mathbf{f}_1) \otimes (V_2, \mathbf{f}_2)$ as described in Table 4.4.A, above. Thus

$$H = \Delta_1 \otimes \Delta_2. \tag{4.4.13}$$

We now give a description of H_I. Suppose for the moment that $\Delta_1 = I_1 \mathbf{F}^*$. Then any element in H_I can be written $\lambda g_1 \otimes g_2$, where $g_1 \in I_1$, $\lambda \in \mathbf{F}^*$ and $g_2 \in \Delta_2$. But this equals $g_1 \otimes \lambda g_2$, and it follows that $\lambda g_2 \in I_2$. Therefore $g_2 \in I_2 \mathbf{F}^*$, whence $H_I \leq (I_1 \otimes I_2)\mathbf{F}^*$. Consequently $H_I = (I_1 \otimes I_2)(I \cap \mathbf{F}^*)$. The same holds of course if we assume that $\Delta_2 = I_2 \mathbf{F}^*$ instead. Note that $\Delta_i = I_i \mathbf{F}^*$ for some i precisely when case $\mathbf{L}^\epsilon$ holds, or when case $\mathbf{O}^+$ holds with q even, or when $I_i \approx O_{n_i}(q)$ with qn_i odd. In all of these cases we see that $\mathbf{F}^* \cap I \leq I_1 \cap I_2$ (for example, when $I_i \approx O_{n_i}(q)$ with qn_i odd, then case $\mathbf{O}$

or **S** holds, and so $\mathbf{F}^* \cap I = \langle -1 \rangle \leq I_1 \cap I_2$). It therefore follows that

$$H_I = I_1 \otimes I_2 \text{ if } H \text{ is of type} \begin{cases} GL^{\pm}_{n_1}(q) \otimes GL^{\pm}_{n_2}(q) \\ Sp_{n_1}(q) \otimes O_{n_2}(q), \ n_2 \text{ odd} \\ Sp_{n_1}(q) \otimes Sp_{n_2}(q), \ q \text{ even} \\ O^{\epsilon_1}_{n_1}(q) \otimes O^{\epsilon_2}_{n_2}(q), \ n_i \text{ odd for some } i. \end{cases} \tag{4.4.14}$$

Now suppose that $\Delta_i \neq I_i\mathbf{F}^*$ for both i. There exists $\delta_i \in \Delta_i$ such that $\tau_i(\delta_i) = \mu$, and we put $z = \delta_1 \otimes \delta_2^{-1}$. If $g_i \in \Delta_i$ and $g_1 \otimes g_2 \in H_I$, then clearly $\tau_1(g_1)\tau_2(g_2) = 1$ by (4.4.6), and so $g_1 \otimes g_2$ times a power of z lies in $I_1 \otimes I_2$. Thus we have

$$H_I = (I_1 \otimes I_2)\langle z \rangle \text{ if } H \text{ is of type} \begin{cases} Sp_{n_1}(q) \otimes O^{\epsilon}_{n_2}(q), \ n_2 \text{ even} \\ Sp_{n_1}(q) \otimes Sp_{n_2}(q), \ q \text{ odd} \\ O^{\epsilon_1}_{n_1}(q) \otimes O^{\epsilon_2}_{n_2}(q), \ n_1, n_2 \text{ both even.} \end{cases} \tag{4.4.15}$$

As a convenience we record several pieces of information concerning this element z.

Lemma 4.4.5. *Assume that situation* (4.4.15) *occurs and that* $z = \delta_1 \otimes \delta_2^{-1}$, *as described above.*

(i) $z \in S$.
(ii) $z \notin I_1 \otimes I_2$.
(iii) $z^2 \in I_1 \otimes I_2$.
(iv) $|H_I : I_1 \otimes I_2| = 2$.

Proof. It follows from Lemmas 2.4.5, 2.7.5 and 2.8.4 that $\det_{V_i}(\delta_i) = \pm\mu^{n_i/2}$, and hence $\det(z) = 1$ in view of (4.4.9). Thus (i) holds. Suppose for a contradiction that $z \in I_1 \otimes I_2$, so that $z = g_1 \otimes g_2$ with $g_i \in I_i$. Then $\delta_1 = \lambda g_1$ for some $\lambda \in \mathbf{F}^*$, which forces $\mu = \tau_i(\delta_i) = \lambda^2$, which is impossible as μ is a non-square. Thus (ii) holds. To prove (iii), observe $\frac{1}{\mu}\delta_i^2 \in I_i$ and hence $z^2 = (\frac{1}{\mu}\delta_1^2) \otimes (\mu\delta_2^{-2}) \in I_1 \otimes I_2$. Part (iv) is immediate from (iii) and the fact that $H_I = (I_1 \otimes I_2)\langle z \rangle$. ■

Lemma 4.4.6. *We have* $H = N_\Delta(\Omega_i)$ *for* $i = 1, 2$. *Moreover, if* $\Omega_i \approx Sp_4(2)$, *then* $H = N_\Delta(\Omega_i')$.

Proof. The proof of the second assertion is exactly the same as the proof of the first, since Ω_i' is absolutely irreducible on V_i when $\Omega_i \approx Sp_4(2)$. So we prove the first only. In light of the restriction $n_i \geq 3$ when I_i is an orthogonal group, it follows from Proposition 2.10.6 that Ω_i is absolutely irreducible on V_i for $i = 1, 2$. Thus by Lemma 4.4.3 and Corollary 2.10.4.ii, we have $N_{GL(V)}(\Omega_1) = \Delta_1 \otimes GL(V_2)$. This proves the assertion in case **L**, so we assume hereafter that case **L** does not hold. Take $g_1 \otimes g_2 \in (\Delta_1 \otimes GL(V_2)) \cap \Delta$, with $g_1 \in \Delta_1$ and $g_2 \in GL(V_2)$. Clearly there exist $v_1, w_1 \in V_1$ such that $\mathbf{f}_1(v_1, w_1) \neq 0$. Thus for all $v_2, w_2 \in V_2$, we have

$$\begin{aligned} \tau(g_1 \otimes g_2)\mathbf{f}_1(v_1, w_1)\mathbf{f}_2(v_2, w_2) &= \tau(g_1 \otimes g_2)\mathbf{f}(v_1 \otimes v_2, w_1 \otimes w_2) \\ &= \mathbf{f}\big((v_1 \otimes v_2)(g_1 \otimes g_2), (w_1 \otimes w_2)(g_1 \otimes g_2)\big) \\ &= \mathbf{f}_1(v_1 g_1, w_1 g_1)\mathbf{f}_2(v_2 g_2, w_2 g_2) \\ &= \tau_1(g_1)\mathbf{f}_1(v_1, w_1)\mathbf{f}_2(v_2 g_2, w_2 g_2). \end{aligned}$$

Consequently $\mathbf{f}_2(v_2g_2, w_2g_2) = \tau(g_1 \otimes g_2)\tau_1(g_1)^{-1}\mathbf{f}_2(v_2, w_2)$ for all $v_2, w_2 \in V_2$, which proves $g_2 \in \Delta_2$. Thus $g_1 \otimes g_2 \in \Delta_1 \otimes \Delta_2 = H$, and hence $N_\Delta(\Omega_1) \leq H$. The reverse inclusion is clear, and the same argument applies to Ω_2. ■

We now prove Theorem 3.1.2 for the groups in $\mathcal{C}_4$.

Proposition 4.4.7. *Theorem 3.1.2 holds for the groups in $\mathcal{C}_4$.*

Proof. Since $\phi_{\mathcal{D}} \in H_\Gamma$ and $\sigma(\phi_{\mathcal{D}}) = \nu$, it follows that $H_\Gamma\Delta = \Gamma$, and so if $H\Omega = \Delta$, then $H_\Gamma\Omega = \Gamma$. Furthermore, if $N_\Delta(H_\Omega) = H$ then $N_\Gamma(H_\Omega) = H_\Gamma$. Thus using Proposition 4.0.1, it suffices to prove that $H = N_\Delta(H_\Omega)$ in case **L**, and that either $H = N_\Delta(H_\Omega)$ or $H\Omega = \Delta$ in the remaining cases.

Suppose first that Ω_i is insoluble for some i, while Ω_{3-i} is soluble. Then it follows from Proposition 2.9.2 that either $\Omega_i = \Omega_i'$ or $\Omega_i \approx Sp_4(2)$, and in both cases we have $\Omega_i' = H_\Omega^\infty$. Using Lemma 4.4.6, $N_\Delta(H_\Omega) \leq N_\Delta(H_\Omega^\infty) = N_\Delta(\Omega_i') = H$, as desired.

Next suppose that Ω_i is insoluble for $i = 1, 2$. If case $\mathbf{L}^\pm$ holds, then $\Omega_1 \approx SL_{n_1}^\pm(q)$ is not isomorphic to $\Omega_2 \approx SL_{n_2}^\pm(q)$ (in view of the restriction $n_1 < n_2$); hence Ω_i is characteristic in H_Ω for each i. Thus arguing as in the previous paragraph, we find that $N_\Delta(H_\Omega) \leq N_\Delta(\Omega_i) \leq H$, as required. Similarly, if case **O** holds, then each Ω_i is characteristic in H_Ω, and the same reasoning applies. Next take case **S**. Here $\tau_1(\Delta_1) = \mathbf{F}^*$ (see (2.4.4)) and hence $\tau(H) = \mathbf{F}^*$ by (4.4.6) and (4.4.13). Thus $H\Omega = \Delta$.

To complete the proof, assume that both Ω_1 and Ω_2 are soluble. From Proposition 2.9.2 we see that H is of type $GU_2(2) \otimes GU_3(2)$, $Sp_2(3) \otimes O_3(3)$, $Sp_2(3) \otimes O_4^+(3)$ or $O_3(3) \otimes O_4^+(3)$. First take H of type $GU_2(2) \otimes GU_3(2)$. Then (4.4.9) and (2.3.1) show that $|\det(H_I)| = q+1$, which means $H_I\Omega = I$. Since $\Delta = I\mathbf{F}^*$ by (2.3.3), we have $H\Omega = \Delta$, as required. If H is of type $Sp_2(3) \otimes O_3(3)$ in $Sp_6(3)$, or of type $Sp_2(3) \otimes O_4^+(3)$ in $Sp_8(3)$, then we deduce that $H\Omega = \Delta$ as in the end of the previous paragraph. Finally, suppose that H is of type $O_3(3) \otimes O_4^+(3)$ in $O_{12}^+(3)$. Since $I_2 \approx O_4^+(3)$, we have $\tau(\Delta_2) = \mathbf{F}^*$, and so $HI = \Delta$. Also $\det(I_2) = \langle -1 \rangle$ by (4.4.9), and so $H_IS = I$. Furthermore, it follows from Lemma 4.4.13 below that if $v, w \in V_2$ have norms $1, -1$ respectively, then $(1 \otimes r_v)(1 \otimes r_w) \in H_S\Omega$. Thus $H_S\Omega = S$. We now conclude $H\Omega = \Delta$, completing the proof. ■

Lemma 4.4.8. *We have $c = |\ddot{I} : \ddot{H}_I|$.*

Proof. In view of Proposition 4.4.7, we may apply Theorem 3.1.2 here. If part (b) of Theorem 3.1.2.i holds, then $H_{\overline{A}}\overline{\Omega} = \overline{A}$, and hence $N_{\overline{A}}(H_{\overline{\Omega}})\overline{\Omega} = \overline{A}$ (since $H_{\overline{\Omega}} \trianglelefteq H_{\overline{A}}$). Therfore $c = 1$ by (3.2.3), and also $|\ddot{I} : \ddot{H}_I| = 1$ since $H_{\overline{I}}\overline{\Omega} = \overline{I}$. Now assume that part (a) of Theorem 3.1.2.i holds. Here $H_{\overline{\Gamma}} = N_{\overline{\Gamma}}(H_{\overline{\Omega}})$, whence $H_\Gamma = N_\Gamma(H_\Omega)$. Therefore equality holds throughout Lemma 4.0.3 (see Lemma 4.0.3.ix). Thus by Lemma 4.0.3.iii, it suffices to show that $HI = \Delta$. Now in cases $\mathbf{L}^\pm$ and $\mathbf{O}^\circ$ we know that $\Delta = I\mathbf{F}^*$, and so $HI = \Delta$ as $\mathbf{F}^* \leq H$. In case **S**, $\mathbf{f}_1$ is symplectic, so $\tau_1(\Delta_1) = \mathbf{F}^*$. Therefore $\tau(H) = \mathbf{F}^*$, which means $HI = \Delta$, as desired. Finally in case $\mathbf{O}^\pm$, n_i is even for some i and $\mathbf{f}_i$ is either symplectic or symmetric; again $\tau_i(\Delta_i) = \mathbf{F}^*$, and the result follows. ■

The next Lemma is a direct consequence of Lemma 4.0.5.

Lemma 4.4.9. *$H_{\overline{\Omega}}$ is local if and only if $P\Omega_i$ is local for some i. If $H_{\overline{\Omega}}$ is non-local, then* $\mathrm{soc}(H_{\overline{\Omega}}) \cong P\Omega_1' \times P\Omega_2'$.

Proof. Simply put $K = \Omega_1' \otimes \Omega_2'$ in Lemma 4.0.5.iv. Note that K is absolutely irreducible on V, and so in the notation of Lemma 4.0.5.iv we have $t = 1$. ∎

We are now ready to investigate the structure of $H_{\overline{\Omega}}$ and determine c. Throughout the rest of this section, part (III) of each Proposition is a direct consequence of Lemma 4.4.9, in view of the fact that $P\Omega_i$ is local if and only if $I_i \approx GL_2^{\pm}(2)$, $GL_2^{\pm}(3)$, $GU_3(2)$, $Sp_2(2)$, $Sp_2(3)$, $O_3(3)$ or $O_4^+(3)$ (see Proposition 2.9.2). Thus we will not offer a proof of (III) for the rest of this section.

Proposition 4.4.10. *Assume that case $\mathbf{L}^\epsilon$ holds so that H is of type $GL_{n_1}^\epsilon(q) \otimes GL_{n_2}^\epsilon(q)$ with $n_1 < n_2$.*

(I) $c = \mathrm{h.c.f.}(q-\epsilon, n_1, n_2)$ *and* $\pi = \pi_1(c)$.

(II) $H_{\overline{\Omega}} \cong (L_{n_1}^\epsilon(q) \times L_{n_2}^\epsilon(q)).[(q-\epsilon, n_1)(q-\epsilon, n_2)c/(q-\epsilon, n)]$.

(III) *$H_{\overline{\Omega}}$ is local if and only if $I_1 \approx GL_2^\epsilon(2)$, $GL_2^\epsilon(3)$ or $GU_3(2)$. When $H_{\overline{\Omega}}$ is non-local,* $\mathrm{soc}(H_{\overline{\Omega}}) \cong L_{n_1}^\epsilon(q) \times L_{n_2}^\epsilon(q)$.

Proof. According to (4.4.14) we have $H_I = I_1 \otimes I_2$, and so (4.4.9) implies that $|\det(H_I)| = \left[\frac{q-\epsilon}{(q-\epsilon, n_1)}, \frac{q-\epsilon}{(q-\epsilon, n_2)}\right]$. Since $|\ddot{H}_I| = |\det(H_I)|/|\det(\mathbf{F}^* \cap I)| = |\det(H_I)|(q-\epsilon, n)/(q-\epsilon)$, the value of c given in (I) is correct by Lemma 4.4.8. And clearly Lemma 4.4.8 and (4.0.3) imply $H_{\overline{\Omega}}$ is a subgroup of $H_{\overline{I}} \cong PGL_{n_1}^\epsilon(q) \times PGL_{n_2}^\epsilon(q)$ of index $(q-\epsilon, n)/c$, whence (II) follows. We now determine π. Let β_i be a basis for V_i, and when $\epsilon = -$ assume that β_i is orthonormal. Then the tensor product basis (described after (4.4.9)) $\beta = \beta_1 \otimes \beta_2$ is a basis of V which is orthonormal when $\epsilon = -$. Let δ be as in (2.2.2) and (2.3.1), so that $\ddot{\Delta} = \ddot{I} = \langle \ddot{\delta} \rangle$. We see that $\ddot{H}_I = \langle \ddot{\delta}^c \rangle$, for $\langle \ddot{\delta}^c \rangle$ is the unique subgroup of index c in $\ddot{I}$. Moveover, $\phi = \phi_\beta(\nu) \in H_\Gamma$, and $\iota = \iota_\beta \in N_A(H)$ when $\epsilon = +$. Therefore $\ddot{H}_\Gamma = \langle \ddot{\delta}^c, \ddot{\phi} \rangle$ in case $\mathbf{U}$, and $\ddot{H}_A = \langle \ddot{\delta}^c, \ddot{\phi}, \ddot{\iota} \rangle$ in case $\mathbf{L}$. Consequently $\pi = \pi_1(c)$, as claimed. ∎

Proposition 4.4.11. *Assume that case $\mathbf{S}$ holds, so that H is of type $Sp_{n_1}(q) \otimes O_{n_2}^\epsilon(q)$, with q odd and $n_2 \geq 3$.*

(I) $c = 1$.

(II) $H_{\overline{\Omega}} \cong (PSp_{n_1}(q) \times PO_{n_2}^\epsilon(q)).(n_2, 2)$.

(III) *$H_{\overline{\Omega}}$ is local if and only if $I_1 \approx Sp_2(3)$, or $I_2 \approx O_3(3)$, or $I_2 \approx O_4^+(3)$. If $H_{\overline{\Omega}}$ is non-local then* $\mathrm{soc}(H_{\overline{\Omega}}) \cong PSp_{n_1}(q) \times P\Omega_{n_2}^\epsilon(q)$.

Proposition 4.4.12. *Assume that case $\mathbf{O}^+$ holds and H is of type $Sp_{n_1}(q) \otimes Sp_{n_2}(q)$, with $n_1 < n_2$.*

(I)
$$c = \begin{cases} 2 & \text{if } q \text{ is even or if } n \equiv 4 \pmod 8 \\ 4 & \text{if } q \text{ is odd and } n \equiv 0 \pmod 8). \end{cases}$$

Also $\pi = \pi_2$ *if* $c = 2$ *and* $\pi = \pi_5$ *if* $c = 4$.

(II) $H_{\overline{\Omega}} \cong (PSp_{n_1}(q) \times PSp_{n_2}(q)).\frac{c}{2}$.

(III) $H_{\overline{\Omega}}$ *is local if and only if* $n_1 = 2$ *and* $q \leq 3$. *If* $H_{\overline{\Omega}}$ *is non-local then* $\mathrm{soc}(H_{\overline{\Omega}}) \cong PSp_{n_1}(q)' \times PSp_{n_2}(q)'$.

Proof. Set $m_i = n_i/2$ and let $\beta_i = \{e_1, \ldots, e_{m_i}, f_1, \ldots, f_{m_i}\}$ be a standard basis for $(V_i, \mathbf{f}_i)$ (see Proposition 2.4.1). Then $W = \langle e_1, \ldots, e_{m_1} \rangle \otimes V_2$ is a totally singular $n/2$-space in V. Now if $g \in I_2$, then $1 \otimes g$ fixes W and $\det_W(1 \otimes g) = \det_{V_2}(g)^{m_1} = 1$, and so by Lemma 4.1.9.iii, $1 \otimes g \in \Omega$. Consequently $1 \otimes I_2 \leq \Omega$. Similarly $I_1 \otimes 1 \leq \Omega$, and so
$$I_1 \otimes I_2 \leq \Omega. \tag{4.4.16}$$
Now when q is odd, set $\delta_i = \delta_{\mathbf{f}_i, \beta_i}(\mu)$ as in (2.4.3), and put $z = \delta_1 \otimes \delta_2^{-1}$, as in the discussion just before (4.4.15). We know that $z \in S$ by Lemma 4.4.5.i, and evidently z fixes W. Furthermore $\det_W(z) = (\mu^{m_1})^{n_2}(\mu^{-m_2})^{m_1} = \mu^{n/4}$. Therefore by Lemma 4.1.9.iii, for odd q we have
$$z \in \begin{cases} \Omega & \text{if } n \equiv 0 \pmod 8) \\ S \backslash \Omega & \text{if } n \equiv 4 \pmod 8). \end{cases} \tag{4.4.17}$$
So combining (4.4.14), (4.4.15), (4.4.16), (4.4.17) and Proposition 2.7.4 we obtain
$$\ddot{H}_I = \begin{cases} \ker_{\ddot{I}}(\ddot{\gamma}) & \text{if } q \text{ is even or if } n \equiv 4 \pmod 8) \\ 1 & \text{if } q \text{ is odd and } n \equiv 0 \pmod 8). \end{cases} \tag{4.4.18}$$

Thus Lemma 4.4.8 and (4.4.18) indicate that the value of c given in (I) is correct. Since $H_{\overline{I}} \cong (PSp_{n_1}(q) \times PSp_{n_2}(q)).(2, q-1)$, it is also clear that (II) holds by Lemma 4.4.8 and (4.0.3).

We now determine π. Observe that $\beta = \beta_1 \otimes \beta_2$ is a standard basis of V, in the sense of Proposition 2.5.3.i. Choose $\phi_i = \phi_{\mathbf{f}_i, \beta_i}$ as in (2.4.5), and put $\phi_{\mathcal{D}} = \phi_1 \otimes \phi_2$, as in the discussion just before (4.4.7). Then $\phi_{\mathcal{D}} = \phi_\beta(\nu) \in H_\Gamma$, and thus $H_\Gamma = (\Delta_1 \otimes \Delta_2)\langle \phi_\beta(\nu) \rangle$. Since $I_i \leq \Omega$, we have
$$\ddot{H}_\Gamma = \langle \ddot{\delta}_1, \ddot{\delta}_2, \ddot{\phi}_{\mathcal{D}} \rangle \tag{4.4.19}$$
(here we identify δ_1, δ_2 with $\delta_1 \otimes 1, 1 \otimes \delta_2$). Notice that δ_1, δ_2 and $\phi_{\mathcal{D}}$ all fix W, and hence $\ddot{H}_\Gamma \leq \ker(\ddot{\gamma})$. Thus when $c = 2$ we have $\ddot{H}_\Gamma = \ker(\ddot{\gamma})$ and so $\pi = \pi_2$. It remains to consider the case $c = 4$. An easy calculation shows $\delta_1 = \delta_{Q,\beta}(\mu)$, and hence $\ddot{H}_\Gamma = \langle \ddot{\delta}_1, \ddot{\phi}_{\mathcal{D}} \rangle$, which in the notation of Proposition 2.7.3 becomes $\langle \ddot{\delta}, \ddot{\phi} \rangle$. Therefore $\pi = \pi_5$. ∎

Lemma 4.4.13. *Assume that case* **O** *holds,* q *is odd,* $\kappa_i = Q_i$ *is orthogonal for* $i = 1, 2$ *and that* v *is a non-singular vector in* V_1.

(i) $r_v \otimes 1 = r_{v \otimes w_1} \cdots r_{v \otimes w_{n_2}}$, *where* $w_1, \ldots, w_{n_2}$ *is a basis of* (V_2, Q_2) *as in Proposition 2.5.12.*

(ii)

$$r_v \otimes 1 \in \begin{cases} I\backslash S & \text{if } n_2 \text{ is odd} \\ S\backslash\Omega & \text{if } n_2 \text{ is even and } D(Q_2) = \boxtimes \\ \Omega & \text{if } n_2 \text{ is even and } D(Q_2) = \square. \end{cases}$$

(iii) *If n_1 and n_2 are even, then*

(a) $S_1 \otimes S_2 \leq \Omega$ *and* $I_1 \otimes I_2 \leq S$

(b)

$$(I_1 \otimes I_2) \cap \Omega = \begin{cases} I_1 \otimes I_2 & \text{if } D(Q_i) = \square \text{ for } i = 1,2 \\ S_i \otimes I_{3-i} & \text{if } D(Q_i) = \square \text{ but } D(Q_{3-i}) = \boxtimes \\ (S_1 \otimes S_2).2 & \text{if } D(Q_i) = \boxtimes \text{ for } i = 1,2 \end{cases}$$

(c)

$$(I_1 \otimes I_2)\Omega = \begin{cases} \Omega & \text{if } D(Q_i) = \square \text{ for both } i = 1,2 \\ S & \text{otherwise.} \end{cases}$$

Proof. Using (2.5.7) we see that both $r_v \otimes 1$ and $r_{v\otimes w_1} \dots r_{v \otimes w_{n_2}}$ centralize $v^{\perp} \otimes V_2$, and negate $\langle v \rangle \otimes V_2$, and so they are equal. Assertion (ii) is immediate from (i) and the spinor norm formula (2.5.9). Now (iii)(a) follows from (ii) and the facts that I is generated by reflections and S is generated by products of pairs of reflections (see Proposition 2.5.6 and (2.5.13)). To prove (iii)(b), observe that $I_i = S_i\langle r_i \rangle$, where r_i is a reflection. Also $r_i \in \Omega$ if and only if $D(Q_{3-i})$ is a square; otherwise $r_i \in S\backslash\Omega$. Thus $(I_1 \otimes I_2) \cap \Omega$ is generated by $S_1 \otimes S_2$ along with $\langle r_1, r_2 \rangle$, $\langle r_{3-i} \rangle$ or $\langle r_1 \otimes r_2 \rangle$ in the respective cases. Assertion (iii)(c) is by now clear. ∎

Proposition 4.4.14. *Assume that case* $\mathbf{O}^+$ *holds and* H *is of type* $O^+_{n_1}(q) \otimes O^+_{n_2}(q)$, *with* $n_2 > n_1 \geq 4$ *and* q *odd.*

(I)

$$c = \begin{cases} 2 & \text{if } n_i \equiv 2 \pmod 4 \text{ for some } i \text{ and } q \equiv 3 \pmod 4 \\ 2 & \text{if } n \equiv 4 \pmod 8 \\ 4 & \text{otherwise.} \end{cases}$$

If $c = 2$ *then* $\pi = \pi_2$. *If* $c = 4$ *then* $\pi = \pi_5$.

(II) $H_{\overline{\Omega}} \cong (PSO^+_{n_1}(q) \times PSO^+_{n_2}(q)).[2c]$.

(III) $H_{\overline{\Omega}}$ *is local if and only if* $I_1 \approx O^+_4(3)$. *If* $H_{\overline{\Omega}}$ *is non-local then* $\mathrm{soc}(H_{\overline{\Omega}}) \cong P\Omega^+_{n_1}(q) \times P\Omega^+_{n_2}(q)$.

(IV) *If* $n_1 = 4$, *then* H_Ω *interchanges the two factors* $SL_2(q)$ *of* Ω_1 *if and only if* $D(Q_2)$ *is a square.*

Proof. First note that Lemma 4.4.13.iii(a) implies that $S_1 \otimes S_2 \leq H_\Omega$. Second, we have $\overline{H} = \overline{\Delta}_1 \times \overline{\Delta}_2$, and according to Table 2.1.D, $|\overline{\Delta}_i : \overline{S}_i| = 4$. Therefore

$$PSO^+_{n_1}(q) \times PSO^+_{n_2}(q) \cong \overline{S}_1 \times \overline{S}_2 \trianglelefteq H_{\overline{\Omega}} \trianglelefteq \overline{H} \cong PSO^+_{n_1}(q).2^2 \times PSO^+_{n_2}(q).2^2.$$

Therefore $H_{\overline{I}} \cong (PSO^+_{n_1}(q) \times PSO^+_{n_2}(q)).2^3$. And since $|\ddot{I}| = 4$, we see that (II) holds by Lemma 4.4.8 and (4.0.3). So it remains to prove (I) and (IV).

As in the proof of Proposition 4.4.12, put $m_i = n_i/2$ and let $\beta_i = \{e_1, \ldots, f_{m_i}\}$ be a standard basis for V_i. Then by (4.4.15) $H_I = (I_1 \otimes I_2)\langle z \rangle$, where $z = \delta_1 \otimes \delta_2^{-1}$ and $\delta_i = \delta_{Q_i,\beta_i}(\mu)$. Then reasoning as for (4.4.17) we obtain

$$z \in \begin{cases} \Omega & \text{if } n \equiv 0 \ (\text{mod } 8) \\ S\backslash\Omega & \text{if } n \equiv 4 \ (\text{mod } 8). \end{cases} \tag{4.4.20}$$

So combining (4.4.20) and Lemma 4.4.13.iii(b),

$$\ddot{H}_I = \begin{cases} \ddot{S} & \text{if } D(Q_i) = \boxtimes \text{ for some } i \\ \ddot{S} & \text{if } n \equiv 4 \ (\text{mod } 8) \\ 1 & \text{otherwise.} \end{cases} \tag{4.4.21}$$

Thus c is as given in (I) in light of Lemma 4.4.8. The homomorphism π is determined as in the proof of Proposition 4.4.12.

We now prove (IV). Since $n \equiv 0 \ (\text{mod } 8)$ it follows from (4.4.20) that $z \in \Omega$, and since $D(Q_1)$ is a square it follows from Lemma 4.4.13.ii that $I_2 \leq \Omega$. Therefore $H_\Omega = ((I_1 \cap \Omega) \otimes I_2)\langle z \rangle$. Since δ_1 fixes $\langle e_1, e_2 \rangle$, it follows that z fixes each factor $SL_2(q)$ of Ω_1 by Lemma 2.9.4. Consequently H_Ω interchanges the two factors if and only if $I_1 \cap \Omega$ does, and this occurs if and only if $I_1 \leq \Omega$, which is true if and only if $D(Q_2)$ is a square. ■

Proposition 4.4.15. *Assume that case* $\mathbf{O}^+$ *holds and that* H *is of type* $O^+_{n_1}(q) \otimes O^-_{n_2}(q)$, *with* q *odd and* $n_i \geq 4$ *for both* i.

(I)
$$c = \begin{cases} 4 & \textit{if } n_1 \equiv 0 \ (\text{mod } 4),\ n_2 \equiv 2 \ (\text{mod } 4) \textit{ and } q \equiv 3 \ (\text{mod } 4) \\ 2 & \textit{otherwise} \end{cases}$$

If $c = 2$ *then* $\pi = \pi_2$. *If* $c = 4$ *then* $\pi = \pi_5$.

(II) $H_{\overline{\Omega}} \cong (PSO^+_{n_1}(q) \times PSO^-_{n_2}(q)).[2c]$.

(III) $H_{\overline{\Omega}}$ *is local if and only if* $I_1 \approx O^+_4(3)$. *If* $H_{\overline{\Omega}}$ *is non-local then* $\mathrm{soc}(H_{\overline{\Omega}}) \cong P\Omega^+_{n_1}(q) \times P\Omega^-_{n_2}(q)$.

(IV) *If* $n_1 = 4$, *then* H_Ω *interchanges the two factors* $SL_2(q)$ *of* Ω_1 *if and only if* $D(Q_2)$ *is a square.*

Proof. The proof of (II) is analogous to the proof of Proposition 4.4.14(II). Thus we concentrate on (I) and (IV). Let $\beta_1 = \{e_1, \ldots, f_{n_1/2}\}$ be a standard basis of V_1 and let β_2 be a basis of V_2 as given in Proposition 2.5.12. Write $\delta_i = \delta_{Q_i,\beta_i}(\mu)$ as described in (2.7.2) and (2.8.2). Thus $z = \delta_1 \otimes \delta_2^{-1} \in S$ by Lemma 4.4.5.i, and hence $H_I \leq S$ by (4.4.15). If either $D(Q_1)$ or $D(Q_2)$ is a non-square, then $H_I\Omega = S$ by Lemma 4.4.13.iii(c), and in this case $c = 2$ according to Lemma 4.4.8. So assume that $D(Q_i)$ is a square for both i. Quoting Proposition 2.5.10, $n_2 \equiv 2 \ (\text{mod } 4)$, $q \equiv 3 \ (\text{mod } 4)$ and $n_1 \equiv 0 \ (\text{mod } 4)$. Moreover $I_1 \otimes I_2 \leq \Omega$ by Lemma 4.4.13.iii(b). Now as in the proof of Proposition 4.4.12, z fixes the totally singular $n/2$-space $W = \langle e_1, \ldots, e_{n_1/2} \rangle \otimes V_2$ of V and $\det_W(z) = \mu^{n/4} \in (\mathbf{F}^*)^2$ (because $n_1 \equiv 0 \ (\text{mod } 4)$). Thus $H_I \leq \Omega$ by Lemma 4.1.9.iii, and so Lemma 4.4.8

implies $c = |\ddot{I}| = 4$. This establishes that the value of c given in (I) is correct. In particular, $c \neq 1$, and so $|\ddot{\Gamma} : \ddot{H}_\Gamma| = c$, by Proposition 4.4.7 and Lemma 4.0.3.i. Evidently $\delta_i \in N_\Delta(W) \leq \ker(\gamma)$. Moreover, we can choose $\phi_1 = \phi_{Q_1,\beta_1}$ and $\phi_2 = \phi_{Q_2,\beta_2}$ as in (2.7.3) and (2.8.4), and we set $\phi_{\mathcal{D}} = \phi_1 \otimes \phi_2$. Clearly $\phi_{\mathcal{D}} \in N_\Gamma(W) \leq \ker(\gamma)$. Therefore $H_\Gamma \leq \ker(\gamma)$, and so $\ddot{H}_\Gamma = \ker_{\ddot{\Gamma}}(\ddot{\gamma})$ when $c = 2$. Thus $\pi = \pi_2$ when $c = 2$. Assume therefore that $c = 4$. Then $D(Q_2)$ is a square and so β_2 is orthonormal. Consequently $\beta = \beta_1 \otimes \beta_2$ is a standard basis of V and $\phi_{\mathcal{D}} = \phi_{Q,\beta}(\nu)$. Moreover $\delta_1 = \delta_{Q,\beta}(\mu)$, and so in the notation of Proposition 2.7.3, $\langle \ddot{\delta}, \ddot{\phi} \rangle \leq \ddot{H}_\Gamma$. Since $|\ddot{\Gamma} : \ddot{H}_\Gamma| = 4$, we now conclude that $\ddot{H}_\Gamma = \langle \ddot{\delta}, \ddot{\phi} \rangle$, and so $\pi = \pi_5$.

To prove (IV), argue as in the proof of Proposition 4.4.14(IV). ∎

Proposition 4.4.16. *Assume that case* $\mathbf{O}^+$ *holds and that* H *is of type* $O^-_{n_1}(q) \otimes O^-_{n_2}(q)$, *with* q *odd and* $n_2 > n_1 \geq 4$.

(I) $c = 2$ *and* $\pi = \pi_2$.

(II) $H_{\overline{\Omega}} \cong (PSO^-_{n_1}(q) \times PSO^-_{n_2}(q)).[4]$

(III) $H_{\overline{\Omega}}$ *is non-local and* $\mathrm{soc}(H_{\overline{\Omega}}) \cong P\Omega^-_{n_1}(q) \times P\Omega^-_{n_2}(q)$.

Proof. Let β_i be a basis for (V_i, Q_i) as given in Lemma 2.5.12, and define $\delta_i \in \Delta_i$ as in (2.8.2). As in the proof of Proposition 4.4.15, we have $H_I \leq S$, and so to prove that $c = 2$ it remains to show $H_I\Omega = S$. By Lemma 4.4.13.iii(c), we have $H_I\Omega = S$ if $D(Q_i)$ is a non-square for some i. We may assume therefore that $D(Q_i)$ is a square for both i, so that $m_i = n_i/2$ is odd for both i and $q \equiv 3 \pmod 4$. In particular, β_i is orthonormal for $i = 1, 2$. Write $\beta_1 = \{v_1, \ldots, v_{n_1}\}$ and $\beta_2 = \{w_1, \ldots, w_{n_2}\}$, so that $\beta = \beta_1 \otimes \beta_2 = \{v_i \otimes w_j\}$ is also orthonormal. Consider the -2-spaces $W_1 = \{v_1, v_2\}$ and $W_2 = \{w_1, w_2\}$ of V_1 and V_2. In view of (2.8.2), we see that z acts on the $+4$-space $W = W_1 \otimes W_2$ (with respect to $\{v_1 \otimes w_1, v_1 \otimes w_2, v_2 \otimes w_1, v_2 \otimes w_2\}$) as the matrix

$$\frac{1}{\mu}\begin{pmatrix} a^2 & ab & ab & b^2 \\ ab & -a^2 & b^2 & -ab \\ ab & b^2 & -a^2 & -ab \\ b^2 & -ab & -ab & a^2 \end{pmatrix}.$$

This is an involution, and the space negated by z is $\langle x, y \rangle$, where $x = (-b, 2a, 0, b)$, $y = (-b, 0, 2a, b)$. Since $(x, x) = (y, y) = 2b^2 + 4a^2$ and $(x, y) = 2b^2$, the discriminant $D(\langle x, y \rangle) = 16b^2a^2 + 16a^4 \equiv b^2 + a^2 \equiv \mu \pmod{(\mathbf{F}^*)^2}$. In particular, $z_W \in S(W, Q) \backslash \Omega(W, Q)$. Clearly z preserves a decomposition $V = W_1 \perp \cdots \perp W_{m_1 m_2}$, where each W_i, like W, is a $+4$-space which is a tensor product of two -2-spaces. Arguing as we did for W, we see that $z_{W_i} \in S(W_i, Q) \backslash \Omega(W_i, Q)$, and since $m_1 m_2$ is odd, we deduce $z \in S \backslash \Omega$. Therefore $H_I \Omega = S$, and it now follows that $c = 2$. Part (II) follows as in Proposition 4.4.14.

We now show that $\pi = \pi_2$. First of all, note that $D(Q) = \square$, as $n \equiv 0 \pmod 4$. Now $\tau(\delta_1) = \mu$, and so $\ddot{\delta}_1 \in \ddot{\Delta} \backslash \ddot{I}$. Therefore in the notation of Proposition 2.7.3, $\ddot{\delta}_1$ is one of $\ddot{\delta}$, $\ddot{r}_\square \ddot{\delta}$, $\ddot{r}_\boxtimes \ddot{\delta}$ and $\ddot{r}_\square \ddot{r}_\boxtimes \ddot{\delta}$. Now δ_1^2 is the scalar μ, which means $|\ddot{\delta}_1| = 2$. As $\ddot{r}_\square \ddot{\delta}$ and $\ddot{r}_\boxtimes \ddot{\delta}$ have order 4 (see Proposition 2.7.3.iii), it follows that $\ddot{\delta}_1 \in \langle \ddot{r}_\square \ddot{r}_\boxtimes, \ddot{\delta} \rangle =$

$\ker_{\ddot{\Delta}}(\ddot{\gamma})$. And as $\ddot{H}_I = \ddot{S}$, we see that $\ddot{H} = \ker_{\ddot{\Delta}}(\ddot{\gamma})$. Now let $\phi_i = \phi_{Q_i,\beta_i} \in \Gamma_i$ be as in (2.8.4), and put $\phi_{\mathcal{D}} = \phi_1 \otimes \phi_2$, as usual. By Proposition 2.8.2, $\ddot{\delta}_i$ and $\ddot{\phi}_i$ commute, and hence $\ddot{\phi}_{\mathcal{D}}$ commutes with $\ddot{\delta}_1$. Consequently, $\ddot{H}_\Gamma = C_{\ddot{\Gamma}}(\ddot{H}) = C_{\ddot{\Gamma}}(\ker_{\ddot{\Delta}}(\ddot{\gamma}))$, which in the notation of Proposition 2.7.3 is $\langle \ddot{r}_\square \ddot{r}_\boxtimes, \ddot{\delta}, \ddot{\phi}\rangle$, and this group is precisely $\ker_{\ddot{\Gamma}}(\ddot{\gamma})$, by Proposition 2.7.4.iii. This finishes the proof. ■

Proposition 4.4.17. *Assume that case* $\mathbf{O}^\epsilon$ *holds and that* H *is of type* $O^\epsilon_{n_1}(q)\otimes O_{n_2}(q)$, *with* qn_2 *odd and* $n_1 \geq 4$.

(I) $c = 1$.

(II) $H_{\overline{\Omega}} \cong P\Omega^\epsilon_{n_1}(q) \times SO_{n_2}(q)$.

(III) $H_{\overline{\Omega}}$ *is local if and only if* $I_1 \approx O_4^+(3)$ *or* $I_2 \approx O_3(3)$. *If* $H_{\overline{\Omega}}$ *is non-local then* $\mathrm{soc}(H_{\overline{\Omega}}) \cong P\Omega^\epsilon_{n_1}(q) \times \Omega_{n_2}(q)$.

(IV) *If* $I_1 \cong O_4^+(q)$, *then* H_Ω *does not interchange the two factors* $SL_2(q)$ *of* Ω_1.

Proof. Here (4.4.14) and Lemma 4.4.13 imply that $H_I = I_1 \otimes I_2 = I_1 \otimes S_2$ (recall $I_2 = S_2 \times \langle -1\rangle$ as n_2 is odd). Moreover $S_2 \leq \Omega$ by Lemma 4.4.13.ii and so $H_\Omega = (I_1 \cap \Omega) \otimes S_2$. Let $u, v \in V_1$ be non-singular vectors such that $r_u r_v \in S_1 \backslash \Omega_1$. Then it follows easily from Lemma 4.4.13.i that $r_v \otimes 1 \in I \backslash S$ and $(r_u \otimes 1)(r_v \otimes 1) \in S \backslash \Omega$. Consequently $I_1 \Omega = I$. Therefore (I) holds by Lemma 4.4.8, and also $I_1 \cap \Omega = \Omega_1$, which shows $H_\Omega = \Omega_1 \otimes S_2$. Thus (II) and (IV) hold. ■

Proposition 4.4.18. *Assume that case* $\mathbf{O}^\circ$ *holds, so that* H *is of type* $O_{n_1}(q)\otimes O_{n_2}(q)$, *with* q *odd and* $n_1 < n_2$.

(I) $c = 1$.

(II) $H_{\overline{\Omega}} \cong (\Omega_{n_1}(q) \times \Omega_{n_2}(q)).2$.

(III) $H_{\overline{\Omega}}$ *is local if and only* $I_1 \approx O_3(3)$, *and if* $H_{\overline{\Omega}}$ *is non-local then* $\mathrm{soc}(H_{\overline{\Omega}}) \cong \Omega_{n_1}(q) \times \Omega_{n_2}(q)$.

§4.5 The subfield subgroups $\mathcal{C}_5$

Let $\mathbf{F}_\sharp$ be a subfield of index r in $\mathbf{F}$ and let $V_\sharp$ be the $\mathbf{F}_\sharp$-span of an $\mathbf{F}$-basis β of V. Then $V_\sharp$ is an n-dimensional $\mathbf{F}_\sharp$-space. Moreover, if $g \in GL(V_\sharp, \mathbf{F}_\sharp)$, then g extends to a unique element in $GL(V, \mathbf{F})$ and hence there is a natural inclusion

$$GL_n(q^{1/r}) \cong GL(V_\sharp, \mathbf{F}_\sharp) \leq GL(V, \mathbf{F}). \tag{4.5.1}$$

Observe that if $g \in GL(V, \mathbf{F})$, then the following are equivalent: $g \in GL(V_\sharp, \mathbf{F}_\sharp)$; the entries of g_β lie in $\mathbf{F}_\sharp$; g preserves the set $V_\sharp$; g centralizes $\phi_\beta(\nu_\sharp)$, where $\nu_\sharp$ is a generator for $\mathrm{Gal}(\mathbf{F}:\mathbf{F}_\sharp)$. Therefore

$$\begin{aligned} GL(V_\sharp, \mathbf{F}_\sharp) &= N_{GL(V,\mathbf{F})}(V_\sharp) \\ &= C_{GL(V,\mathbf{F})}(\phi_\beta(\nu_\sharp)). \end{aligned} \tag{4.5.2}$$

One may exhibit the inclusion in (4.5.1) in an alternative way by starting with $(V_\sharp, \mathbf{F}_\sharp)$ and forming the tensor product space $(V_\sharp \otimes \mathbf{F}, \mathbf{F})$. By identifying $V_\sharp$ with $\{v \otimes 1 \mid v \in V_\sharp\}$, we may take $V_\sharp \subseteq V$, as before.

Now assume that case **S** holds, so that $\kappa = \mathbf{f}$ is a non-degenerate symplectic form. Let β be a symplectic basis, and let $V_\sharp = \mathrm{span}_{\mathbf{F}_\sharp}(\beta)$. Then it is clear that the restriction $\mathbf{f}_{V_\sharp}$ of $\mathbf{f}$ to $V_\sharp$ is a non-degenerate symplectic form on $V_\sharp$. Also, if $g \in I(V_\sharp, \mathbf{F}_\sharp, \mathbf{f}_{V_\sharp})$, then the (unique) extension of g to $GL(V, \mathbf{F})$ lies in $I = I(V, \mathbf{F}, \mathbf{f})$. Thus we obtain

$$Sp_n(q^{1/r}) \approx I(V_\sharp, \mathbf{F}_\sharp, \mathbf{f}_{V_\sharp}) \leq I(V, \mathbf{F}, \mathbf{f}) \approx Sp_n(q). \tag{4.5.3}$$

Alternatively, if we begin with a symplectic geometry $(V_\sharp, \mathbf{F}_\sharp, \mathbf{f}_\sharp)$, then we obtain a symplectic geometry over $\mathbf{F}$ given by $(V_\sharp \otimes \mathbf{F}, \mathbf{F}, \mathbf{f}_\sharp \otimes \mathbf{F})$, where $\mathbf{f}_\sharp \otimes \mathbf{F}(v \otimes \lambda, w \otimes \eta) = \lambda\eta \mathbf{f}_\sharp(v, w)$ for all $v, w \in V_\sharp$. This also yields an embedding $Sp_n(q^{1/r}) \leq Sp_n(q)$.

In a similar fashion we obtain other embeddings, listed in Table 4.5.A, of an isometry group $I_\sharp = I(V_\sharp, \mathbf{F}_\sharp, \kappa_\sharp)$ in the isometry group $I = I(V, \mathbf{F}, \kappa)$. These embeddings will be discussed explicitly in the rest of this section. Throughout, we write $\mathrm{T} = \mathrm{T}^{\mathbf{F}}_{\mathbf{F}_\sharp}$.

Definition. The members of $\mathcal{C}_5(\Gamma)$ are the groups $N_\Gamma(V_\sharp)\mathbf{F}^*$, where $V_\sharp \subset V$ and $(V_\sharp, \mathbf{F}_\sharp, \kappa_\sharp)$ is a classical geometry over $\mathbf{F}_\sharp$ with $\kappa, \kappa_\sharp$ as in Table 4.5.A below, and where $r = |\mathbf{F} : \mathbf{F}_\sharp|$ is prime.

Table 4.5.A

case	type	description of $\kappa_\sharp$	conditions
L	$GL_n(q^{1/r})$	$\kappa = \kappa_\sharp = 0$	
S	$Sp_n(q^{1/r})$	$\kappa_\sharp = \kappa_{V_\sharp}$	
$\mathbf{O}^\epsilon$	$O_n^{\epsilon_\sharp}(q^{1/r})$	$\kappa_\sharp = \kappa_{V_\sharp}$	$\epsilon = \epsilon_\sharp^r$
U	$GU_n(q^{1/r})$	$\kappa_\sharp = \kappa_{V_\sharp}$	r odd
U	$O_n^\epsilon(q)$	$\kappa_\sharp = \kappa_{V_\sharp}$	q odd, $r = 2$
U	$Sp_n(q)$	$\kappa_\sharp = \zeta\kappa_{V_\sharp}$, $\mathrm{T}(\zeta) = 0$, $\zeta \in \mathbf{F}^*$	n even, $r = 2$

Remarks on the conditions. When H is of type $O_n^\epsilon(q)$ in $GU_n(q)$, the condition q odd is imposed since if q were even, such a subgroup would normalize a symplectic group $Sp_n(q)$.

For the rest of this section we assume that $H_\Gamma = N_\Gamma(V_\sharp)\mathbf{F}^* \in \mathcal{C}_5(\Gamma)$, where $(V_\sharp, \mathbf{F}_\sharp, \kappa_\sharp)$ is a classical geometry as described in the definition and in Table 4.5.A. As usual, put $X_\sharp = X(V_\sharp, \mathbf{F}_\sharp, \kappa_\sharp)$ as X ranges over the symbols in (2.1.15), and set $\tau_\sharp = \tau_{(V_\sharp, \mathbf{F}_\sharp, \mathbf{f}_\sharp)}$ and $\sigma_\sharp = \sigma_{(V_\sharp, \mathbf{F}_\sharp, \mathbf{f}_\sharp)}$. Since $\kappa_\sharp = \lambda\kappa_{V_\sharp}$ for some $\lambda \in \mathbf{F}^*$, it follows that $N_\Delta(V_\sharp) = \Delta_\sharp$ and that

$$\tau_{\Delta_\sharp} = \tau_\sharp. \tag{4.5.4}$$

Consequently

$$H = \Delta_\sharp \mathbf{F}^*. \tag{4.5.5}$$

Furthermore, if $g \in N_\Delta(V_\sharp)$ then $\det_{V_\sharp}(g) = \det_V(g)$. Therefore

$$I_\sharp = N_I(V_\sharp) \text{ and } \Omega_\sharp \leq \Omega.$$

The inclusion $\Omega_\sharp \leq \Omega$ holds in case **O** because $n \geq 7$ and so $\Omega_\sharp$ is perfect.

Proposition 4.5.1. *We have $H_\Gamma = N_\Gamma(H_\Omega)$, and hence Theorem 3.1.2 holds.*

Proof. Note that if $H_\Gamma = N_\Gamma(H_\Omega)$, then Theorem 3.1.2 holds by Proposition 4.0.1. Thus it suffices to prove that $H_\Gamma = N_\Gamma(H_\Omega)$.

We first show that $\Omega_\sharp$ is characteristic in H_Ω. This is clear if $\Omega_\sharp$ is perfect, for then $\Omega_\sharp = H^\infty$. If $\Omega_\sharp$ is not perfect, then it is $SL_2(2)$, $SL_2(3)$, $SU_3(2)$ or $Sp_4(2)$. In these cases $\mathbf{F}^* \cap \Omega \leq \Omega_\sharp$, so $\Omega_\sharp = H_\Omega$, $O^2(H_\Omega)$, $O^3(H_\Omega)$ or H_Ω, respectively. Thus $\Omega_\sharp$ is characteristic in H_Ω in all cases.

By Proposition 2.10.6, $\Omega_\sharp$ is absolutely irreducible on $(V_\sharp, \mathbf{F}_\sharp)$. As we remarked earlier in this section, we may regard V as the tensor product space $V_\sharp \otimes \mathbf{F}$, identifying $V_\sharp$ with $V_\sharp \otimes 1$. Now $\mathbf{F}$ is an r-dimensional space over $\mathbf{F}_\sharp$, and so $V_\sharp \otimes \mathbf{F}$ is in fact an nr-dimensional $\mathbf{F}_\sharp$-space. Thus $(V, \mathbf{F}_\sharp) = (V_\sharp, \mathbf{F}_\sharp) \otimes (\mathbf{F}, \mathbf{F}_\sharp)$ is a tensor decomposition of $\mathbf{F}_\sharp$-spaces. Since $\Omega_\sharp$ acts absolutely irreducibly on $(V_\sharp, \mathbf{F}_\sharp)$, Lemma 4.4.3.iv implies that the irreducible $\mathbf{F}_\sharp\Omega_\sharp$-submodules of V are precisely the subspaces $V_\sharp \otimes \lambda$, with $\lambda \in \mathbf{F}^*$. Now let $g \in N_\Gamma(H_\Omega)$. Then g normalizes $\Omega_\sharp$ (since $\Omega_\sharp$ is characteristic), and hence $V_\sharp g = (V_\sharp \otimes 1)g = V_\sharp \otimes \lambda$, for some $\lambda \in \mathbf{F}^*$. Then $g\lambda^{-1} \in N_\Gamma(V_\sharp)$, so $g \in H_\Gamma$, as required. ■

Proposition 4.5.2.
(i) *$H_{\overline{\Omega}}$ is local if and only if $P\Omega_\sharp$ is local.*
(ii) *If $P\Omega_\sharp$ is non-local then $\mathrm{soc}(H_{\overline{\Omega}}) \cong P\Omega_\sharp'$.*

Proof. Clearly $P\Omega_\sharp \trianglelefteq H_{\overline{\Omega}}$ and so $H_{\overline{\Omega}}$ is local if $P\Omega_\sharp$ is local. Assume therefore that $P\Omega_\sharp$ is non-local. The result now follows from Lemma 4.0.5.iv, putting $K = \Omega_\sharp$. ■

Thus as in §4.4, part (III) of each Proposition to follow is a direct consequence of Proposition 4.5.2, so no proof will be offered.

We begin with the linear and unitary groups. We have already exhibited embeddings of type $GL_n(q^{1/r})$ in $GL_n(q)$ at the beginning of this section. To obtain the unitary embedding, simply take an orthonormal basis of $(V, \mathbf{F}, \kappa)$, let $V_\sharp$ be its $\mathbf{F}_\sharp$-span and $\kappa_\sharp = \kappa_{V_\sharp}$. Then provided r is odd, $\kappa_\sharp$ is unitary, and this yields the embedding $GU_n(q^{1/r}) \leq GU_n(q)$. Note, however, that if $r = 2$, then $\kappa_\sharp$ is bilinear, rather than unitary; hence the condition r odd in Table 4.5.A.

Proposition 4.5.3. *Assume that case $\mathbf{L}^\epsilon$ holds and that H is of type $GL_n^\epsilon(q^{1/r})$, so that r is odd if $\epsilon = -$.*
(I) $c = (q-\epsilon)/\left[q^{1/r} - \epsilon, \dfrac{q-\epsilon}{(q-\epsilon, n)}\right]$ and $\pi = \pi_1(c)$.
(II) $H_{\overline{\Omega}} \cong \frac{c}{(q-\epsilon,n)} PGL_n^\epsilon(q^{1/r})$.
(III) *$H_{\overline{\Omega}}$ is local if and only if $\Omega_\sharp \approx SL_2(2)$, $SL_2(3)$ or $SU_3(2)$, and when $H_{\overline{\Omega}}$ is non-local, $\mathrm{soc}(H_{\overline{\Omega}}) \cong L_n^\epsilon(q^{1/r})$.*

Proof. Here $\Delta_\sharp = I_\sharp \mathbf{F}_\sharp^*$, and so $H = I_\sharp \mathbf{F}^*$. Therefore $H_I = I_\sharp(\mathbf{F}^* \cap I)$ and hence $|\det(H_I)| = \left[q^{1/r} - \epsilon, \frac{q-\epsilon}{(q-\epsilon, n)}\right]$. Consequently the value of c given in (I) is correct in view of Lemma 4.0.3.v,ix and Proposition 4.5.1. Clearly $\overline{H} = \overline{\Delta}_\sharp \cong PGL_n^\epsilon(q^{1/r})$, and so (II) is a consequence of Lemma 4.0.3.ii,ix. It remains to determine π. For $\epsilon = +$ let β be any $\mathbf{F}_\sharp$-basis of $V_\sharp$, and when $\epsilon = -$ let β be an orthormal basis of $(V_\sharp, \kappa_\sharp)$. Then β is also an $\mathbf{F}$-basis of V, and β is orthonormal when $\epsilon = -$. Evidently $\phi_\beta(\nu) \in H_\Gamma$, and when $\epsilon = +$ we have $\iota_\beta \in N_A(H_\Omega)$. Therefore, mimicking the argument in the proof of Proposition 4.4.10(I), we see that $\pi = \pi_1(c)$. ∎

Proposition 4.5.4. *Assume that case* **S** *holds, so that H is of type $Sp_n(q^{1/r})$.*

(I) $c = \text{h.c.f.}(2, q-1, r)$, *and* $\pi = \pi_1$ *if* $c = 2$.

(II) $H_{\overline{\Omega}} \cong PSp_n(q^{1/r}).c$.

(III) $H_{\overline{\Omega}}$ *is non-local and* $\text{soc}(H_{\overline{\Omega}}) \cong PSp_n(q^{1/r})'$.

To obtain an embedding of type $O_n^\epsilon(q)$ in $GU_n(q)$, assume that $\kappa = \mathbf{f}$ is unitary with q odd and let $\mathbf{F}_\sharp$ be the subfield of index 2 in $\mathbf{F}$. Choose a basis $\beta = \{v_1, \dots, v_n\}$ of V such that $\mathbf{f}(v_i, v_j) \in \mathbf{F}_\sharp$ for all i, j and set $V_\sharp = \text{span}_{\mathbf{F}_\sharp}(\beta)$ as usual. Then observe that $\mathbf{f}(v, w) \in \mathbf{F}_\sharp$ for all $v, w \in V_\sharp$, and therefore $\mathbf{f}_{V_\sharp}$ is a non-degenerate symmetric bilinear form. Put $\mathbf{f}_\sharp = \mathbf{f}_{V_\sharp}$ and note that if β is an orthonormal basis, then $D(\mathbf{f}_\sharp)$ is a square. And if we replace v_1 by μv_1, then $D(\mathbf{f}_\sharp) \equiv \mu^{q+1} \pmod{(\mathbf{F}_\sharp^*)^2}$, which is a non-square in $\mathbf{F}_\sharp$ (recall μ is a generator for $\mathbf{F}^*$). Therefore when n is even we obtain embeddings $O_n^\epsilon(q) < GU_n(q)$ for both $\epsilon = +$ and $\epsilon = -$.

Proposition 4.5.5. *Assume that case* **U** *holds and that H is of type $O_n^\epsilon(q)$, so that q is odd and $r = 2$.*

(I) $c = (q+1, n)/(2, n)$, *and* $\pi = \pi_1(c)$ *if n is odd or $D(\kappa_\sharp)$ is a square, while* $\pi = \pi_2(c)$ *if n is even and $D(\kappa_\sharp)$ is a non-square.*

(II) $H_{\overline{\Omega}} \cong PSO_n^\epsilon(q).(2, n)$.

(III) $H_{\overline{\Omega}}$ *is local if and only if* $I_\sharp \approx O_3(3)$ *or* $O_4^+(3)$, *and when $H_{\overline{\Omega}}$ is non-local* $\text{soc}(H_{\overline{\Omega}}) \cong P\Omega_n^\epsilon(q)$.

Proof. First take n odd. Then $\Delta_\sharp = S_\sharp \times \mathbf{F}_\sharp^*$ and so $H = S_\sharp \mathbf{F}^*$. Therefore $H_{\overline{\Omega}} = \overline{H} = \overline{S}_\sharp \cong SO_n(q)$, and $c = (q+1, n)$ by Lemma 4.0.3.v,ix and Proposition 4.5.1 (recall from §2.3 that $|\ddot{\Delta}| = (q+1, n)$). With no loss, we may assume that $V_\sharp$ is the $\mathbf{F}_\sharp$-span of an orthonormal basis β of $(V, \mathbf{F}, \kappa)$, and hence $\phi_\beta(\nu) \in H_\Gamma$. Thus in the notation of Proposition 2.3.5, we have $\ddot{\phi} \in \ddot{H}_\Gamma$. And since $|\ddot{\Gamma} : \langle \ddot{\phi} \rangle| = c = |\ddot{\Gamma} : \ddot{H}_\Gamma|$, it follows that $\ddot{H}_\Gamma = \langle \ddot{\phi} \rangle$, which proves $\pi = \pi_1(c)$.

For the rest of this proof we take n even. Set $\mu_\sharp = \mu^{q+1}$, so that $\mu_\sharp$ generates $\mathbf{F}_\sharp^*$. Thus $\Delta_\sharp = I_\sharp \langle \delta_\sharp \rangle$, where $\tau_\sharp(\delta_\sharp) = \mu_\sharp$, as given in (2.7.2) and (2.8.2). Also $\det(\delta_\sharp) = \pm \mu_\sharp^{n/2}$ by Lemmas 2.7.5 and 2.8.4. In view of (4.5.4) and (2.1.21), we have $\tau(\delta_\sharp) = \mu_\sharp = \tau(\mu \mathbf{I})$, and hence $y = \mu^{-1}\delta_\sharp \in I$. We now argue that

$$H_I = I_\sharp \langle y \rangle. \tag{4.5.6}$$

For notice that $H = I_\sharp\langle\delta_\sharp\rangle\mathbf{F}^*$, and hence

$$H_I = I_\sharp\big(\langle\delta_\sharp\rangle\mathbf{F}^* \cap I\big).$$

Thus to prove (4.5.6) it suffices to show that $\langle\delta_\sharp\rangle\mathbf{F}^* \cap I \leq I_\sharp\langle y\rangle$. So take $x \in \langle\delta_\sharp\rangle\mathbf{F}^* \cap I$ and write $x = \delta_\sharp^i\rho$ with $\rho \in \mathbf{F}^*$. Evidently $y^{-i}x = \mu^i\rho \in Z(I)$, which shows $x \in Z(I)\langle y\rangle$. However $y^{q-1} = \mu^{1-q}\delta_\sharp^{q-1}$, and since $\delta_\sharp^{q-1} \in I_\sharp$ it follows that $\mu^{1-q} \in I_\sharp\langle y\rangle$. But μ^{1-q} generates $Z(I)$, and so $Z(I) \leq I_\sharp\langle y\rangle$. Therefore $x \in Z(I)\langle y\rangle \leq I_\sharp\langle y\rangle$, proving (4.5.6).

Now

$$\det(y) = \det(\mu^{-1}\delta_\sharp) = \pm\mu^{-n}\mu_\sharp^{n/2} = \pm\mu^{n(q-1)/2}.$$

And since $|\det(I_\sharp)| = 2$, (4.5.6) yields

$$|\det(H_I)| = \left[2, \frac{q^2-1}{(q^2-1, \frac{n(q-1)}{2})}\right] = \frac{2(q+1)}{(q+1,n)}.$$

Therefore by Proposition 4.5.1 and Lemma 4.0.3.v,ix the value of c given in (I) is correct. Moreover, Lemma 4.0.3.ii,ix yields $|\overline{H} : H_{\overline{\Omega}}| = (q+1,n)/c = 2$. Therefore $PSO_n^\epsilon(q) \cong \overline{S}_\sharp \trianglelefteq H_{\overline{\Omega}} = \frac{1}{2}\overline{H} = \frac{1}{2}\Delta_\sharp \cong PSO_n^\epsilon(q).2$, and thus (II) holds.

We now determine π. Assume first that $D(\kappa_\sharp)$ is a square. Then we can take $V_\sharp$ to be the $\mathbf{F}_\sharp$-span of an orthonormal basis, and so, just as in the proof of Proposition 4.4.10(I), we see that $\pi = \pi_1(c)$. Next suppose that $D(\kappa_\sharp)$ is a non-square. Then we can take $V_\sharp$ to be the $\mathbf{F}_\sharp$-span of the basis $\{\mu v_1, v_2, \ldots, v_n\}$, where $\beta = \{v_1, \ldots, v_n\}$ is orthonormal. Now set $\phi = \phi_\beta(\nu)$ and $\delta = \delta_\beta(\omega)$ as in (2.3.1). Then $(\mu v_1)(\phi\delta^{(p-1)/2}) = \mu_\sharp^{(p-1)/2}(\mu v_1)$, and so $\phi\delta^{(p-1)/2} \in N_\Gamma(V_\sharp)$. As in the proof of Proposition 4.4.10(I), $\ddot{H} = \langle\ddot{\delta}^c\rangle$, and therefore

$$\ddot{H}_\Gamma = \langle\ddot{\delta}^c, \ddot{\phi}\ddot{\delta}^{(p-1)/2}\rangle.$$

Thus $\pi = \pi_2(c)$. ■

The embedding of type $Sp_n(q)$ in $GU_n(q)$ is not as apparent as the previous one. We begin with a symplectic geometry $(V_\sharp, \kappa_\sharp, \mathbf{F}_\sharp)$ and we let $\zeta \in \mathbf{F}^*$ satisfy $\zeta^q = -\zeta$. Then $\mathbf{F} = \mathbf{F}_\sharp \oplus \mathbf{F}_\sharp\zeta$, and so $\{(\lambda + \eta\zeta)v \mid \lambda, \eta \in \mathbf{F}_\sharp,\ v \in V_\sharp\} = V$. Now define the form $\kappa : V \times V \to \mathbf{F}$ by

$$\kappa\big((\lambda_1 + \eta_1\zeta)v_1, (\lambda_2 + \eta_2\zeta)v_2\big) = (\lambda_1 + \eta_1\zeta)(\lambda_2 - \eta_2\zeta)\zeta\kappa_\sharp(v_1, v_2).$$

Then κ is a non-degenerate unitary form on V and $\kappa_{V_\sharp} = \zeta\kappa_\sharp$. It follows that

$$Sp_n(q) \approx I(V_\sharp, \mathbf{F}_\sharp, \kappa_\sharp) \leq I(V, \mathbf{F}, \zeta^{-1}\kappa) = I(V, \mathbf{F}, \kappa) \approx GU_n(q). \tag{4.5.7}$$

Evidently (4.5.7) indicates that the embedding does not depend on the choice of ζ. In particular, when q is even we can take $\zeta = 1$.

Proposition 4.5.6. *Assume that case* **U** *holds and that H is of type $Sp_n(q)$, so that $r = 2$ and n is even.*

(I) $c = (q+1, n/2)$ *and* $\pi = \pi_1(c)$.

(II) $H_{\overline{\Omega}} \cong PSp_n(q).((2, q-1)c/(q+1, n))$.

(III) $H_{\overline{\Omega}}$ *is non-local and* $\mathrm{soc}(H_{\overline{\Omega}}) \cong PSp_n(q)'$.

Proof. Following closely the argument in Proposition 4.5.5, one may show that $H_I = I_\sharp\langle y\rangle$, where $\det(y) = \mu^{n(q-1)/2}$. Taking δ as in (2.3.1), we have $\det(y) = \det(\delta^{n/2})$, and hence $\ddot{y} = \ddot{\delta}^{n/2} = \ddot{\delta}^{(q+1,n/2)}$. Moreover $I_\sharp \le \Omega$ since $\det(I_\sharp) = 1$, and so

$$\ddot{H} = \ddot{H}_I = \langle \ddot{\delta}^{(q+1,n/2)} \rangle. \tag{4.5.8}$$

Note that $|\ddot{\delta}| = (q+1, n)$, and so $|\ddot{\delta}^{(q+1,n/2)}| \le 2$, with equality if and only if the power of 2 dividing $q+1$ is strictly greater than that dividing $\frac{n}{2}$. The proof of (II) and the calculation of c proceeds as in Proposition 4.5.5.

To determine π, let $\beta_1 = \{e_1', \ldots, e_{n/2}', f_1, \ldots, f_{n/2}\}$ be a symplectic basis of $(V_\sharp, \mathbf{F}_\sharp, \kappa_\sharp)$. Then as $\kappa_\sharp = \zeta\kappa_{V_\sharp}$, we know that $\beta_2 = \{e_1, \ldots, e_{n/2}, f_1, \ldots, f_{n/2}\}$ is a unitary basis of $(V, \mathbf{F}, \kappa)$, where $e_i = \zeta^{-1}e_i'$. Next observe that $\phi_{\beta_2}(\nu) \in N_\Gamma(V_\sharp)$. For $e_i'\phi_{\beta_2}(\nu) = (\zeta e_i)\phi_{\beta_2}(\nu) = \zeta^{p-1}e_i'$; and because $\zeta^{q-1} = -1$, it follows easily that $\zeta^{p-1} \in \mathbf{F}_\sharp$. Consequently by (4.5.8)

$$\ddot{H}_\Gamma = \langle \ddot{\delta}^c, \ddot{\phi}_{\beta_2}(\nu) \rangle. \tag{4.5.9}$$

First we consider q even. Here we may take $\zeta = 1$. Choose $\lambda \in \mathbf{F}$ such that $\mathrm{T}(\lambda) = 1$, and for $1 \le i \le \frac{n}{2}$, set $v_{2i-1} = \lambda e_i + f_i$ and $v_{2i} = (\lambda+1)e_i + f_i$. Then $\beta = \{v_1, \ldots, v_n\}$ is orthonormal, and we put $\phi = \phi_\beta(\nu)$. A straightforward calculation shows that $\phi_{\beta_2}(\nu) = \phi g$, where g satisfies

$$g_{\{v_{2i-1}, v_{2i}\}} = \begin{pmatrix} 1+\lambda+\lambda^2 & \lambda+\lambda^2 \\ \lambda+\lambda^2 & 1+\lambda+\lambda^2 \end{pmatrix}.$$

Evidently $g \in \Omega$, which means $\ddot{\phi}_{\beta_2}(\nu) = \ddot{\phi}$. We deduce from (4.5.9) that $\ddot{H}_\Gamma = \langle \ddot{\delta}^c, \ddot{\phi} \rangle$, and so $\pi = \pi_1(c)$.

For the rest of this proof we take q odd. Let $\lambda \in \mathbf{F}$ satisfy $\lambda^{q+1} = \frac{1}{2}$, and put $\eta = \mu^{(q-1)/2}$, so that $\eta^{q+1} = -1$. This time set $v_{2i-1} = \lambda(e_i + f_i)$ and $v_{2i} = \eta\lambda(e_i - f_i)$, so that $\beta = \{v_1, \ldots, v_n\}$ is an orthonormal basis. Putting $\phi = \phi_\beta(\nu)$, one checks that $\phi_{\beta_2}(\nu) = \phi\lambda^{p-1}g$, where $g = \mathrm{diag}_\beta(1, \eta^{p-1}, 1, \eta^{p-1}, \ldots, 1, \eta^{p-1}) \in I$. Thus $\det(g) = \eta^{(p-1)n/2} = \mu^{(q-1)(p-1)n/4} = \det(\delta^{(p-1)n/4})$. Consequently

$$\ddot{g} = \ddot{\delta}^{(p-1)n/4} = \begin{cases} 1 & \text{if } p \equiv 1 \pmod 4 \\ \ddot{\delta}^{(q+1,n/2)} = \ddot{\delta}^c & \text{if } p \equiv 3 \pmod 4. \end{cases}$$

Thus

$$\ddot{\phi}_{\beta_2}(\nu) = \ddot{\phi}\ddot{g} = \begin{cases} \ddot{\phi} & \text{if } p \equiv 1 \pmod 4 \\ \ddot{\phi}\ddot{\delta}^c & \text{if } p \equiv 3 \pmod 4. \end{cases}$$

Using (4.5.9) we deduce $\ddot{H}_\Gamma = \langle \ddot{\delta}^c, \ddot{\phi} \rangle$, whence $\pi = \pi_1(c)$, as claimed. ∎

We now turn to the orthogonal groups, and we write $Q = \kappa$ and $Q_\sharp = \kappa_\sharp$. These embeddings can be exhibited as follows. Start with an orthogonal geometry $(V_\sharp, \mathbf{F}_\sharp, Q_\sharp)$ with associated bilinear form $\mathbf{f}_\sharp = \mathbf{f}_{Q_\sharp}$, and let β be a standard basis for $V_\sharp$ as described in Proposition 2.5.3. Now form the tensor product space $V_\sharp \otimes \mathbf{F}$ (an n-dimensional space over $\mathbf{F}$) and indentify $V_\sharp$ with $V_\sharp \otimes 1$, so that β is also a basis of $V_\sharp \otimes \mathbf{F}$. Next, let $\mathbf{f}_\sharp \otimes \mathbf{F}$ be the bilinear form on $V_\sharp \otimes \mathbf{F}$ such that $(\mathbf{f}_\sharp \otimes \mathbf{F})_\beta = (\mathbf{f}_\sharp)_\beta$. Then there is a unique quadratic form $Q_\sharp \otimes \mathbf{F}$ on $V_\sharp \otimes \mathbf{F}$ such that its associated bilinear form is $\mathbf{f}_\sharp \otimes \mathbf{F}$ and such that $Q_\sharp \otimes \mathbf{F}$ and $Q_\sharp$ agree on $V_\sharp$. This yields an embedding $I(V_\sharp, \mathbf{F}_\sharp, Q_\sharp) \leq I(V_\sharp \otimes \mathbf{F}, \mathbf{F}, Q_\sharp \otimes \mathbf{F})$. We now claim that $\mathrm{sgn}(Q_\sharp \otimes \mathbf{F}) = \mathrm{sgn}(Q_\sharp)^r$. When n is odd this is trivial, and when $\mathrm{sgn}(Q_\sharp) = +$ this is also easy, for $\mathrm{span}_{\mathbf{F}}\{e_1, \ldots, e_{n/2}\}$ is a totally singular space in $V_\sharp \otimes \mathbf{F}$. Assume now that $\mathrm{sgn}(Q_\sharp) = -$. When $r = 2$, the embedding $I(V_\sharp, \mathbf{F}_\sharp, Q_\sharp) \leq I(V_\sharp \otimes \mathbf{F}, \mathbf{F}, Q_\sharp \otimes \mathbf{F})$ was already constructed in §2.8 (see the discussion preceding Proposition 2.8.1), and there we have seen that $\mathrm{sgn}(Q_\sharp \otimes \mathbf{F}) = +$. Thus we are left with the case where r is odd. Here we observe that the polynomial $\mathbf{x}^2 + \mathbf{x} + \zeta$ appearing in Lemma 2.5.2.ii remains irreducible over $\mathbf{F}$, and so the 2-space $\langle x, y \rangle \otimes \mathbf{F}$ is a -2-space in $V \otimes \mathbf{F}$. Therefore $\mathrm{span}_{\mathbf{F}}\{e_1, \ldots, e_{n/2-1}\}$ is a maximal totally singular subspace in $V_\sharp \otimes \mathbf{F}$, which means $\mathrm{sgn}(Q_\sharp \otimes \mathbf{F}) = -$, as claimed.

Putting $\epsilon_\sharp = \mathrm{sgn}(Q_\sharp)$, the preceding discussion establishes the embeddings of type $O_n^{\epsilon_\sharp}(q^{1/r})$ in $O_n^{\epsilon}(q)$, where $\epsilon = \epsilon_\sharp^r$.

Lemma 4.5.7. *Suppose that case* $\mathbf{O}^{\pm}$ *holds, so that n is even.*

(i) *If q is odd, then* $I_\sharp \backslash S_\sharp \subseteq I \backslash S$.
(ii) *If q is even, then* $I_\sharp \backslash \Omega_\sharp \subseteq I \backslash \Omega$.
(iii) *If r and q are odd, then* $S_\sharp \backslash \Omega_\sharp \subseteq S \backslash \Omega$.
(iv) *If $r = 2$ and q is odd, then* $S_\sharp \leq \Omega$.
(v) *If r and q are odd, then* $\Delta_\sharp \backslash I_\sharp \mathbf{F}^* \subseteq \Delta \backslash I\mathbf{F}^*$.
(vi) *If $r = 2$ and q is odd, then* $\Delta_\sharp \leq I\mathbf{F}^*$.

Proof. (i). This is trivial, for elements in $I_\sharp \backslash S_\sharp$ have determinant -1.

(ii). Since $\Omega_\sharp \leq \Omega$, it suffices to exhibit one element of $I_\sharp \backslash \Omega_\sharp$ which does not lie in Ω. So let $v \in V_\sharp$ satisfy $Q_\sharp(v) \neq 0$. Then $Q(v) \neq 0$, and so we may let $g \in I \backslash \Omega$ be the reflection in v. Clearly g is also a reflection in the $(V_\sharp, \mathbf{F}_\sharp, Q_\sharp)$-geometry, and so $g \in I_\sharp \backslash \Omega_\sharp$, as desired.

(iii) and (iv). Take $g \in S_\sharp \backslash \Omega_\sharp$, and write $g = r_1 \ldots r_k$, where $r_i = r_{v_i}$ and v_i is a non-singular vector in $(V_\sharp, \mathbf{F}_\sharp, Q_\sharp)$. Then the spinor norm (in the $(V_\sharp, \mathbf{F}_\sharp, Q_\sharp)$-geometry) of g is $\lambda = \prod_{i=1}^k Q_\sharp(v_i)$, and by assumption λ is a non-square in $\mathbf{F}_\sharp$. However, since $Q_\sharp = Q_{V_\sharp}$, the spinor norm of g in the $(V, \mathbf{F}, Q)$-geometry is also λ, and so (iii) and (iv) are now clear from Description 1 in §2.5.

(v) and (vi). These follow directly from (4.5.4). ■

Proposition 4.5.8. *Assume that case* $\mathbf{O}^{\circ}$ *holds, so that H is of type $O_n(q^{1/r})$ with qn odd.*

(I) $c = (2, r)$, *and if* $c = 2$, *then* $\pi = \pi_1$.
(II) $H_{\overline{\Omega}} \cong \Omega_n(q^{1/r}).c$.
(III) $H_{\overline{\Omega}}$ *is non-local and* $\mathrm{soc}(H_{\overline{\Omega}}) \cong \Omega_n(q^{1/r})$.

Lemma 4.5.9. *Suppose that q is odd and that $(n, \epsilon, r) = (2, +, 2)$. Define λ to be a generator for $\mathbf{F}_\sharp^*$ if $q^{1/2} \equiv 1 \pmod 4$ and define $\lambda = -1$ if $q^{1/2} \equiv 3 \pmod 4$. Also let $\eta \in \mathbf{F}^*$ satisfy $\eta^2 = \lambda^{-1}$. Then there exists $k \in \Delta_\sharp$ such that $\tau_\sharp(k) = \lambda$ and*

$$\eta k \in \begin{cases} \Omega & \text{if } D(Q_\sharp) = \boxtimes \\ S \backslash \Omega & \text{if } D(Q_\sharp) = \square. \end{cases}$$

Proof. *Case $\epsilon_\sharp = +$ and $q^{1/2} \equiv 1 \pmod 4$.* Let β be a standard basis of $V_\sharp$ and define $k = \mathrm{diag}_\beta(\lambda, 1)$. Thus $\tau_\sharp(k) = \lambda$ and $\eta k = \mathrm{diag}(\eta\lambda, \eta) = \mathrm{diag}(\eta^{-1}, \eta) \in S$. Furthermore $|\eta| = 2|\lambda| = 2(q^{1/2} - 1)$, which does not divide $(q-1)/2$ as $q^{1/2} \equiv 1 \pmod 4$. Therefore $\eta k \in S \backslash \Omega$, as $\Omega \cong \mathbf{Z}_{(q-1)/2}$.

Case $\epsilon_\sharp = +$ and $q^{1/2} \equiv 3 \pmod 4$. Let β be as in the previous case and this time define $k = \mathrm{diag}_\beta(-1, 1)$. As before we have $\eta k \in S$, but now $|\eta k| = 4$, which divides $(q-1)/2$. Consequently $\eta k \in \Omega$, as $S \cong \mathbf{Z}_{q-1}$ and $\Omega \cong \mathbf{Z}_{(q-1)/2}$.

Case $\epsilon_\sharp = -$ and $q^{1/2} \equiv 1 \pmod 4$. By Proposition 2.5.10 $D(Q_\sharp)$ is a non-square and so Proposition 2.5.12 implies that there is a basis β of $V_\sharp$ such that $(\mathbf{f}_\sharp)_\beta = \mathrm{diag}_\beta(1, \lambda)$. Now define $k = \left(\begin{smallmatrix} 0 & -1 \\ \lambda & 0 \end{smallmatrix}\right)$, so that $\tau_\sharp(k) = \lambda$. Then it is easily seen that $\eta k \in S$ and that $|\eta k| = 4$. Thus the result follows as in the previous case.

Case $\epsilon_\sharp = -$ and $q^{1/2} \equiv 3 \pmod 4$. Here $D(Q_\sharp)$ is a square and so $(V_\sharp, Q_\sharp)$ has an orthonormal basis β. Choose $a, b \in \mathbf{F}_\sharp$ such that $a^2 + b^2 = -1$, and put $k = \left(\begin{smallmatrix} a & -b \\ b & a \end{smallmatrix}\right)$, so that $\eta k \in S$. Now ηk has eigenvalues $\eta a \pm b$, which means by changing basis (over $\mathbf{F}$) we may write $\eta k = \mathrm{diag}(\eta a + b, \eta a - b)$. If $\eta a + b$ is a square in $\mathbf{F}$, then $\eta a + b = (x + \eta y)^2$ for some $x, y \in \mathbf{F}_\sharp$. But then $b = x^2 - y^2$ and $a = 2xy$, which forces $-1 = a^2 + b^2 = (x^2 + y^2)^2$, contradicting the fact that -1 is a non-square in $\mathbf{F}_\sharp$. Thus $\eta a + b$ is a non-square in $\mathbf{F}$, and hence $\eta k \in S \backslash \Omega$, completing the proof. ∎

Proposition 4.5.10. *Assume that case $\mathbf{O}^\epsilon$ holds with n even, so that H is of type $O_n^{\epsilon_\sharp}(q^{1/r})$ and $\epsilon = \epsilon_\sharp^r$.*

(I)

$$c = \begin{cases} 1 & \text{if } r \text{ is odd or if } q \text{ is even} \\ 2 & \text{if } r = 2,\ q \text{ is odd},\ n \equiv 0 \pmod 4, \text{ and } D(Q_\sharp) = \boxtimes \\ 2 & \text{if } r = 2,\ q \text{ is odd},\ n \equiv 2 \pmod 4, \text{ and } D(Q_\sharp) = \square \\ 4 & \text{if } r = 2,\ q \text{ is odd},\ n \equiv 0 \pmod 4, \text{ and } D(Q_\sharp) = \square \\ 4 & \text{if } r = 2,\ q \text{ is odd},\ n \equiv 2 \pmod 4, \text{ and } D(Q_\sharp) = \boxtimes. \end{cases}$$

$$\pi = \begin{cases} \pi_1 & \text{if } c = 2 \\ \pi_4 & \text{if } c = 4. \end{cases}$$

(II)

$$H_{\overline{\Omega}} \cong \begin{cases} P\Omega_n^{\epsilon_\sharp}(q^{1/r}) & \text{if } r \text{ is odd or } q \text{ is even} \\ PSO_n^{\epsilon_\sharp}(q^{1/2}).[c/2] & \text{otherwise.} \end{cases}$$

(III) *$H_{\overline{\Omega}}$ is non-local and* $\mathrm{soc}(H_{\overline{\Omega}}) \cong P\Omega_n^{\epsilon_\sharp}(q^{1/r})$.

Proof. First suppose that q is even. By Lemma 4.0.3.viii,ix we have $c = |\ddot{I} : \ddot{H}_I|$. It follows from Lemma 4.5.7.ii that $H_I\Omega = I$, and so $c = 1$. As $\Delta_\sharp = I_\sharp \times \mathbf{F}_\sharp^*$, we see from (4.5.5) that $H = I_\sharp \times \mathbf{F}^*$, whence $H_I = I_\sharp(\mathbf{F}^* \cap I) = I_\sharp$. Consequently Lemma 4.0.3.iv gives $H_{\overline{\Omega}} = \frac{1}{2}\overline{H}_I \cong \frac{1}{2}O_n^{\epsilon_\sharp}(q^{1/r}) \cong \Omega_n^{\epsilon_\sharp}(q^{1/r})$, as desired.

Next suppose that qr is odd. It follows from Lemma 4.5.7.i,iii,v that $HI = \Delta$, $I_\sharp S = I$, and $S_\sharp\Omega = S$, whence $H\Omega = \Delta$. Consequently $c = 1$ by Lemma 4.0.3.i. Note that $D(Q_\sharp)$ is a square if and only if $D(Q)$ is a square, and hence $|P\Delta_\sharp : P\Omega_\sharp| = |P\Delta : P\Omega|$. Therefore by Lemma 4.0.3.ii and the fact that $\overline{H} \cong P\Delta_\sharp$, we see that $H_{\overline{\Omega}} \cong P\Omega_\sharp \cong P\Omega_n^{\epsilon_\sharp}(q^{1/r})$.

For the rest of this proof we suppose that q is odd and $r = 2$. Thus $I \approx O_n^+(q)$ and $D(Q)$ is a square. Moreover, by Lemma 4.5.7.iv we have $S_\sharp \leq \Omega$, and since $P\Delta_\sharp \cong PS_\sharp.2^2$ (see Table 2.1.D), we obtain

$$PSO_n^{\epsilon_\sharp}(q^{1/2}) \cong \overline{S}_\sharp \trianglelefteq H_{\overline{\Omega}} \trianglelefteq \overline{H} \cong PSO_n^{\epsilon_\sharp}(q^{1/2}).2^2. \tag{4.5.10}$$

Therefore by Lemma 4.0.3.ii,ix and Proposition 4.5.1, $H_{\overline{\Omega}} = \frac{c}{8}\overline{H} \cong PSO_n^{\epsilon_\sharp}(q).[\frac{c}{2}]$, and so (II) holds. It remains to establish (I). If $g \in I_\sharp$ is a reflection in a vector $v \in V_\sharp$ of $\mathbf{f}_\sharp$-norm $\lambda \in \mathbf{F}_\sharp^*$, then as $\mathbf{F}_\sharp^* \leq (\mathbf{F}^*)^2$, we have $\mathbf{f}(v, v) \in (\mathbf{F}^*)^2$, which means $\ddot{g} \in \langle \ddot{r}_\square \rangle$. Thus as $I_\sharp = S_\sharp\langle g \rangle$, we deduce

$$\ddot{I}_\sharp = \langle \ddot{r}_\square \rangle. \tag{4.5.11}$$

Now according to Lemma 4.5.7.vi we have $H \leq I\mathbf{F}^*$, and hence $\ddot{H} \leq \ddot{I} = \langle \ddot{r}_\square, \ddot{r}_\boxtimes \rangle$. Therefore from (4.5.11) we obtain

$$\langle \ddot{r}_\square \rangle \leq \ddot{H} \leq \langle \ddot{r}_\square, \ddot{r}_\boxtimes \rangle. \tag{4.5.12}$$

We will establish whether equality holds on the left or the right of (4.5.12).

Write $V_\sharp = W_1 \perp \cdots \perp W_{n/2}$, where W_i is an $\epsilon_i 2$-space for $1 \leq i \leq \frac{n}{2}$. Then $V = V_1 \perp \cdots \perp V_{n/2}$, where $V_i = W_i\mathbf{F}$, and V_i is a $+2$-space in V for all i. Now let λ and η be as in Lemma 4.5.9. By applying Lemma 4.5.9 to each of the spaces W_i, we see that there exists $k \in \Delta_\sharp$ such that k fixes each space W_i, $\tau_\sharp(k) = \lambda$, and ηk^{W_i} satisfies the conclusion of the Lemma. Since λ is a non-square in $\mathbf{F}_\sharp$, we have $\Delta_\sharp = I_\sharp\langle k \rangle\mathbf{F}_\sharp^*$, and hence $\ddot{H} = \ddot{I}_\sharp\langle \ddot{k} \rangle$. Now $\eta k^{V_i} \in S(V_i, Q)$ for all $i \leq n/2$, and so $\eta k \in S$. Therefore $\ddot{k} \in \ddot{S} = \langle \ddot{r}_\square \ddot{r}_\boxtimes \rangle$, and since $\langle \ddot{S}, \ddot{r}_\square \rangle = \ddot{I}$ we know that equality holds on the left in (4.5.12) if $\eta k \in \Omega$ and equality holds on the right if $\eta k \in S \backslash \Omega$. Thus equality holds on the

$$\begin{cases} \text{left} & \text{if } \eta k^{V_i} \in S(V_i, Q) \backslash \Omega(V_i, Q) \text{ for evenly many } i \\ \text{right} & \text{if } \eta k^{V_i} \in S(V_i, Q) \backslash \Omega(V_i, Q) \text{ for oddly many } i. \end{cases} \tag{4.5.13}$$

It is now straightforward to determine whether equality holds on the right or the left. First take $n \equiv 0 \pmod 4$ and $D(Q_\sharp)$ a non-square. Then $\epsilon_\sharp = -$ and so we can take $\epsilon_i = +$ for $2 \leq i \leq \frac{n}{2}$ and $\epsilon_1 = -$. Hence according to Lemma 4.5.9, ηk^{V_i} lies in

$\Omega(V_i, Q)$ precisely once or precisely $\frac{n}{2} - 1$ times, and so oddly many ηk^{V_i} lie in $\Omega(V_i, Q)$. Consequently equality holds on the right in (4.5.12). Therefore by Lemma 4.0.3.i,ix and Proposition 4.5.1, $c = |\ddot{\Delta} : \ddot{I}| = 2$. Furthermore observe that the norm of every vector in $V_\sharp$ is a square in $\mathbf{F}$, and so if $h \in H_\Gamma$, then $\tau(h) \in (\mathbf{F}^*)^2$. Therefore $H_{\overline{\Gamma}} \leq \ker(\overline{\tau})$, and so $\ddot{H}_\Gamma = \ker_{\ddot{\Gamma}}(\ddot{\tau})$, which means $\pi = \pi_1$.

Next take the case where $n \equiv 2 \pmod 4$ and $D(Q_\sharp)$ is a non-square. Then we can choose the W_i so that $D(W_i)$ is a square for $2 \leq i \leq \frac{n}{2}$ and $D(W_1)$ is a non-square. Then by Lemma 4.5.9, $\eta k^{V_i} \in S(V_i, Q) \backslash \Omega(V_i, Q)$ only when $2 \leq i \leq \frac{n}{2}$, and hence for evenly many i. Therefore equality holds on the left in (4.5.12), and hence $c = |\ddot{\Delta} : \langle \ddot{r}_\square \rangle| = 4$. We now determine π in this situation. Let $\beta = \{e_1, \ldots, f_{n/2}\}$ be a standard basis of V and set $\phi = \phi_\beta(\nu)$. For $1 \leq j \leq \frac{n}{2}$ put $v_{2j-1} = (\frac{1}{\sqrt{2}})(e_j + f_j)$ and $v_{2j} = (\frac{i}{\sqrt{2}})(e_j - f_j)$, where $i = \sqrt{-1}$. Thus $\beta_1 = \{v_1, \ldots, v_n\}$ is an orthonormal basis of (V, Q). Now choose $\xi \in \mathbf{F}$ such that $\xi^2 = \mu_\sharp$, where $\mu_\sharp$ is a generator for $\mathbf{F}_\sharp$. Then if $\beta_2 = \{\xi v_1, v_2, \ldots, v_n\}$, we see that $\mathbf{f}_{\beta_2} = \mathrm{diag}_{\beta_2}(\mu_\sharp, 1, \ldots, 1)$, which has determinant a non-square in $\mathbf{F}_\sharp$. Hence we may take $V_\sharp$ to be the $\mathbf{F}_\sharp$-span of β_2. Now set $\phi_1 = \phi_{\beta_1}(\nu) \in \Gamma$. Evidently $(\xi v_1)\phi_1 = \xi^p v_1 = \mu_\sharp^{(p-1)/2}(\xi v_1) \in V_\sharp$, and hence $\phi_1 \in N_\Gamma(V_\sharp) \leq H_\Gamma$. Consequently $\ddot{H}_\Gamma = \langle \ddot{r}_\square, \ddot{\phi}_1 \rangle$. Now it is clear that $e_j = (\frac{1}{\sqrt{2}})(v_{2j-1} - iv_{2j})$ and $f_j = (\frac{1}{\sqrt{2}})(v_{2j-1} + iv_{2j})$. Thus if $p \equiv 1 \pmod 4$, we have $e_j\phi_1 = (\frac{1}{2})^{(p-1)/2} e_j = \pm e_j$ and similarly $f_j\phi_1 = \pm f_j$. Therefore $\phi_1 = \pm\phi$, and hence $\ddot{\phi}_1 = \ddot{\phi}$. Therefore $\pi = \pi_4$. On the other hand, if $p \equiv 3 \pmod 4$, then $e_j\phi_1 = \pm f_j$ and $f_j\phi_1 = \pm e_j$. Define $r_j = r_{e_j - f_j}$, so that r_j interchanges e_j and f_j, and set $s = r_1 \ldots r_{n/2}$. Thus $\phi_1 = \pm s\phi$, and furthermore $\ddot{s} = \ddot{r}_1$ because $r_2 \ldots r_{n/2} \in \Omega$. Since $e_1 - f_1$ has norm $-2 \in (\mathbf{F}^*)^2$ (note $q \equiv 1 \pmod 8$), we know that $\ddot{r}_1 = \ddot{r}_\square$, and so $\ddot{\phi}_1 = \ddot{r}_\square\ddot{\phi}$. Thus again $\ddot{H}_\Gamma = \langle \ddot{r}_\square, \ddot{\phi} \rangle$, whence $\pi = \pi_4$.

The other two cases are treated similarly, and we leave them as an exercise. ■

We conclude this section with a result to be used in Chapter 6, where overgroups of members of $\mathcal{C}$ are described.

Proposition 4.5.11. *If case* $\mathbf{O}^+$ *holds, then* $\Omega_\sharp$ *centralizes an element of order* r *in* Γ.

Proof. Here H is of type $O_n^{\epsilon_\sharp}(q^{1/r})$, where $(\epsilon_\sharp)^r = +$. First assume that $\epsilon_\sharp = +$ and let $\beta = \{e_1, \ldots, e_{n/2}, f_1, \ldots, f_{n/2}\}$ be a standard basis for $(V, \mathbf{F}, Q)$. We may take $V_\sharp = \mathrm{span}_{\mathbf{F}_\sharp}(\beta)$, and so for $g \in \Omega_\sharp$, all entries in the matrix g_β lie in $\mathbf{F}_\sharp$. Therefore $\Omega_\sharp$ centralizes $\phi_\beta(\nu_\sharp)$, where $\nu_\sharp$ generates $\mathrm{Gal}(\mathbf{F} : \mathbf{F}_\sharp)$. When $\epsilon_\sharp = -$, it follows from Proposition 2.8.1.ii (beware that the roles of V and $V_\sharp$ are reversed in Proposition 2.8.1) that an involution in $\Gamma \backslash \Delta$ centralizes $V_\sharp$. Therefore this involution centralizes $\Omega_\sharp$, as required. ■

§4.6 The symplectic-type normalizers $\mathcal{C}_6$

The members of $\mathcal{C}_6(\Gamma)$ are the normalizers in Γ of certain absolutely irreducible symplectic-type r-groups R, with $r \neq p$. We first give a brief description of such groups for those readers who are unfamiliar with them.

Definition. Let R be an r-group for some prime r. Then R is *extraspecial* if $Z(R) =$

Table 4.6.A			
structure	$\|R\|$, $\|Z(R)\|$	notation	$C_{Aut(R)}(Z(R))$
$\overbrace{R_o \circ \cdots \circ R_o}^{m}$	r^{1+2m}, r	r^{1+2m}	$r^{2m}.Sp_{2m}(r)$
$\overbrace{D_8 \circ \cdots \circ D_8}^{m}$	2^{1+2m}, 2	2_+^{1+2m}	$2^{2m}.O_{2m}^+(2)$
$\overbrace{D_8 \circ \cdots \circ D_8}^{m-1} \circ Q_8$	2^{1+2m}, 2	2_-^{1+2m}	$2^{2m}.O_{2m}^-(2)$
$\mathbf{Z}_4 \circ \overbrace{D_8 \circ \cdots \circ D_8}^{m}$	2^{2+2m}, 4	$4 \circ 2^{1+2m}$	$2^{2m}.Sp_{2m}(2)$

$\Phi(R) = R' \cong \mathbf{Z}_r$. Furthermore, R is of *symplectic-type* if every characteristic abelian subgroup of R is cyclic.

This first Proposition is an old result of P. Hall, describing the structure of extraspecial r-groups. A proof can be found in [Su$_2$, p.69].

Proposition 4.6.1. *Let r be prime.*

(i) *There are, up to isomorphism, just two extraspecial groups of order r^3. When $r = 2$, the two groups are Q_8 and D_8. When r is odd, one group has exponent r^2, while the other has exponent r.*

(ii) *When r is odd, the extraspecial r-group with exponent r is isomorphic to*

$$R_o = \langle x, y, z \mid x^r = y^r = z^r = [x, z] = [y, z] = z^{-1}[x, y] = 1\rangle.$$

(iii) *Let R be an extraspecial r-group. Then R is a central product of extraspecial r-groups of order r^3, and $|R| = r^{1+2m}$ for some m. If $r = 2$, then R is either a central product of m copies of D_8, or $m-1$ copies of D_8 and one Q_8. If r is odd and R has exponent r, then R is a central product of m copies of R_o, where R_o is as in (ii).*

The structure of symplectic-type r-groups is also well understood, and is closely related to that of extraspecial groups (see [Su$_2$, pp.75-76]). We will be concerned only with symplectic-type r-groups with minimal exponent — that is, exponent r when r is odd and exponent 4 when $r = 2$. In the next Proposition, we provide the information we shall need about such groups.

Proposition 4.6.2. *If R is an r-group of symplectic-type with exponent $r(2, r)$, then R appears in Table 4.6.A. We also give the structure of $C_{Aut(R)}(Z(R))$. In the first row, r is odd and R_o is as given in Proposition 4.6.1.ii.*

The assertions concerning structure in the first column are a corollary to the structure theorem for symplectic-type groups (see [As$_8$, (23.9)] and [Su$_2$, pp.75-76]). The groups in the first three rows are in fact extraspecial. The information in the right hand column can be read off from [As$_8$, Ex. 8.5]. Note that no sign $\pm$ appears in the 2^{1+2m} in the last row since $\mathbf{Z}_4 \circ 2_+^{1+2m} \cong \mathbf{Z}_4 \circ 2_-^{1+2m}$.

The next Proposition describes the representation theory of the symplectic-type groups.

Proposition 4.6.3. *Assume that R is a symplectic-type r-group of exponent $r(2,r)$ as in Table 4.6.A, and that $r \neq p$.*

(i) *R has precisely $|Z(R)|-1$ inequivalent faithful absolutely irreducible representations over an algebraically closed field of characteristic p. Denote these by $\rho_1, \ldots, \rho_k$, where $k = |Z(R)| - 1$.*

(ii) *The ρ_i are quasiequivalent, they have degree r^m, and the smallest field over which they can be realized is $\mathbf{F}_{p^e}$, where e is the smallest integer for which $p^e \equiv 1 \pmod{|Z(R)|}$.*

(iii) *If $i \neq j$, then ρ_i and ρ_j differ on $Z(R)$.*

Proof. The analogous result for representations of R over the complex numbers is proved in [Su$_2$, p.335]. The result in characteristic p follows immediately in view of [Is, Th. 15.13]. ∎

The representations ρ_i will be constructed at various points in the ensuing discussion. We are now ready to define the family $\mathcal{C}_6$.

Definition. The members of $\mathcal{C}_6(\Gamma)$ are the groups $N_\Gamma(R)$ such that

(a) $R \leq \Delta$ and R is an r-group ($r \neq p$) of symplectic-type of exponent $r(2,r)$,

(b) R acts absolutely irreducibly on V (so that $n = r^m$),

(c) $\mathbf{F} = \mathbf{F}_{p^e}$, where e is the smallest integer for which $p^e \equiv 1 \pmod{|Z(R)|}$,

(d) R appears in Table 4.6.B.

Table 4.6.B

case	type	description of R	conditions
L	$r^{1+2m}.Sp_{2m}(r)$	$R \cong r^{1+2m}$	er odd
L	$(4 \circ 2^{1+2m}).Sp_{2m}(2)$	$R \cong 4 \circ 2^{1+2m}$	$e=1,\ n \geq 4$
L	$2_-^{1+2}.O_2^-(2)$	$R \cong 2_-^{1+2}$	$e=1,\ n=2$
U	$r^{1+2m}.Sp_{2m}(r)$	$R \cong r^{1+2m}$	r odd, e even
U	$(4 \circ 2^{1+2m}).Sp_{2m}(2)$	$R \cong 4 \circ 2^{1+2m}$	$e=2$
S	$2_-^{1+2m}.O_{2m}^-(2)$	$R \cong 2_-^{1+2m}$	$e=1$
O$^+$	$2_+^{1+2m}.O_{2m}^+(2)$	$R \cong 2_+^{1+2m}$	$e=1$

Remarks on the conditions. The condition on e in part (c) of the Definition is imposed to ensure that $N_\Gamma(R)$ is contained in no subfield subgroup in $\mathcal{C}_5$ (see Proposition 4.6.3.ii). Furthermore, when e is even, R fixes a non-degenerate unitary form as described below; thus we demand that e is odd in case **L** to guarantee that $N_\Gamma(R)$ is not contained in a unitary group in $\mathcal{C}_8$. We will see below that the representation ρ_1 of 2_-^{1+2m} or 2_+^{1+2m} fixes a non-degenerate symplectic or quadratic form, respectively, and hence these groups occur in case **S** or **O**.

For the remainder of this section we shall assume that R is one of the groups appearing in Table 4.6.B and $H_\Gamma = N_\Gamma(R)$. According to [As$_1$, Theorem A(4)]

$$\overline{H} \cong C_{Aut(R)}(Z(R)). \tag{4.6.1}$$

Thus the structure of $\overline{H}$ is given in Table 4.6.A. Note further that

$$n = r^m \text{ and } e = uf. \tag{4.6.2}$$

These equalities hold in view of parts (b) and (c) of the Definition, and by Proposition 4.6.3.ii.

We begin by proving Theorem 3.1.2 for the subgroups in $\mathcal{C}_6$.

Proposition 4.6.4. *We have $H_\Gamma = N_\Gamma(H_\Omega)$. Hence Theorem 3.1.2 holds for the members of $\mathcal{C}_6$.*

Proof. The second assertion follows from the first according to Proposition 4.0.1. Thus it suffices to show that $H_\Gamma = N_\Gamma(H_\Omega)$. According to (4.6.1), we may write $H = (\mathbf{F}^* \circ R).C\ell_{2m}(r)$, where $C\ell_{2m}(r)$ denotes either $Sp_{2m}(r)$ or $O^{\pm}_{2m}(2)$. Thus for $n \geq 4$, we know that $C\ell_{2m}(r)'$ is quasisimple and so $H_\Omega^\infty = H^\infty = (\langle\lambda\rangle \circ R).C\ell_{2m}(r)'$, for some $\lambda \in \mathbf{F}^* \cap \Omega$. Consequently R is the group generated by the elements of order r in the soluble radical of H_Ω^∞, and hence R is characteristic in H_Ω. Therefore $N_\Gamma(H_\Omega) \leq N_\Gamma(R) = H_\Gamma$, as claimed. For $n = 3$, we have $H_\Omega \geq H' = (\langle\lambda\rangle \circ R).Sp_2(3)'$, where $\lambda \in \mathbf{F}^* \cap \Omega$. But $Z(R) = \mathbf{F}^* \cap \Omega$, and so $\lambda \in R$. Therefore $R.Q_8 \leq H_\Omega \leq R.Sp_2(3)$, and so $R = O_3(H_\Omega)$. So again R is characteristic in H_Ω and the result follows. Finally take $n = 2$, so that case **L** holds and H is of type $2^{1+2}_{-}.O_2^-(2)$. Here we check that $R = O_2(H_\Omega)$, and so we reach the conclusion as before. ∎

Now we turn to the structure results. First we handle case $\mathbf{L}^{\pm}$ with $R \cong r^{1+2m}$ and r odd. To describe the faithful irreducible representations of R, let $\mathbf{F} = \mathbf{F}_{p^e}$ where e is the smallest integer for which $p^e \equiv 1 \pmod r$, and let $\omega \in \mathbf{F}^*$ be a primitive r^{th} root of unity. Assume that W is an r-dimensional space over $\mathbf{F}$ with basis $\beta = \{w_1, \ldots, w_r\}$, and let x, y, R_o be as in Proposition 4.6.1.ii. Define $y\rho$ as the element in $GL(W, \mathbf{F})$ which satisfies $w_i(y\rho) = w_{i+1}$ (subscripts modulo r) and $x\rho = \text{diag}_\beta(1, \omega, \omega^2, \ldots, \omega^{r-1})$. One checks that ρ yields an absolutely irreducible representation of R_o. Thus by tensoring (see Lemma 4.4.3.vi) one obtains an absolutely irreducible representation of $R \cong R_o \otimes \cdots \otimes R_o$ of degree r^m on $W \otimes \cdots \otimes W$. Observe that when e is even, $(R_o)\rho \leq I(W, \mathbf{F}, \mathbf{f})$, where $\mathbf{f}$ is the unitary form for which β is an orthonormal basis; therefore R embeds in $I(W \otimes \cdots \otimes W, \mathbf{F}, \mathbf{f} \otimes \cdots \otimes \mathbf{f})$ (see (4.4.4)). Hence the condition e odd in case **L** and e even in case **U**.

Proposition 4.6.5. *Assume that case $\mathbf{L}^\epsilon$ holds and H is of type $r^{1+2m}.Sp_{2m}(r)$ with r odd.*

(I)

$$c = \begin{cases} 1 & \text{if } n = 3 \text{ and } q \equiv \epsilon 4 \text{ or } \epsilon 7 \pmod 9 \\ (q - \epsilon, n) & \text{otherwise} \end{cases}$$

and $\pi = \pi_1(c)$.

(II)

$$H_{\overline{\Omega}} \cong \begin{cases} 3^2.Q_8 & \text{if } n = 3 \text{ and } c = 1 \\ r^{2m}.Sp_{2m}(r) & \text{otherwise.} \end{cases}$$

(III) $H_{\overline{\Omega}}$ *is* r*-local.*

Proof. First assume that $\overline{H}$ is perfect. Then $H_{\overline{\Omega}} = \overline{H}$ and so $c = |\ddot{\Delta}| = (q-\epsilon, n)$ by Lemma 4.0.3.ii,ix and Proposition 4.6.4. Observe that part (c) of the Definition above ensures that p has order e in the multiplicative group of the field $\mathbf{F}_r$. Thus $e \mid r-1$, and hence $f \mid r-1$. And as $n = r^m$, we see that $|A : \Delta| = 2f$ is relatively prime to $(q-\epsilon, n)$. Therefore $\ddot{A}$ contains a unique conjugacy class of subgroups of order $2f$. Using the notation of Propositions 2.2.3 and 2.3.5, $\langle \ddot{\phi}, \ddot{\iota} \rangle$ and $\langle \ddot{\phi} \rangle$ have order $2f$ in cases **L** and **U**, respectively. Hence these groups must be $\ddot{H}_\Gamma$, and so $\pi = \pi_1(c)$.

Now take the case where $\overline{H}$ is not perfect — namely, $r = n = 3$. Here we may take $R = R_o = \langle x, y \rangle$ as in Proposition 4.6.1.ii, and following the discussion after Proposition 4.6.4, we may take y to be the map $v_i \mapsto v_{i+1}$ (subscripts modulo 3) and $x_\beta = \mathrm{diag}(1, \omega, \omega^2)$, where $\beta = \{v_1, v_2, v_3\}$ is a basis of V and ω is a primitive cube root of 1. Moreover, if case **U** holds (so that $e = 2$ and $p \equiv 2 \pmod 3$), then β is an orthonormal basis for $(V, \mathbf{f})$. As we mentioned in the proof of Proposition 4.6.4, $R.Q_8 \le H_\Omega \le R.Sp_2(3)$. If $p \equiv \epsilon 4$ or $\epsilon 7 \pmod 9$, then 27 does not divide $L_3^\epsilon(p)$, and hence $H_{\overline{\Omega}} = \frac{1}{3}\overline{H} \cong 3^2.Q_8$. In this case $|\ddot{H}| = 3 = |\ddot{\Delta}|$, and hence $c = 1$ by Lemma 4.0.3.ii,ix. Now suppose that $p \equiv \epsilon \pmod 9$. Then there exists $\lambda \in \mathbf{F}^* \cap I$ such that $\lambda^3 = \omega^{-1}$. Define $g \in I$ by $g_\beta = \mathrm{diag}_\beta(\lambda, \lambda, \lambda\omega) \in \Omega$. Then $x^g = x$ and $y^g = xy$, and hence g induces an outer automorphism of order 3 on R. Consequently $H_{\overline{\Omega}} \cong 3^2.Sp_2(3)$, which shows $H_{\overline{\Omega}} = \overline{H}$. Therefore as in the first part of the proof we deduce $c = 3$ and $\pi = \pi_1(3)$. ∎

Continue to assume that case $\mathbf{L}^\epsilon$ holds. Here we exhibit the embedding of $4 \circ 2^{1+2m}$ in $I = I(V, \mathbf{F}, \kappa)$ (with $m \ge 2$). For $j = 1, \ldots, m$, let V_j be a 2-dimensional vector space over $\mathbf{F}$. When $\epsilon = -$ assume that $\mathbf{f}_j$ is a non-degenerate unitary form on V_j, and when $\epsilon = +$ let $\mathbf{f}_j$ be the zero form. Let β_j be a basis for V_j, and when $\epsilon = -$ assume that β_j is orthonormal. Now let $x_j, y_j \in GL(V_j, \mathbf{F})$ satisfy

$$(x_j)_{\beta_j} = \begin{pmatrix} i & 0 \\ 0 & -i \end{pmatrix} \quad (y_j)_{\beta_j} = \begin{pmatrix} 0 & 1 \\ -1 & 0 \end{pmatrix}, \tag{4.6.3}$$

where $i = \sqrt{-1}$. Then it is clear that $R_j = \langle x_j, y_j \rangle \cong Q_8$ and that $R_j \le I(V_j, \mathbf{F}, \mathbf{f}_j)$. Forming the tensor product as before, we may take $R = \langle i \rangle \circ (R_1 \otimes \cdots \otimes R_m) \le I(V, \mathbf{F}, \mathbf{f})$, where $V = V_1 \otimes \cdots \otimes V_m$ and $\mathbf{f} = \mathbf{f}_1 \otimes \cdots \otimes \mathbf{f}_m$. We use this notation in the proof of the following Proposition.

Proposition 4.6.6. *Assume that case* $\mathbf{L}^\epsilon$ *holds with* H *of type* $(4 \circ 2^{1+2m}).Sp_{2m}(2)$*, so that* $n = 2^m$ *with* $m \ge 2$ *and* $q = p \equiv \epsilon \pmod 4$*.*

(I)

$$c = \begin{cases} 2 & \text{if } n = 4 \text{ and } p \equiv \epsilon 5 \pmod 8 \\ (p-\epsilon, n) & \text{otherwise} \end{cases}$$

and $\pi = \pi_1(c)$.

(II)
$$H_{\overline{\Omega}} \cong \begin{cases} 2^4.A_6 & \text{if } n = 4 \text{ and } p \equiv \epsilon 5 \pmod 8 \\ 2^{2m}.Sp_{2m}(2) & \text{otherwise.} \end{cases}$$

(III) $H_{\overline{\Omega}}$ *is 2-local.*

Proof. First suppose that $n \geq 8$, so that $\overline{H}$ is perfect. Then as in Proposition 4.6.5, $H_{\overline{\Omega}} = \overline{H}$ and $c = (p-\epsilon, n)$. When $\epsilon = +$, it is clear from (4.6.3) that $\iota_{\beta_j} \in N_{A(V_j,\mathbf{F},\mathbf{f}_j)}(R_j)$, and hence $\iota = \iota_\beta \in N_A(R)$, where $\beta = \beta_1 \otimes \cdots \otimes \beta_j$. Therefore $\ddot{H}_A = \langle \ddot{\iota} \rangle$, which proves $\pi = \pi_1(c)$. Similarly, when $\epsilon = -$, we see that $\phi_{\beta_j}(\nu) \in N_{\Gamma(V_j,\mathbf{F},\mathbf{f}_j)}(R_j)$, and so $\phi = \phi_\beta(\nu) \in H_\Gamma$, where $\beta = \beta_1 \otimes \cdots \otimes \beta_j$. Thus $\ddot{H}_\Gamma = \langle \ddot{\phi} \rangle$, and once again $\pi = \pi_1(c)$.

Next assume that $n = 4$. Here $\overline{H} \cong 2^4.Sp_4(2)$, which has a perfect subgroup K of index 2. Thus $2^4.A_6 \cong K \trianglelefteq H_{\overline{\Omega}}$. First suppose that $p \equiv \epsilon 5 \pmod 8$. Then 256 does not divide $|\overline{\Omega}|$, and hence $H_{\overline{\Omega}} = K$. Therefore $|\ddot{H}| = 2$, while $|\ddot{\Delta}| = 4$. Thus $c = 2$ by Lemma 4.0.3.ii,ix, and arguing as before we find that $\pi = \pi_1(2)$. Now take the case where $p \equiv \epsilon \pmod 8$, and let $\lambda \in \mathbf{F}^*$ satisfy $\lambda^4 = -1$. Take $g_1 \in I(V_1, \mathbf{F}, \mathbf{f}_1)$ satisfying $(g_1)_{\beta_1} = \mathrm{diag}_{\beta_1}(1, i)$. Then $x_1^{g_1} = x_1$ and $y_1^{g_1} = x_1 y_1$. Therefore $g = g_1 \otimes \lambda \in N_\Omega(R)$, and since g centralizes $1 \otimes x_2$ and $1 \otimes y_2$, we know that g induces a transvection on the 4-dimensional space $(\overline{R}, \mathbf{F}_2)$. Now the image of $\overline{H}$ in $GL(\overline{R}, \mathbf{F}_2) \cong GL_4(2)$ is $Sp_4(2) \cong S_6$, and transvections do not lie in the derived group $Sp_4(2)' \cong A_6$ (this can be verified using the isomorphism explicitly described in proof of Proposition 2.9.1.xvi). Hence $\overline{g} \in H_{\overline{\Omega}} \backslash K$, which proves $H_{\overline{\Omega}} = \overline{H}$. Thus as in the previous paragraph we deduce that $c = 4$ and that $\pi = \pi_1(4)$. ■

The preceding Proposition does not address case **L** with $n = 2$ and H of type $2_-^{1+2}.O_2^-(2)$. The structure and conjugacy of these subgroups is given in the next result. Its proof is most readily obtained by exploiting the isomorphism $SL_2(q) \cong Sp_2(q)$, and so Proposition 4.6.7 is actually subsumed by Proposition 4.6.9, below.

Proposition 4.6.7. *Assume that case* **L** *holds with* $n = 2$, *so that* H *is of type* $2_-^{1+2}.O_2^-(2)$ *with* $q = p \geq 3$.

(I) *We have*
$$c = \begin{cases} 1 & \text{if } p \equiv \pm 3 \pmod 8 \\ 2 & \text{if } p \equiv \pm 1 \pmod 8 \end{cases}$$

and $\pi = \pi_1(2)$ *if* $c = 2$.

(II) $H_{\overline{\Omega}} \cong A_4.c$.

(III) $H_{\overline{\Omega}}$ *is 2-local.*

We now turn to cases **S** and **O**, and we begin by describing the faithful irreducible representations of $2_\pm^{1+2m}$. Throughout this discussion, $\mathbf{F} = \mathbf{F}_p$ with p odd. Let j run through the symbols $+, 1, 2, \ldots, m-1$ and let $(V_j, \mathbf{F}, \mathbf{f}_j)$ be a 2-dimensional orthogonal geometry such that $D(\mathbf{f}_j)$ is a square. Let β_j be an orthonormal basis of V_j and let x_j, y_j satisfy

$$(x_j)_{\beta_j} = \begin{pmatrix} 0 & 1 \\ -1 & 0 \end{pmatrix} \quad (y_j)_{\beta_j} = \begin{pmatrix} 1 & 0 \\ 0 & -1 \end{pmatrix} \tag{4.6.4}$$

so that $R_j := \langle x_j, y_j \rangle \cong D_8$ and $R_j \leq I(V_j, \mathbf{F}, \mathbf{f}_j)$. Furthermore, let $(V_-, \mathbf{F}, \mathbf{f}_-)$ be a 2-dimensional symplectic geometry and let β_- be a symplectic basis. Choose $a, b \in \mathbf{F}$ such that $a^2 + b^2 = -1$ and let x_-, y_- satisfy

$$(x_-)_{\beta_-} = \begin{pmatrix} 0 & 1 \\ -1 & 0 \end{pmatrix} \quad (y_-)_{\beta_-} = \begin{pmatrix} a & b \\ b & -a \end{pmatrix}. \tag{4.6.5}$$

Thus $R_- = \langle x_-, y_- \rangle \cong Q_8$ and $R_- \leq I(V_-, \mathbf{F}, \mathbf{f}_-)$. Thus for $\epsilon = \pm$ we obtain, using (4.4.2), the embedding

$$\begin{aligned} 2_\epsilon^{1+2m} &\cong R_1 \circ \cdots \circ R_{m-1} \circ R_\epsilon \\ &\leq I(V_1 \otimes \cdots \otimes V_{m-1} \otimes V_\epsilon, \mathbf{F}, \mathbf{f}_1 \otimes \cdots \otimes \mathbf{f}_{m-1} \otimes \mathbf{f}_\epsilon) \\ &\approx \begin{cases} O_{2^m}^+(p) & \text{if } \epsilon = + \text{ and } m \geq 2 \\ Sp_{2^m}(p) & \text{if } \epsilon = -. \end{cases} \end{aligned}$$

It is convenient to define $g_\epsilon \in \Delta(V_\epsilon, \mathbf{F}, \mathbf{f}_\epsilon)$ by

$$(g_\epsilon)_{\beta_\epsilon} = \begin{pmatrix} 1 & 1 \\ -1 & 1 \end{pmatrix}.$$

It is easy to check that

$$\begin{aligned} x_\epsilon^{g_\epsilon} &= x_\epsilon \\ y_\epsilon^{g_\epsilon} &= y_\epsilon x_\epsilon \\ \tau_{(V_\epsilon, \mathbf{f}_\epsilon)}(g_\epsilon) &= 2. \end{aligned} \tag{4.6.6}$$

Proposition 4.6.8. *Assume that case* $\mathbf{O}^+$ *holds, so that* H *is of type* $2_+^{1+2m}.O_{2m}^+(2)$, *with* $n = 2^m$, $m \geq 3$ *and* $q = p \geq 3$.

(I)

$$c = \begin{cases} 4 & \text{if } p \equiv \pm 3 \pmod 8 \\ 8 & \text{if } p \equiv \pm 1 \pmod 8). \end{cases}$$

If $c = 4$ *then* $\pi = \pi_5$ *and if* $c = 8$ *then* $\pi = \pi_6$.

(II) $H_{\overline{\Omega}} \cong 2^{2m}.\Omega_{2m}^+(2).\frac{c}{4}$.

(III) $H_{\overline{\Omega}}$ *is 2-local.*

Proof. Clearly $2^{2m}.\Omega_{2m}^+(2) \cong O^2(\overline{H}) \leq H_{\overline{\Omega}}$, and so we must determine whether $H_{\overline{\Omega}}$ is isomorphic to $2^{2m}.\Omega_{2m}^+(2)$ or $2^{2m}.O_{2m}^+(2)$. Regard $\overline{R}$ as a $2m$-dimensional vector space over $\mathbf{F}_2$, and observe that $\overline{R}$ admits a non-degenerate quadratic form P such that $P(\overline{x}) = 0$ or 1 according as $|x| = 2$ or 4 (see [Su₂, p.97]). Notice that if $x, y \in R$ then $\mathbf{f}_P(\overline{x}, \overline{y}) = 0$ if and only if x and y commute. According to the discussion before this Proposition, we may write $R = R_1 \otimes \cdots \otimes R_{m-1} \otimes R_+$ and we set $x = 1 \otimes \cdots \otimes 1 \otimes x_+ \in R$, where $x_+ \in R_+ < I(V_+, \mathbf{F}, \mathbf{f}_+)$. Since $|x| = 4$, we know that $\overline{x}$ is non-singular in $(\overline{R}, P)$, and thus $H_{\overline{\Omega}} \cong 2^{2m}.O_{2m}^+(2)$ if and only if the image of $H_{\overline{\Omega}}$ in $I(\overline{R}, \mathbf{F}_2, P)$ contains the reflection $r_{\overline{x}}$. It follows from (4.6.6) that $g = 1 \otimes \cdots \otimes 1 \otimes g_+ \in \Delta$ and $\tau(g) = 2$, and furthermore g induces the reflection $r_{\overline{x}} \in I(\overline{R}, \mathbf{F}_2, P)$. Therefore $\overline{H} = \langle O^2(\overline{H}), \overline{g} \rangle$, and hence

$$\ddot{H} = \langle \ddot{g} \rangle.$$

If $p \equiv \pm 3 \pmod 8$, then $\tau(g) = 2$ is a non-square, and hence $\ddot{g} \in \ddot{\Delta} \backslash \ddot{I}$. Thus $H_{\overline{\Omega}} = O^2(\overline{H})$, proving (II). Moreover $\overline{g}^2$ lies in the perfect group $O^2(\overline{H})$, and hence $|\ddot{g}| = 2$. Therefore, using the notation of Proposition 2.7.3, $\ddot{g} = \ddot{\delta}$ or $\ddot{r}_\square \ddot{r}_\boxtimes \ddot{\delta}$. These two elements are conjugate in $\ddot{\Delta}$, and so $\pi = \pi_5$, as claimed. Suppose on the other hand that $p \equiv \pm 1 \pmod 8$, so that $\sqrt{2} \in \mathbf{F} = \mathbf{F}_p$. Evidently $\frac{1}{\sqrt{2}} g_+ \in S(V_+, \mathbf{F}, \mathbf{f}_+)$, and so by Lemma 4.4.13.iii(a), $\frac{1}{\sqrt{2}} g \in \Omega$, which shows $H_{\overline{\Omega}} = \overline{H}$. Therefore $c = |\ddot{\Delta}| = 8$ and $\ddot{H}_\Gamma = 1$, which means $\pi = \pi_6$. ■

Proposition 4.6.9. *Assume that case* **S** *holds, so that H is of type $2_-^{1+2m}.O_{2m}^-(2)$, with $n = 2^m$ and $q = p \geq 3$. Include the case $n = 2$.*

(I)

$$c = \begin{cases} 1 & \text{if } p \equiv \pm 3 \pmod 8 \\ 2 & \text{if } p \equiv \pm 1 \pmod 8 \end{cases}$$

and $\pi = \pi_1$ if $c = 2$.

(II) $H_{\overline{\Omega}} \cong 2^{2m}.\Omega_{2m}^-(2).c$.

(III) *$H_{\overline{\Omega}}$ is 2-local.*

Proof. Argue as in Proposition 4.6.8, using x_-, y_-, R_-, $(V_-, \mathbf{F}, \mathbf{f}_-)$ etc. instead of x_+, y_+, R_+, $(V_+, \mathbf{F}, \mathbf{f}_+)$. ■

§4.7 The tensor product subgroups $\mathcal{C}_7$

Let V_1 be an m-dimensional vector space over $\mathbf{F}$, and assume that V_1 comes equipped with a form $\mathbf{f}_1$, which is either 0, a non-degenerate bilinear form, or a non-degenerate unitary form. For $i = 1, \ldots, t$, let $(V_i, \mathbf{f}_i)$ be a classical geometry which is similar (this term is defined in §2.1) to $(V_1, \mathbf{f}_1)$. Thus for each i there exists a similarity $\eta_i : (V_1, \mathbf{f}_1) \to (V_i, \mathbf{f}_i)$ which satisfies $\mathbf{f}_i(v\eta_i, w\eta_i) = \lambda_i \mathbf{f}_1(v, w)$ for all $v, w \in V_1$, where $\lambda_i \in \mathbf{F}^*$ is independent of v and w. We thus obtain a tensor decomposition $\mathcal{D}$ given by

$$(V, \kappa) = (V_1, \mathbf{f}_1) \otimes \cdots \otimes (V_t, \mathbf{f}_t)$$

as described in §4.4 (see (4.4.5)). Here $V = V_1 \otimes \cdots \otimes V_t$ and

$$\kappa = \begin{cases} Q(\mathbf{f}_1, \ldots, \mathbf{f}_t) & \text{if } q \text{ is even and each } \mathbf{f}_i \text{ is symplectic} \\ \mathbf{f}_1 \otimes \cdots \otimes \mathbf{f}_t & \text{otherwise.} \end{cases}$$

As in §4.4, let $X_i = X(V_i, \mathbf{f}_i)$ and $x_i = x_{(V_i, \mathbf{f}_i)}$, where X ranges over the symbols in (2.1.15) and x is σ or τ. Since the spaces V_i all have the same dimension, we may write

$$\dim(V_i) = m \quad \text{and} \quad n = \dim(V) = m^t.$$

Clearly V is spanned by the vectors $v_1\eta_1 \otimes \cdots \otimes v_t\eta_t$ with $v_i \in V_1$. Now define $\alpha_i : \Gamma_1 \to \Gamma L(V_i)$ by

$$(v\eta_i)(g\alpha_i) = (vg)\eta_i \qquad (g \in \Gamma_1,\ v \in V_1).$$

Then for $v, w \in V_1$, we have

$$\begin{aligned} \mathbf{f}_i\big((v\eta_i)(g\alpha_i),(w\eta_i)(g\alpha_i)\big) &= \mathbf{f}_i\big((vg)\eta_i,(wg)\eta_i\big) \\ &= \lambda_i \mathbf{f}_1(vg,wg) = \lambda_i\tau_1(g)\mathbf{f}_1(v,w)^{\sigma_1(g)} \\ &= \tau_1(g)\lambda_i^{1-\sigma_1(g)}\big(\lambda_i\mathbf{f}_1(v,w)\big)^{\sigma_1(g)}. \end{aligned}$$

Therefore $g\alpha_i \in \Gamma_i$ with $\tau_i(g\alpha_i) = \tau_1(g)\lambda_i^{1-\sigma_1(g)}$ and $\sigma_i(g\alpha_i) = \sigma_1(g)$. Hence α_i is an isomorphism from Γ_1 to Γ_i, and α_i takes Δ_1 to Δ_i. Pick $\phi_1 \in \Gamma_1$ such that $\sigma_1(\phi_1) = \nu$. Then set $\phi_{\mathcal{D}} = \phi_1\alpha_1 \otimes \cdots \otimes \phi_1\alpha_t$, so that $\Gamma_{(\mathcal{D})} = \Delta_{(\mathcal{D})}\langle\phi_{\mathcal{D}}\rangle$ as described in (4.4.7).

Consider now a permutation $\rho \in S_t$, and define g_ρ by

$$(v_1\eta_1 \otimes \cdots \otimes v_t\eta_t)g_\rho = v_{1\rho^{-1}}\eta_1 \otimes \cdots \otimes v_{t\rho^{-1}}\eta_t.$$

Evidently $g_\rho \in I$, and so

$$S_t \cong J := \{g_\rho \mid \rho \in S_t\} \leq I. \tag{4.7.1}$$

Further calculations show that for all $g \in \Delta_1$,

$$(g\alpha_i)^{g_\rho} = g\alpha_{i\rho}.$$

Here, as in §4.4, we identify Δ_i with $1 \otimes \cdots \otimes \Delta_i \otimes \cdots \otimes 1$. Hence J permutes naturally the subgroups Δ_i. Also $[J, \phi_{\mathcal{D}}] = 1$, and hence J normalizes $\Gamma_{(\mathcal{D})}$. We define

$$\Gamma_{\mathcal{D}} := \Gamma_{(\mathcal{D})}J = (\Delta_1 \otimes \cdots \otimes \Delta_t)(\langle\phi_{\mathcal{D}}\rangle \times J), \tag{4.7.2}$$

and we set $G_{\mathcal{D}} = G \cap \Gamma_{\mathcal{D}}$ for all subgroups $G \leq \Gamma$. In particular, we have

$$\begin{aligned} \Delta_{\mathcal{D}} &= \Delta_{(\mathcal{D})}J \\ I_{\mathcal{D}} &= I_{(\mathcal{D})}J \end{aligned} \tag{4.7.3}$$

For the remainder of this section, we shall abuse notation, writing $v_1 \otimes \cdots \otimes v_t$ instead of $v_1\eta_1 \otimes \cdots \otimes v_t\eta_t$, and $g_1 \otimes \cdots \otimes g_t$ instead of $g_1\alpha_1 \otimes \cdots \otimes g_t\alpha_t$ (here $v_i \in V_1$ and $g_i \in \Delta_1$).

Definition. The members of $\mathcal{C}_7(\Gamma)$ are the groups $\Gamma_{\mathcal{D}}$ given in (4.7.2) (with $t \geq 2$) such that

(a)

$$\begin{cases} (V,Q) \cong (V_1 \otimes \cdots \otimes V_t, Q(\mathbf{f}_1 \otimes \cdots \otimes \mathbf{f}_t)) & \text{in case } \mathbf{O}^+ \text{ with } q \text{ even} \\ (V,\mathbf{f}) \cong (V_1 \otimes \cdots \otimes V_k, \mathbf{f}_1 \otimes \cdots \otimes \mathbf{f}_t) & \text{otherwise,} \end{cases}$$

(b) $\Omega(V_i, \mathbf{f}_i)'$ is quasisimple, and

(c) the forms $\mathbf{f}_i$ are as described in the Table 4.7.A.

Table 4.7.A

case	type	description of $\mathcal{D}$	conditions
L	$GL_m(q) \wr S_t$	$\mathrm{f}_i = 0$	$m \geq 3$
U	$GU_m(q) \wr S_t$	f_i unitary	$m \geq 3$
S	$Sp_m(q) \wr S_t$	f_i symplectic	qt odd
$\mathbf{O}^+$	$O_m^{\pm}(q) \wr S_t$	f_i symmetric	q odd
$\mathbf{O}^+$	$Sp_m(q) \wr S_t$	f_i symplectic	qt even
$\mathbf{O}^\circ$	$O_m(q) \wr S_t$	f_i symmetric	qm odd

Remarks on the conditions. Our definition here differs from that given in [As$_1$], in that we impose the condition $m \geq 3$ in case $\mathbf{L}^\pm$. This is justified in the discussion before Proposition 4.7.3. Observe that case $\mathbf{O}^-$ does not arise here, in view of Lemma 4.4.2. We also comment on condition (b) in the Definition (which *is* imposed in [As$_1$]). First note that if I_1 is soluble, then $\overline{\Gamma}_{\mathcal{D}}$ contains a non-trival abelian normal subgroup, which forces $\Gamma_{\mathcal{D}}$ to be contained in a member of $\mathcal{C}_1 \cup \mathcal{C}_2 \cup \mathcal{C}_3 \cup \mathcal{C}_6$. For example, if I_1 is $Sp_2(3)$ or $SU_3(2)$, then $\Gamma_{\mathcal{D}}$ is contained in a member of $\mathcal{C}_6(\Gamma)$ in view of the isomorphisms $Sp_2(3) \cong 2_-^{1+2}.O_2^-(2)$ and $GU_3(2) \cong 3_+^{1+2}.Sp_2(3)$. And if I_1 is $Sp_2(2)$, $O_2^{\pm}(q)$ or $O_3(3)$, then $\Gamma_{\mathcal{D}}$ lies in a member of $\mathcal{C}_2 \cup \mathcal{C}_3$. Second, condition (b) also excludes the case in which Ω_1' is insoluble but not quasisimple — namely, $\Omega_1 \approx \Omega_4^+(q)$. Here, we have $\Omega_1 \cong Sp_2(q) \otimes Sp_2(q)$, and hence $\Gamma_{\mathcal{D}}$ is contained in a member of $\mathcal{C}_7(\Gamma)$ of type $Sp_2(q) \wr S_{2t}$.

For the remainder of this section we assume that $H_\Gamma \in \mathcal{C}_7(\Gamma)$, so that $H_\Gamma = \Gamma_{\mathcal{D}}$, where $\mathcal{D}$ is a tensor decomposition as given Table 4.7.A. Thus $H = \Delta_{\mathcal{D}}$ is as in (4.7.3). As a matter of convenience we set

$$X_\sharp = X_1 \otimes \cdots \otimes X_t$$

as X ranges over the symbols Ω, S, I and Δ. Reasoning as in the proof of (4.4.14), we see that

$$I_{(\mathcal{D})} = I_\sharp \text{ when } H \text{ is of type} \begin{cases} GL_m^{\pm}(q) \wr S_t \\ Sp_m(q) \wr S_t \text{ in } O_{m^t}^+(q) \text{ with } q \text{ even} \\ O_m(q) \wr S_t \text{ with } qm \text{ odd.} \end{cases} \tag{4.7.4}$$

When none of the embeddings in (4.7.4) arises, there exists $\delta_i \in \Delta_i$ such that $\tau_i(\delta_i) = \mu$. We define $z_{i,j} = \delta_i \otimes \delta_j^{-1}$, so that analogously to (4.4.15), we have

$$I_{(\mathcal{D})} = I_\sharp \langle z_{i,i+1} \mid i < t \rangle \text{ if } H \text{ is of type} \begin{cases} Sp_m(q) \wr S_t \ (q \text{ odd}) \\ O_m^{\pm}(q) \wr S_t \ (q \text{ odd, } m \text{ even}). \end{cases} \tag{4.7.5}$$

Furthermore $z_{i,i+1}^2 \in I_i \otimes I_{i+1}$ (see Lemma 4.4.5.iii). Therefore $I_{(\mathcal{D})}/I_\sharp \cong 2^{t-1}$.

The discussion in the previous two paragraphs, together with (4.7.3), yields

$$\overline{H}_I \cong \begin{cases} \overline{I}_1^t.S_t & \text{in situation (4.7.4)} \\ \overline{I}_1^t.2^{t-1}.S_t & \text{in situation (4.7.5).} \end{cases} \tag{4.7.6}$$

In situation (4.7.5), note that $z_{i,i+1} \in S(V_i \otimes V_{i+1}, \mathbf{f}_i \otimes \mathbf{f}_{i+1}) \leq S$. Thus $\ddot{z}_{i,i+1} = 1$ when H is of type $Sp_m(q) \wr S_t$ in $Sp_{m^t}(q)$ with qt odd. Also, when $t \geq 3$, Lemma 4.4.13.iii(a) implies that $S(V_i \otimes V_{i+1}, \mathbf{f}_i \otimes \mathbf{f}_{i+1}) \leq S(V_i \otimes V_{i+1}, \mathbf{f}_i \otimes \mathbf{f}_{i+1}) \otimes S(V_{i+2}, \mathbf{f}_{i+2}) \leq \Omega$ (subscripts modulo t), and so in this case $\ddot{z}_{i,i+1} = 1$, too. Therefore, $\ddot{z}_{i,i+1} \neq 1$ only if H is of type $Sp_m(q) \wr S_2$ or $O_m^{\pm}(q) \wr S_2$ in $O_{m^2}^{+}(q)$ with q odd and $t = 2$. Consequently, putting $z = z_{1,2}$, we have (in situation (4.7.5))

$$\ddot{I}_{(\mathcal{D})} = \langle \ddot{I}_\sharp, \ddot{z} \rangle. \tag{4.7.7}$$

Since $J \leq I$, we have $J' \leq \Omega$. Now put $g = g_{(12)}$, so that $(v_1 \otimes v_2 \otimes v_3 \otimes \cdots)g = v_2 \otimes v_1 \otimes v_3 \otimes \cdots$. Thus $\ddot{J} = \langle \ddot{g} \rangle$ and it is clear that

$$\det(g) = (-1)^{\frac{1}{2}m(m-1)m^{t-2}} = \begin{cases} -1 & \text{if } t = 2 \text{ and } m \equiv 2 \pmod 4 \\ -1 & \text{if } m \equiv 3 \pmod 4 \\ 1 & \text{otherwise.} \end{cases} \tag{4.7.8}$$

Therefore, based on our discussion so far, we have established

$$\ddot{H}_I = \begin{cases} \langle \ddot{I}_\sharp, \ddot{z}, \ddot{g} \rangle & \text{if } q \text{ is odd and } H \text{ is of type} \\ & Sp_m(q) \wr S_2 \text{ or } O_m^{\pm}(q) \wr S_2 \text{ in } O_{m^2}^{+}(q) \\ \langle \ddot{I}_\sharp, \ddot{g} \rangle & \text{otherwise.} \end{cases} \tag{4.7.9}$$

An application of Lemma 4.0.5.iv yields

Lemma 4.7.1. *The group $H_{\overline{\Omega}}$ is non-local and* $\mathrm{soc}(H_{\overline{\Omega}}) = \overline{\Omega}'_\sharp \cong (P\Omega'_1)^t$.

Thus we will make no mention of (III) in the proofs below. Here we establish Theorem 3.1.2 for the members of $\mathcal{C}_7$.

Proposition 4.7.2. *We have $H_\Gamma = N_\Gamma(H_\Omega)$, and hence Theorem 3.1.2 holds.*

Proof. The second statement follows from the first along with Proposition 4.0.1. Thus we need only prove the first. Since $\phi_{\mathcal{D}} \in H_\Gamma$ and $\sigma(\phi_{\mathcal{D}}) = \nu$, we have $H_\Gamma \Delta = \Gamma$, so it suffices to prove that $H_\Delta = N_\Delta(H_\Omega)$. So take $g \in N_\Delta(H_\Omega)$. It follows from Lemma 4.7.1 that $\Omega'_\sharp$ is characteristic in H_Ω, and so g permutes the components Ω'_i. Multiplying g by some element in J if necessary, we can assume that $g \in N_{GL(V)}(\Omega'_i)$ for all i. By Lemma 4.4.3.iii and Corollaries 2.10.4.ii and 2.10.7, $g \in \bigcap_{i=1}^{t} N_{GL(V)}(\Omega_i) = \Delta_1 \otimes \cdots \otimes \Delta_t \leq H$. The result now follows. ■

As in the proof of Lemma 4.4.8, we deduce from Proposition 4.7.2 and Lemma 4.0.3.iii,ix that

$$c = |\ddot{I} : \ddot{H}_I| = |\bar{I}|/|H_{\bar{I}} : H_{\overline{\Omega}}|. \tag{4.7.10}$$

We begin the analysis of the groups H by considering case $\mathbf{L}^\epsilon$, so that H is of type $GL_m^\epsilon(q) \wr S_t$. To justify the extra condition $m \geq 3$ in Table 4.7.A, let us suppose for the moment that $m = 2$. First take case $\mathbf{L}^+$, so that $\Omega_i = SL(V_i)$. In view of the isomorphism $SL_2(q) \cong Sp_2(q)$, we know that $\Omega_i = \Omega(V_i, \mathbf{g}_i)$ for some non-degenerate symplectic from $\mathbf{g}_i$. Consequently $\Omega_\sharp \leq \Omega(V, \mathbf{g})$, where $\mathbf{g} = \mathbf{g}_1 \otimes \cdots \otimes \mathbf{g}_t$, a non-degenerate bilinear form. Arguing along the lines of the proof of Corollary 2.10.4, one checks that $N_\Gamma(\Omega_\sharp) \leq \Gamma(V, \mathbf{g})$, whence $H_\Gamma \leq \Gamma(V, \mathbf{g}) \in \mathcal{C}_8(\Gamma)$ (see §4.8, below). Next consider case $\mathbf{L}^-$, so that $\Omega_i = \Omega(V_i, \mathbf{f}_i)$ with $\mathbf{f}_i$ unitary. Let $\mathbf{F}_\sharp$ be the subfield $\mathbf{F}_q$ of $\mathbf{F}$. The embedding (4.5.7) yields a subgroup $Sp_2(q)$ of Ω_i fixing an $\mathbf{F}_\sharp$-space $V_{i,\sharp}$ in V_i. Moreover, this $Sp_2(q)$ preserves the symplectic form $\mathbf{f}_{i,\sharp} = (\zeta \mathbf{f}_i)_{V_{i,\sharp}}$ on $V_{i,\sharp}$, where $\zeta \in \mathbf{F}^*$ and $\mathrm{T}^{\mathbf{F}}_{\mathbf{F}_\sharp}(\zeta) = 0$. Since $|Sp_2(q)| = |SU_2(q)|$, we have $Sp_2(q) = \Omega_i$ here. Thus $\Omega_\sharp \leq \Omega(V_\sharp, \mathbf{F}_\sharp, \mathbf{f}_\sharp)$, where $V_\sharp = V_{1,\sharp} \otimes \cdots \otimes V_{t,\sharp}$ and $\mathbf{f}_\sharp = \mathbf{f}_{1,\sharp} \otimes \cdots \otimes \mathbf{f}_{t,\sharp}$. We deduce that $\Omega_\sharp$ lies in the subgroup $N_\Gamma(V_\sharp)\mathbf{F}^* \in \mathcal{C}_5(\Gamma)$, which is of type $O_n^{\pm}(q)$ or $Sp_n(q)$. As in the proof of Proposition 4.5.1, we see that $H_\Gamma \leq N_\Gamma(\Omega_\sharp) \leq N_\Gamma(V_\sharp)\mathbf{F}^* \in \mathcal{C}_5(\Gamma)$.

Proposition 4.7.3. *Assume that case $\mathbf{L}^\epsilon$ holds, so that H is of type $GL_m^\epsilon(q) \wr S_t$.*

(I)

$$c = \begin{cases} (q-\epsilon, m)/2 & \text{if } t = 2,\ m \equiv 2 \pmod 4 \text{ and } q \equiv -\epsilon \pmod 4 \\ (q-\epsilon, n/m) & \text{otherwise,} \end{cases}$$

and $\pi = \pi_1(c)$.

(II)

$$H_{\overline{\Omega}} \cong \begin{cases} L_m^\epsilon(q)^2.\left[\frac{(q-\epsilon,m)^3}{(q-\epsilon,n)}\right] & t = 2,\ m \equiv 2 \pmod 4,\ q \equiv -\epsilon \pmod 4 \\ L_m^\epsilon(q)^t.\left[\frac{c(q-\epsilon,m)^t}{(q-\epsilon,n)}\right].S_t & \text{otherwise.} \end{cases}$$

(III) *$H_{\overline{\Omega}}$ is non-local and $\mathrm{soc}(H_{\overline{\Omega}}) \cong L_m^\epsilon(q)^t$.*

Proof. Using (4.7.3) and (4.7.4), we have $H_I = I_{(\mathcal{D})}J$, and so $\det(H_I) = \langle \det(I_{(\mathcal{D})}), \det(J) \rangle$. And according to (4.4.9),

$$|\det(I_{(\mathcal{D})})| = |\det(I_1)^{n/m}| = \frac{q-\epsilon}{(q-\epsilon, \frac{n}{m})}. \tag{4.7.11}$$

Suppose in this paragraph that $|\det(H_I)| > |\det(I_{(\mathcal{D})})|$. Thus $\det(g) \notin \det(I_{(\mathcal{D})})$, where g is as in (4.7.8). By (4.7.8), $\det(g) = -1$, and q is odd, and $|\det(I_{(\mathcal{D})})|$ is also odd. Thus (4.7.11) ensures that n (and hence m) is even, and so by (4.7.8), $t = 2$ and $m \equiv 2 \pmod 4$. Because $|\det(I_{(\mathcal{D})})|$ is odd and $\frac{n}{m} = m \equiv 2 \pmod 4$, it follows from (4.7.11) that $q \equiv -\epsilon \pmod 4$.

Thus we have shown

$$|\det(H_I)| = \begin{cases} 2(q-\epsilon)/(q-\epsilon, m) & \text{if } t = 2,\ m \equiv 2 \pmod 4,\ q \equiv -\epsilon \pmod 4 \\ (q-\epsilon)/(q-\epsilon, \frac{n}{m}) & \text{otherwise.} \end{cases}$$

Hence the value of c given in (I) is correct in view of Lemma 4.0.3.v,ix and Proposition 4.7.2. To determine the structure of $H_{\overline{\Omega}}$, first observe that $|\ddot{I}_{(\mathcal{D})}| = |\det(I_{(\mathcal{D})})|(q-\epsilon, n)/(q-\epsilon)$, which by (4.7.11) equals $(q-\epsilon, n)/(q-\epsilon, \frac{n}{m})$, and so

$$\overline{\Omega}_{(\mathcal{D})} \cong \frac{(q-\epsilon, \frac{n}{m})}{(q-\epsilon, n)} \overline{I}_{(\mathcal{D})} \cong L_m^\epsilon(q)^t.\left[\frac{(q-\epsilon, m)^t (q-\epsilon, \frac{n}{m})}{(q-\epsilon, n)}\right].$$

Now provided $t \geq 3$ or $m \not\equiv 2 \pmod 4$ or $q \not\equiv -\epsilon \pmod 4$, the previous paragraph implies $\det(J) \leq \det(I_{(\mathcal{D})})$, and hence $H_{\overline{\Omega}} \cong \overline{\Omega}_{(\mathcal{D})}.S_t$. If however $t = 2$, $m \equiv 2 \pmod 4$ and

$q \equiv -\epsilon \pmod 4$, then $\det(J) \not\leq \det(I_{(\mathcal{D})})$, and it follows that $H_{\overline{\Omega}} = \overline{\Omega}_{(\mathcal{D})}$. This completes the proof of (II). It remains to determine π. Let β_i be a basis of V_i and assume that β_i is orthonormal if $\epsilon = -$. Then $\beta = \beta_1 \otimes \cdots \otimes \beta_t$ is a basis of V, and β is orthonormal if $\epsilon = -$. Thus in the notation of Proposition 2.2.3 and Proposition 2.3.5, we have $\langle \ddot{\phi}, \ddot{\iota} \rangle \leq \ddot{H}_A$ in case **L** and $\langle \ddot{\phi} \rangle \leq \ddot{H}_\Gamma$ in case **U**. Thus as in the proof of Proposition 4.4.10 we see that $\pi = \pi_1(c)$. ■

Proposition 4.7.4. *Assume that case* **S** *holds, so that H is of type $Sp_m(q) \wr S_t$ with qt odd and $(q,m) \neq (3,2)$.*

(I) $c = 1$.

(II) $H_{\overline{\Omega}} \cong PSp_m(q)^t.2^{t-1}.S_t$.

(III) $H_{\overline{\Omega}}$ *is non-local and* $\mathrm{soc}(H_{\overline{\Omega}}) \cong PSp_m(q)^t$.

Proposition 4.7.5. *Assume that case* $\mathbf{O}^+$ *holds and* $\mathbf{f}_i$ *is symplectic, so that H is of type $Sp_m(q) \wr S_t$, with m even, qt even, and $(q,m) \notin \{(2,2),(3,2)\}$.*

(I)

$$c = \begin{cases} 1 & \text{if } t = 2 \text{ and } m \equiv 2 \pmod 4 \\ 2(2,q-1) & \text{otherwise.} \end{cases}$$

If $c = 2$, *then* $\pi = \pi_2$. *If* $c = 4$, *then* $\pi = \pi_5$.

(II)

$$H_{\overline{\Omega}} \cong \begin{cases} PSp_m(q)^2 & \text{if } t = 2 \text{ and } m \equiv 2 \pmod 4 \\ PSp_m(q)^t.(2,q-1)^{t-1}.S_t & \text{otherwise} \end{cases}$$

(III) $H_{\overline{\Omega}}$ *is non-local and* $\mathrm{soc}(H_{\overline{\Omega}}) \cong PSp_m(q)'^{\,t}$.

(IV) $H_{\overline{\Omega}}$ *is intransitive on its components if and only if* $t = 2$ *and* $m \equiv 2 \pmod 4$.

Proof. Choose a symplectic basis $\beta_i = \{e_1, \ldots, e_{m/2}, f_1, \ldots, f_{m/2}\}$ of V_i and write $W_o = \langle e_1, \ldots, e_{m/2} \rangle$. Furthermore, set $W_i = V_1 \otimes \cdots \otimes V_{i-1} \otimes W_o \otimes V_{i+1} \otimes \cdots \otimes V_t$. Clearly

$$W_i \text{ is a maximal totally singular subspace of } V \text{ for all } i. \tag{4.7.12}$$

When q is odd, I_1 is quasisimple, and so $I_\sharp \leq \Omega$. And when q is even, $I_1 \leq N_\Omega(W_2) \leq \Omega$ according to Description 4 in §2.5. Thus for all q we have $I_\sharp \leq \Omega$, which is to say

$$\ddot{I}_\sharp = 1. \tag{4.7.13}$$

Thus in view of (4.7.9) and (4.7.10), to determine c we must find $\ddot{z}$ (when q is odd and $t = 2$) and $\ddot{g}$ (for all q, t).

Here we calculate $\ddot{z}$ when q is odd and $t = 2$ (see (4.7.9)). We may choose $\delta_i = \delta_{\mathbf{f}_i,\beta_i}(\mu)$, so that z fixes W_1. Observe $\det_{W_1}(z) = \mu^{n/2}\mu^{-n/4} = \mu^{n/4}$, which is a square if and only if $4 \mid m$. Thus using Lemma 4.1.9.iii, we have

$$\langle \ddot{z} \rangle = \begin{cases} \ddot{S} & \text{if } q \text{ is odd, } t = 2,\ m \equiv 2 \pmod 4 \\ 1 & \text{otherwise.} \end{cases} \tag{4.7.14}$$

We now determine where $\ddot{g}$ lies. Now $W_1^g = W_2$ and $\dim(W_1 \cap W_2) = \frac{n}{4}$. Thus when q is even, $g \in I \backslash \Omega$ if and only if $t = 2$ and $m \equiv 2 \pmod 4$ (see Description 4 in §2.5). For the rest of this paragraph we take q odd, so that t is even. In view of (4.7.8) we have $g \in I \backslash S$ if and only if $t = 2$ and $m \equiv 2 \pmod 4$. Thus it remains to examine the case in which either $t \geq 4$ or $m \equiv 0 \pmod 4$. If $t \geq 4$, then g fixes W_3 and $\det_{W_3}(g) = 1$, which proves $g \in \Omega$ (see Lemma 4.1.9.iii). Finally consider the case in which $t = 2$ and $m \equiv 0 \pmod 4$. Then g fixes the decomposition $V = U \perp U^\perp$, where U is the $+(m^2 - m)$-space spanned by

$$\begin{array}{ll} e_i \otimes e_j,\ f_i \otimes f_j & 1 \leq i,j \leq \frac{m}{2} \\ e_i \otimes f_j,\ f_j \otimes e_i & 1 \leq i,j \leq \frac{m}{2},\ i \neq j \end{array} \tag{4.7.15}$$

and $U^\perp$ is the $+m$-space spanned by

$$e_i \otimes f_i,\ f_i \otimes e_i \quad 1 \leq i \leq \tfrac{m}{2}. \tag{4.7.16}$$

Furthermore g fixes the totally singular $\frac{1}{2}(m^2 - m)$-space U_o in U, spanned by

$$\begin{array}{ll} e_i \otimes e_j & 1 \leq i,j \leq \frac{m}{2} \\ e_i \otimes f_j,\ f_j \otimes e_i & 1 \leq i < j \leq \frac{m}{2}, \end{array}$$

and $\det_{U_o}(g) = (-1)^{m/2(m/2-1)} = 1$. Thus according to Lemma 4.1.9.iii, the image of g in $I(U,Q)$ lies in $\Omega(U,Q)$. Moreover on $U^\perp$, the element g acts as $\prod_{i=1}^{m/2} r_i$, where r_i is the reflection in $e_i \otimes f_i - f_i \otimes e_i$. So as $m/2$ is even, the spinor norm (see Description 1 in §2.5) is a square, whence the image of g in $I(U^\perp, Q)$ lies in $\Omega(U^\perp, Q)$. It now follows that $g \in \Omega$, and we have shown

$$\ddot{g} \in \begin{cases} \ddot{I} \backslash \ker_{\ddot{I}}(\ddot{\gamma}) & \text{if } t = 2 \text{ and } m \equiv 2 \pmod 4 \\ \ddot{\Omega} = 1 & \text{otherwise.} \end{cases} \tag{4.7.17}$$

So combining (4.7.13), (4.7.14) and (4.7.17) we obtain

$$\ddot{H}_I = \begin{cases} \ddot{I} & \text{if } t = 2 \text{ and } m \equiv 2 \pmod 4 \\ 1 & \text{otherwise.} \end{cases} \tag{4.7.18}$$

Thus by (4.7.10), the value of c given in (I) is correct, since $|\ddot{I}| = 2(2, q-1)$ (note when q is odd, $D(Q)$ is a square by Lemma 2.5.10, and so $|\ddot{S}| = 2$ and $|\ddot{I}| = 4$). Now $\overline{H}_I \cong PSp_m(q)^t.(2, q-1)^{t-1}.S_t$ by (4.7.6). Thus when $t = 2$ and $m \equiv 2 \pmod 4$, it follows from (4.7.18) and (4.0.3) that $|H_{\overline{\Omega}}| = |PSp_m(q)^2.[2(2, q-1)]|/|\ddot{I}|$; and since $|\ddot{I}| = 2(2, q-1)$ and $PSp_m(q)^2 \leq H_{\overline{\Omega}}$ by (4.7.13), we deduce that $H_{\overline{\Omega}} \cong PSp_m(q)^2$. And in the other cases, we have $H_I \leq \Omega$, and hence $H_{\overline{\Omega}} = \overline{H}$. Therefore (II) holds.

It remains to determine π when $c = 2(2, q-1)$. Here $H_I \leq \Omega$, which means $\ddot{H}_I = 1$. Evidently $\phi_{\beta_1}(\nu) \otimes \cdots \otimes \phi_{\beta_t}(\nu) = \phi_\beta(\nu) \in H_\Gamma$, where $\beta = \beta_1 \otimes \cdots \otimes \beta_t$ is a standard basis of (V, Q). Thus in the notation of Proposition 2.7.3, $\ddot{\phi} \in \ddot{H}_\Gamma$, and hence $\ddot{H}_\Gamma = \ddot{H}\langle \ddot{\phi} \rangle$. Now when q is even, $\ddot{H} = \ddot{H}_I = 1$, whence $\ddot{H}_\Gamma = \langle \ddot{\phi} \rangle = \ker_{\ddot{\Gamma}}(\ddot{\gamma})$. Therefore $\pi = \pi_2$, as claimed. Suppose therefore that q is odd. Then $\delta_{Q,\beta}(\mu) = \delta_{\mathbf{f}_1,\beta_1}(\mu) \otimes 1 \otimes \cdots \otimes 1 \in H$, and it follows (since $\ddot{H}_I = 1$) that $\ddot{H}_\Gamma = \langle \ddot{\delta}, \ddot{\phi} \rangle$, and so $\pi = \pi_5$. The proof is now complete. ■

Proposition 4.7.6. *Assume that case* $\mathbf{O}^+$ *holds and* H *is of type* $O_m^+(q) \wr S_t$, *so that* q *is odd and* $m \geq 6$.

(I)

$$c = \begin{cases} 1 & \text{if } t = 2,\ m \equiv 2 \pmod 4 \\ 2 & \text{if } t = 3,\ m \equiv 2 \pmod 4,\ q \equiv 3 \pmod 4 \\ 4 & \text{otherwise.} \end{cases}$$

If $c = 2$, *then* $\pi = \pi_2$. *If* $c = 4$, *then* $\pi = \pi_5$.

(II)

$$H_{\overline{\Omega}} \cong \begin{cases} PSO_m^+(q)^2.[4] & \text{if } t = 2 \text{ and } m \equiv 2 \pmod 4 \\ \Omega_m^+(q)^3.[2^5].3 & \text{if } t = 3,\ m \equiv 2 \pmod 4 \text{ and } q \equiv 3 \pmod 4 \\ PSO_m^+(q)^t.[2^{2t-1}].S_t & \text{otherwise.} \end{cases}$$

(III) $H_{\overline{\Omega}}$ *is non-local and* $\mathrm{soc}(H_{\overline{\Omega}}) \cong P\Omega_m^+(q)^t$.

(IV) $H_{\overline{\Omega}}$ *is intransitive on its components if and only if* $t = 2$ *and* $m \equiv 2 \pmod 4$.

Proof. It follows from Lemma 4.4.13.iii(a) that $S_\sharp \leq \Omega$ and so

$$PSO_m^+(q)^t \cong \overline{S}_\sharp \trianglelefteq H_{\overline{\Omega}}. \tag{4.7.19}$$

Moreover, from (4.4.9) we know $I_\sharp \leq S$, and by Lemma 4.4.13.iii(b) $I_1 \leq \Omega$ if and only if $D(\mathbf{f}_2 \otimes \cdots \otimes \mathbf{f}_t)$ is a square. Thus we obtain from Lemma 4.4.1,

$$\ddot{I}_\sharp = \begin{cases} \ddot{S} & \text{if } t = 2 \text{ and } D(\mathbf{f}_2) = \boxtimes \\ 1 & \text{otherwise.} \end{cases} \tag{4.7.20}$$

And reasoning as in the proof of Proposition 4.7.5 (see the argument just before (4.7.14)), we deduce

$$\langle \ddot{z} \rangle = \begin{cases} \ddot{S} & \text{if } t = 2 \text{ and } m \equiv 2 \pmod 4 \\ 1 & \text{otherwise.} \end{cases} \tag{4.7.21}$$

Next we determine where g lies. If $t = 2$ and $m \equiv 2 \pmod 4$ then by (4.7.8) we have $g \in I \backslash S$. Thus we assume that either $t \geq 3$ or $4 \mid m$. Suppose first that $4 \mid m$. Then the argument used in (4.7.15)-(4.7.17) shows that $g \in \Omega$, although here we must assume that $\{e_1, \ldots, f_{m/2}\}$ is a standard basis of V_i, rather than a symplectic basis. So we take $m \equiv 2 \pmod 4$ and $t \geq 3$. If $t \geq 4$, then as in the proof of Proposition 4.7.5 we find g fixes W_3 and $\det_{W_3}(g) = 1$, proving $g \in \Omega$. So it remains to consider the case $t = 3$ and $m \equiv 2 \pmod 4$. Here $g \in I(V_1 \otimes V_2, \mathbf{f}_1 \otimes \mathbf{f}_2) \backslash S(V_1 \otimes V_2, \mathbf{f}_1 \otimes \mathbf{f}_2)$. Thus by Lemma 4.4.13.iii, if $D(\mathbf{f}_3)$ is a non-square then $g \in S \backslash \Omega$, and if $D(\mathbf{f}_3)$ is a square then $g \in \Omega$. Hence we have shown

$$\ddot{g} \in \begin{cases} \ddot{I} \backslash \ddot{S} & \text{if } t = 2 \text{ and } m \equiv 2 \pmod 4 \\ \ddot{S} \backslash \ddot{\Omega} & \text{if } t = 3,\ m \equiv 2 \pmod 4,\ q \equiv 3 \pmod 4 \\ \ddot{\Omega} = 1 & \text{otherwise.} \end{cases} \tag{4.7.22}$$

So combining (4.7.9), (4.7.20), (4.7.21) and (4.7.22) we deduce

$$\ddot{H}_I = \begin{cases} \ddot{I} & \text{if } t = 2 \text{ and } m \equiv 2 \pmod 4 \\ \ddot{S} & \text{if } t = 3,\ m \equiv 2 \pmod 4 \text{ and } q \equiv 3 \pmod 4 \\ 1 & \text{otherwise.} \end{cases} \tag{4.7.23}$$

Thus the value of c given in (I) is correct by (4.7.10). Moreover by (4.7.6), $\overline{H}_I \cong PO_m^+(q)^t.2^{t-1}.S_t \cong PSO_m^+(q)^t.2^{2t-1}.S_t$. Since $|\ddot{I}| = 4$ and $|\ddot{S}| = 2$, we see that (II) follows from (4.0.3), (4.7.19) and (4.7.23). Note that when $t = 3$, $m \equiv 2 \pmod 4$ and $q \equiv 3 \pmod 4$, we have $g \in S\backslash\Omega$ by (4.7.22) and $I_{(D)} \leq \Omega$ by (4.7.20) and (4.7.21); therefore $H_{\overline{\Omega}}$ does not induce an odd permutation on its three components.

To determine π, we observe that just as in the proof of Proposition 4.7.5, $\phi_\beta(\nu), \delta_\beta(\mu) \in H_\Gamma$ for some standard basis β of V. Hence $\pi = \pi_5$ or π_2 according as $c = 4$ or $c = 2$. ∎

Proposition 4.7.7. *Assume that case* $\mathbf{O}^+$ *holds and that H is of type $O_m^-(q) \wr S_t$, so that q is odd and $m \geq 4$.*

(I)

$$c = \begin{cases} 1 & \text{if } t = 2 \text{ and } m \equiv 2 \pmod 4 \\ 2 & \text{if } t = 2 \text{ and } m \equiv 0 \pmod 4 \\ 2 & \text{if } t = 3,\ m \equiv 2 \pmod 4 \text{ and } q \equiv 1 \pmod 4 \\ 4 & \text{otherwise} \end{cases}$$

If $c = 2$, then $\pi = \pi_2$. If $c = 4$, then $\pi = \pi_5$.

(II)

$$H_{\overline{\Omega}} \cong \begin{cases} PSO_m^-(q)^2.[4] & \text{if } t = 2 \text{ and } m \equiv 2 \pmod 4 \\ \Omega_m^-(q)^2.[8] & \text{if } t = 2 \text{ and } m \equiv 0 \pmod 4 \\ \Omega_m^-(q)^3.[2^5].3 & \text{if } t = 3,\ m \equiv 2 \pmod 4 \text{ and } q \equiv 1 \pmod 4 \\ PSO_m^-(q)^t.[2^{2t-1}].S_t & \text{otherwise.} \end{cases}$$

(III) *$H_{\overline{\Omega}}$ is non-local and* $\mathrm{soc}(H_{\overline{\Omega}}) \cong P\Omega_m^-(q)^t$.

(IV) *$H_{\overline{\Omega}}$ is intransitive on its components if and only if $t = 2$ and $m \equiv 2 \pmod 4$.*

Proof. Reasoning as at the beginning of the proof of Proposition 4.7.6 we see that (4.7.19) (with $+$ replaced by $-$) and (4.7.20) hold. Thus

$$\ddot{I}_\sharp = \begin{cases} \ddot{S} & \text{if } t = 2 \text{ and } m = 0 \pmod 4 \\ \ddot{S} & \text{if } t = 2,\ m \equiv 2 \pmod 4 \text{ and } q \equiv 1 \pmod 4 \\ 1 & \text{otherwise.} \end{cases} \tag{4.7.24}$$

We now want to determine $\langle \ddot{I}_\sharp, \ddot{z} \rangle$ when $t = 2$. We know already that $\ddot{z} \in \ddot{S}$, and so in view of (4.7.20), we may assume that that $D(\mathbf{f}_2)$ is a square, so that $m \equiv 2 \pmod 4$ and $q \equiv 3 \pmod 4$. But now the argument in the proof of Proposition 4.4.16 shows that $z \in S\backslash\Omega$. It now follows that

$$\langle \ddot{I}_\sharp, \ddot{z} \rangle = \begin{cases} 1 & \text{if } t \geq 3 \\ \ddot{S} & \text{if } t = 2 \end{cases} \tag{4.7.25}$$

As in the proof of Proposition 4.7.6, $g \in I\backslash S$ if and only if $t = 2$ and $m \equiv 2 \pmod 4$. Thus to find where g lies, it remains to consider the case $t \geq 3$ or $m \equiv 0 \pmod 4$. Our goal is to determine $\ddot{H}_I$, and in view of (4.7.25) we see that $\ddot{H}_I = \ddot{S}$ if $t = 2$ and $m \equiv 0 \pmod 4$. Thus we can assume that $t \geq 3$. Now if $t \geq 4$, then $g \in I(V_1 \otimes V_2, \mathbf{f}_1 \otimes \mathbf{f}_2) \otimes 1 \otimes 1 \otimes \cdots \leq S(V_1 \otimes V_2 \otimes V_3, \mathbf{f}_1 \otimes \mathbf{f}_2 \otimes \mathbf{f}_3) \otimes 1 \otimes \cdots \leq \Omega$, by Lemma 4.4.13. Thus we can take $t = 3$.

If $m \equiv 0 \pmod 4$, then $g \in S(V_1 \otimes V_2, \mathbf{f}_1 \otimes \mathbf{f}_2) \otimes 1 \leq \Omega$ by Lemma 4.4.13.iii(a). And if $D(\mathbf{f}_3) = \square$ then $g \in I(V_1 \otimes V_2, \mathbf{f}_1 \otimes \mathbf{f}_2) \otimes 1 \leq \Omega$ by Lemma 4.4.13.iii(b). Thus it remains to consider the case where $t = 3$, $m \equiv 2 \pmod 4$ and $D(\mathbf{f}_3)$ is a non-square. Here $g \in I(V_1 \otimes V_2, \mathbf{f}_1 \otimes \mathbf{f}_2) \otimes 1 \backslash S(V_1 \otimes V_2, \mathbf{f}_1 \otimes \mathbf{f}_2) \otimes 1 \subseteq S \backslash \Omega$, according to Lemma 4.4.13.iii(b).

Combining the results in the previous paragraph with (4.7.9) and (4.7.25) gives

$$\ddot{H}_I = \begin{cases} \ddot{I} & \text{if } t = 2 \text{ and } m \equiv 2 \pmod 4 \\ \ddot{S} & \text{if } t = 2 \text{ and } m \equiv 0 \pmod 4 \\ \ddot{S} & \text{if } t = 3,\ m \equiv 2 \pmod 4 \text{ and } q \equiv 1 \pmod 4 \\ 1 & \text{otherwise.} \end{cases} \tag{4.7.26}$$

Thus the value of c given in (I) is correct. And one may now verify (II) using (4.0.3), (4.7.19) (with + replaced by −) and (4.7.26).

We now determine π. Let β_i be a basis for (V_i, Q_i) as described in Proposition 2.5.12, and let $\delta_i = \delta_{Q_i,\beta_i}$, $\phi_i = \phi_{Q_i,\beta_i}$ be as described in §2.8. Also put $\phi_\mathcal{D} = \phi_1 \otimes \cdots \otimes \phi_t$, so that $\ddot{H}_\Gamma = \langle \ddot{H}, \ddot{\phi}_\mathcal{D} \rangle$. Now $\ddot{H} = \langle \ddot{H}_I, \ddot{\delta}_1 \rangle$, and the argument appearing in the last paragraph of the proof of Proposition 4.4.16 shows

$$\ddot{H} = \begin{cases} \ker_{\ddot{\Delta}}(\ddot{\gamma}) & \text{if } c = 2 \\ \langle \ddot{\delta} \rangle \text{ or } \langle \ddot{r}_\square \ddot{r}_\boxtimes \ddot{\delta} \rangle & \text{if } c = 4. \end{cases} \tag{4.7.27}$$

Since $\ddot{\delta}_i$ and $\ddot{\phi}_i$ commute in the (V_i, Q_i)-geometry (see Proposition 2.8.2.i), it follows that $\ddot{\phi}_\mathcal{D} \leq C_{\ddot{\Gamma}}(\ddot{H})$. Thus when $c = 2$, we see that $\ddot{H}_\Gamma = \ker_{\ddot{\Gamma}}(\ddot{\gamma})$, as in the proof of Proposition 4.4.16. So we can asssume for the rest of this proof that $c = 4$. Now in the notation of Proposition 2.7.3, we observe that $\ddot{\delta}$ and $\ddot{r}_\square \ddot{r}_\boxtimes \ddot{\delta}$ are conjugate in $\ddot{\Delta}$, and so in view of (4.7.27) we can assume without loss that $\ddot{H} = \langle \ddot{\delta} \rangle$. Thus $\ddot{\phi}_\mathcal{D} \leq C_{\ddot{\Gamma}}(\ddot{\delta}) = \langle \ddot{r}_\square \ddot{r}_\boxtimes \rangle \times \langle \ddot{\delta} \rangle \times \langle \ddot{\phi} \rangle \cong \mathbf{Z}_2 \times \mathbf{Z}_2 \times \mathbf{Z}_f$. Thus if f is odd, it is clear that $\langle \ddot{\phi}_\mathcal{D} \rangle = \langle \ddot{\phi} \rangle$, which proves $\ddot{H}_\Gamma = \langle \ddot{\delta}, \ddot{\phi} \rangle$, which in turn implies $\pi = \pi_5$. Hence we take f even. Here $q \equiv 1 \pmod 4$, and so $D(\mathbf{f}_i)$ is a non-square. Therefore $(\mathbf{f}_i)_{\beta_i} = \mathrm{diag}_{\beta_i}(\mu, 1, \ldots, 1)$ and $\phi_i = \phi_{\beta_i}(\nu)\mathrm{diag}_{\beta_i}(\mu^{(p-1)/2}, 1, \ldots, 1)$. Now form the tensor product basis $\beta = \beta_1 \otimes \cdots \otimes \beta_t$. Thus each element $v \in \beta$ satisfies $\mathbf{f}(v, v) = \mu^k$ for some $k = 0, \ldots, t$. Replacing v by $\mu^{-[k/2]}v$, we obtain a basis β' such that $\mathbf{f}(w, w) \in \{1, \mu\}$ for all $w \in \beta'$. And as $n \equiv 0 \pmod 4$, we know that $D = \square$. Thus we may write $\beta' = \{w_1, \ldots, w_n\}$, where $\mathbf{f}(w_j, w_j) = 1$ for $1 \leq j \leq s$, and $\mathbf{f}(w_j, w_j) = \mu$ for $s + 1 \leq j \leq n$, with s even. One checks that

$$\phi_\mathcal{D} = \phi_{\beta'}(\nu)\mathrm{diag}_{\beta'}(\overbrace{1, \ldots, 1}^{s}, \overbrace{\mu^{(p-1)/2}, \ldots, \mu^{(p-1)/2}}^{n-s}).$$

Now let $i = \sqrt{-1} \in \mathbf{F}$ and set

$$\begin{aligned} e_j &= \tfrac{1}{\sqrt{2}}(w_{2j-1} + iw_{2j}) \quad 1 \leq j \leq \tfrac{n}{2} \\ f_j &= \tfrac{1}{\sqrt{2}}(w_{2j-1} - iw_{2j}) \quad 1 \leq j \leq \tfrac{s}{2} \\ f_j &= \tfrac{1}{\mu\sqrt{2}}(w_{2j-1} - iw_{2j}) \quad \tfrac{s}{2} + 1 \leq j \leq n, \end{aligned}$$

so that $\beta'' = \{e_1, f_1, \ldots, e_{n/2}, f_{n/2}\}$ is a standard basis of V. Also let $\zeta = +1$ if $p \equiv \pm 1 \pmod 8$ and $\zeta = -1$ if $p \equiv \pm 3 \pmod 8$. Then putting $\beta''_j = \{e_j, f_j\}$, we have

$$\phi_{\mathcal{D}}^{\langle e_j, f_j \rangle} = \begin{cases} \zeta\phi_{\beta''_j}(\nu) & 1 \le i \le \frac{s}{2},\ p \equiv 1 \pmod 4 \\ \zeta\phi_{\beta''_j}(\nu) r_j & 1 \le i \le \frac{s}{2},\ p \equiv 3 \pmod 4 \\ \zeta\phi_{\beta''_j}(\nu) h_j & \frac{s}{2} + 1 \le i \le \frac{n}{2},\ p \equiv 1 \pmod 4 \\ \zeta\phi_{\beta''_j}(\nu) h_j r_j & \frac{s}{2} + 1 \le i \le \frac{n}{2},\ p \equiv 3 \pmod 4), \end{cases}$$

where $r_j = r_{e_j - f_j}$ (interchanging e_j and f_j), and $(h_j)_{\beta''_j} = \mathrm{diag}(\mu^{(p\mp 1)/2}, \mu^{(1\mp p)/2})$ with $p \equiv \pm 1 \pmod 4$. In light of the fact that $\Omega_2^+(q) \cong \mathbf{Z}_{(q-1)/2}$ and $SO_2^+(q) \cong \mathbf{Z}_{q-1}$, it follows that $h_j \in \Omega$. Moreover, if $p \equiv 3 \pmod 4$, then $\prod_{i=1}^{n/2} r_i \in \Omega$ since $n \equiv 0 \pmod 4$. It now follows that $\ddot{\phi}_{\mathcal{D}} = \ddot{\phi}_{\beta''}(\nu) = \ddot{\phi}$ for all p, and so $\ddot{H}_\Gamma = \langle \ddot{\delta}, \ddot{\phi} \rangle$, which means $\pi = \pi_5$, as claimed. ■

Proposition 4.7.8. *Assume that case* $\mathbf{O}^\circ$ *holds so that* H *is of type* $O_m(q) \wr S_t$ *with* qm *odd and* $(q, m) \neq (3, 3)$.

(I) $c = 1$.

(II) $H_{\overline{\Omega}} \cong \Omega_m(q)^t.2^{t-1}.S_t$.

(III) $H_{\overline{\Omega}}$ *is non-local and* $\mathrm{soc}(H_o) \cong \Omega_m(q)^t$.

§4.8 The classical subgroups $\mathcal{C}_8$

We now come to the final collection of subgroups in $\mathcal{C}$.

Definition. The members of $\mathcal{C}_8(\Gamma)$ are the groups $\Gamma(V, \mathbf{F}, \kappa_\sharp)$ such that $\kappa_\sharp$ is a non-degenerate form as described in Table 4.8.A.

Table 4.8.A			
case	type	description of $\kappa_\sharp$	conditions
L	$Sp_n(q)$	$\kappa_\sharp = \mathbf{f}_\sharp$ symplectic	n even, $n \ge 4$
L	$U_n(q^{1/2})$	$\kappa_\sharp = \mathbf{f}_\sharp$ unitary	f even, $n \ge 3$
L	$O_n^\epsilon(q)$	$\kappa_\sharp = Q_\sharp$ quadratic	q odd, $n \ge 3$, $\epsilon = \mathrm{sgn}(Q_\sharp)$
S	$O_n^\epsilon(q)$	$\kappa_\sharp = Q_\sharp$ quadratic, $\mathbf{f}_{Q_\sharp} = \mathbf{f}$	q even, $\epsilon = \mathrm{sgn}(Q_\sharp)$

Remarks on the conditions. The condition $n \ge 4$ in the first row of the table appears due to the isomorphism $SL_2(q) \cong Sp_2(q)$. As for the condition $n \ge 3$ in the second row, observe that $SU_2(q^{1/2}) \cong SL_2(q^{1/2})$, and so groups of type $U_2(q^{1/2})$ are identical with subfield groups in $\mathcal{C}_5$. In the third row we impose the condition $n \ge 3$ since the subgroups $O_2^+(q)$ and $O_2^-(q)$ coincide with the members of $\mathcal{C}_2$ and $\mathcal{C}_3$, respectively.

For the remainder of this section we assume that $H_\Gamma = \Gamma(V, \mathbf{F}, \kappa_\sharp) \in \mathcal{C}_8(\Gamma)$, as described in Table 4.8.A. As usual, let $X_\sharp = X(V, \mathbf{F}, \kappa_\sharp)$ as X ranges over the symbols in (2.1.15). It is clear by Lemma 2.1.2.iv that

$$H = \Delta_\sharp. \tag{4.8.1}$$

This first Lemma is easy, using Proposition 2.9.2, Proposition 2.10.6 and Lemma 4.0.5.iv.

Lemma 4.8.1.
(i) *$H_{\overline{\Omega}}$ is local if and only if $P\Omega_\sharp$ is local.*
(ii) *If $H_{\overline{\Omega}}$ is non-local then $\mathrm{soc}(H_{\overline{\Omega}}) \cong P\Omega_\sharp'$.*

Thus as usual we make no mention of part (III) in the Propositions which follow. In light of the work in the previous sections, the following Proposition completes the proof of Theorem 3.1.2.

Proposition 4.8.2. *We have $H_\Gamma = N_\Gamma(H_\Omega)$, and hence Theorem 3.1.2 holds.*

Proof. As in the proof of Proposition 4.7.2, we need only establish the first statement. Since $\sigma(\Gamma) = Aut(\mathbf{F})$, we see that $\Gamma_\sharp\Delta = \Gamma$, and so it suffices to show that $N_\Delta(H_\Omega) = H$. Now if $\Omega_\sharp$ is perfect, then $\Omega_\sharp = H_\Omega^\infty$, and so by Corollary 2.10.4.ii and Proposition 2.10.6, $N_\Delta(H_\Omega) \le N_{GL(V)}(\Omega_\sharp) = \Delta_\sharp = H$, as desired. If $\Omega_\sharp$ is not perfect, then H is of type $U_3(2)$ in $GL_3(4)$, $O_3(3)$ in $GL_3(3)$, $O_4^+(2)$ in $Sp_4(2)$, $O_4^+(3)$ in $GL_4(3)$ or $Sp_4(2)$ in $GL_4(2)$. Easy arguments apply to each case. For example, if H is of type $U_3(2)$, then $SU_3(2) \approx \Omega_\sharp = O^3(\Delta_\sharp)$, and so $\Omega_\sharp = O^3(H_\Omega)$; thus $N_\Delta(H_\Omega) \le N_{GL(V)}(\Omega_\sharp) = \Delta_\sharp$. And if H is of type $O_3(3)$ or $O_4^+(3)$, then $\Omega_\sharp = O^2(\Delta_\sharp)$, and a similar argument applies. The remaining two cases are left to the reader. ∎

Proposition 4.8.3. *Assume that case* **L** *holds and that H is of type $Sp_n(q)$, so that n is even and $n \ge 4$.*
(I) *$c = (q-1, \frac{n}{2})$ and $\pi = \pi_1(c)$.*
(II) *$H_{\overline{\Omega}} \cong PSp_n(q).[(q-1,2)c/(q-1,n)]$.*
(III) *$H_{\overline{\Omega}}$ is non-local and $\mathrm{soc}(H_{\overline{\Omega}}) \cong PSp_n(q)'$.*

Proof. Proposition 4.8.2 and Lemma 4.0.3.v,ix imply

$$c = \frac{q-1}{|\det(H)|}. \tag{4.8.2}$$

And as $\det(I_\sharp) = 1$, it follows from Lemma 2.4.5 that $|\det(H)| = |(\mathbf{F}^*)^{n/2}| = (q-1)/(q-1, \frac{n}{2})$, and so $c = (q-1, \frac{n}{2})$, as required. Moreover Lemma 4.0.3.ii yields $|\overline{H} : H_{\overline{\Omega}}| = \frac{(q-1,n)}{c}$, and so (II) holds because $\overline{H} = \overline{\Omega}_\sharp.(q-1,2) \cong PSp_n(q).(q-1,2)$. It remains to show that $\pi = \pi_1(c)$. Let β be a symplectic basis for $(V, \mathbf{f}_\sharp)$. Clearly $\phi_\beta(\nu) \in N_\Gamma(H)$, and so $\ddot{\phi} \in \ddot{H}_\Gamma$ (in the notation of Proposition 2.2.3). Now take $h \in H$ and write $h_\beta(\mathbf{f}_\sharp)_\beta h_\beta^t = \tau_\sharp(h)(\mathbf{f}_\sharp)_\beta$ (see Lemma 2.1.8). Thus $h_\beta^{-t} = \tau_\sharp(h)^{-1}(\mathbf{f}_\sharp)_\beta^{-1}h_\beta(\mathbf{f}_\sharp)_\beta$. However $\mathbf{F}^* \le H$ and $(\mathbf{f}_\sharp)_\beta \in H$, and hence $h^{\iota_\beta} \in H$, and so $\iota = \iota_\beta$ normalizes H. Hence $\iota \in N_A(H_\Omega)$, which proves $\ddot{\iota} \in \ddot{H}_A$. The usual argument — first appearing in the proof of Proposition 4.4.10 — shows that $\pi = \pi_1(c)$. ∎

Proposition 4.8.4. *Assume that case* **L** *holds and that* H *is of type* $O_n^\epsilon(q)$*, so that* q *is odd and* $n \geq 3$.

(I) $c = (q-1,n)/(n,2)$, and

$$\pi = \begin{cases} \pi_1(c) & \text{if } \epsilon \in \{\circ, +\} \text{ or } D(Q_\sharp) = \square \\ \pi_2(c) & \text{if } \epsilon = - \text{ and } D(Q_\sharp) = \boxtimes. \end{cases}$$

(II) $H_{\overline{\Omega}} \cong PSO_n^\epsilon(q).(n,2)$.

(III) $H_{\overline{\Omega}}$ *is local if and only if* $I_\sharp \approx O_3(3)$ *or* $O_4^+(3)$*, and if* $H_{\overline{\Omega}}$ *is non-local then* $\mathrm{soc}(H_{\overline{\Omega}}) \cong P\Omega_n^\epsilon(q)$.

Proof. First notice that (4.8.2) holds as before. Take n odd. Then by (2.6.2) $H = S_\sharp \times \mathbf{F}^*$, and hence $H_{\overline{\Omega}} = \overline{H} \cong PSO_n(q)$, proving (II). Also $|\det(H)| = (q-1)/(q-1,n)$, and hence $c = (q-1,n)$ by (4.8.2). We may replace $Q_\sharp$ by $\lambda Q_\sharp$ for any $\lambda \in \mathbf{F}^*$ in order to assume that $D(Q_\sharp) = \square$. Thus $(V, Q_\sharp)$ has an orthonormal basis β, which means $H = \{g \in GL(V) \mid g_\beta g_\beta^{\mathrm{t}} \in \mathbf{F}^*\}$. Thus it is clear that $\phi_\beta(\nu) \in H_\Gamma$ and $\iota_\beta \in H_A$, and hence $\pi = \pi_1(c)$ as in the proof Proposition 4.4.10.

Now take n even. Using (4.8.1) and Lemmas 2.7.5 and 2.8.4, we deduce

$$|\det(H)| = |\langle -1, (\mathbf{F}^*)^{n/2}\rangle| = \left[2, \frac{q-1}{(q-1,n/2)}\right] = \frac{2(q-1)}{(q-1,n)}. \tag{4.8.3}$$

Thus the value of c given in (I) is correct in light of (4.8.2). Moreover it follows from Lemma 4.0.3.iv,ix that $H_{\overline{\Omega}} = \frac{1}{2}\overline{H}$, and since $\overline{H} = PSO_n^\epsilon(q).[4]$, we see that (II) holds.

We now determine π. If $\mathrm{sgn}(Q_\sharp) = +$, then mimicking the proof of Proposition 4.8.3.I (using a standard orthogonal basis instead of a symplectic basis) one may show that $\pi = \pi_1(c)$. Thus we assume that $\mathrm{sgn}(Q_\sharp) = -$. If $D(Q_\sharp) = \square$, then V has an orthonormal basis, and so the argument used for n odd shows that $\pi = \pi_1(c)$, once again. So it remains to consider the case where $D(Q_\sharp) = \boxtimes$. Here V has a basis β such that $(\mathbf{f}_\sharp)_\beta = \mathrm{diag}_\beta(\mu, 1, \ldots, 1)$ (see Proposition 2.5.12). As in §2.2, set $\delta = \delta_\beta(\mu)$ and $\phi = \phi_\beta(\nu)$. Let $\phi_\sharp = \phi_{\beta, Q_\sharp}$ be as given in (2.8.4), so that $\phi_\sharp = \phi\delta^{(p-1)/2}$. Thus $H_\Gamma = \Delta_\sharp\langle\phi_\sharp\rangle$. Put $\iota = \iota_\beta$ and $g = \iota\delta^{-1}$. Now $H = \{h \in GL(V) \mid h_\beta\delta_\beta h_\beta^{\mathrm{t}} = \lambda\delta_\beta$ for some $\lambda \in \mathbf{F}^*\}$, and so for all $h \in H$ we have (supressing the subscript β)

$$(h^g)\delta(h^g)^{\mathrm{t}} = \delta h^{-\mathrm{t}}\delta^{-1}\delta\delta^{-\mathrm{t}}h^{-1}\delta^{\mathrm{t}} = \delta h^{-\mathrm{t}}\delta^{-1}h^{-1}\delta = \lambda^{-1}\delta.$$

Therefore g normalizes H and hence also normalizes H_Ω. So $g \in H_A$, whence

$$\ddot{H}_A = \langle \ddot{H}, \ddot{\phi}_\sharp, \ddot{g}\rangle = \langle \ddot{H}, \ddot{\phi}\ddot{\delta}^{(p-1)/2}, \ddot{\iota}\ddot{\delta}^{-1}\rangle.$$

We already computed in (4.8.3) that $|\ddot{H}| = 2$, and so $\ddot{H} = \langle\ddot{\delta}^{(q-1,n)/2}\rangle$. Consequently $\ddot{H}_A = \langle\ddot{\delta}^{(q-1,n)/2}, \ddot{\phi}\ddot{\delta}^{(p-1)/2}, \ddot{\iota}\ddot{\delta}^{-1}\rangle$, which means $\pi = \pi_2(c)$. The proof is now complete.

■

Proposition 4.8.5. *Assume that case* **L** *holds and that H is of type $U_n(q^{1/2})$, so that q is a square and $n \geq 3$.*

(I) $c = (q-1)/\left[q^{1/2}+1, q-1/(q-1,n)\right]$ and $\pi = \pi_1(c)$.

(II) $H_{\overline{\Omega}} \cong U_n(q^{1/2}).\left[(q^{1/2}+1,n)c/(q-1,n)\right]$.

(III) $H_{\overline{\Omega}}$ *is local if and only if* $I_\sharp \approx GU_2(2)$, $GU_2(3)$ *or* $GU_3(2)$, *and if* $H_{\overline{\Omega}}$ *is non-local, then* $\mathrm{soc}(H_{\overline{\Omega}}) \cong U_n(q^{1/2})$.

Proof. Assertion (II) is proved in the usual way using Lemma 4.0.3.ii,x so we concentrate on (I). Since $H = I_\sharp \mathbf{F}^*$, we have $|\det(H)| = \left[q^{1/2}+1, q-1/(q-1,n)\right]$, and so the value of c given in (I) is correct by (4.8.2). To determine π, let β be an orthonormal basis for $(V, \mathbf{f}_\sharp)$. Then $\phi_\beta(\nu)$ and $\iota_\beta \in N_A(H_\Omega)$, and so the result follows as in the proof of Proposition 4.4.10. ∎

Proposition 4.8.6. *Assume that case* **S** *holds and that H is of type $O_n^\epsilon(q)$, so that q is even.*

(I) $c = 1$.

(II) $H_{\overline{\Omega}} \cong O_n^\epsilon(q)$.

(III) $H_{\overline{\Omega}}$ *is local if and only if* $I_\sharp \approx O_4^+(2)$, *and if* $H_{\overline{\Omega}}$ *is non-local then* $\mathrm{soc}(H_{\overline{\Omega}}) \cong \Omega_n^\epsilon(q)$.

This concludes Chapter 4, completing the proof of parts (A) and (B) of the Main Theorem in §3.1.

Chapter 5
PROPERTIES OF THE FINITE SIMPLE GROUPS

The rest of this book is devoted to proving part (C) of the Main Theorem stated in §3.1. That is, we shall determine precisely when the members of $\mathcal{C}(G)$ are maximal in G, where G is any group whose socle is a classical simple group. In Chapters 6 and 7 we find all overgroups of members of $\mathcal{C}$ which themselves lie in $\mathcal{C}$, and in Chapter 8 we find those overgroups which lie in $\mathcal{S}$. The results in Chapter 8 depend on the classification of the finite simple groups, while those in Chapters 6 and 7 do not. Nevertheless, all three chapters require a fair amount of information about the simple groups, and in this chapter we survey various results from the literature concerning the finite simple groups which will be needed. The material here falls into three broad areas:

(1) basic properties of the simple groups (§5.1);

(2) subgroups of simple groups (§5.2);

(3) representations of the simple groups (§§5.3,5.4).

In §5.5 we include some further results on representations of direct products of simple groups and on extensions of soluble groups by simple groups. Throughout this chapter, L will usually denote a non-abelian simple group.

§5.1 Basic properties of the simple groups

As we mentioned in Chapter 1, the recent Classification Theorem asserts that the non-abelian simple groups fall into four categories: the alternating groups, the classical groups, the exceptional groups, and the sporadic groups. Of course, the alternating group A_n has order $\frac{1}{2}n!$. We display the orders of the other simple groups in Tables 5.1.A, 5.1.B and 5.1.C. The simple group L appears in the left hand column, and its order appears in the right hand column. We also give the order of the outer automorphism group $Out(L) = Aut(L)/L$, and an associated integer d, which is usually the order of the *multiplier* of L (see Theorem 5.1.4 below). The simple classical groups (Table 5.1.A) and the simple exceptional groups (Table 5.1.B) together comprise the *simple groups of Lie type*. Each simple group of Lie type has an associated *Lie rank*, usually denoted ℓ, which is also given in Tables 5.1.A and 5.1.B. The classical groups in some sense have two interpretations — a classical interpretation and a Lie theoretic interpretation. We give Lie notation for the classical groups in Table 5.1.A. In Tables 5.1.A and 5.1.B it is to be understood that

$$q = p^f,$$

where p is prime. Thus the groups in Tables 5.1.A and 5.1.B comprise the *simple groups of Lie type over* $\mathbf{F}_q$.

Recall from Chapter 2 that we use the convention $L_n^+(q) = L_n(q)$ and $L_n^-(q) = U_n(q)$.

Table 5.1.A
The Simple Classical Groups

L	Lie notation Lie rank ℓ	d	$\lvert Out(L)\rvert$	$\lvert L\rvert$
$L_n(q)$	$A_{n-1}(q)$ $n-1$	$(n,q-1)$	$2df,\ n\geq 3$ $df,\ n=2$	$\frac{1}{d}q^{n(n-1)/2}\prod_{i=2}^{n}(q^i-1)$
$U_n(q)$	${}^2A_{n-1}(q)$ $[\frac{n}{2}]$	$(n,q+1)$	$2df,\ n\geq 3$ $df,\ n=2$	$\frac{1}{d}q^{n(n-1)/2}\prod_{i=2}^{n}(q^i-(-1)^i)$
$PSp_{2m}(q)$	$C_m(q)$ m	$(2,q-1)$	$df,\ m\geq 3$ $2f,\ m=2$	$\frac{1}{d}q^{m^2}\prod_{i=1}^{m}(q^{2i}-1)$
$\Omega_{2m+1}(q)$ q odd	$B_m(q)$ m	2	$2f$	$\frac{1}{2}q^{m^2}\prod_{i=1}^{m}(q^{2i}-1)$
$P\Omega^+_{2m}(q)$ $m\geq 3$	$D_m(q)$ m	$(4,q^m-1)$	$2df,\ m\neq 4$ $6df,\ m=4$	$\frac{1}{d}q^{m(m-1)}(q^m-1)\prod_{i=1}^{m-1}(q^{2i}-1)$
$P\Omega^-_{2m}(q)$ $m\geq 2$	${}^2D_m(q)$ $m-1$	$(4,q^m+1)$	$2df$	$\frac{1}{d}q^{m(m-1)}(q^m+1)\prod_{i=1}^{m-1}(q^{2i}-1)$

Table 5.1.B
The Simple Exceptional Groups

L	ℓ	d	$\lvert Out(L)\rvert$	$\lvert L\rvert$
$G_2(q)$	2	1	f if $p\neq 3$ $2f$ if $p=3$	$q^6(q^2-1)(q^6-1)$
$F_4(q)$	4	1	$(2,p)f$	$q^{24}(q^2-1)(q^6-1)(q^8-1)(q^{12}-1)$
$E_6(q)$	6	$(3,q-1)$	$2df$	$\frac{1}{d}q^{36}\prod_{i\in\{2,5,6,8,9,12\}}(q^i-1)$
$E_7(q)$	7	$(2,q-1)$	df	$\frac{1}{d}q^{63}\prod_{i\in\{2,6,8,10,12,14,18\}}(q^i-1)$
$E_8(q)$	8	1	f	$q^{120}\prod_{i\in\{2,8,12,14,18,20,24,30\}}(q^i-1)$
${}^2B_2(q),\ q=2^{2m+1}$	1	1	f	$q^2(q^2+1)(q-1)$
${}^2G_2(q),\ q=3^{2m+1}$	1	1	f	$q^3(q^3+1)(q-1)$
${}^2F_4(q),\ q=2^{2m+1}$	2	1	f	$q^{12}(q^6+1)(q^4-1)(q^3+1)(q-1)$
${}^3D_4(q)$	2	1	$3f$	$q^{12}(q^8+q^4+1)(q^6-1)(q^2-1)$
${}^2E_6(q)$	4	$(3,q+1)$	$2df$	$\frac{1}{d}q^{36}\prod_{i\in\{2,5,6,8,9,12\}}(q^i-(-1)^i)$

Table 5.1.C
The Simple Sporadic Groups

L	d	$\lvert Out(L)\rvert$	$\lvert L\rvert$
M_{11}	1	1	$2^4.3^2.5.11$
M_{12}	2	2	$2^6.3^3.5.11$
M_{22}	12	2	$2^7.3^2.5.7.11$
M_{23}	1	1	$2^7.3^2.5.7.11.23$
M_{24}	1	1	$2^{10}.3^3.5.7.11.23$
J_1	1	1	$2^3.3.5.7.11.19$
J_2	2	2	$2^7.3^3.5^2.7$
J_3	3	2	$2^7.3^5.5.17.19$
J_4	1	1	$2^{21}.3^3.5.7.11^3.23.29.31.37.43$
HS	2	2	$2^9.3^2.5^3.7.11$
Suz	6	2	$2^{13}.3^7.5^2.7.11.13$
McL	3	2	$2^7.3^6.5^3.7.11$
Ru	2	1	$2^{14}.3^3.5^3.7.13.29$
$He = F_7$	1	2	$2^{10}.3^3.5^2.7^3.17$
Ly	1	1	$2^8.3^7.5^6.7.11.31.37.67$
$O'N$	3	2	$2^9.3^4.5.7^3.11.19.31$
Co_1	2	1	$2^{21}.3^9.5^4.7^2.11.13.23$
Co_2	1	1	$2^{18}.3^6.5^3.7.11.23$
Co_3	1	1	$2^{10}.3^7.5^3.7.11.23$
Fi_{22}	6	2	$2^{17}.3^9.5^2.7.11.13$
Fi_{23}	1	1	$2^{18}.3^{13}.5^2.7.11.13.17.23$
Fi'_{24}	3	2	$2^{21}.3^{16}.5^2.7^3.11.13.17.23.29$
$HN = F_5$	1	2	$2^{14}.3^6.5^6.7.11.19$
$Th = F_3$	1	1	$2^{15}.3^{10}.5^3.7^2.13.19.31$
$BM = F_2$	2	1	$2^{41}.3^{13}.5^6.7^2.11.13.17.19.23.31.47$
$M = F_1$	1	1	$2^{46}.3^{20}.5^9.7^6.11^2 \times$ $13^3.17.19.23.29.31.41.47.59.71$

Using the Lie notation, this becomes

$$A_n^\epsilon(q) = \begin{cases} A_n(q) & \text{if } \epsilon = + \\ {}^2A_n(q) & \text{if } \epsilon = -. \end{cases}$$

In the same manner, we define

$$X_\ell^\epsilon(q) = \begin{cases} X_\ell(q) & \text{if } \epsilon = + \\ {}^2X_\ell(q) & \text{if } \epsilon = - \end{cases} \tag{5.1.1}$$

where X_ℓ is one of the symbols G_2, F_4, and E_6.

The groups of Lie type in Tables 5.1.A and 5.1.B account for sixteen families of simple groups as q and n vary. However, when q (and/or n) is small, the groups of Lie type behave somewhat atypically, and for our purposes this behaviour is important to record. For example, we saw in Proposition 2.9.2 that certain classical groups are not simple. Analogously, for the exceptional groups, we have the following (see [Ca$_1$, Theorems 11.1.2, 14.4.1]).

Theorem 5.1.1. *The exceptional groups in Table 5.1.B are simple except for*

$$\begin{aligned} {}^2B_2(2) &\cong \mathbf{Z}_5{:}\mathbf{Z}_4 \\ G_2(2) &\cong Aut(U_3(3)) \cong U_3(3).2 \\ {}^2G_2(3) &\cong Aut(L_2(8)) \cong L_2(8).3 \\ {}^2F_4(2) &\cong {}^2F_4(2)'.2. \end{aligned}$$

The group ${}^2F_4(2)'$ is simple (and is called the Tits group).

Remark. In view of the isomorphisms stated in Theorem 5.1.1, we regard $G_2(2)$ and ${}^2G_2(3)$ as classical groups rather than exceptional groups of Lie type.

Theorem 5.1.2. *The isomorphisms among the groups in Tables 5.1.A, 5.1.B, 5.1.C and the alternating groups are precisely those given in Proposition 2.9.1 and Theorem 5.1.1.*

This result can be proved using the observation (due essentially to Artin — see [K-L-S]) that the only coincidences among the orders of simple groups (apart from those given by the isomorphisms in Proposition 2.9.1 and Theorem 5.1.1) are $|PSp_{2m}(q)| = |\Omega_{2m+1}(q)|$ and $|A_8| = |L_3(4)|$. Details can be found in [K-L-S].

We define

$$Lie(p)$$

to be the set of *simple* groups of Lie type over fields of characteristic p. We include $PSp_4(2)'$ and ${}^2F_4(2)'$ in $Lie(2)$. Also we set

$$Lie(p') = \bigcup_{r \neq p} Lie(r).$$

In view of Proposition 2.9.1 and Theorem 5.1.1, some simple groups belong to $Lie(p) \cap Lie(r)$ for distinct primes p, r, and may have different Lie ranks with respect to these different primes. For instance $L_2(9) = Sp_4(2)' \in Lie(2) \cap Lie(3)$, and so $L_2(9)$ has Lie rank 2 as a member of $Lie(2)$ and Lie rank 1 as a member of $Lie(3)$. A member of $Lie(p)$ is said to be *twisted* if it lies in one of the families ${}^2A_n$, ${}^2B_2$, ${}^2D_n$, ${}^3D_4$, ${}^2E_6$, ${}^2F_4$ or ${}^2G_2$. Otherwise L is *untwisted.* The *untwisted Lie rank* of a twisted group of type ${}^aX_\ell$ is just the Lie rank of the corresponding untwisted group X_ℓ, namely ℓ. Thus the untwisted Lie rank of ${}^2F_4(q)$ is 4, while its Lie rank is 2.

At times we will use the notation

$$C\ell_d(q)$$

to denote a classical group of dimension d over the field $\mathbf{F}_q$. Recall that the dimension is just the dimension of the vector space naturally associated to the classical group. In view of the isomorphisms appearing in Proposition 2.9.1, it may be the case that $C\ell_d(q) \cong C\ell_{d'}(q')$ with $(d,q) \neq (d',q')$. We will usually choose d minimal in these situations. For example, if $C\ell_d(q) \cong P\Omega_4^-(r)$ with d minimal, then $d = 2$ and $q = r^2$, in view of the isomorphism $P\Omega_4^-(r) \cong L_2(r^2)$.

We have given in Tables 5.1.A,B,C the orders of the outer automorphism groups of the groups of Lie type and the sporadic groups. References for this information can be found in [Ca$_1$, Chapter 12] and [St$_1$, p.195] for the groups of Lie type and [At] for the sporadic groups. Note that Proposition 2.9.1.xii and Table 5.1.A imply that $|Out(A_6)| = 4$. ¿From the point of view of $L_2(9)$ this is natural behaviour. However, from the point of view of A_6 this is a sporadic phenomenon, as the following Theorem indicates; a proof can be found in [Pas, Th. 5.7].

Theorem 5.1.3. *If $n \geq 5$ and $n \neq 6$, then $Aut(A_n) = S_n$ and hence $|Out(A_n)| = 2$. However $|Out(A_6)| = 4$.*

We now consider the multipliers of the finite simple groups. Let G be a finite perfect group. A *central extension* of G is a group H satisfying $H/Z(H) \cong G$. A central extension of G which is perfect is called a *covering group* of G. It was shown by Schur [Sch$_1$] that all covering groups of G are finite, and that there is a unique covering group of maximal order, called the *full covering group* of G. Letting H be the full covering group of G, define the (*Schur*) *multiplier* of G, written $M(G)$, to be the centre of H. If H_1 is any other covering group of G, then there is a surjective homomorphism from H to H_1, with kernel contained in $Z(H)$. Moreover, any covering group of H_1 is still a covering group of G. These results, as well as further information about Schur multipliers, can be found in [As$_8$] and [Go, §4.15]. The following Theorem describes the multipliers of the simple groups. Again, we see the atypical behaviour of the groups of Lie type when q and n are small.

Theorem 5.1.4.

(i) *For $n \geq 5$ we have*

$$M(A_n) \cong \begin{cases} \mathbf{Z}_2 & \text{if } n \neq 6,7 \\ \mathbf{Z}_6 & \text{if } n = 6 \text{ or } 7. \end{cases}$$

(ii) *Assume that L is a simple group of Lie type or a sporadic simple group, and that d is as given in Tables 5.1.A,B,C. Then provided L does not appear in Table 5.1.D, we have*

$$M(L) \cong \begin{cases} \mathbf{Z}_2 \times \mathbf{Z}_2 & \text{if } L \cong P\Omega_{2m}^+(q) \text{ with } q \text{ odd and } m \text{ even} \\ \mathbf{Z}_d & \text{otherwise.} \end{cases}$$

Table 5.1.D Exceptional Schur multipliers	
L	$M(L)$
$L_2(4)$, $L_3(2)$, $L_4(2)$, $U_4(2)$, $Sp_6(2)$, $G_2(4)$, $F_4(2)$	$\mathbf{Z}_2$
$G_2(3)$	$\mathbf{Z}_3$
$L_2(9)$, $Sp_4(2)'$, $\Omega_7(3)$	$\mathbf{Z}_6$
$L_3(4)$	$\mathbf{Z}_4 \times \mathbf{Z}_{12}$
$U_4(3)$	$\mathbf{Z}_3 \times \mathbf{Z}_{12}$
$U_6(2)$, ${}^2E_6(2)$	$\mathbf{Z}_2 \times \mathbf{Z}_6$
$\Omega_8^+(2)$, ${}^2B_2(8)$	$\mathbf{Z}_2 \times \mathbf{Z}_2$

Part (i) was proved originally by Schur [Sch$_2$] and a discussion of (ii) appears in [Go, §4.15] and [Col, pp.20-21].

§5.2 Subgroups of the simple groups

We begin our study of the subgroups of the simple groups by considering their large subgroups. Clearly large subgroups correspond to permutation representations of small degree, and for this reason the following definitions are useful. For any group G, define

$$P(G) = \min\{n \mid G \text{ has a non-trivial permutation representation of degree } n\},$$
$$P_f(G) = \min\{n \mid G \text{ has a faithful permutation representation of degree } n\},$$
$$P_f^t(G) = \min\{n \mid G \text{ has a faithful transitive permutation representation of degree } n\}.$$

Thus $P(G)$ is the index of the largest proper subgroup of G, and $P_f^t(G)$ is the index of its largest core-free subgroup. In particular, for the finite simple group L, the integer $P(L)$ is just the index of the largest proper subgroup, and this is now known for all the simple groups (see for instance [At] for the sporadic groups, [Co$_1$] for the classical groups and [L-S$_1$] for the exceptional groups). Also observe

Proposition 5.2.1. *Assume that L is simple and that L_1 is a covering group of L.*
(i) $P(L_1) = P(L)$.
(ii) $P_f(L_1) \geq P_f(L)$.
(iii) $P_f^t(L_1) \geq P_f^t(L)$.

The problem of determining $P(L)$ for the simple groups L has quite an interesting history. For example, in Galois' famous letter to Chevalier in 1832 [Ga$_1$], he stated that $P(L_2(p)) = p + 1$ for all primes $p \geq 13$. The first published proof of this theorem appeared in a paper of Jordan in 1868 [Jo]. However, even before Galois was born, the Italian mathematician and physician Paolo Ruffini in 1799 [Ru] performed various calculations with permutations of degree 5 and in effect gave a proof of the fact that $P(A_5) = 5$. (Of course, Ruffini did not even dream of such notation for the concept of a 'group' was not fully developed until many decades later!) Then in 1815, Cauchy [Cau],

Table 5.2.A
$P(L)$ for L a classical simple group

L	$P(L)$
$L_n(q)$ $(n,q) \neq (2,5),(2,7),(2,9),(2,11),(4,2)$	$(q^n-1)/(q-1)$
$L_2(5)$, $L_2(7)$, $L_2(9)$, $L_2(11)$, $L_4(2)$	$5,7,6,11,8$
$PSp_{2m}(q)$, $m \geq 2$, $q > 2$, $(m,q) \neq (2,3)$	$(q^{2m}-1)/(q-1)$
$Sp_{2m}(2)$, $m \geq 3$	$2^{m-1}(2^m-1)$
$Sp_4(2)'$, $PSp_4(3)$	6, 27
$\Omega_{2m+1}(q)$, $m \geq 3$, q odd, $q \geq 5$	$(q^{2m}-1)/(q-1)$
$\Omega_{2m+1}(3)$, $m \geq 3$	$\frac{1}{2}3^m(3^m-1)$ *
$P\Omega_{2m}^+(q)$, $m \geq 4$, $q \geq 3$	$(q^m-1)(q^{m-1}+1)/(q-1)$
$P\Omega_{2m}^+(2)$, $n \geq 4$	$2^{m-1}(2^m-1)$
$P\Omega_{2m}^-(q)$, $n \geq 4$	$(q^m+1)(q^{m-1}-1)/(q-1)$
$U_3(q)$, $q \neq 5$	q^3+1
$U_3(5)$	50
$U_4(q)$	$(q+1)(q^3+1)$
$U_n(q)$, $n \geq 5$, $(n,q) \neq (6m,2)$	$\frac{\left(q^n-(-1)^n\right)\left(q^{n-1}-(-1)^{n-1}\right)}{q^2-1}$
$U_n(2)$, $6 \mid n$	$2^{n-1}(2^n-1)/3$ *

gave lower bounds for $P(A_n)$ for all n. Of course, it follows immediately from the fact that A_n is simple (for $n \geq 5$) that $P(A_n) = n$. The calculation of $P(L)$ for the simple groups L has continued throughout the course of finite group theory, and here we present some of the more modern results.

When L is a classical simple group, $P(L)$ has been determined in certain special cases by many mathematicians, including most notably Galois, Jordan, Burnside and Dickson. The complete solution has been obtained by Cooperstein [Co_1] and Patton [Pat].

Theorem 5.2.2. [Co_1, Pat] *The values of $P(L)$ for L a classical simple group are as in Table 5.2.A.*

Remark. There are actually two errors in [Co_1], so we take this opportunity here to state the full result with corrections marked with a $*$ in Table 5.2.A. These corrections can be verified using [Ka_1] or [Lie_1].

We draw several consequences of Theorem 5.2.2 which will serve as a convenience in later chapters.

Corollary 5.2.3. *Assume that L is a classical simple group of dimension n.*

(i) $P(L) \geq n+1$.

(ii) *If $n \geq 4$, then $P(L) \geq n^2+3$, except when $L \approx L_4(2)$, $Sp_4(2)'$, $\Omega_4^-(2)$, $\Omega_4^-(3)$, $\Omega_4^-(4)$, $\Omega_5(3)$ or $\Omega_6^{\pm}(2)$.*

Most of the minimal actions in Cooperstein's Theorem 5.2.2 are actions on either singular or non-singular 1-spaces in the classical geometry associated with L. Thus the corresponding stabilizers are reducible subgroups. In [Ka$_1$], Kantor obtains strong upper bounds for the orders of irreducible subgroups of the classical groups. Using the classification of simple groups and Aschbacher's Theorem 1.2.1, Kantor's results have been improved in the following Theorem. Recall the definition of u given in (2.1.17).

Theorem 5.2.4 [Lie$_1$]. *Let L be a classical simple group, with associated geometry of dimension n over $\mathbf{F}$, where $\mathbf{F} = \mathbf{F}_{q^u}$. Also let G satisfy $L \trianglelefteq G \leq Aut(L)$, and let H be any subgroup of G not containing L. Then either H is contained a member of $\mathcal{C}(G)$ or one of the following holds:*

(i) *H is A_m or S_m with $n+1 \leq m \leq n+2$, and V is the fully deleted permutation module for H, described in §5.3, below;*

(ii) $|H| < |\mathbf{F}|^{3n}$.

We now present a few elementary results concerning the values of $P(G)$ and $P_f(G)$ for various groups G.

Proposition 5.2.5. *Let $L_1, \ldots, L_n$ be non-abelian simple groups and put $G = L_1 \times \cdots \times L_n$.*

(i) *Any normal subgroup of G is of the form $L_I = \prod_{i \in I} L_i$ for some subset $I \subseteq \{1, \ldots, n\}$.*

(ii) *If C is a covering group of G then $C = C_1 \circ \cdots \circ C_n$, where C_i is a covering group of L_i.*

(iii) *With C as in* (ii), *we have $P_f(C) \geq P_f(G)$ and $P_f^t(C) \geq P_f^t(G)$.*

Proof. Assertions (i) and (ii) are elementary facts. Part (iii) is an easy generalization of Proposition 5.2.1. ∎

Lemma 5.2.6. *Let $L_1, \ldots, L_n$ be non-abelian simple groups and let X be a core-free subgroup of $L_1 \times \cdots \times L_n$ which projects surjectively onto each L_i. Then $|X| \leq \prod_{i=1}^n |L_i|^{1/2}$.*

Proof. When $n = 2$, it is clear that $L_1 \cong L_2$ and that X is a diagonal subgroup isomorphic to L_1. Thus the result holds here, and we may now proceed by induction. Let $\pi_i : X \mapsto L_i$ be the ith projection and put $X_i = \ker(\pi_i)$. For $x = (x_1, \ldots, x_n) \in X$, define $\text{supp}(x)$ to be the number of non-trivial x_i. Amongst all members of $X \backslash X_1$, choose x such that $s = \text{supp}(x)$ is minimal. We may relabel the indices so that $\pi_i(x) \neq 1$ for $1 \leq i \leq s$ while $\pi_i(x) = 1$ for $s+1 \leq i \leq n$ (possibly $s = n$). Thus $x = (x_1, \ldots, x_s, 1, \ldots, 1)$ with $x_i \neq 1$. Since $\pi_1(X) = L_1$, we have $\pi_1(\langle x^X \rangle) = L_1$. It follows that $s \geq 2$, for if $s = 1$, then $L_1 = \langle x^X \rangle \leq X$, contrary to our assumption that X is core-free. We now claim that $X_1 = X_2$. For assume otherwise. Now $X/X_i \cong L_i$, and so each X_i is a maximal normal subgroup of X. Thus if $X_1 \neq X_2$, then $X = X_1X_2$, whence there exists $y \in X_2 \backslash X_1$. Evidently $y = (y_1, 1, y_3, \ldots, y_n)$ with $y_1 \neq 1$. Since $\pi_1(\langle y^X \rangle) = L_1$, it is clear that y may be chosen so that x_1 and y_1 do not commute, and hence $[x, y] \in X \backslash X_1$. However $1 \leq \mathrm{supp}([x, y]) < \mathrm{supp}(x)$, which contradicts the minimality of s. Thus we have established $X_1 = X_2$. Relabel the indices again so that $X_i = X_1$ for $1 \leq i \leq t$ and $X_i \neq X_1$ for $t + 1 \leq i \leq n$ (possibly $t = n$). Then $X_1 \leq L_{t+1} \times \cdots \times L_n$, and since $X_1X_i = X$ for $t + 1 \leq i \leq n$, it follows that $\pi_i(X_1) = L_i$ for $t + 1 \leq i \leq n$. Thus by induction, we have $|X_1| \leq \prod_{i=t+1}^n |L_i|^{1/2}$. Since $X/X_1 = X/X_i \cong L_i$ for $1 \leq i \leq t$, we have $|X/X_1| = |L_1| = \prod_{i=1}^t |L_i|^{1/t} \leq \prod_{i=1}^t |L_i|^{1/2}$ since $t \geq 2$, and the proof is complete. ■

Proposition 5.2.7. *Let $L_1, \ldots, L_n$ be non-abelian simple groups and let C be a covering group of $L_1 \times \cdots \times L_n$.*

(i) $P_f(C) \geq \sum_{i=1}^n P(L_i)$.

(ii) $P_f^t(C) \geq \prod_{i=1}^n m_i$, *where* $m_i = \min\{|L_i|^{1/2}, P(L_i)\}$.

(iii) *If each L_i is classical, then* $P_f^t(C) \geq \prod_{i=1}^n P(L_i)$.

Proof. In view of Proposition 5.2.5.iii we may assume that $C = L_1 \times \cdots \times L_n$. The proof goes by induction on n and on $|C|$. The case $n = 1$ is true by definition, so suppose that $n > 1$. Let X be a set of size $P_f(C)$ on which C acts faithfully. We aim to show that $|X| \geq \sum_{i=1}^n P(L_i)$.

Suppose first that C is intransitive on X, and write $X = X_1 \cup X_2$, where X_i is a C-invariant proper subset of X. Thus for $i = 1, 2$, we have $C^{X_i} = \prod_{a \in I_i} L_a$ for some subset I_i of $\{1, \ldots, n\}$, where $I_1 \cup I_2 = \{1, \ldots, n\}$. By the minimality of $P_f(C)$, we know that C does not act faithfully on X_i and so by induction on $|C|$ we deduce $|X_i| \geq P_f(C^{X_i}) \geq \sum_{a \in I_i} P(L_a)$. It now follows that $|X| \geq |X_1| + |X_2| \geq \sum_{a=1}^n P(L_a)$, as required.

Now suppose that C is transitive on X. Fix $x \in X$ and let $\pi_i : C_x \to L_i$ be the ith projection map. By relabelling, we may assume that π_i is surjective for $1 \leq i \leq s$, and is not surjective for $s + 1 \leq i \leq n$ (possibly s is 0 or n). Set $R_i = \pi_i(C_x)$ for $s + 1 \leq i \leq n$ and $H = L_1 \ldots L_s$. Thus we have $C_x \leq HR_{s+1} \ldots R_n$. Clearly there is a homomorphism from C_x to $R_{s+1} \ldots R_n$ with kernel $H_x = H \cap C_x$, and hence

$$|C_x| \leq |H_x||R_{s+1} \ldots R_n|. \tag{5.2.1}$$

Now any normal subgroup of H is also normal in C, and so H_x is core-free in H. Therefore $|H{:}H_x| \geq P_f^t(H)$. Thus if $H \neq C$, we deduce by induction that $|H{:}H_x| \geq \max\{\sum_{i=1}^s P(L_i), \prod_{i=1}^s m_i\}$. And for $s + 1 \leq i \leq n$ we have $|L_i{:}R_i| \geq P(L_i) \geq m_i$, whence $|L_{s+1} \ldots L_n : R_{s+1} \ldots R_n| \geq \prod_{i=s+1}^n P(L_i) \geq \max\{\sum_{i=s+1}^n P(L_i), \prod_{i=s+1}^n m_i\}$. Thus it now follows from (5.2.1) that $|X| = |C{:}C_x| \geq \max\{\sum_{i=1}^n P(L_i), \prod_{i=1}^n m_i\}$, as required. We have therefore reduced to the case in which $s = n$. Thus by Lemma 5.2.6, $|C : C_x| \geq \prod_{i=1}^n |L_i|^{1/2} \geq \prod_{i=1}^n m_i$, proving (ii). Note that (iii) follows from (ii), since every classical simple group $|L|$ has a subgroup of order at least $|L|^{1/2}$. To complete the proof we must check that $\prod_{i=1}^{n}|L_i|^{1/2} \geq \sum_{i=1}^{n} P(L_i)$ when $n \geq 2$. Now it follows from [Is, Cor. 4.13] that L_i has a subgroup of order at least $|L_i|^{1/3}$, which shows $P(L_i) \leq |L_i|^{2/3}$. It is clear that $\prod_{i=1}^{n}|L_i|^{1/2} \geq \sum_{i=1}^{n}|L_i|^{2/3}$ if $n \geq 2$, and hence (i) holds. ■

It follows from the classification theorem that every finite simple group $|L|$ has a proper subgroup of order at least $|L|^{1/2}$, and hence in practice $m_i = P(L_i)$.

Proposition 5.2.8. *Let p be a prime.*

(i) $P_f(\mathbf{Z}_p^t) = pt$ *for all t.*

(ii) *The p-rank of* S_n *satisfies* $m_p(S_n) = \left[\frac{n}{p}\right]$.

(iii) *Let H be a group with structure* $p^t.A_n$, *where* $n \geq 5$, *and* $O_p(H) = p^t$ *is either the deleted permutation module or the fully deleted permutation module for* $H/O_p(H) \cong A_n$ *over* $\mathbf{F}_p$, *as described in §5.3, below. Then* $P_f(H) \geq np$.

Proof. (i) As in the proof of Proposition 5.2.7 we reduce to the case where $\mathbf{Z}_p^t$ is transitive on a minimal set X. But a transitive abelian group is regular [Wie, 4.4], so $|X| = p^t \geq pt$. Obviously $\mathbf{Z}_p^t$ has a faithful permutation representation of degree pt, and so we deduce (i).

(ii) This is immediate from (i).

(iii) Again, let X be a minimal set on which H acts faithfully. Put $V = O_p(H)$ and let Y be a V-orbit in X upon which V acts non-trivially. Let $Y = Y_1, \ldots, Y_k$ be the orbit of Y under $H/O_p(H) \cong A_n$. If $k > 1$, then $k \geq P(A_n) = n$, and since $|Y| \geq p$, we deduce $|X| \geq k|Y| \geq np$, as desired. Otherwise $k = 1$, which is to say Y is H-invariant. Thus $V_{(Y)} \trianglelefteq V.A_n$, and in view of Lemma 5.3.4 below, $|V^Y| = |V : V_{(Y)}| = p^{n-1}$ or p^{n-2}. But V^Y acts regularly, and so $|X| \geq |Y| \geq p^{n-2} \geq np$, except possibly when $p = 2$ and $n = 5$. But here the fully deleted permutation module has order 2^4, and so the argument above shows $|X| \geq 16 > np$, as claimed. ■

Subgroups of sporadic groups

Complete lists of the maximal subgroups of all but three of the 26 sporadic simple groups have been determined; the three exceptions are the largest Fischer group Fi_{24}', the Baby Monster BM and the Monster M. These lists appear complete in [At], except for J_4, Fi_{22}, Fi_{23} and Th, which appear in [K-W$_1$,K-W$_2$,K-P-W,Lin]. From these lists, together with the information in [At, p.231], we obtain information concerning the large alternating and classical groups which are involved in the sporadic groups. This is recorded in Proposition 5.2.9. It is worth remarking, however, that the complete lists of maximal subgroups are not needed to derive this result. Indeed, there is a host of elementary techniques which can be used to show that suitably large simple groups cannot embed in a given sporadic group. To illustrate, it is shown in [Wi$_4$] that A_8 is not contained in the Suzuki group Suz, since there is no compatible restriction of the 12-dimensional character of $6.Suz$ to A_8 or $2.A_8$. Nevertheless, we use the above references for the sake of convenience.

Proposition 5.2.9. *Let L be a sporadic simple group, and define*

$$a(L) = \max\{m \mid A_m \text{ is involved in } L\}$$
$$\mathcal{X}(L) = \text{the set of classical simple groups of dimension at least 5 which are involved in } L.$$

(In the definition of $\mathcal{X}(L)$ we ignore $\Omega_5(q)$ and $P\Omega_6^{\pm}(q)$ in view of their isomorphisms with $PSp_4(q)$ and $L_4^{\pm}(q)$, respectively.)

(i) *The values of $a(L)$ are given in the middle column of Table 5.2.B.*

(ii) *If the symbol '–' appears in the right hand column then $\mathcal{X}(L)$ is empty. Otherwise, the maximal members of $\mathcal{X}(L)$ appear in the right hand column.*

Table 5.2.B

L	$a(L)$	maximal members of $\mathcal{X}(L)$
J_1	5	–
M_{11}, M_{12}, J_2, J_3	6	–
$M_{22}, He, O'N$	7	–
$M_{23}, M_{24}, HS, McL, Ru$	8	–
J_4	8	$L_5(2)$
Suz	7	$U_5(2)$
Ly	11	–
Co_1	9	$\Omega_8^+(2), U_6(2)$
Co_2	8	$U_6(2)$
Co_3	8	$Sp_6(2)$
Fi_{22}	10	$\Omega_8^+(2), \Omega_7(3), U_6(2)$
Fi_{23}	12	$P\Omega_8^+(3), Sp_8(2), U_6(2)$
Fi_{24}', BM, M	12	$\Omega_{10}^{\pm}(2), P\Omega_8^{\pm}(3), U_6(2)$
HN	12	–
Th	9	$L_5(2)$

Groups of Lie Type

Here we collect a few results concerning the subgroups of the groups of Lie type. Assume that $L \in Lie(p)$, so that L is a simple group of Lie type in characteristic p. For the time being, assume also that L is untwisted and of rank ℓ. As described in [Ca_1], there is a simple Lie algebra over $\mathbf{C}$ corresponding to L, with root system Φ and fundamental root system $\{\alpha_1, \ldots, \alpha_\ell\}$ giving one of the following Dynkin diagrams:

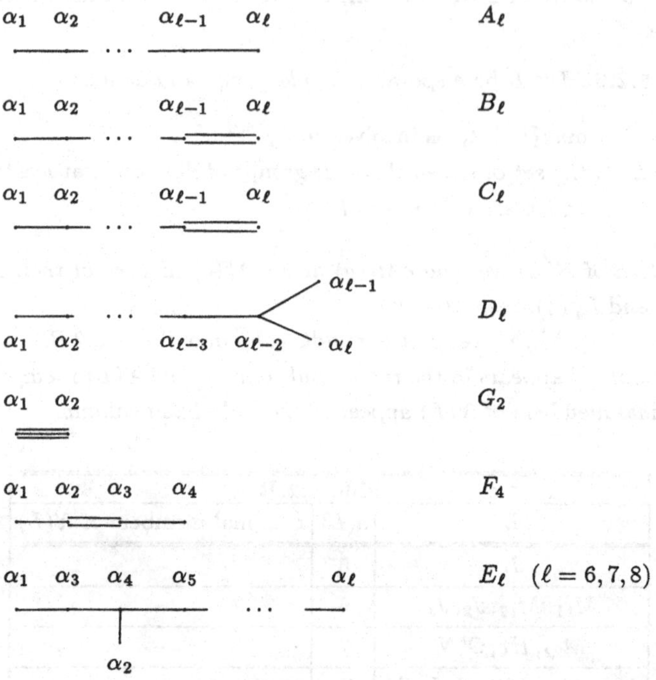

(5.2.2)

The group L is generated by *root* subgroups X_α $(\alpha \in \Phi)$ as described in [Ca$_1$, Ch. 6]. Denote by Φ^+ the set of positive linear combinations of $\alpha_1, \ldots, \alpha_\ell$ lying in Φ. Then $U = \langle X_\alpha \mid \alpha \in \Phi^+ \rangle$ is a Sylow p-subgroup of L, and if H is a *Cartan* subgroup normalizing U, then $B = HU$ is a *Borel* subgroup of L. The subgroups of L which contain B are called *parabolic* subgroups. They correspond to subsets of $\{\alpha_1, \ldots, \alpha_\ell\}$. If J is such a subset then the corresponding parabolic subgroup P_J is of the form $Q_J L_J H$, where

$$Q_J = \langle X_\alpha \mid \alpha \in \Phi^+ \backslash \mathbf{Z}J \rangle,$$
$$L_J = \langle X_\alpha \mid \alpha \in \mathbf{Z}J \cap \Phi \rangle.$$

Here Q_J is called the ***unipotent radical*** of P_J, and $L_J H$ the ***Levi factor***. Moreover, $L_J = O^{p'}(L_J H)$, and L_J is a central product of groups of Lie type corresponding to the subdiagram on J. For example, the Levi factors of the maximal parabolic subgroups of $F_4(q)$ are of types $B_3(q)$, $C_3(q)$ and $A_1(q) \times A_2(q)$.

When L is a twisted group, it consists of elements centralized by an automorphism γ of the corresponding untwisted group which induces a non-trivial symmetry ρ on the Dynkin diagram (see [Ca$_1$, Chs. 12,13]). Such symmetries exist for A_ℓ, B_2, D_ℓ, E_6, F_4, G_2 (with ρ of order 2) and for D_4 (with ρ of order 3). For twisted groups, parabolic subgroups P_J exsits only when J is a ρ-invariant subset of the Dynkin diagram. The Levi factor of P_J is obtained by taking the fixed points of the automorphism γ on the

Levi factor L_J for the corresponding untwisted group. For example, the Levi factors of the maximal parabolic subgroups of ${}^2E_6(q)$ are of types ${}^2A_5(q)$, ${}^2D_4(q)$, $A_1(q) \times A_2(q^2)$ and $A_1(q^2) \times A_2(q)$. We shall require two basic results concerning parabolic subgroups. The first is a celebrated result of Borel and Tits ([B-T, 3.12] — see also [B-W]).

Proposition 5.2.10. *Let $L \in Lie(p)$, and let X be a p-local subgroup of L. Then X lies in a parabolic subgroup of L.*

In fact, [B-T] does much more than this — for instance, their arguments provide a 'canonical' parabolic subgroup containing X.

The second result concerning parabolic subgroups is an argument taken from [L-S$_2$], we shall call the 'parabolic argument'. This consists of the following result, together with its proof.

Proposition 5.2.11. (Parabolic argument) *Let $L, M \in Lie(p)$ have Lie ranks ℓ, m, respectively, and suppose that $L \preceq M$. Let X be a Levi factor of L with $O^{p'}(X)$ of Lie rank $\ell - u$, where $u \leq \min\{\ell, m\}$. Then there is a Levi factor Y of M such that $O^{p'}(X) \preceq O^{p'}(Y)$ and $O^{p'}(Y)$ has Lie rank $m - u$.*

We sketch a proof of this — details can be found in [L-S$_2$, p.244]. First let $L_1 = O^{p'}(X)$ have Lie rank $\ell - 1$. Now L_1 normalizes some non-trivial p-subgroup of L (for example, the unipotent radical of a parabolic subgroup corresponding to X), and hence by Proposition 5.2.10, $L_1 \preceq M_1 = O^{p'}(Y)$ for some Levi factor Y of M. Clearly the Lie rank of M_1 is at most $m - 1$. Applying the same reasoning to the pair L_1, M_1, we see that L_1 has a subgroup L_2 of Lie rank $\ell - 2$ which embeds in a subgroup M_2 of M_1 of Lie rank at most $m - 2$. Continuing in this way, we obtain the Lemma.

Before stating the next result, which is a consequence of the parabolic argument, we need to extend the definition of Lie rank to include various soluble groups of Lie type (the Lie rank was defined only for simple groups in Tables 5.1.A,B). Let $\mathcal{E}$ be the collection of groups $\mathcal{E}_o \cup \mathcal{E}_1 \cup \mathcal{E}_2$, where

$$\begin{aligned} \mathcal{E}_o &= \{P\Omega_2^{\pm}(q) \mid \text{any } q\} \\ \mathcal{E}_1 &= \{L_2(q), PSp_2(q) \mid q \leq 3\} \cup \{\Omega_3(3), U_3(2)\} \\ \mathcal{E}_2 &= \{P\Omega_4^{+}(q) \mid \text{any } q\}. \end{aligned} \tag{5.2.3}$$

For $X \in \mathcal{E}_i$ define the Lie rank of X to be i. The parabolic argument gives

Proposition 5.2.12.

(i) [L-S$_2$, Fact 1, p.244] *Let $L, M \in Lie(p) \cup \mathcal{E}$ have Lie ranks ℓ, m, respectively, and suppose that $L \preceq M$. Then $\ell \leq m$.*

(ii) *Let $L, L_i \in Lie(p) \cup \mathcal{E}$ for $1 \leq i \leq t$, and suppose that a covering group of $\prod_{i=1}^{t} L_i$ embeds in L. Let ℓ, ℓ_i be the Lie ranks of L and L_i, respectively. Then $\sum_{i=1}^{t} \ell_i \leq \ell$.*

Finally, we state a result taken from [G-L, 10.1,10.2] concerning the r-rank $m_r(L)$, where $r \neq p$.

Proposition 5.2.13. *Assume that $L \in Lie(p)$ and that L has untwisted Lie rank ℓ_1 and Lie rank ℓ_2. Let r be an odd prime distinct from p.*

(i) $m_r(L) \leq \ell_1$.

(ii) *If L is exceptional and not of type ${}^2E_6$, then $m_r(L) \leq \ell_2$.*

Primitive prime divisors

Primitive prime divisors were first discovered by Zsigmondy [Zs] (see also [Dic$_3$]) at the end of the 19th century. Zsigmondy's theorem may be stated as follows.

Theorem 5.2.14 (Zsigmondy [Zs]). *Let q and n be integers with $q \geq 2$ and $n \geq 3$. Provided $(q,n) \neq (2,6)$, there is a prime s such that $s \mid q^n - 1$ but s does not divide $q^i - 1$ for $i < n$.*

We call such a prime s a *primitive prime divisor of* $q^n - 1$, and denote such a prime by q_n. Of course, there may be more than one primitive prime divisor of $q^n - 1$. For example, the primitive prime divisors of $17^6 - 1$ are 7, 13 and 307, and thus the symbol 17_6 denotes any one of these three primes.

When n is odd and $(q,n) \neq (2,3)$, then there is a primitive prime divisor q_{2n} of $q^{2n} - 1$, and we sometimes write $q_{-n} = q_{2n}$ in this situation.

The following properties of q_n are rather easy to prove.

Proposition 5.2.15. *Assume that $q \geq 2$, $n \geq 3$ and $(q,n) \neq (2,6)$.*

(i) *If $q_n \mid q^m - 1$ then $n \mid m$.*

(ii) $q_n \equiv 1 \pmod{n}$.

Let p be a prime, r a prime distinct from p, and m an integer which is not a power of p. Also let X be a group which is not a p-group. Then we define

$$\begin{aligned} \zeta_p(r) &= \min\{z \mid z \geq 1 \text{ and } p^z \equiv 1 \pmod{r}\} \\ \zeta_p(m) &= \max\{\zeta_p(r) \mid r \text{ prime}, r \neq p,\ r \mid m\} \qquad (5.2.4) \\ \zeta_p(X) &= \zeta_p(|X|). \end{aligned}$$

Using Zsigmondy's Theorem 5.2.14 and Tables 5.1.A and 5.1.B, we may derive the following result.

Proposition 5.2.16. *Assume that L is a simple group of Lie type over $\mathbf{F}_q$, where $q = p^f$. Then $\zeta_p(L)$ is given in Table 5.2.C.*

Table 5.2.C

L	$\zeta_p(L)$	exceptions
$L_n(p^f)$	nf	$\zeta_p(L_2(p)) = 1$ if $p+1 = 2^k$ $\zeta_2(L_6(2)) = 5$
$PSp_n(p^f)$, n even, $n \geq 4$ $P\Omega_n^-(p^f)$, n even, $n \geq 8$		$\zeta_2(Sp_6(2)) = 4$
$U_n(p^f)$, $n \geq 3$	$2fn$, n odd $2f(n-1)$, n even	$\zeta_2(U_4(2)) = 4$
$P\Omega_n^+(p^f)$, n even, $n \geq 8$	$f(n-2)$	$\zeta_2(\Omega_8^+(2)) = 4$
$\Omega_n(p^f)$, qn odd, $n \geq 7$	$f(n-1)$	
${}^2B_2(2^f)$	$4f$	
$G_2(p^f)$ $(p^f \geq 3)$, ${}^2G_2(3^f)$ $(3^f \geq 27)$	$6f$	
$F_4(p^f)$, ${}^2F_4(2^f)$, $E_6(p^f)$, ${}^3D_4(p^f)$	$12f$	
${}^2E_6(p^f)$	$18f$	
$E_7(p^f)$	$18f$	
$E_8(p^f)$	$30f$	

Finally, we record the following result.

Proposition 5.2.17. *Let L be a simple group and p a prime.*

(i) *If $L \in Lie(p)$, then $\zeta_p(L) = \zeta_p(Aut(L))$.*

(ii) *If L is alternating or sporadic, then $\zeta_p(L) = \zeta_p(Aut(L))$.*

Proof. This follows from the values of $|Out(L)|$ given in Tables 5.1.A,B,C, together with Proposition 5.2.15.ii. ■

§5.3 Representations of simple groups

In our proofs in Chapters 7,8 we shall require a fair amount information about representations of simple groups. A good deal of this information is concerned with lower bounds for the degrees of non-trivial irreducible modular projective representations.

Let G be a group and p a prime. Of course, a p-modular representation of G is simply a homomorphism from G to $GL_n(\mathbf{F})$ for some n and some field $\mathbf{F}$ of characteristic p. Similarly, a *projective p-modular representation* of G is a homomorphism from G to $PGL_n(\mathbf{F})$. Obviously G embeds in $PGL_n(\mathbf{F})$ for some n and $\mathbf{F}$ if and only if G has a faithful projective p-modular representation of degree n, and often we will be concerned with determining the smallest values of n for which this occurs. We define

$$\begin{aligned} R_{\mathbf{F}}(G) &= \min\{n \mid G \preceq PGL_n(\mathbf{F})\}, \\ R_p(G) &= \min\{R_{\mathbf{F}}(G) \mid \mathbf{F} \text{ a field of characteristic } p\}, \\ R_{p'}(G) &= \min\{R_s(G) \mid s \text{ prime}, s \neq p\}, \\ R(G) &= \min\{R_p(G) \mid \text{all primes } p\}. \end{aligned} \tag{5.3.1}$$

Observe that if $\overline{\mathbf{F}}_p$ denotes the algebraic closure of $\mathbf{F}_p$, then

$$R_p(G) = \min\{n \mid G \preceq PGL_n(\overline{\mathbf{F}}_p)\}. \tag{5.3.2}$$

This first Proposition relates minimal projective representations of a simple group to minimal representations of its covering group.

Proposition 5.3.1. *Assume that L is a simple group and that L_1 is the full covering group of L.*

(i) $R_{\mathbf{F}}(L) \geq \min\{n \mid L_1$ *has a non-trivial representation in* $GL_n(\mathbf{F})\}$.

(ii) $R_p(L) = \min\{n \mid L_1$ *has a non-trivial representation in* $GL_n(\overline{\mathbf{F}}_p)\}$.

Proof. Choose n minimal such that $L \leq PGL_n(\mathbf{F})$. Let $\widetilde{L}$ be a minimal preimage of L in $GL_n(\mathbf{F})$. Then $\widetilde{L}$ is a covering group of L, and so there is a non-trivial homomorphism from L_1 to $GL_n(\mathbf{F})$, proving (i). The same argument also gives an inequality in (ii). To establish the reverse inequality, choose n minimal such that there is a non-trivial homomorphism from L_1 to $GL_n(\overline{\mathbf{F}}_p)$. The minimality of n ensures that this representation is absolutely irreducible, and so $Z(L_1)$ acts as scalars by Schur's Lemma. Hence L embeds in $PGL_n(\overline{\mathbf{F}}_p)$, as required. ∎

Usually we will apply the function R_p to perfect groups, and the following Proposition will help us in certain situations.

Proposition 5.3.2. *Assume that G is a perfect group and that p is prime. Further assume that $G/Z(G)$ has a unique minimal normal subgroup $N/Z(G)$ and that $N/Z(G)$ is not a p-group.*

(i) $R_p(G) \geq R_p(G/Z(G))$.

(ii) *If* $Z(G) \neq 1$, *then* $R_p(G) > R_p(G/Z(G))$.

Proof. Let $G \leq PGL(V, \mathbf{F})$, with $\mathbf{F} = \overline{\mathbf{F}}_p$ and $\dim(V)$ minimal. Let $\widetilde{G}$ be the minimal preimage of G in $GL(V, \mathbf{F})$. For any subgroup $X \leq G$, let $\widetilde{X}$ denote the full preimage of X in $\widetilde{G}$. Observe that $\widetilde{G}$ is perfect since G is. Put $Z = Z(G)$, and note that $[\widetilde{Z}, \widetilde{G}] \leq \mathbf{F}^*$. Therefore $[\widetilde{Z}, \widetilde{G}, \widetilde{G}] = 1$, and similarly $[\widetilde{G}, \widetilde{Z}, \widetilde{G}] = 1$. So as $\widetilde{G}$ is perfect, the Three Subgroup Lemma implies that $\widetilde{Z} \leq Z(\widetilde{G})$. The reverse inclusion is clear, and hence $\widetilde{Z} = Z(\widetilde{G})$. Now write $0 = V_o < V_1 < \cdots < V_k = V$ for a $\widetilde{G}$-composition series, and let $C = \bigcap_i C_{\widetilde{G}}(V_i/V_{i-1})$. Thus C is a p-group and $C \trianglelefteq \widetilde{G}$, and so our assumption on G/Z implies $C \leq \widetilde{Z}$. Suppose for a contradiction that $\widetilde{N}$ acts as scalars on V_i/V_{i-1} for all $i = 1, \ldots, k$. Then $[\widetilde{G}, \widetilde{N}] \leq C \leq \widetilde{Z} = Z(\widetilde{G})$. But then the Three Subgroup Lemma shows

$\widetilde{N} \leq Z(\widetilde{G})$, a contradiction. Thus there exists i such that $\widetilde{N}$ does not act as scalars on $W = V_i/V_{i-1}$. Since $\widetilde{G}$ is absolutely irreducible on W, it follows that $\widetilde{Z}$ acts as scalars on W, and so the fact that N/Z is the unique minimal normal subgroup of $G/Z(G)$ ensures that the subgroup of $\widetilde{G}$ which acts as scalars on W is precisely $\widetilde{Z}$. Thus $\widetilde{G}/\widetilde{Z}$ embeds in $PGL(W, \mathbf{F})$, and part (i) now follows. To prove (ii), we assume that $Z \neq 1$. Thus $\widetilde{Z} \not\leq \mathbf{F}^*$, which means $Z(\widetilde{G})$ does not act as scalars. Therefore $\widetilde{G}$ cannot be irreducible, whence $k > 1$. Therefore $\dim(V) > \dim(W)$, and hence (ii) is proved. ∎

As a direct consequence of Lemma 5.3.2 we obtain

Corollary 5.3.3. *If L is quasisimple, then $R_p(L) \geq R_p(L/Z(L))$ for all primes p.*

An irreducible representation is of course one which leaves no non-zero proper subspace invariant. And since $PGL_n(\mathbf{F})$ also acts on the set of subspaces of the associated n-dimensional vector space, we can make exactly the same definition for a projective representation. At times it will be useful to draw the distinction between reducible and irreducible projective representations, and so we make another definition:

$$\begin{aligned} R^i_{\mathbf{F}}(G) &= \min\{n \mid G \text{ embeds irreducibly in } PGL_n(\mathbf{F})\} \\ R^i_p(G) &= \min\{R^i_{\mathbf{F}}(G) \mid \mathbf{F} \text{ a field of characteristic } p\} \\ &= \min\{n \mid G \text{ embeds irreducibly in } PGL_n(\overline{\mathbf{F}}_p)\} \qquad (5.3.3) \\ R^i_{p'}(G) &= \min\{R^i_s(G) \mid s \text{ prime, } s \neq p\} \\ R^i(G) &= \min\{R^i_p(G) \mid \text{all primes } p\}. \end{aligned}$$

Alternating groups

Let $n \geq 5$, let p be prime, and let S_n act on the permutation module $\mathbf{F}_p^n$ by permuting the coordinates naturally. Define submodules U, W of $\mathbf{F}_p^n$ by

$$U = \{(a_1, \ldots, a_n) \,\Big|\, \sum_{i=1}^n a_i = 0\}, \quad W = \{(a, \ldots, a) \mid a \in \mathbf{F}_p\}. \qquad (5.3.4)$$

The $(n-1)$-dimensional space U is called the *deleted permutation module* for A_n and S_n. A direct calculation shows that

Lemma 5.3.4. *The subspaces U and W are the only non-zero proper A_n-invariant submodules of $\mathbf{F}_p^n$.*

Now define

$$M = U/(U \cap W). \qquad (5.3.5)$$

Clearly W is contained in U if and only if $p \mid n$, and so

$$\dim(M) = \begin{cases} n-1 & \text{if } p \text{ does not divide } n \\ n-2 & \text{if } p \mid n. \end{cases} \qquad (5.3.6)$$

We call M the ***fully deleted permutation module*** for A_n over $\mathbf{F}_p$. For any field extension $\mathbf{F}$ of $\mathbf{F}_p$, the fully deleted permutation module for A_n over $\mathbf{F}$ is defined as $M \otimes \mathbf{F}$. Clearly M is irreducible by Lemma 5.3.4, and it is elementary to check that $C_{GL(M)}(A_n)$ consists only of scalars. Thus by Lemma 2.10.1, M is absolutely irreducible, and so the fully deleted permutation module over $\mathbf{F}$ is always irreducible. As the following Proposition indicates, the fully deleted permutation module is usually the smallest A_n-module.

Proposition 5.3.5 ([Dic$_2$,Wa$_1$,Wa$_2$] and [Ja$_1$, Th. 6]). *Assume that $n \geq 10$, and let V be a non-trivial irreducible A_n-module in characteristic p. Suppose further that $\dim(V) \leq n$. Then V is isomorphic to the fully deleted permutation module.*

To determine $R(A_n)$, we must obtain some information about the representations of the covering groups of A_n (see Proposition 5.3.1). Recall from Theorem 5.1.4.i that $M(A_n) \cong \mathbf{Z}_2$ if $n \geq 8$.

Proposition 5.3.6 [Wa$_3$]. *Let $n \geq 9$ and let s be the number of non-zero terms in the 2-adic expansion of n (so that $n = 2^{w_1} + \cdots + 2^{w_s}$ with $w_1 > \cdots > w_s \geq 0$). Then the degree of any faithful representation of the full covering group of S_n or A_n in odd characteristic is divisible by $2^{[(n-s)/2]}$ or $2^{[(n-s-1)/2]}$, respectively.*

Using Propositions 5.3.5 and 5.3.6, along with some well-known information concerning the modular representations of A_n for small n, we obtain Proposition 5.3.7. Part (ii) of Proposition 5.3.7 is true for $n = 8$ by Theorem 5.3.9 (recall $A_8 \cong L_4(2)$). For $n = 5, 6, 7$, Proposition 5.3.7.ii is easily deduced from calculations with ordinary character tables, the theory of Brauer trees, and restrictions to smaller alternating groups. We provide no details here; the complete set of modular character tables for these groups can be found in [Pa].

Proposition 5.3.7. *For $n \geq 5$ we have $R(A_n) \geq n - 4$. More specifically:*
(i) *for $n \geq 9$ we have $R(A_n) = n - 2$;*
(ii) *for $5 \leq n \leq 8$, the values of $R_p(A_n)$ are as follows:*

n	$R_2(A_n)$	$R_3(A_n)$	$R_5(A_n)$	$R_p(A_n)$, $p \geq 7$
5	2	2	2	2
6	3	2	3	3
7	4	4	3	4
8	4	7	7	7

We now discuss various forms on the fully deleted permutation module M over $\mathbf{F}_p$ which are preserved by A_n. Clearly A_n preserves the natural symmetric bilinear form $\mathbf{f}: U \times U \to \mathbf{F}_p$ given by

$$\mathbf{f}((a_1, \ldots, a_n), (b_1, \ldots, b_n)) = \sum_{i=1}^{n} a_i b_i.$$

Evidently **f** induces a symmetric bilinear form on the fully deleted permutation module M, which we again call **f**, and so we obtain the embedding $A_n \leq \Omega(M, \mathbf{F}_p, \mathbf{f})$. Now let $p = 2$ and define $Q : U \to \mathbf{F}_2$ by

$$Q(a_1, \ldots, a_n) = \begin{cases} 1 & \text{if the number of non-zero } a_i \text{ is congruent to 2 modulo 4} \\ 0 & \text{if the number of non-zero } a_i \text{ is congruent to 0 modulo 4.} \end{cases} \tag{5.3.7}$$

Then Q is an A_n-invariant quadratic form on U with associated bilinear form **f**. Provided $n \not\equiv 2 \pmod 4$, Q induces a quadratic form on M, again called Q, and this yields an embedding $A_n \leq \Omega(M, \mathbf{F}_2, Q)$. Thus it may be checked that the representation of A_n on its fully deleted permutation module in characteristic 2 gives rise to the following inclusions:

$$A_{2m+2} \leq \begin{cases} \Omega^+_{2m}(2) & \text{if } m \equiv 3 \pmod 4 \\ \Omega^-_{2m}(2) & \text{if } m \equiv 1 \pmod 4 \\ Sp_{2m}(2) & \text{if } m \text{ is even} \end{cases} \tag{5.3.8}$$

$$A_{2m+1} \leq \begin{cases} \Omega^+_{2m}(2) & \text{if } m \equiv 0 \pmod 4 \\ \Omega^-_{2m}(2) & \text{if } m \equiv 2 \pmod 4 \\ \Omega^\mp_{2m}(2) & \text{if } m \equiv \pm 1 \pmod 4. \end{cases} \tag{5.3.9}$$

Of course, if $m \equiv \pm 1 \pmod 4$, then $A_{2m+1} < A_{2m+2} \leq \Omega^\mp_{2m}(2)$. Furthermore, when m is even, A_{2m+2} fixes no quadratic form on its minimal module. This last fact is easy to prove, and is left as an exercise for the reader.

Sporadic groups

The following result is taken from [L-P-S$_3$, Theorem 2.3.2]. In many cases, the result is proved by considering restrictions of representations to suitably chosen subgroups.

Proposition 5.3.8. *Let L be a sporadic simple group. Then a lower bound for $R(L)$ is given as follows.*

L	M_{11}	M_{12}	M_{22}	M_{23}	M_{24}	J_1	J_2	J_3	J_4	HS	McL	He	Ru
$R(L) \geq$	5	6	6	11	11	7	6	9	110	20	21	18	28

L	Suz	$O'N$	Co_1	Co_2	Co_3	Fi_{22}	Fi_{23}	Fi'_{24}	HN	Ly	Th	BM	M
$R(L) \geq$	12	31	24	22	22	27	234	702	56	110	48	234	729

Groups of Lie type: representations in coprime characteristic

We now turn to modular representations of simple groups in $Lie(p)$. Not surprisingly, the representation theory in characteristic p is substantially different from that in characteristic coprime to p, and we divide our discussions accordingly. Here we are concerned with the representations of the groups in $Lie(p)$ in characteristic coprime to p, whereas § 5.4 is devoted to representations in characteristic p. The basic result which we employ is the main theorem of [La-Se], which asserts that in general, $R_{p'}(L)$ is much larger than $R_p(L)$ for any $L \in Lie(p)$. However, there is some atypical behaviour for low n and q, as described in an earlier part of this chapter.

Table 5.3.A

L	$e(L)$	exceptions
$L_2(q)$	$(q-1)/(2,q-1)$	$e(L_2(4))=2$, $e(L_2(9))=3$
$L_n(q)$, $n\geq 3$	$q^{n-1}-1$	$e(L_3(2))=2$, $e(L_3(4))=4$
$PSp_{2m}(q)$, $m\geq 2$	$\frac{1}{2}(q^m-1)$, q odd $\frac{1}{2}q^{m-1}(q^{m-1}-1)(q-1)$, q even	$e(Sp_4(2)')=2$ $e(PSp_6(2))=7$
$U_n(q)$, $n\geq 3$	$q(q^{n-1}-1)/(q+1)$, n odd $(q^n-1)/(q+1)$, n even	$e(U_4(2))=4$ $e(U_4(3))=6$
$P\Omega^+_{2m}(q)$, $m\geq 4$	$(q^{m-1}-1)(q^{m-2}+1)$, $q\neq 2,3,5$ $q^{m-2}(q^{m-1}-1)$, $q=2,3,5$	$e(\Omega_8^+(2))=8$
$P\Omega^-_{2m}(q)$, $m\geq 4$	$(q^{m-1}+1)(q^{m-2}-1)$	
$\Omega_{2m+1}(q)$ $m\geq 3$, q odd	$q^{2(m-1)}-1$, $q>5$ $q^{m-1}(q^{m-1}-1)$, $q=3,5$	$e(\Omega_7(3))=27$
$E_6(q)$	$q^9(q^2-1)$	
$E_7(q)$	$q^{15}(q^2-1)$	
$E_8(q)$	$q^{27}(q^2-1)$	
$F_4(q)$	$q^6(q^2-1)$, q odd * $\frac{1}{2}q^7(q^3-1)(q-1)$, q even	 $e(F_4(2))\geq 44$
${}^2E_6(q)$	$q^9(q^2-1)$ *	
$G_2(q)$	$q(q^2-1)$	$e(G_2(3))=14$ $e(G_2(4))=12$ *
${}^3D_4(q)$	$q^3(q^2-1)$	
${}^2F_4(q)$	$q^4\sqrt{q/2}(q-1)$	
$Sz(q)$	$\sqrt{q/2}(q-1)$	$e(Sz(8))=8$
${}^2G_2(q)$	$q(q-1)$	

Theorem 5.3.9. *Assume that L is a simple group in $Lie(p)$. Then $R_{p'}(L)\geq e(L)$, where $e(L)$ is as in Table 5.3.A.*

Remarks. There are actually three slight errors in the main theorem of [La-Se], and we have presented a corrected version; the corrections are indicated with the symbol *. The bound 16 given for $R_{2'}({}^2F_4(2))$ also holds for the Tits simple group ${}^2F_4(2)'$. In fact, it follows from the theorems in [Hi] that $R_{2'}({}^2F_4(2)')=26$.

As a convenience in later arguments, we draw a few consequences of Theorem 5.3.9.

Corollary 5.3.10. *Let L be a classical simple group of dimension d over $\mathbf{F}_{p^e}$, with d chosen minimally (see the discussion before Theorem 5.1.3).*

(i) *If $d\geq 9$, then $R_{p'}(L)>d^2$.*

(ii) *If $d\geq 8$, then $R_{p'}(L)>d^2$, except for $R_{2'}(\Omega_8^+(2))=8$, $R_{2'}(\Omega_8^-(2))\geq 27$, $R_{2'}(Sp_8(2))\geq 28$ and $R_{3'}(PSp_8(3))\geq 40$.*

(iii) *If $d\geq 7$, then $R_{p'}(L)\geq 28$, except for $R_{2'}(\Omega_8^+(2))=8$, $R_{3'}(\Omega_7(3))\geq 27$ and $R_{2'}(\Omega_8^-(2))\geq 27$.*

(iv) *If $d\geq 6$, then $R_{p'}(L)\geq\frac{1}{2}(p^{ed/2}-1)$.*

Corollary 5.3.11. *Assume that $\Omega = \Omega(V, \mathbf{F}, \kappa)$ is a classical group of dimension d in characteristic p, and that $P\Omega'$ is non-abelian simple.*

(i) *If $R_{p'}(P\Omega') = d$, then $\Omega \approx SL_2^{\pm}(4)$, $SL_2^{\pm}(5)$, $Sp_2(4)$, $Sp_2(5)$, $\Omega_3(7)$, $\Omega_3(9)$, $SU_4(2)$, $Sp_4(3)$, $\Omega_6^-(3)$ or $\Omega_8^+(2)$.*

(ii) *If $d \geq 4$ and $R_{p'}(P\Omega') \leq \frac{1}{2}d^2$, then $\Omega \approx \Omega_4^-(2)$, $\Omega_4^-(3)$, $Sp_4(2)$, $Sp_4(3)$, $SL_4^{\pm}(2)$, $SU_4(3)$, $\Omega_5(3)$, $\Omega_5(5)$, $SU_5(2)$, $\Omega_6^{\pm}(2)$, $\Omega_6^-(3)$, $\Omega_8^{\pm}(2)$ or $Sp_8(2)$.*

(iii) *$R_{p'}(P\Omega') \geq d$, apart from the following exceptions.*

Ω	$L_3(2)$	$\Omega_3(5)$	$\Omega_4^-(2)$	$Sp_4(2)$	$\Omega_4^-(3)$	$\Omega_5(3)$	$\Omega_6^-(2)$
$R_{p'}(P\Omega')$	2	2	2	2	3	4	4

§5.4 Groups of Lie type: representations in the natural characteristic

In this section we discuss some of the theory of representations in characteristic p of the groups in $Lie(p)$. The theory is rather advanced, and we present a brief survey, using [Ca$_2$,Hu$_1$,Hu$_2$,St$_2$,St$_3$] as our main references.

The representation theory of groups in $Lie(p)$ depends largely on the theory for the corresponding simple algebraic groups, so we begin with a discussion of these groups.

Simple algebraic groups

Let k be an algebraically closed field of characteristic p. By a *simple algebraic group* over k we mean a linear algebraic group over k which has no proper, closed, connected, normal subgroups. The simple algebraic groups over k were classifed by Chevalley [Ch$_1$], and they are the groups of types $A_\ell(k)$, $B_\ell(k)$, $C_\ell(k)$, $D_\ell(k)$, $E_\ell(k)$ ($\ell = 6,7,8$), $F_4(k)$ and $G_2(k)$. For each fixed type (e.g., $A_\ell(k)$ with ℓ fixed), there may be several pairwise non-isomorphic simple algebraic groups, which can usually be distinguished by their centers. For example, both the groups $SL_{\ell+1}(k)$ and $PSL_{\ell+1}(k)$ are of type $A_\ell(k)$, but the former has center of order $\ell+1$, while the latter has trivial centre. (Note, however, that when ℓ is even there are two non-isomorphic groups of type $D_\ell(k)$, both having a centre of order 2.) For each type, there is always a unique *simply connected* group whose centre Z is as large as possible, and an *adjoint* group with trivial centre. We summarize the relevant details in the following table.

Type	$\lvert Z\rvert$	simply connected group	adjoint group
A_ℓ	$\ell+1$	$SL_{\ell+1}(k)$	$PSL_{\ell+1}(k)$
B_ℓ, p odd	2	$Spin_{2\ell+1}(k)$	$\Omega_{2\ell+1}(k)$
C_ℓ	$(2, p-1)$	$Sp_{2\ell}(k)$	$PSp_{2\ell}(k)$
D_ℓ	$(4, p^2-1)$	$Spin_{2\ell}(k)$	$P\Omega_{2\ell}(k)$
E_6	$(3, p^2-1)$		
E_7	$(2, p-1)$		
E_8, F_4, G_2	1		

As described in [Ca$_2$, §1.17], the groups in $Lie(p)$ arise as fixed point groups of suitable *Frobenius* maps on simple algebraic groups. Let G be a simple algebraic group over k, and let $q = p^f$. Regard G as a subgroup of $GL_n(k)$ for some n. If the map σ_q sending $(a_{ij}) \mapsto (a_{ij}^q)$ maps G into itself, it is called a *standard* Frobenius map on G. Generally, a map $\sigma: G \to G$ is said to be a *Frobenius map* if some power of σ is a standard Frobenius map. By [St₂], any surjective homomorphism $\sigma: G \to G$, such that the fixed point group G_σ is finite, is a Frobenius map. The finite groups $O^{p'}(G_\sigma)$ which arise in this way are precisely the finite groups of Lie type in characteristic p. Untwisted groups arise from standard Frobenius maps, and twisted groups from non-standard ones. For example, let $G = SL_n(k)$ and define $\sigma_1 = \sigma_q$ and $\sigma_2:(a_{ij}) \mapsto ((a_{ij}^q)^t)^{-1}$ (so that $\sigma_2^2 = \sigma_{q^2}$). Then $G_{\sigma_1} = SL_n(q)$ and $G_{\sigma_2} = SU_n(q)$.

A *torus* of the simple algebraic group G over k is a subgroup which is isomorphic to a direct product of copies of k^*. Every torus lies in a maximal torus, and and all maximal tori are conjugate in G. Each maximal torus T is isomorphic to $(k^*)^\ell$, where ℓ is the Lie rank of G, and $C_G(T) = T$. Moreover, the group $W = W(G) = N_G(T)/T$ is finite and is called the *Weyl group* of G. For all this see [Hu₁, Chs.VIII,IX].

Modules and weights

Let L be simply connected of Lie type over $\mathbf{F}_q$, where $q = p^f$ as described above. Thus $L = G_\sigma$, where G is a simply connected, simple algebraic group over k, and σ is a suitable Frobenius map.

Fix a maximal torus T of G, and let $X = X(T)$ be the *character group of* T, that is, the set of algebraic group homomorphisms from T to k^*. If M is any finite-dimensional *rational* kG-module (that is, the corresponding map $G \to GL(M)$ is an algebraic group homomorphism), then

$$M = \bigoplus_{\mu \in X} M_\mu,$$

where for $\mu \in X$, we have

$$M_\mu = \{v \in M \mid vt = \mu(t)v \text{ for all } t \in T\}.$$

If $M_\mu \neq 0$, then μ is said to be a *weight* of M, and M_μ the *μ-weight space* of M. Clearly the Weyl group $W = N_G(T)/T$ acts on X and hence induces a group of permutations on the weights of M. Obviously X is an abelian group, and hence a $\mathbf{Z}$-module. Thus we may form the tensor product $E = \mathbf{R} \otimes_{\mathbf{Z}} X$, and the action of W on X yields an action of W on E. Choose a positive definite, bilinear, symmetric W-invariant $\mathbf{R}$-form $(\ ,\ )$ on E.

Assume here that $\mathcal{L}$ is the *adjoint module* for G, that is, $\mathcal{L}$ is the Lie algebra of G over k, with the natural G-action. The set Φ of weights of $\mathcal{L}$ is called the set of *roots* of G (see [Hu₁, 16.4]). Next, select a system Φ^+ of *positive* roots from Φ, with corresponding set $\Pi = \{\alpha_1, \ldots, \alpha_\ell\}$ of *fundamental* roots ([Hu₁, §27,]), giving Dynkin diagram as described in §5.2. Define a partial order on X by writing $\mu \leq \lambda$ if and only if $\lambda - \mu$ is a sum of positive roots. For $\alpha \in \Phi$, define $\alpha^* = 2\alpha/(\alpha, \alpha) \in E$, the *co-root* corresponding to

α. Then there is a basis $\{\lambda_1, \ldots, \lambda_\ell\}$ of E which is dual to $\{\alpha_1^*, \ldots, \alpha_\ell^*\}$, that is, with $(\lambda_i, \alpha_j^*) = \delta_{ij}$. The λ_i form a $\mathbf{Z}$-basis for X, and they are called the *fundamental dominant weights*. Now define

$$X^+ = \left\{ \sum_{i=1}^{\ell} c_i \lambda_i \mid c_i \in \mathbf{Z},\, c_i \geq 0 \right\},$$

$$X_q = \left\{ \sum_{i=1}^{\ell} c_i \lambda_i \mid c_i \in \mathbf{Z},\, 0 \leq c_i \leq q-1 \right\}.$$

As described in [Hu$_2$] (originally, [St$_3$]), for each $\lambda \in X^+$ there is a unique irreducible kG-module $M(\lambda)$ with highest weight λ (relative to the above partial order). The relationship between the kG-modules and the kL-modules is summarized in the following result, taken from [St$_2$, §13].

Theorem 5.4.1. *Suppose that L is either an untwisted group over $\mathbf{F}_q$, or a twisted group of type ${}^2A_\ell$, ${}^2D_\ell$, ${}^2E_6$ or ${}^3D_4$ over $\mathbf{F}_q$. Then for $\lambda \in X_q$, the modules $M(\lambda)$ remain irreducible and inequivalent upon restriction to L, and exhaust the irreducible kL-modules.*

There is an analogous result in [St$_2$] for groups of type ${}^2B_2$, ${}^2G_2$ and ${}^2F_4$ which we state here for completeness. Let L be one of these groups, over $\mathbf{F}_{p^{2a+1}}$, where p is 2 or 3. For a root α, set $q(\alpha) = p^a$ if α is long, $q(\alpha) = p^{a+1}$ if α is short. Define

$$X_q' = \left\{ \sum_i c_i \lambda_i \mid 0 \leq c_i \leq q(\alpha_i) - 1 \right\}.$$

Then Theorem 5.4.1 holds with X_q replaced by X_q'.

Remark. Theorem 5.4.1 has been stated only for simply connected groups. However, it also applies to all projective representations of the simple groups $\overline{L} = L/Z(L)$, apart from $Sp_4(2)'$. For if M is an irreducible projective module for $\overline{L}$ over k, then M is a module for $C/O_p(C)$, where C is the full covering group of $\overline{L}$; however, by Theorem 5.1.4, $C/O_p(C) = L$ (except when $L = Sp_4(2)$), so M is an irreducible kL-module, and hence is given by Theorem 5.4.1.

We now discuss the action of the automorphisms of L on the modules $M(\lambda)$. Let M be a kL-module affording a representation $\rho : L \to GL_n(k)$, and let ν be the automorphism of $GL_n(k)$ induced by the action of the field automorphism $t \mapsto t^p$ on matrix entries. For $r \geq 1$ we denote by $M^{(r)}$ the space M with L-action given by the representation $\nu^r \rho$ (writing maps on the left). If $M = M(\lambda)$, then $M^{(r)} = M(p^r \lambda)$. The group L possesses field automorphisms arising from the maps $x_\alpha(t) \mapsto x_\alpha(t^{p^r})$ ($\alpha \in \Phi$, $t \in k$) (see [Ca$_1$, Ch. 12]) or [St$_1$]).

When L is of type A_ℓ, D_ℓ, E_6 or D_4 over $\mathbf{F}_q$, L possesses a graph automorphism τ_o of order 2, 2, 2 or 3, respectively, which induces a symmetry τ on the Dynkin diagram (see [Ca$_1$, Ch. 12]). If L also has a field automorphism ϕ with the same order as τ_o and commuting with τ_o, then $C_L(\tau_o\phi)$ gives twisted groups of types ${}^2A_\ell$, ${}^2D_\ell$, ${}^2E_6$ and ${}^3D_4$ [Ca₁, Ch. 13]. We denote also by τ_o the restriction of τ_o to the corresponding twisted group. Of course, the symmetry τ induces a permutation on the set of weights for L.

For a kL-module M affording the representation ρ, and an automorphism γ of L, we denote by M^γ the space M with L-action given by the representation $\rho\gamma$. Thus M^γ is quasiequivalent to M (see §2.10).

Proposition 5.4.2.

(i) *Let ϕ be a field automorphism of L induced by the map $t \mapsto t^p$ on k. Then $M(\lambda)^{\phi^r} \cong M(\lambda)^{(r)}$ as kL-modules.*

(ii) *Let L be of type A_ℓ, D_ℓ, E_6 or D_4, with graph automorphism τ_o as above. Then $M(\lambda)^{\tau_o} \cong M(\tau(\lambda))$ as kL-modules.*

(iii) *Let L be of type ${}^2A_\ell$, ${}^2D_\ell$, ${}^2E_6$ or ${}^3D_4$ over $\mathbf{F}_q$, where $q = p^e$. Then as kL-modules, $M(\lambda)^{(e)} \cong M(\tau^{-1}(\lambda))$.*

Parts (i) and (ii) of this can easily be justified by considering the action of the automorphisms ϕ and τ_o, extended to G, on a suitable maximal torus. Part (iii) is also elementary; a proof can be found in [Lie₁, p. 437].

Duality

We continue with the notation of the previous section. In order to discuss duality among the modules $M(\lambda)$, we introduce the *longest element* w_o of the Weyl group W of G (see [Ca₁, p. 20]). The action of w_o on the root system is as follows [Bou, p.250-275]:

$$w_o = \begin{cases} -1 & \text{for types } B_\ell, C_\ell, D_\ell\ (\ell \text{ even}), G_2, F_4, E_7, E_8 \\ -\tau & \text{for types } A_\ell, D_\ell\ (\ell \text{ odd}), E_6. \end{cases}$$

Proposition 5.4.3 [Hu₁, 3.1.6]. *We have $M(\lambda)^* \cong M(-w_o(\lambda))$.*

In particular, for types B_ℓ, C_ℓ, D_ℓ (ℓ even), G_2, F_4, E_7 and E_8, all the kG-modules $M(\lambda)$ are self dual.

Fields of definition

We continue with the previous notation. Let H be any subgroup of $GL(V, \mathbf{F})$, where V and $\mathbf{F}$ are arbitrary, and recall from §2.10 what it means for H to be realized over a subfield $\mathbf{F}_o$ of $\mathbf{F}$. We extend this notion to a representation, as follows: if ρ is a representation of H to $GL(V, \mathbf{F})$, then we say ρ can be realized over $\mathbf{F}_o$ provided $\rho(H)$ can. Furthermore, if it is understood that V is a module for H, then we say that V can be realized over $\mathbf{F}_o$ provided the same holds of image of H in $GL(V, \mathbf{F})$. In our proofs in Chapter 8, it will be important for us to be able to determine the smallest fields over which the kL-modules $M(\lambda)$ are realized. We describe methods for doing this at the present juncture.

First we require a result on splitting fields for characteristic p representations for groups $L \in Lie(p)$, taken from [St₁, p. 241]

Proposition 5.4.4. *Let L be a group of Lie type over $\mathbf{F}_q$.*
(i) *If L is untwisted or of type ${}^2B_2$, ${}^2G_2$ or ${}^2F_4$, then $\mathbf{F}_q$ is a splitting field for L.*
(ii) *If L is of type ${}^2A_\ell$, ${}^2D_\ell$ or ${}^2E_6$, then $\mathbf{F}_{q^2}$ is a splitting field.*
(iii) *If L is of type ${}^3D_4$, then $\mathbf{F}_{q^3}$ is a splitting field.*

To prove a more detailed result on fields of definition we require Steinberg's twisted tensor product theorem (see [St₃] or [Bo]).

Theorem 5.4.5 (Steinberg's twisted tensor product theorem). *Let λ be a weight in X_q, where $q = p^e$, and write $\lambda = \mu_o + p\mu_1 + \cdots + p^{e-1}\mu_{e-1}$ with $\mu_i \in X_p$ for all i. Then*

$$M(\lambda) \cong M(\mu_o) \otimes M(\mu_1)^{(1)} \otimes \cdots \otimes M(\mu_{e-1})^{(e-1)}.$$

Proposition 5.4.6. *Let L be simply connected of Lie type over $\mathbf{F}_{p^e}$, and suppose that V is an absolutely irreducible $\mathbf{F}_{p^f}L$-module which is realized over no proper subfield of $\mathbf{F}_{p^f}$.*
(i) *If L is untwisted, then $f \mid e$ and there is an irreducible kL-module M such that*

$$V \otimes k \cong M \otimes M^{(f)} \otimes \cdots \otimes M^{(e-f)}.$$

In particular, $\dim(V) = \dim(M)^{e/f}$.
(ii) *If L is of type ${}^2A_\ell$, ${}^2D_\ell$ or ${}^2E_6$, then one of the following occurs.*
(a) *$f \mid e$, $V \cong V^{\tau_o}$ there is an irreducible kL-module M such that $M \cong M^{\tau_o}$ and*

$$V \otimes k \cong M \otimes M^{(f)} \otimes \cdots \otimes M^{(e-f)}.$$

In particular, $\dim(V) = \dim(M)^{e/f}$.
(b) *$f \mid 2e$ but f does not divide e, $V \not\cong V^{\tau_o}$, $V^{(f/2)} \cong V^{\tau_o}$, there is an irreducible kL-module M such that $M \not\cong M^{\tau_o}$ and*

$$V \otimes k \cong M \otimes (M^{\tau_o})^{(f/2)} \otimes M^{(f)} \otimes (M^{\tau_o})^{(3f/2)} \otimes \cdots \otimes (M^{\tau_o})^{(e-f)} \otimes M^{(e-(f/2))}.$$

In particular, $\dim(V) = \dim(M)^{2e/f}$.

Proof. Let $\overline{V} = V \otimes k$. By Theorems 5.4.1 and 5.4.5, there are weights $\mu_i \in X_p$ such that

$$\overline{V} \cong M(\mu_o) \otimes M(\mu_1)^{(1)} \otimes \cdots \otimes M(\mu_{e-1})^{(e-1)}.$$

Since V is realized over $\mathbf{F}_{p^f}$, we have $\overline{V} \cong \overline{V}^{(f)}$.

First consider case (i), in which L is untwisted. Here $\overline{V} \cong \overline{V}^{(e)}$ by Proposition 5.4.4, so $\overline{V} \cong \overline{V}^{((e,f))}$. As V is realized over no proper subfield of $\mathbf{F}_{p^f}$, we deduce that $(e,f) = f$, that is, $f \mid e$. Moreover, $\overline{V} \cong \overline{V}^{(f)}$ gives

$$M(\mu_o)\otimes M(\mu_1)^{(1)}\otimes\cdots\otimes M(\mu_{e-1})^{(e-1)} \cong M(\mu_o)^{(f)}\otimes M(\mu_1)^{(f+1)}\otimes\cdots\otimes M(\mu_{e-1})^{(f-1)}.$$

Hence for $0 \le i \le f-1$ we have

$$\mu_i = \mu_{f+i} = \cdots = \mu_{e-f+i}.$$

Consequently if we put $M = M(\mu_o)\otimes M(\mu_1)^{(1)}\otimes\cdots\otimes M(\mu_{f-1})^{(f-1)}$, then

$$\overline{V} \cong M \otimes M^{(f)} \otimes \cdots \otimes M^{(e-f)},$$

as required.

Now consider case (ii), so that L is of type ${}^2A_\ell$, ${}^2D_\ell$ or ${}^2E_6$. Here $\overline{V} \cong \overline{V}^{(2e)}$ by Proposition 5.4.4, so we see as above that $f \mid 2e$. Suppose first that $f \mid e$. Using Proposition 5.4.2.iii, we have

$$\overline{V} \cong M(\mu_o)^{(f)}\otimes\cdots\otimes M(\mu_{e-f-1})^{(e-1)}\otimes M(\tau(\mu_{e-f}))\otimes\cdots\otimes M(\tau(\mu_{e-1}))^{(f-1)}. \quad (5.4.1)$$

Consequently, for $0 \le i \le f-1$,

$$\mu_i = \mu_{f+i} = \cdots = \mu_{e-f+i} = \tau(\mu_i).$$

Put $M = M(\mu_o)\otimes M(\mu_1)^{(1)}\otimes\cdots\otimes M(\mu_{f-1})^{(f-1)}$. Then $M \cong M^{\tau_o}$ and $\overline{V} \cong M \otimes\cdots\otimes M^{(e-f)}$, as required.

Finally, suppose that f does not divide e (but $f \mid 2e$). Then $f = 2e/m$ with m odd. Here (5.4.1) holds again, giving, for $0 \le i \le \frac{f}{2} - 1$,

$$\mu_i = \mu_{f+i} = \cdots = \mu_{e-(f/2)+i} = \tau(\mu_{(f/2)+i}) = \tau(\mu_{(3f/2)+i}) = \cdots = \tau(\mu_{e-f+i}).$$

Put $M = M(\mu_o)\otimes M(\mu_1)^{(1)}\otimes\cdots\otimes M(\mu_{f/2-1})^{(f/2-1)}$. Then

$$\overline{V} \cong M \otimes (M^{\tau_o})^{(f/2)} \otimes \cdots \otimes (M^{\tau_o})^{(e-f)} \otimes M^{(e-f/2)}$$

and $\overline{V}^{(f/2)} \cong \overline{V}^{\tau_o} \not\cong V$. ∎

Remark 5.4.7. (a) There is a result entirely similar to Proposition 5.4.6.ii for the groups of type ${}^3D_4$, which is a little more cumbersome to state. Here, either $f \mid e$ and Proposition 5.4.6.ii.a holds, or $f \mid 3e$ and f does not divide e and $V \otimes k$ is a tensor product of $3e/f$ modules $M \otimes (M^{\tau_o})^{(f/3)} \otimes (M^{\tau_o^2})^{(2f/3)} \otimes M^{(f)} \otimes \cdots$.

(b) The situation of Proposition 5.4.6 is rather more complicated for groups L of type ${}^2B_2$, ${}^2G_2$ or ${}^2F_4$ over $\mathbf{F}_q$, where $q = p^{2a+1}$ and p is 2, 3 or 2, respectively. Here we state a result which follows directly from the proof of [Lie$_1$, Theorem 2.3]. If V is an absolutely irreducible $\mathbf{F}_{p^f}L$-module which is realized over no proper subfield of $\mathbf{F}_{p^f}$, then $f \mid 2a+1$ and $\dim(V) \ge R_p(L)^{(2a+1)/f}$. Note that $R_p(L)$ is 4, 7 and 26 for types ${}^2B_2$, ${}^2G_2$ and ${}^2F_4$, respectively (see Proposition 5.4.12, below).

The next result is taken from [Lie$_1$, Theorem 2.2].

Proposition 5.4.8. *Let L be of type ${}^2A_\ell$, ${}^2D_\ell$, ${}^2E_6$ or ${}^3D_4$, and suppose that V is an irreducible kL-module such that $V \cong V^{\tau_0}$. Then $\dim(V) \geq m$, where m is defined as follows:*

type of L	m
${}^2A_\ell$	$\ell(\ell+1)$, $\ell \geq 6$
	20, $\ell = 4, 5$
	6, $\ell = 2, 3$
${}^2D_\ell$	2ℓ
${}^2E_6$	72
${}^3D_4$	24

Spin modules

A critical role in our proofs in Chapter 8 will be played by the spin modules for the symplectic and orthogonal groups. Let $L = G_\sigma$ be a simply connected group of type $B_\ell(q)$ or $D_\ell(q)$. With the labelling of Dynkin diagrams as in (5.2.2), we refer to the modules $M(\lambda_\ell)$ for type B_ℓ, and $M(\lambda_{\ell-1})$ and $M(\lambda_\ell)$ for type D_ℓ, as *spin modules* for L. Their algebraic conjugates $M(\lambda_\ell)^{(r)}$ and $M(\lambda_{\ell-1})^{(r)}$ will also be called spin modules.

The basic theory of spin modules is expounded by Chevalley in [Ch$_2$]. We shall outline some of this theory, and use it to describe the weights of the spin modules relative to a specific maximal torus of G. We shall only need this information when q is even, so we suppose in the discussion below that $p = 2$. The description for p odd is similar, but slightly more complicated at certain points. Thus $L = G_\sigma = Sp_{2\ell}(q)$ or $\Omega^+_{2\ell}(q)$ with q even.

We first describe the spin module for $\Omega^+_{2\ell}(k')$, where k' is either $\mathbf{F}_q$ or the algebraically closed field k. Let V be the natural 2ℓ-dimensional module over k' for $\Omega_{2\ell}(k')$, with quadratic form Q, associated bilinear form $(\ ,\ )$ and standard basis $\beta = \{e_1, \ldots, e_\ell, f_1, \ldots, f_\ell\}$ as in Proposition 2.5.3.i. The *Clifford algebra* $C = C(Q)$ of (V, k', Q) is defined to be T/I, where T is the tensor algebra of V over k' (see [La, Ch. 16, §15]), and I is the ideal generated by all elements $x \otimes x - Q(x).1$ for $x \in V$. We identify V with the set of elements of degree 1 in C. Then $x^2 = Q(x)$ and $xy + yx = (x, y)$ for $x, y \in V$. Also $\dim(C) = 2^{2\ell}$ and C has a basis consisting of elements $b_S = \prod_{s \in S} s$ for subsets S of β. Let C_+ (respectively, C_-) be the subspace of C spanned by all b_S with $|S|$ even (respectively, all b_S with $|S|$ odd).

Fix $v \in V$ with $Q(v) \neq 0$, so that $v^{-1} = Q(v)^{-1}v$. Then for $u \in V$, we have $v^{-1}uv = -r_v(u)$, where r_v is the reflection defined in (2.5.7). The *Clifford group* X of C is defined to be

$$X = \{s \in C \mid s \text{ invertible}, s^{-1}us \in V \text{ for all } u \in V\}.$$

Clearly X has a natural representation on V (with $s \in X$ acting as $u \mapsto s^{-1}us$ for $u \in V$), and in fact in our situation with q even, we have

$$X \cong O^+_{2\ell}(k') \times (k')^*.$$

Let θ be the natural homomorphism $X \to (k')^*$. The *special Clifford group* is $X^+ = X \cap C_+ \cap \ker(\theta)$. We have

$$X^+ \cong \Omega^+_{2\ell}(k').$$

Define $E = \langle e_1, \ldots, e_\ell \rangle$, $F = \langle f_1, \ldots, f_\ell \rangle$, maximal totally singular subspaces of V. Let $C(E)$ be the subalgebra of C generated by E, and set $C_+(E) = C(E) \cap C_+$. As E is totally singular, $C(E)$ is isomorphic to the exterior algebra of E. By [Ch$_2$, II.2.2], if we put $f = f_1 \ldots f_\ell$, then

$$Cf = C(E)f.$$

Hence we can define a representation $\rho : C \to \mathrm{End}(C(E))$ by putting

$$\rho(c)u = u' \qquad (u, u' \in C(E),\ c \in C),$$

where

$$u'f = cuf.$$

For $c \in C_+$, $\rho(c)$ leaves $C_+(E)$ invariant, and thus gives a representation $\rho_+ : C_+ \to \mathrm{End}(C_+(E))$. The restriction of ρ_+ to the group X^+ is called the *spin representation* of X^+, and has dimension $2^{\ell-1}$; The space $C_+(E)$ with this action is called the *spin module* for X^+.

We now calculate the weights of the spin representation relative to a suitable maximal torus of X^+, taking $k' = k$. Pick $\mu_1, \ldots, \mu_\ell \in k^*$, and let $\mu \in k$ satisfy $\mu^2 = \mu_1 \ldots \mu_\ell$. Define

$$s_{\mu_1,\ldots,\mu_\ell} = \mu^{-1} \prod_{i=1}^{\ell} (e_i + f_i)(e_i + \mu_i f_i) \in C_+.$$

In the natural representation on V, the element $s_{\mu_1,\ldots,\mu_\ell}$ corresponds to the product $\prod_{i=1}^{\ell} r_{e_i+f_i} r_{e_i+\mu_i f_i}$ of reflections, which as a matrix relative to the basis β is

$$\mathrm{diag}(\mu_1, \ldots, \mu_\ell, \mu_1^{-1}, \ldots, \mu_\ell^{-1}) \in G.$$

Thus $T = \{s_{\mu_1,\ldots,\mu_\ell} \mid \mu_i \in k^*\}$ is a maximal torus of X^+.

For a subset $S = \{i_1, \ldots, i_t\}$ of $\{1, \ldots, \ell\}$ with $i_1 < \cdots < i_t$, set $e_S = e_{i_1} \ldots e_{i_t} \in C(E)$. We consider the action of T on the basis $\{e_S \mid |S| \text{ even}\}$ of $C_+(E)$. Calculation shows that

$$\rho(s_{\mu_1,\ldots,\mu_\ell}).(e_S) = \left(\prod_{i \in S} \mu_i \prod_{i \notin S} \mu_i^{-1} \right)^{1/2} e_S.$$

Hence the 1-spaces $\langle e_S \rangle$ are the weight spaces for T, and the weights are the $2^{\ell-1}$ maps $\chi_{\epsilon_1,\ldots,\epsilon_\ell} : T \to k^*$, where $\epsilon_i = \pm 1$, $\prod_{i=1}^{\ell} \epsilon_i = 1$ and

$$\chi_{\epsilon_1,\ldots,\epsilon_\ell}(s_{\mu_1,\ldots,\mu_\ell}) = \left(\prod_{i=1}^{\ell} \mu_i^{\epsilon_i} \right)^{1/2}.$$

Let $\Delta = \{\chi_{\epsilon_1,\ldots,\epsilon_\ell} \mid \epsilon_i = \pm 1, \prod_{i=1}^{\ell} \epsilon_i = 1\}$, the set of weights. The Weyl group $W = W(D_\ell) \cong 2^{\ell-1}.S_\ell$ acts transitively on Δ as follows. The normal subgroup $2^{\ell-1}$ induces all sign changes of even weight on $\epsilon_1, \ldots, \epsilon_\ell$; and the complement S_ℓ permutes $\epsilon_1, \ldots, \epsilon_\ell$ naturally. When ℓ is even, the action of W on Δ is imprimitive, with a block system $\mathcal{B}$ consisting of blocks $\{\chi_{\epsilon_1,\ldots,\epsilon_\ell}, \chi_{-\epsilon_1,\ldots,-\epsilon_\ell}\}$ of size 2. Here the kernel $W_{(\mathcal{B})}$ of the action of W on $\mathcal{B}$ is $\langle w_o \rangle$, of order 2, and the action $W^{\mathcal{B}}$ is the primitive affine group $2^{\ell-2}.S_\ell$. When ℓ is odd, W^Δ is primitive.

The stabilizer in X^+ of a nonsingular 1-space of V is a group $B_{\ell-1}(k) \cong Sp_{2\ell-2}(k)$ (see Proposition 4.1.7). The spin representation for this group $Sp_{2\ell-2}(k)$ is defined to be the restriction of ρ_+. Working relative to the maximal torus $\{s_{\mu_1,\ldots,\mu_{\ell-1},1} \mid \mu_i \in k^*\}$ of $B_{\ell-1}(k)$, we obtain the set Δ of weights as before. The Weyl group $W(B_{\ell-1}) \cong 2^{\ell-1}.S_{\ell-1}$ is a subgroup of $W(D_\ell)$ and is still transitive on Δ. The action $W(B_{\ell-1})^\Delta$ is the standard wreath action of $S_2 \wr S_{\ell-1}$ on the Cartesian product of $\ell - 1$ sets of size 2.

Finally, we define the spin module for $\Omega_{2\ell}^-(q)$ to be the restriction of the spin module for a group $\Omega_{2\ell}^+(q^2)$ containing $\Omega_{2\ell}^-(q)$.

As we remarked before, when q is odd, the above description of spin modules for $D_\ell(q)$ goes through with some minor changes. In particular, $X^+ \cong Spin_{2\ell}(q)$ here, and there are some slight sign problems in defining the elements $s_{\mu_1,\ldots,\mu_\ell}$.

Proposition 5.4.9.

(i) *The spin representation embeds*

$$B_\ell(q) \text{ in } \begin{cases} \Omega_{2^\ell}^+(q) & q \text{ even} \\ \Omega_{2^\ell}^+(q) & q \text{ odd}, \ell \equiv 0,3 \pmod 4 \\ Sp_{2^\ell}(q) & q \text{ odd}, \ell \equiv 1,2 \pmod 4. \end{cases}$$

The representation cannot be realised over a proper subfield of $\mathbf{F}_q$.

(ii) *The spin representation embeds*

$$D_\ell(q) \text{ in } \begin{cases} \Omega_{2^{\ell-1}}^+(q) & \ell \text{ even}, q \text{ even} \\ \Omega_{2^{\ell-1}}^+(q) & \ell \equiv 0 \pmod 4, q \text{ odd} \\ Sp_{2^{\ell-1}}(q) & \ell \equiv 2 \pmod 4, q \text{ odd}. \end{cases}$$

If ℓ is odd then the representation is not self-dual. The representation cannot be realized over a proper subfield of $\mathbf{F}_q$.

(iii) *The spin representation embeds*

$$^2D_\ell(q) \text{ in } \begin{cases} SU_{2^{\ell-1}}(q) & \ell \text{ odd} \\ \Omega_{2^{\ell-1}}^+(q^2) & \ell \text{ even}, q \text{ even} \\ \Omega_{2^{\ell-1}}^+(q^2) & \ell \equiv 0 \pmod 4, q \text{ odd} \\ Sp_{2^{\ell-1}}(q^2) & \ell \equiv 2 \pmod 4, q \text{ odd}. \end{cases}$$

The representation cannot be realized over a proper subfield of $\mathbf{F}_{q^2}$.

Proof. (i). When q is odd, this follows from [Ch₂, p.103]; and when q is even it follows from our proof of Proposition 8.1.2.i in Chapter 8. That the spin representation cannot be realized over a proper subfield of $\mathbf{F}_q$ is immediate from the fact that for $f < e$ (where $q = p^e$), $M(\lambda_\ell)^{(f)} \cong M(p^f\lambda_\ell) \ncong M(\lambda_\ell)$.

(ii). If ℓ is odd then by Proposition 5.4.3, $M(\lambda_{\ell-1})^* \cong M(\lambda_\ell)$, and hence the spin module for $D_\ell(q)$ is not self dual. When ℓ is even, the result follows from [Ch$_2$, p.78] for q odd, and from [Ch$_2$, pp.81-83] for q even. The last sentence in (ii) is proved as for (i).

(iii). Let $q = p^e$. By Proposition 5.4.2.iii, we have $M(\lambda_{\ell-1})^{(e)} \cong M(\lambda_\ell)$ as ${}^2D_\ell(q)$-modules. Hence if ℓ is odd then by Proposition 5.4.3, $M(\lambda_{\ell-1})^{(e)} \cong M(\lambda_{\ell-1})^*$, so the spin representation is unitary. Now let ℓ be even. As before, $M(\lambda_{\ell-1})$ is not realized over a proper subfield fo $\mathbf{F}_{q^2}$. The embeddings of ${}^2D_\ell(q)$ follow from part (ii). ∎

Lemma 5.4.10. *Let k be an algebraically closed field of characteristic 2, let G be a group of type B_ℓ ($\ell \geq 3$) or D_ℓ ($\ell \geq 5$) over k, and let $W = W(G)$ be the Weyl group of G. Then the kG-module $M = M(\lambda)$ is a spin module for G provided the following two conditions hold.*

(i)

$$\dim(M) = \begin{cases} 2^\ell & \text{if } G = B_\ell(k), \\ 2^{\ell-1} & \text{if } G = D_\ell(k), \end{cases}$$

(ii) $|\lambda^W| = \dim(M)$.

Proof. We give the proof for $G = D_\ell(k)$ with $\ell \geq 5$, and leave the similar and easier proof for $B_\ell(k)$ to the reader. Let $T = \{s_{\mu_1,\ldots,\mu_\ell} \mid \mu_i \in k^*\}$ be the maximal torus of G described in the discussion before Proposition 5.4.9 and for $1 \leq i \leq \ell$ let $e_i : T \to k^*$ be the element of $X(T)$ sending $s_{\mu_1,\ldots,\mu_\ell}$ to μ_i. Then, as summarized in [Bou, pp.256-7], the fundamental roots $\alpha_1, \ldots, \alpha_\ell$ and the fundamental dominant weights $\lambda_1, \ldots, \lambda_\ell$ are given by

$$\begin{aligned}
\alpha_i &= e_i - e_{i+1} \quad (1 \leq i \leq \ell - 1) \\
\alpha_\ell &= e_{\ell-1} + e_\ell \\
\lambda_i &= e_1 + \cdots + e_i \quad (1 \leq i \leq \ell - 2) \\
\lambda_{\ell-1} &= \tfrac{1}{2}(e_1 + \cdots + e_{\ell-1} - e_\ell) \\
\lambda_\ell &= \tfrac{1}{2}(e_1 + \cdots + e_{\ell-1} + e_\ell).
\end{aligned}$$

The normal subgroup $2^{\ell-1}$ of $W \cong 2^{\ell-1}.S_\ell$ induces all sign changes of even weight on $e_1, \ldots, e_\ell$, while the complement S_ℓ permutes $e_1, \ldots, e_\ell$ naturally.

Now suppose that $M = M(\lambda)$ satisfies $\dim(M) = 2^{\ell-1} = |\lambda^W|$. Write $\lambda = \sum_{i=1}^{\ell} a_i e_i$, and let $x = |\{i \mid a_i \neq 0\}|$. From the action of the complement S_ℓ we see that $|\lambda^W|$ is divisible by $\binom{\ell}{x}$, which must therefore be a power of 2. However [Ja$_2$, Lemma 22.3] shows that the exponent of the full power of 2 dividing $\binom{\ell}{x}$ is less than $\log_2(\ell) + 1$, and hence $\binom{\ell}{x} < 2\ell$. Consequently $x = 1$, ℓ or $\ell - 1$. If $x = 1$, then $|\lambda^W| = 2\ell < 2^{\ell-1}$, a contradiction; and if $x = \ell - 1$, then $|\lambda^W|$ is divisible by $2^{\ell-1}\ell$, which is also impossible. Hence $x = \ell$. Moreover, as $|\lambda^W| = 2^{\ell-1}$, all the coefficients a_i are equal up to sign. It follows that λ^W contains $a\lambda_{\ell-1}$ or $a\lambda_\ell$ for some positive integer a. As $\lambda \in X^+$, we have $\lambda = a\lambda_{\ell-1}$ or $a\lambda_\ell$. If the 2-adic expansion of a is $a = 2^{i_1} + \cdots + 2^{i_s}$, then by Steinberg's tensor product theorem, $M(\lambda) \cong M(\lambda_{\ell-\delta})^{(i_1)} \otimes \cdots \otimes M(\lambda_{\ell-\delta})^{(i_s)}$ with $\delta = 0$ or 1. As $\dim(M) = 2^{\ell-1}$, we deduce that $a = 2^i$. Consequently M is $M(\lambda_{\ell-1})^{(i)}$ or $M(\lambda_\ell)^{(i)}$; in other words, M is a spin module, as required. ■

Small modules in characteristic p

We shall need results describing modules in characteristic p of relatively small dimension for groups in *Lie*(p). These are taken from [Lie$_1$, Th. 1.1] and [Lie$_2$, Ths. 2.2,2.7,2.10]. Recall that k is an algebraically closed field of characteristic p.

Proposition 5.4.11. *Let L be one of the classical simple groups $L_d^{\pm}(q)$, $PSp_d(q)'$ or $P\Omega_d^{\epsilon}(q)$, and assume that M is a non-trivial irreducible projective kL-module satisfying $\dim_k(M) \leq \frac{1}{2}d(d+1)$, $\frac{1}{2}d^2$ or $\frac{1}{2}d^2 - 1$ in the respective cases. Let V be the natural projective kL-module of dimension d. Then M is quasiequivalent either to V or to one of those modules appearing in Table 5.4.A.*

Table 5.4.A

type of L	M	$\dim(M)$
$A_\ell^{\pm}(q)$, $d = \ell + 1$	$\Lambda^2 V$	$\frac{1}{2}d(d-1)$
	$S^2 V$	$\frac{1}{2}d(d+1)$
	$\Lambda^3 V$, $d = 6$	20
$B_\ell(q)$,	$\Lambda^2 V$	$\frac{1}{2}d(d-1)$
$\ell \geq 3$, q odd, $d = 2\ell + 1$	spin module, $\ell \leq 6$	2^ℓ
$C_\ell(q)$, $d = 2\ell$	section of $\Lambda^2 V$	$\frac{1}{2}d(d-1) - 1$ if $(p, \ell) = 1$
		$\frac{1}{2}d(d-1) - 2$ if $p \mid \ell$
	spin module, q even, $\ell \leq 6$	2^ℓ
	section of $\Lambda^3 V$, $\ell = 3$, q odd	14
	$L \cong Sp_4(2)'$	3
$D_\ell^{\pm}(q)$, $d = 2\ell$	section of $\Lambda^2 V$	$\frac{1}{2}d(d-1)$, q odd
		$\frac{1}{2}d(d-1) - (2, \ell)$, q even
	spin module, $\ell \leq 7$	$2^{\ell-1}$

The 3-dimensional module for $Sp_4(2)'$ occurs because of the exceptional multiplier — see Table 5.1.D. The corresponding representation is actually faithful on the 3-fold cover $3.Sp_4(2)'$.

The following result for the exceptional groups is a slight extension of [Lie$_2$, Th. 2.10].

Proposition 5.4.12. *Let $L \in Lie(p)$ be an exceptional group, not of type ${}^3D_4$ or ${}^2B_2$, and define e_L as in Table 5.4.B, below. If M is any irreducible projective kL-module with $1 < \dim(M) < e_L$, then M is quasiequivalent to either the minimal module $M(\mu)$ or the adjoint module $M(\lambda)$, as given in Table 5.4.B.*

Table 5.4.B					
type of L	e_L	μ	$\dim(M(\mu))$	λ	$\dim(M(\lambda))$
$E_6^{\pm}(q)$	82	λ_1, λ_6	27	λ_2	$78 - \delta_{p,3}$
$E_7(q)$	244	λ_7	56	λ_1	$133 - \delta_{p,2}$
$E_8(q)$	1024	λ_8	248	λ_8	248
$F_4^{\pm}(q)$	96	λ_4	$26 - \delta_{p,3}$	λ_1	$52 - 26\delta_{p,2}$
$G_2^{\pm}(q)$	18	λ_1	$7 - \delta_{p,2}$	λ_2	$14 - 7\delta_{p,3}$

The extension of [Lie_2, Th. 2.10] needed to establish above involves increasing e_L from 80 to 82 for $E_6^{\pm}(q)$, and from 192 to 244 for $E_7(q)$. The proof of this extension is entirely straightforward, following the lines of the proof of [Lie_2, Th. 2.10]. The values of $\dim(M(\mu))$ and $\dim(M(\lambda))$ given in Table 5.4.B are well-known, and can be found in [G-S], for example.

By a *minimal module* for a group $L \in Lie(p)$ we mean a non-trivial projective irreducible module in characteristic p of minimal dimension. Letting M be a minimal module, we give in Proposition 5.4.13 the dimension $\dim(M)$, and we also give some facts about forms on M which are preserved by L. Proposition 5.4.13 may be proved using the previous two results. In Table 5.4.C, the term 'proper' appearing in the right hand column for $E_6^{\pm}$ and E_7 means that the corresponding representation is a faithful representation of $L_1/O_p(L_1)$, where L_1 is the full covering group of L. The minimal modules for $E_6^{\pm}$ and E_7 will be discussed in further detail at the end of this section.

Proposition 5.4.13. *Let L be a simple group in $Lie(p)$. Then $R_2(Sp_4(2)') = 3$, and for the remaining groups, $R_p(L)$ is given in Table 5.4.C.*

Table 5.4.C		
type of L	$R_p(L)$	facts about minimal module
$A_\ell^{\pm}(q)$	$\ell+1$	natural module
$B_\ell(q)$, $\ell \geq 3$, q odd	$2\ell+1$	natural module
${}^2B_2(q)$, q even	4	symplectic
$C_\ell(q)$	2ℓ	natural module
$D_\ell^{\pm}(q)$, $\ell \geq 4$	2ℓ	natural module
${}^3D_4(q)$	8	orthogonal
$E_6^{\pm}(q)$	27	proper
$E_7(q)$	56	proper
$E_8(q)$	248	orthogonal
$F_4(q)$	$26 - \delta_{p,3}$	orthogonal
${}^2F_4(q)'$	26	orthogonal
$G_2(q)$	$7 - \delta_{p,2}$	orthogonal if q odd symplectic if q even
${}^2G_2(q)$	7	orthogonal

The following results are immediate from Theorem 5.3.9 and Proposition 5.4.13.

Corollary 5.4.14.

(i) *Assume that L is a classical group in Lie(p). Then $R_p(L) < R_{p'}(L)$, except when L is one of $L_2(4)$, $L_2(5)$, $L_3(2)$, $Sp_4(2)'$, $PSp_4(3)$, $U_4(2)$, $\Omega_3(5)$, $\Omega_4^-(2)$, $\Omega_5(3)$, $\Omega_6^-(2)$ and $\Omega_8^+(2)$.*

(ii) *If L is an exceptional group in Lie(p), then $R_p(L) < R_{p'}(L)$, except when L is one of ${}^2G_2(3)'$, $G_2(2)'$ and ${}^2F_4(2)'$. Moreover, $R_p(L) = R_{p'}(L)$ only when $L = {}^2F_4(2)'$.*

Proposition 5.4.15. *Assume that $L = C\ell_d(q)$ is non-abelian simple. Then $R_p(L) = d$, except in the following cases.*

L	$R_p(L)$
$P\Omega_6^\pm(q)$	4
$\Omega_5(q)$ (q odd)	4
$\Omega_4^-(q)$	2
$\Omega_3(q)$ (q odd)	2
$Sp_4(2)'$	3

In several of the arguments in Chapter 7 we will encounter a situation in which one classical group embeds in another, both in characteristic p. The following Proposition describes when this can occur.

Proposition 5.4.16.

(i) *Assume that $m \geq 3$, and let L be one of the simple classical groups $PSp_m(q^2)$, $U_m(q)$ or $P\Omega_m^\epsilon(q^2)$. Then a covering group of $L_m(q)$ does not embed in a covering group of L.*

(ii) *Assume that L, L_1 are classical simple groups over $\mathbf{F}_q$ of dimension m, m_1, respectively. Suppose that $L \precsim L_1$. Then $m \leq m_1$, except when (L, L_1) is one of $(P\Omega_6^\pm(q), L_4^\pm(q))$, $(P\Omega_6^\pm(q), L_5^\pm(q))$, $(\Omega_5(q), PSp_4(q))$, $(\Omega_5(q), L_4^\pm(q))$, $(\Omega_3(q), L_2^\pm(q))$ and $(\Omega_3(q), PSp_2(q))$.*

(iii) *Assume in (ii) that $L < L_1$, with $m = m_1 \geq 3$, and $L_1 \ncong L_m^\pm(q)$. Then (L, L_1) is either $(\Omega_m^\epsilon(q), Sp_m(q))$ with m and q even, or $(\Omega_4^-(q), PSp_4(q))$ with q odd.*

(iv) *For q odd, $\Omega_4^-(q) \not\precsim Sp_4(q)$.*

Proof. (i). Write $L = P\Omega(V, \mathbf{F}_{q^2}, \kappa)$, where V is the natural projective module for L over $\mathbf{F}_{q^2}$ and κ the non-degenerate form on V preserved by L. Let $\Omega = \Omega(V, \mathbf{F}_{q^2}, \kappa)$. Suppose (i) is false; then some covering group K of $L_m(q)$ embeds in Ω. By Proposition 5.4.4, as an $\mathbf{F}_{q^2}K$-module, V is realized over $\mathbf{F}_q$, and so there is an $\mathbf{F}_qK$-module W such that $V \cong W \otimes \mathbf{F}_{q^2}$. Clearly the form κ restricts to a non-degenerate form on W, which is preserved by K. But by Proposition 5.4.13, W is the natural m-dimensional module for K, and clearly K preserves no such form on W. This contradiction proves (i).

(ii). Suppose that $m > m_1$. Then $R_p(L) < m$, and so L is one of the groups occurring in Proposition 5.4.15. If $L = P\Omega_6^\pm(q)$, then L_1 is a classical group over $\mathbf{F}_q$ of dimension 4 or 5 containing $L_4^\pm(q)$, and so L_1 is $L_4^\pm(q)$ or $L_5^\pm(q)$. The other possibilities for L are handled in the same way.

(iii). Here L_1 is $PSp_m(q)$ or $P\Omega^\epsilon_m(q)$. Since $L^\pm_m(q)$ does not embed in L_1 by Lagrange's Theorem, L is also symplectic or orthogonal. Further, application of Lagrange's Theorem shows that the only possibility is $(L, L_1) = (P\Omega^\epsilon_m(q), PSp_m(q))$. Suppose here that q is odd and $m \geq 6$. Let V be the natural module of dimension m over $\mathbf{F}_q$ for $Sp_m(q)$. Using Proposition 5.4.11 we see that V must be a natural irreducible orthogonal module for the preimage of L in $Sp_m(q)$. Consequently L preserves both a non-degenerate symplectic and a non-degenerate symmetric form on V. This is easily seen to be impossible by the method of proof of Lemma 2.10.3. Thus either q is even, or q is odd and $m = 4$, giving (iii).

(iv). For a contradiction, assume that $L \approx \Omega^-_4(q) \cong L_2(q^2)$ embeds in $Sp_4(q)$. Let V be the natural 4-dimensional module for $Sp_4(q)$ over $\mathbf{F}_q$. If L is not absolutely irreducible on V, then using Lemma 2.10.1 we find that L embeds in $GL_2(q^2)$, which is not so. Therefore L acts absolutely irreducibly on V. Thus as $\mathbf{F}_{q^2}L$-modules, $V \otimes \mathbf{F}_{q^2} \cong W \otimes W^\psi$, where W is a natural 2-dimensional $\mathbf{F}_{q^2}SL_2(q^2)$ module, and ψ is the involutory automorphism of $\mathbf{F}_{q^2}$. Since $SL_2(q^2)$ preserves a non-degenerate symplectic form on W and on W^ψ, it follows that L preserves a non-degenerate symmetric form on $V \otimes \mathbf{F}_{q^2}$. This form restricts to a non-degenerate L-invariant symmetric form on V, and we reach the same contradiction arising in the previous paragraph. ∎

Minimal modules for $E^\pm_6$ and E_7

In some of the proofs in Chapter 8, we shall require additional information concerning the minimal modules for the groups $E^\epsilon_6(q)$ and $E_7(q)$. We present this information here.

Proposition 5.4.17. *The action of $E^\epsilon_6(q)$ on its minimal module $M(\lambda_6)$ embeds the simply connected group $E^\epsilon_6(q)$ faithfully in $SL^\epsilon_{27}(q)$. The representation is not self-dual, and is realized over no proper subfield of $\mathbf{F}_q$ if $\epsilon = +$, and of $\mathbf{F}_{q^2}$ if $\epsilon = -$.*

Proof. If $\epsilon = +$, then $M(\lambda_6)^* \cong M(\lambda_1)$ by Proposition 5.4.3, so $M(\lambda_6)$ is not self-dual. And if $\epsilon = -$, then $M(\lambda_6)^* \cong M(\lambda_1) \cong M(\lambda_6)^{(e)}$ by Proposition 5.4.2.iii (here $q = p^e$); thus the representation is unitary by Lemma 2.10.15.ii. That the simply connected group is faithful in this representation follows from the description of the module $M(\lambda_6)$ given below. The statements concerning the fields of definition are established as in the proof of Proposition 5.4.10. ∎

We now give an explicit description of this module for E^ϵ_6, which enables us to perform calculations with it. This is most conveniently done within an E_6-parabolic subgroup of $E_7(q)$. With the E_7 Dynkin diagram labelled as in (5.2.2), let $J = \{\alpha_1, \ldots, \alpha_6\}$.

Then $P_J = Q_J L_J H$, where

$$Q_J = \langle X_\alpha \mid \alpha = \sum_{i=1}^{7} m_i\alpha_i,\ m_i \geq 0,\ m_7 > 0\rangle$$

$$L_J = \langle X_\alpha \mid \alpha = \sum_{i=1}^{7} m_i\alpha_i,\ m_7 = 0\rangle \cong E_6(q)$$

and L_J is simply connected. There are precisely 27 positive roots α with $m_7 > 0$ (see [Bou, p.264]). Using the Chevalley commutator relations [Ca_1, 5.2.2], we find that Q_J is elementary abelian of order q^{27}. The action of L_J on Q_J (by conjugation) is the action of $E_6(q)$ on $M(\lambda_6)$ (realized over $\mathbf{F}_q$). And of course ${}^2E_6(q)$ on $M(\lambda_6)$ may be regarded as a subgroup of $E_6(q^2)$ in this action.

The action of $E_7(q)$ on its minimal module has a similar description, in terms of a E_7-parabolic subgroup of $E_8(q)$. Let $J = \{\alpha_1, \ldots, \alpha_7\}$ here, so that in $E_8(q)$, we have $P_J = Q_J L_J H$, where $Q_J \cong q^{1+56}$ (a special group with $Z(Q_J) = Q'_J$ elementary abelian of order q and Q_J/Q'_J elementary abelian of order q^{56}), and $L_J \cong E_7(q)$ (simply connected). The action of L_J on Q_J/Q'_J is the action of $E_7(q)$ on $M(\lambda_7)$ (realized over $\mathbf{F}_q$). Write $\overline{}$ for reduction modulo Q'_J in Q_J. Then for $u, v \in Q_J$,

$$(\overline{u}, \overline{v}) = [u, v] \in Z(Q_J)$$

defines an L_J-invariant symplectic form (,) on $\overline{Q}_J$ (here we identify $Z(Q_J)$ with $\mathbf{F}_q$ as they are both elementary abelian of order q). Moreover, if $p = 2$, then

$$Q(\overline{u}) = u^2 \in Z(Q_J)$$

defines an L_J-invariant quadratic form on $\overline{Q}_J$ with associated bilinear form (,). Thus we have

Proposition 5.4.18. *The action of $E_7(q)$ on its minimal module $M(\lambda_7)$ embeds the simply connected group $E_7(q)$ in $Sp_{56}(q)$, and in $\Omega^\epsilon_{56}(q)$ if q is even. The representation is realized over no proper subfield of $\mathbf{F}_q$.*

Further scrutiny of this module for $E_7(q)$ shows that $E_7(q)$ embeds in $\Omega^+_{56}(q)$ when q is even (as opposed to $\Omega^-_{56}(q)$). This can be seen by restricting to a subgroup $L_8(q)$, which acts on the 56-space as on $V \oplus V^*$, where V is the exterior square of the natural 8-dimensional module. Since $L_8(q)$ fixes no non-degenerate quadratic form on V, the 28-space V must be totally singular with respect to the quadratic form, and hence $\epsilon = +$.

§5.5 Further results on representations

We conclude this chapter with various further results on representation theory which we shall need. Recall $\overline{\mathbf{F}}_r$ is the algebraic closure of the field $\mathbf{F}_r$. Observe that if G is any finite subgroup of $PGL(V, \overline{\mathbf{F}}_r)$, then there is a finite preimage of G in $GL(V, \overline{\mathbf{F}}_r)$. Thus there is a minimal finite preimage.

Lemma 5.5.1. *Let p be prime and assume that $P \leq GL(V, \overline{\mathbf{F}}_r)$, where P is a finite p-group and $r \neq p$. Assume further that P acts irreducibly on V and that $P' \leq \overline{\mathbf{F}}_r^*$. Then $|P : P \cap \overline{\mathbf{F}}_r^*| = p^m$ for some even integer m and $\dim(V) = p^{m/2}$.*

Proof. By [Is, 15.13], the Brauer character χ of the representation of P on V is a complex irreducible character of P. Moreover, $P \cap \overline{\mathbf{F}}_r^* = Z(\chi)$ (which is by definition $\{g \in P \mid |\chi(G)| = \chi(1)\}$). Hence the conclusion follows from [Is, 2.31]. ∎

Lemma 5.5.2. *Let p be prime.*
(i) *$m_{p'}(GL_n(p^f)) \leq n$ for all integers f.*
(ii) *$m_{p'}(PGL_n(p^f)) \leq n$ for integers f.*

Proof. Assertion (i) follows readily from Maschke's Theorem, along with the fact that the image of an irreducible representation of an abelian p'-group in characteristic p is cyclic.

To prove (ii), write $U = \mathbf{Z}_r^t$ for some prime r distinct from p, and assume that $U \leq PGL(V, \mathbf{E})$, where $\mathbf{E} = \overline{\mathbf{F}}_p$. Let $\widetilde{U}$ be a minimal finite preimage of U in $GL(V, \mathbf{E})$; observe that $\widetilde{U}$ is an r-group. If $\widetilde{U}$ is abelian, then $m_r(\widetilde{U}) \geq m_r(U) = t$. Consequently by (i) we obtain $\dim(V) \geq m_r(\widetilde{U}) \geq t$, as desired.

Assume therefore that $\widetilde{U}$ is non-abelian and put $Z = \widetilde{U} \cap \mathbf{E}^*$, so that $U = \widetilde{U}/Z$. Set $K = Z(\widetilde{U})$, so that

$$U = K/Z \times M/Z$$

for some $M \leq \widetilde{U}$. Clearly $Z(M) \leq Z(\widetilde{U}) = K$, and hence $Z(M) = Z$. Our assumption that $\widetilde{U}$ is non-abelian ensures that $Z \neq M$, and we set

$$m = m_r(M/Z) \geq 1.$$

By Maschke's Theorem, $V = \bigoplus_{i=1}^k V_i$, where each V_i is an irreducible $\widetilde{U}$-module. By Schur's Lemma, K acts as scalars on each V_i, and so K embeds in k copies of $\mathbf{E}^*$. Therefore,

$$k \geq m_r(K) \geq m_r(K/Z) = t - m. \tag{5.5.1}$$

Now each V_i is an irreducible M-module. Since Z is a group of scalars, $C_Z(V_i) = 1$, and hence $C_M(V_i) \cap Z = 1$. However $Z = Z(M)$ and $C_M(V_i) \trianglelefteq M$, and thus $C_M(V_i) = 1$. Therefore M acts faithfully on V_i. So defining M_i to be the subgroup of M which acts as scalars on V_i, we see that $M_i \leq Z(M) = Z$. Therefore $M_i = Z$ and consequently by Lemma 5.5.1, m is even and $\dim(V_i) = r^{m/2}$. Therefore

$$\dim(V) = kr^{m/2} \tag{5.5.2}$$

Combining (5.5.1) and (5.5.2), we see that $\dim(V) \geq t$, except possibly when $k = t - m = 1$. But this situation is impossible, for when $k = 1$, we know that $\widetilde{U}$ is absolutely irreducible on V (since $\mathbf{E}$ is algebraically closed). Hence $Z(\widetilde{U}) \leq \mathbf{E}^*$, which implies $t - m = 0$. ∎

Lemma 5.5.3. *Let K be a finite perfect group with a unique minimal normal subgroup N. If $N \cong \mathbf{Z}_p^t$ and $t \geq 2$, then $R_{p'}(K) \geq \min\{P(K/N), p^{t/2}\}$.*

Proof. Write $K \leq PGL(V, \mathbf{E})$, where $\mathbf{E} = \overline{\mathbf{F}}_r$, with r a prime distinct from p, and with $\dim(V)$ minimal. Let $\widetilde{K}$ be a minimal finite preimage of K in $GL(V, \mathbf{E})$ and for any subgroup X of K write $\widetilde{X}$ for the preimage of X in $\widetilde{K}$. Note that $\widetilde{K}$ is perfect since K is. Now let $0 = V_o < V_1 < \cdots < V_s = V$ be a $\widetilde{K}$-composition series of V, and as in the proof of Proposition 5.3.2, put $C = \bigcup_{i=1}^{s} C_{\widetilde{K}}(V_i/V_{i-1})$. Suppose for a contradiction that $\widetilde{N}$ acts as scalars on each V_i/V_{i-1}. Then $[\widetilde{K}, \widetilde{N}] \leq C \cap \widetilde{N}$, and since $\widetilde{N}$ is an r'-group while C is an r-group, we have $[\widetilde{K}, \widetilde{N}] = 1$. But this contradicts the fact that N is non-central in K. Therefore $\widetilde{N}$ does not act as scalars on $W = V_i/V_{i-1}$ for some i. Since N is the unique minimal normal subgroup, it is clear that the subgroup of $\widetilde{K}$ inducing scalars on W is precisely $\widetilde{K} \cap \mathbf{E}^*$, and hence $K \leq PGL(W, \mathbf{E})$. Thus by minimality, $V = W$, which is to say $\widetilde{K}$ is irreducible in $GL(V, \mathbf{E})$. Obviously K/N acts on the Wedderburn components of $\widetilde{N}$, and if this action is non-trivial, we obtain $n \geq P(K/N)$, as required. It may be assumed therefore that $\widetilde{N}$ acts homogeneously on V. Thus $\widetilde{N}$ acts faithfully on an irreducible $\widetilde{N}$-submodule U of V. Since N is a p-group, $\widetilde{N} = N_p \times N_{p'}$, where N_p is the Sylow p-subgroup of $\widetilde{N}$ and $N_{p'}$ is a p'-group. Clearly $N_{p'} \leq \mathbf{E}^*$, and so N_p also acts faithfully and irreducibly on W. The subgroup N_o of N_p which acts as scalars on W must act as scalars on all of V (since $\widetilde{N}$ is homogeneous), and hence $N_o = N_p \cap \mathbf{E}^*$. Therefore by Lemma 5.5.1, $\dim(W) = |N_p : N_p \cap \mathbf{E}^*|^{1/2} = |N|^{1/2}$, completing the proof. ∎

Lemma 5.5.3 has the following Corollary, which will be used in Chapters 7 and 8 to give lower bounds for R_p for certain members of $\mathcal{C}_2$ and $\mathcal{C}_6$. See, for example, the proof Lemma 8.2.3.

Corollary 5.5.4.

(i) *Assume that H has structure $p^t.A_n$, where $n \geq 5$ and $O_p(H)$ is either the deleted or fully deleted permutation module for $H/O_p(H)$. Then either $R_{p'}(H) \geq n$ or else $H \cong 2^4.A_5$, $2^4.A_6$ or $2^5.A_6$, in which case $R_{2'}(H) \geq 4$, 4 or 5, respectively.*

(ii) *Assume that H is an insoluble member of $\mathcal{C}_6(\Omega)$ for some classical group Ω, so that $H^\infty \cong r^{1+2m}.Sp_{2m}(r)$, $(4 \circ 2^{1+2m}).Sp_{2m}(2)$ or $2_\pm^{1+2m}.\Omega_{2m}^\pm(2)$ (see §4.6). Then $R_{2'}(H^\infty/Z(H^\infty)) \geq 2^m$.*

Proof. Assertion (ii) is immediate from Lemma 5.5.3, since $P(Sp_{2m}(r)) \geq r^m$ and $P(\Omega_{2m}^\pm(2)) \geq 2^m$ (see Theorem 5.2.2). Part (i) is also clear, provided $Z(H) = 1$. When $Z(H) \neq 1$ (so that $O_p(H)$ is the deleted but not fully deleted permutation module), we appeal to Proposition 5.3.2, which asserts that $R_{p'}(H) \geq R_{p'}(H/Z(H)) + 1$. ∎

Lemma 5.5.5. *Assume that $M = M_1 \circ \cdots \circ M_t$ is a central product of groups M_i, and that M is an irreducible subgroup of $GL(V, \mathbf{F})$, with $\mathbf{F}$ finite. Define $\mathbf{E} = \mathrm{End}_{\mathbf{F}M}(V)$ and write $(V, \mathbf{E})$ for V regarded as a vector space over $\mathbf{E}$. Then there is a tensor decomposition $(V, \mathbf{E}) = (V_1, \mathbf{E}) \otimes \cdots \otimes (V_t, \mathbf{E})$ such that M_i is embedded absolutely irreducibly in*

$1 \otimes \cdots \otimes GL(V_i, \mathbf{E}) \otimes \cdots \otimes 1$.

Proof. We prove the Lemma for $t = 2$, and the general result will follow by induction. Observe that M may be regarded as an absolutely irreducible subgroup of $GL(V, \mathbf{E})$ (see Lemmas 2.10.1, 2.10.2) and for the remainder of this proof, all vector spaces are to be regarded as $\mathbf{E}$-spaces. Write $V = W_1 \oplus \cdots \oplus W_k$, where the W_i are the Wedderburn components of M_1 on V. Evidently $M_1 C_{GL(V)}(M_1)$ fixes W_1, and hence so does M. Consequently $k = 1$, which is to say M_1 is homogeneous on V. As we remarked after the proof of Lemma 4.4.3, there is a tensor decomposition $V = V_1 \otimes V_2$ such that $M_1 \leq GL(V_1) \otimes 1$. Moreover, the proof of Lemma 4.4.3.i shows that $C_{GL(V)}(M_1)$ is isomorphic to $GL_m(\mathbf{E}_o)$, where $m = \dim_{\mathbf{E}}(V_2)$ and $\mathbf{E}_o = \mathrm{End}_{\mathbf{E}M_1}(V_1)$. Thus $Z(C_{GL(V)}(M_1)) \cong \mathbf{E}_o^*$, and this group clearly lies in $C_{GL(V)}(M)$. Since V is an absolutely irreducible $\mathbf{E}M$-module, we have $\mathbf{E}_o = \mathbf{E}$, which shows M_1 is embedded absolutely irreducibly in $GL(V_1)$. Moreover, $C_{GL(V)}(M_1) = GL(V_2)$ (Lemma 4.4.3.i), and the argument before shows that M_2 is embedded absolutely irreducibly in $GL(V_2)$. The Lemma is proved. ■

Corollary 5.5.6. *Assume that $G = L_1 \times \cdots \times L_t$, with each L_i a non-abelian simple group. Further suppose that G is an irreducible subgroup of $PGL(V, \mathbf{F})$, with $\mathbf{F}$ finite. Then there is a field $\mathbf{E}$ satsifying $\mathbf{F} \subseteq \mathbf{E} \subseteq \mathrm{End}_{\mathbf{F}}(V)$ and a tensor decomposition $(V, \mathbf{E}) = (V_1, \mathbf{E}) \otimes \cdots \otimes (V_t, \mathbf{E})$, such that $L_i \leq 1 \otimes \cdots \otimes PGL(V_i, \mathbf{E}) \otimes \cdots \otimes 1$.*

Proof. Let $\widetilde{G}$ be a minimal preimage of G in $GL(V, \mathbf{F})$. Then $\widetilde{G}$ is a covering group of G, and so $\widetilde{G}$ is a central product of groups C_i, where C_i is a covering group of L_i (see Proposition 5.2.5). Now apply Lemma 5.5.5. ■

Proposition 5.5.7. *Assume that $L_1, \ldots, L_t$ are non-abelian simple groups. Put $G = L_1 \times \cdots \times L_t$, $n_i = R_p(L_i)$ and $n = \sum_{i=1}^t n_i$.*

(i) *$R_p(G) \geq n$.*

(ii) *$R_p^i(G) = \prod_{i=1}^t n_i$.*

Proof. First of all, observe that (ii) is immediate from Corollary 5.5.6, and so it remains to prove (i). We proceed by induction on t. There is nothing to prove when $t = 1$, so we assume hereafter that $t \geq 2$. Write $G \leq PGL(V, \mathbf{E})$, where $\mathbf{E}$ is the algebraic closure of $\mathbf{F}_p$ and $\dim(V)$ is minimal. Our goal is to show that $\dim(V) \geq n$. Let $\widetilde{G}$ be a minimal preimage of G in $GL(V, \mathbf{E})$, so that $\widetilde{G} = C_1 \circ \cdots \circ C_t$, a central product of quasisimple groups with C_i a covering group of L_i. We put $Z = \widetilde{G} \cap \mathbf{E}^*$, so that $G = \widetilde{G}/Z$. It follows from Proposition 5.2.5.i that any normal subgroup N of $\widetilde{G}$ satisfies $\prod_{i \in I} C_i \leq N \leq Z \prod_{i \in I} C_i$ for some subset $I \subseteq \{1, \ldots, t\}$. If G is irreducible, then by (ii) we have $\dim(V) \geq \prod_{i=1}^t n_i \geq n$, as desired. Now suppose that $\widetilde{G}$ is reducible on V, and pick an irreducible submodule W. Then $C_{\widetilde{G}}(W) \trianglelefteq \widetilde{G}$ and hence by our observation before, we have (relabelling if necessary) $\prod_{i=1}^s C_i \leq C_{\widetilde{G}}(W) \leq Z \prod_{i=1}^s C_i$ for some $s \leq t$. It now follows that $\prod_{i=s+1}^t C_i$ acts faithfully on W, for if $x \in \prod_{i=s+1}^t C_i$ centralizes W, then $x \in Z$ and hence $x = 1$ as $C_Z(W) = 1$ (since $Z \leq \mathbf{E}^*$). Therefore by (ii),

$$\dim(W) \geq \prod_{i=s+1}^{t} n_i \geq \sum_{i=s+1}^{t} n_i. \tag{5.5.3}$$

Moreover, $C_{\widetilde{G}}(W) \cap C_{\widetilde{G}}(V/W) \leq O_p(\widetilde{G}) = 1$, and so $\prod_{i=1}^{s} C_i$ acts faithfully on V/W. Furthermore, by the minimality of $\dim(V)$, we know that $\widetilde{G}$ does not act trivally on W (or else $G \leq PGL(V/W)$), and so $s < t$. Thus $L_1 \times \cdots \times L_s < PGL(V/W)$, and so $\dim(V/W) \geq \sum_{i=1}^{s} n_i$ by induction. The result now follows. ■

Corollary 5.5.8. *Let L_i, G, n_i, n and t be as in Proposition 5.5.7, and assume that $n_i \geq 3$ for all i. Further suppose that $G \leq PGL(V, \mathbf{F})$ for some n-dimensional space V over $\mathbf{F}$ in characteristic p. Then there is G-invariant series $0 = V_o < V_1 < \cdots < V_t = V$ such that the image of L_i in $PGL(V_j/V_{j-1}, \mathbf{F})$ is non-trivial if and only if $i = j$.*

Proof. Clearly we can assume that $t \geq 2$. Since $\prod_{i=1}^{t} n_i > n$, it follows from Proposition 5.5.7.ii that G is reducible on V. So let W be an irreducible $\widetilde{G}$-submodule of V, where $\widetilde{G}$ is a minimal preimage of G in $GL(V)$. Arguing as in the previous proof, we see that equality must hold in (5.5.3). And since $n_i \geq 3$, we have $s = t - 1$. Evidently $L_1 \times \cdots \times L_{t-1} \leq PGL(V/W)$, and so the result now follows by induction on t. ■

The proof of Corollary 5.5.8 has the following technical Corollary, which will be used in §8.6.

Corollary 5.5.9. *Retain the notation L_i, G, n_i, n, t of Corollary 5.5.8, and assume that $n_i \geq 3$ for all i. Further suppose that $G \leq P\Omega(V, \mathbf{F}, \kappa)$ for some n-dimensional space V over $\mathbf{F}$ in characteristic p, with κ a non-degenerate symplectic, quadratic or unitary form. Then there is a subspace decomposition $V = W_1 \perp \cdots \perp W_t$, with W_i a non-degenerate n_i-space, such that the image of L_i in $P\Omega(W_j, \mathbf{F}, \kappa)$ is non-trivial if and only if $i = j$.*

Proof. Let V_1 be as in Corollary 5.5.8. Clearly L_1 is irreducible in $PGL(V_1)$, and so V_1 is either totally singular or non-degenerate. If V_1 is totally singular, then L_1 also acts irreducibly in $PGL(V/V_1^{\perp})$. But according to Corollary 5.5.8, L_1 acts trivially on V/V_1. Thus V_1 is non-degenerate, and so $V = V_1 \perp V_1^{\perp}$. The group $L_2 \times \cdots \times L_t$ embeds in $P\Omega(V_1^{\perp}, \mathbf{F}, \kappa)$, and the result now follows by induction. ■

Proposition 5.5.10. *No non-abelian finite simple group has a non-trivial representation of degree 2 in odd characteristic.*

Proof. Assume for a contradiction that $L \leq GL_2(\mathbf{F})$, with L a non-abelian finite simple group and $\mathbf{F}$ a field of odd characteristic. Obviously $L \leq SL_2(\mathbf{F})$, and it follows that $|L|$ is odd, for the unique involution in $SL_2(\mathbf{F})$ is central. (Here one may quote the Feit-Thompson Theorem, but we continue with an elementary proof.) Let r be a prime divisor of $|L|$ distinct from the characteristic of $\mathbf{F}$ and pick an element z of order r in the centre of a Sylow r-subgroup R of L. Then we may write $z = \mathrm{diag}(\lambda, \lambda^{-1})$, where λ is a primitive r^{th} root of unity in $\overline{\mathbf{F}}$. Obviously $C_L(z)$ acts on each of the two eigenspaces for z, and hence $R = \langle \mathrm{diag}(\mu, \mu^{-1}) \rangle$ for some $\mu \in \overline{\mathbf{F}}^*$. Since $|L|$ is odd, L contains no element interchanging the two eigenspaces of $\mathrm{diag}(\mu, \mu^{-1})$, and hence $R \leq Z(N_L(R))$. This contradicts Burnside's Normal p-complement Theorem (see [As8, (39.1)]). ∎

Proposition 5.5.11. *Assume that L is a non-abelian simple group embedded in $PGL(V, \mathbf{F})$, where $\mathbf{F}$ is a field of characteristic p. Also suppose that $C_{PGL(V,\mathbf{F})}(L)$ contains an elementary abelian group of order r^2, with r prime and $r \neq p$. Then* $\dim(V) \geq R_p(L) + 2$.

Proof. Write E for the elementary abelian group of order r^2 in the centralizer of L, and let $\widetilde{E}$ and $\widetilde{L}$ be minimal preimages of E and L in $GL(V, \mathbf{F})$. Thus $\widetilde{L}$ is quasisimple and $[\widetilde{E}, \widetilde{L}] = 1$. Assume for a contradiction that $\dim(V) \leq R_p(L) + 1$. Since $\widetilde{E}$ is not cyclic, $\widetilde{L}$ acts reducibly on V, and so $\widetilde{L}$ has just two irreducible constituents on V, one trivial and the other non-trivial. In addition, $\widetilde{L}$ must act absolutely irreducibly on the non-trivial constituent (or else $\dim(V) \geq 2R_p(L)$). Consequently $\widetilde{E}$ acts as scalars on these two constituents; but this means $\widetilde{E}$ modulo scalars is cyclic, a contradiction. ∎

Chapter 6
NON-MAXIMAL SUBGROUPS IN $\mathcal{C}$: THE EXAMPLES

Let $(V, \mathbf{F}, \kappa)$ be a classical geometry of dimension n, and let $X = X(V, \mathbf{F}, \kappa)$ as X ranges over the symbols in (2.1.15). In this chapter we describe those classes of triples (H, K, Ω) appearing in Table 3.5.H — that is, the triples $H < K < \Omega$ with $H, K \in \mathcal{C}(\Omega)$. In Chapter 7, we prove that the triples occurring here in fact exhaust all such triples, thereby showing that Table 3.5.H is complete. As described in §3.4, one may read off from Table 3.5.H all triples (H, K, G) where $H < K < G$, such that $H, K \in \mathcal{C}(G)$ and G is any classical group satisfying $\overline{\Omega} \le G \le \overline{A}$ (as usual $\overline{}$ denotes reduction modulo scalars).

For the rest of this chapter we assume that $H, K \in \mathcal{C}(\Omega)$ and that

$$H < K < \Omega. \tag{6.0.1}$$

Recall the definition of the overgroup sets $\mathcal{G}_{\mathcal{C}}(X)$ given in (3.4.2), where $X \le \overline{\Omega}$. Here we make anologous definitions for groups $X \le \Omega$. For a subgroup $X \le \Omega$, define

$$\begin{aligned} \mathcal{G}_i(X) &= \{Y \in \mathcal{C}_i(\Omega) \mid X < Y < \Omega\}, \\ \mathcal{G}_{\mathcal{C}}(X) &= \bigcup_{i=1}^{8} \mathcal{G}_i(X) \end{aligned} \tag{6.0.2}$$

Thus we have $K \in \mathcal{G}_{\mathcal{C}}(H)$. It turns out that triples arise only when $H \in \mathcal{C}_1 \cup \mathcal{C}_2 \cup \mathcal{C}_4$. We treat these three cases in the remaining sections of this chapter. As mentioned in §3.4, we are concerned with the overgroup sets only when the dimension is at least 13. Thus for the remainder of our work we adopt the blanket assumption

$$n \ge 13. \tag{6.0.3}$$

The overgroup sets when $n \le 12$ are determined in [Kl$_2$].

§6.1 The case $H \in \mathcal{C}_1$

In this section assume that $H \in \mathcal{C}_1(\Omega)$, so that $H = N_\Omega(W)$ or $N_\Omega(U, W)$ for suitable subspaces W, U of V as described in §4.1. The triples described here appear in those rows of Table 3.5.H for which $\mathcal{C}_1$ appears in the second column. We first consider row 12, with H of type $P_{n/2-1}$ in case $\mathbf{O}^+$. Here W is a totally singular $(\frac{n}{2} - 1)$-space, and by Witt's Lemma 2.1.6, we may take $W = \langle e_1, \ldots, e_{n/2-1} \rangle$, where $\{e_1, \ldots, f_{n/2}\}$ is a standard basis of V as described in Proposition 2.5.3.i. Evidently $W = W_1 \cap W_2$, where $W_1 = \langle W, e_{n/2} \rangle$ and $W_2 = \langle W, f_{n/2} \rangle$, and each W_i is a totally singular $\frac{n}{2}$-space. Obviously W_1 and W_2 lie in distinct families $\mathcal{U}^i_{n/2}$ (see Description 4 in §2.5) and we may write $W_i \in \mathcal{U}^i_{n/2}$. Also by Lemma 2.5.8.iii, W_i is the unique member of $\mathcal{U}^i_{n/2}$ which contains W. Consequently $H \le N_\Omega(W_i)$, since H fixes each set $\mathcal{U}^i_{n/2}$. Now $N_\Omega(W_i)^{W_i} \ge SL(W_i)$ by Corollary 4.1.10 and hence $H < N_\Omega(W_i)$. Therefore $\{N_\Omega(W_1), N_\Omega(W_2)\} \subseteq \mathcal{G}_1(H)$. We are now in position to prove the first result of this chapter.

Proposition 6.1.1. *Assume that case* $\mathbf{O}^+$ *holds and that* H *is of type* $P_{n/2-1}$, *so that* W *is a totally singular* $(\frac{n}{2}-1)$*-space. Then* $\mathcal{G}_1(H) = \{N_\Omega(W_1), N_\Omega(W_2)\}$, *where* W_i *is the unique member of* $\mathcal{U}^i_{n/2}$ *containing* W.

Proof. Write $K = N_\Omega(U) \in \mathcal{G}_1(H)$ for some subspace U of V. Now according to Proposition 4.1.14.ii, there exists $P \in Syl_p(\Omega)$ the stabilizing the maximal flag in V given by $0 = V_o < V_1 < \cdots < V_{(n-4)/2} < W < W_1$, where $V_i = \langle e_1, \ldots, e_i\rangle$ for $i = 1, \ldots, \frac{1}{2}(n-4)$. Obviously $P \leq H$. Now the stabilizers of non-degenerate subspaces or of non-singular 1-spaces do not contain a Sylow p-subgroup of Ω. Therefore U is totally singular. Now according to Proposition 4.1.14.ii, P fixes precisely two maximal flags; the second is obtained from the one above by replacing W_1 by W_2. Therefore U must be one of the spaces V_i $(i = 1, \ldots, (n-4)/2)$ or W or W_j $(j = 1, 2)$. However $U \neq W$ as $H \neq K$, and $U \neq V_i$ as $H^W \geq SL(W)$. Therefore $U = W_j$, and the proof is complete. ■

If W, W_1 and W_2 are as above, notice that $N_I(W)$ contains an element interchanging W_1 and W_2. Thus using the terminology introduced in §3.4, $N_I(W)$ is an I-novelty with respect to $N_\Omega(W_i)$.

The examples in the next Proposition are based on the fact that in the $O_2^+(2)$-geometry there is a unique non-singular vector, and in the $O_2^+(3)$-geometry there are exactly two non-degenerate 1-spaces, one □1-space $W_\square$ and and one ⊠1-space $W_\boxtimes$. Obviously $W_\square = \langle e_1 - f_1\rangle$ and $W_\boxtimes = \langle e_1 + f_1\rangle$, where $\{e_1, f_1\}$ is a standard basis for the $O_2^+(3)$-geometry.

Proposition 6.1.2. *Assume that case* $\mathbf{O}$ *holds and that* H *is of type* $O_2^+(q) \perp O_{n-2}^\xi(q)$, *so that* W *is a* $+2$*-space in* V.
 (i) *If* $q = 2$ *then* $\mathcal{G}_1(H) = \{N_\Omega(v)\}$, *where* v *is the non-singular vector in* W.
 (ii) *If* $q = 3$, *then* $\mathcal{G}_1(H) = \{N_\Omega(W_\square), N_\Omega(W_\boxtimes)\}$, *where* $W_\square$ *is the* □1*-space in* W *and* $W_\boxtimes$ *is the* ⊠1*-space in* W.

Proof. According to Proposition 2.10.6, H is irreducible on the non-degenerate $(n-2)$-space $W^\perp$. Moreover H interchanges the two singular 1-spaces in W. The result is now clear. ■

Observe that if $I \approx O_n^+(3)$ with n even, then Δ contains an element δ satisfying $\tau(\delta) = -1$ and interchanging $W_\square$ and $W_\boxtimes$. Therefore $N_\Delta(W)$ is a Δ-novelty with respect to $N_I(W_\square)$ and $N_I(W_\boxtimes)$.

The final example in this section arises because we have included the groups in Aschbacher's collection $\mathcal{C}_1'$ in our collection $\mathcal{C}_1$.

Proposition 6.1.3. *Assume that case* $\mathbf{L}$ *holds and that* H *is of type* $GL_m(q) \oplus GL_{n-m}(q)$ *or of type* $P_{m,n-m}$, *so that* $H = N_\Omega(W, U)$. *Then* $\mathcal{G}_1(H) = \{N_\Omega(W), N_\Omega(U)\}$.

Proof. Exercise. ■

Evidently $N_\Omega(W, U)$ will give rise to an A-novelty with respect to $N_\Gamma(W)$ and $N_\Gamma(U)$.

§6.2 The case $H \in \mathcal{C}_2$

In this section assume that $H \in \mathcal{C}_2(\Omega)$, so that $H = \Omega_{\mathcal{D}}$, where $\mathcal{D}$ is an m-space decomposition of V given by $V = V_1 \oplus \cdots \oplus V_t$, as described in §4.2. Thus $n = mt$ where $m = \dim(V_i)$. The triples described here appear in those rows of Table 3.5.H for which $\mathcal{C}_2$ occurs in the second column.

We begin with totally singular $\frac{n}{2}$-decompositions in case $\mathbf{O}^+$. Recall the families $\mathcal{U}^i_{n/2}$ mentioned in Description 4 in §2.5.

Proposition 6.2.1. *Assume that case* $\mathbf{O}^+$ *holds and that* H *is of type* $GL_{n/2}(q).2$ *so that* $\mathcal{D}$ *is a singular* $\frac{n}{2}$*-space decomposition. Suppose further that* $\frac{n}{2}$ *is odd. Then* $\mathcal{G}_1(H) = \{N_\Omega(V_1), N_\Omega(V_2)\}$.

Proof. Since $\frac{n}{2}$ is odd, V_1 and V_2 lie in different Ω-orbits $\mathcal{U}^i_{n/2}$ (see Lemma 2.5.8), and hence there is no element of Ω interchanging V_1 and V_2. Therefore $N_\Omega(V_i) \in \mathcal{G}_1(H)$ for $i = 1, 2$. Now $SL(V_i) \leq H^{V_i}$ (see Corollary 4.1.10), and so H is irreducible on V_i. Moreover $V_1 \cong V_2^*$ as H-modules, and hence $V_1 \not\cong V_2$ as H-modules. The Proposition now follows from Lemma 2.10.11. ■

For the rest of this section we shall assume that either case $\mathbf{L}$ holds, so that H is of type $GL_m(q) \wr S_t$, or that κ and $\mathcal{D}$ are non-degenerate and $\mathcal{D}$ is isometric, so that H is of type $GU_m(q) \wr S_t$, $Sp_m(q) \wr S_t$ or $O^\xi_m(q) \wr S_t$ (see Table 4.2.A.). As in (4.2.8), define

$$X_i = X(V_i, \kappa), \tag{6.2.1}$$

where X ranges over the symbols Ω, S and I. According to Proposition 4.2.8.iii, we have $I_{\mathcal{D}} = I_{(\mathcal{D})} J$, where $J \cong S_t$ acts naturally on the spaces V_i. Clearly $A_t \cong J' \leq I' \leq \Omega$, and hence $J' \leq H$. It is convenient to define

$$L = (J')^H, \tag{6.2.2}$$

the normal closure of J' in H. The following result is clear.

Lemma 6.2.2.
(i) $L^{\mathcal{D}} = Alt(\mathcal{D}) \cong A_t$, *and hence* $L^{\mathcal{D}}$ *is transitive on* $\mathcal{D}$, *provided* $t \geq 3$.
(ii) L *is perfect provided* $t \geq 5$.

In order to perform calculations with elements in the stabilizer of $\mathcal{D}$, we identify $I_{\mathcal{D}}$ with the wreath product $I_1 \wr S_t$. So with a slight abuse of notation, elements of H may be written $(g_1, \ldots, g_t)\rho$, where $g_i \in I_1$ and $\rho \in S_t$ acts naturally on the t coordinates. For $g \in I_1$, denote by $g_{i,j}$ the element of $I_{(\mathcal{D})}$ whose i^{th} coordinate is g, whose j^{th} coordinate is g^{-1}, and whose remaining coordinates are 1. We claim that

$$g_{i,j} \in \Omega_{(\mathcal{D})} \tag{6.2.3}$$

for all i, j and for all $g \in I_1$. This claim is trivial in case $\mathbf{S}$ (for here $\Omega = I$) and it is also clear in case $\mathbf{L}^\pm$, for $\det(g_1, \ldots, g_t) = \prod_{i=1}^t \det_{V_1}(g_i)$. Finally consider case $\mathbf{O}$. Then by Propositions 2.5.6 and 2.5.9, g may be written as the product of an element of Ω_1 with a product of reflections. Now $\Omega_1 \times \cdots \times \Omega_t \leq \Omega$ by Lemma 4.1.1.ii. Thus we may write $g_{i,j} \equiv h \pmod{\Omega}$, with $h = r_1 \ldots r_k r'_k \ldots r'_1$, where r_ℓ $(1 \leq \ell \leq k)$ is a reflection in a non-singular vector in V_i and r'_ℓ is a reflection in a non-singular vector in V_j, and the two vectors have the same norm. Thus h is the product of an even number of reflections, and if q is odd the spinor norm of h is a square. Hence $h \in \Omega$, from which it follows that $g_{i,j} \in \Omega$, as claimed.

Lemma 6.2.3. *Assume that* $t \geq 3$ *and that* L *is as in* (6.2.2).

(i) *We have* $L_{(\mathcal{D})}^{V_i} = I_i$.

(ii) I_1^{t-2} *embeds in* $L_{(\mathcal{D})}$.

(iii) *If* I_i *is irreducible on* V_i *and* $I_i \neq 1$, *then* L *is absolutely irreducible on* V.

(iv) *The group* L *is reducible on* V *if and only if* $I_i \approx GL_1(2)$ *in case* **L** *or* $I_i \approx O_2^+(q)$ *with* $q \leq 3$ *in case* $\mathbf{O}^+$.

Proof. (i) and (ii). Whenever H contains an element of the form $x = (g, h, 1, 1, \ldots, 1)$, then $L_{(\mathcal{D})}$ contains

$$[x^{-1}, (123)] = (g, hg^{-1}, h^{-1}, 1, 1, \ldots, 1). \tag{6.2.4}$$

Thus for all $g \in I_1$, we deduce from (6.2.3) that $(g, g^{-2}, g, 1, \ldots, 1) = [g_{1,2}^{-1}, (123)] \in L_{(\mathcal{D})}$, which shows $L_{(\mathcal{D})}^{V_1} = I_1$. Assertion (i) now follows as $L^{\mathcal{D}}$ is transitive. Moreover, as $L^{\mathcal{D}} = Alt(\mathcal{D})$, it follows that $L_{(\mathcal{D})}$ contains all elements of the form $(1, \ldots, 1, g, 1, \ldots, 1, g^{-2}, g)$, and so (ii) follows easily.

(iii). Assume here that I_1 is irreducible on V_1. We first show that the spaces V_i are mutually non-isomorphic as $L_{(\mathcal{D})}$-modules. Consider the case $t \geq 4$. By assumption $I_1 \neq 1$, and we pick $g \in I_1 \backslash 1$. Then the element $(g, g^{-2}, g, 1, \ldots, 1)$ (appearing in the first paragraph) acts trivially on V_4 and non-trivially on V_1; thus V_1 and V_4 are non-isomorphic, and hence the spaces V_i are pair-wise non-isomorphic by the 2-transitivity of $Alt(\mathcal{D})$. If $t = 3$, then $m \geq 5$ (since $n \geq 13$ by (6.0.3)), and hence $\Omega_i \neq 1$. Now $\Omega_1 \leq \Omega$ by Lemma 4.1.1.ii and so we can take $g \in \Omega_1 \backslash 1$ and $h = 1$ in (6.2.4). Thus $V_1 \not\cong V_3$ as $L_{(\mathcal{D})}$-modules; using $Alt(\mathcal{D})$ again we see that the spaces V_i are mutually non-isomorphic, as desired. To complete the proof of (iii), observe that I_i is in fact absolutely irreducible on V_i according to Proposition 2.10.6.iii. The result now follows by appealing to Proposition 2.10.13.

(iv). The 'only if' is immediate from part (iii) and Proposition 2.10.6. Now if H is of type $GL_1(2) \wr S_n$, then H is reducible, for it fixes $v = v_1 + \ldots + v_n$ with $v_i \in V_i \backslash 0$. And if H is of type $O_2^+(q) \wr S_{n/2}$ with $q \leq 3$, then I_i is reducible on V_i, and hence H is reducible on V. Thus (iv) is proved. ∎

The triples which we describe in the rest of this section all occur because of the exceptional behaviour of the classical groups with m small. We begin with the case $m = 1$, and then work our way up to examples with $m = 2, 3, 4$.

Proposition 6.2.4. *Assume that case* **L** *holds and that H is of type $GL_1(2) \wr S_n$. Then there is a 1-space W and an $(n-1)$-space U such that the following hold.*
(i) $\mathcal{G}_1(H) = \{N_\Omega(U), N_\Omega(W), N_\Omega(U,W)\}$.
(ii) *When n is even,* $|\mathcal{G}_8(H)| = 1$.

Proof. Here $\dim(V_i) = 1$, and we let v_i be the non-zero vector in V_i. Thus H acts naturally as S_n on the vectors v_i, and hence V is a permutation module for S_n. Part (i) follows directly from Lemma 5.3.4. To prove (ii), assume that n is even and note that H preserves the non-degenerate symmetric bilinear form $\mathbf{f}$ given by $\mathbf{f}(v_i, v_j) = 1 + \delta_{ij}$. Therefore $H < I(V, \mathbf{f}) \in \mathcal{C}_8$. Now let $\mathbf{g}$ be any non-degenerate symmetric bilinear form on V satisfying $H < I(V, \mathbf{g}) \in \mathcal{C}_8$. Since $\mathbf{g}$ is non-degenerate, there exists $i \geq 2$ such that $\mathbf{g}(v_1, v_i) = 1$. Thus as H is 2-transitive on the vectors v_i, we have $\mathbf{g}(v_i, v_j) = 1 + \delta_{ij}$ for all i, j, which is to say $\mathbf{g} = \mathbf{f}$. Thus (ii) holds, and the proof is finished. ■

The next triples occur because $GL_1(3) = O_1(3)$ and $GL_1(4) = GU_1(2)$.

Proposition 6.2.5. *Assume that case* **L** *holds and that H is of type $GL_1(q) \wr S_n$.*
(i) *If $q = 3$, then $\mathcal{G}_8(H) = \{N_\Omega(\mathbf{f})\}$ for some non-degenerate symmetric bilinear form* $\mathbf{f}$. *Moreover if n is even, then* $\mathrm{sgn}(\mathbf{f}) = (-)^{n/2}$.
(ii) *If $q = 4$, then $\mathcal{G}_8(H) = \{N_\Omega(\mathbf{f})\}$ for some non-degenerate unitary form* $\mathbf{f}$.

Proof. Choose $v_i \in V_i \backslash 0$ and define $\mathbf{f}(v_i, v_j) = \delta_{ij}$. Then for $q = 3$ (respectively, $q = 4$), $\mathbf{f}$ extends to a unique non-degenerate bilinear (respectively, unitary) form on V. An easy calculation shows that $H < I(V, \mathbf{f}) \in \mathcal{C}_8$, and the uniqueness follows from the fact that H is absolutely irreducible (see Lemma 2.10.3 and Lemma 6.2.3.iii). Note that when $q = 3$ and n is even, $D(\mathbf{f})$ is a square, and so the final statement of (i) follows from Proposition 2.5.10. ■

We now examine some examples when $m = 2$. Observe that $O_2^-(2) \leq Sp_2(2) \leq GL_2(2)$, and that these groups are all isomorphic to S_3. Therefore equality holds throughout.

Proposition 6.2.6. *If case* **L** *or* **S** *holds and H is of type $GL_2(2) \wr S_{n/2}$ or $Sp_2(2) \wr S_{n/2}$, respectively, then $|\mathcal{G}_8(H)| = 1$. Moreover in case* **S**, *the quadratic form fixed by H has sign* $(-)^{n/2}$.

Proof. As we observed above, $GL(V_i) = I(V_i, Q_i)$, where Q_i is a non-degenerate quadratic form on V_i satisfying $\mathrm{sgn}(Q_i) = -$. Now define the non-degenerate quadratic form Q on V by $Q(v_1 + \cdots + v_{n/2}) = \sum_{i=1}^{n/2} Q_i(v_i)$ (here $v_i \in V_i$). It is easy to check that $H \leq I(V, Q) < I(V, \mathbf{f}_Q)$. Since H is absolutely irreducible (Lemma 6.2.3.iii), Q is the unique H-invariant non-degenerate quadratic form on V, and $\mathbf{f}_Q$ is the unique H-invariant non-degenerate symmetric bilinear form on V. Thus in case **L**, $\mathcal{G}_8(H) = \{I(V, \mathbf{f}_Q)\}$ and in case **S**, $\mathcal{G}_8(H) = \{I(V, Q)\}$. Finally we note that $(V, Q) = (V_1, Q_1) \perp \cdots \perp (V_{n/2}, Q_{n/2})$, and so $\mathrm{sgn}(Q) = (-)^{n/2}$ by Proposition 2.5.11. ■

Now consider the $GU_2(2)$-geometry. Here $\mathbf{F} = \mathbf{F}_4$ and for all $\lambda \in \mathbf{F}$ we have $\lambda\lambda^\alpha = \lambda^3 = 1$ (recall α is the field automorphism of $\mathbf{F}$ of order 2). Thus if $\{x, y\}$ is an orthonormal basis, we find that $\lambda x + \mu y$ is singular if $\lambda\mu \neq 0$. Therefore $\langle x \rangle$ and $\langle y \rangle$ are the only non-singular 1-spaces. Consequently $GU_2(2)$ fixes a unique non-degenerate 1-decomposition, and this leads to the following result.

Proposition 6.2.7. *If case* **U** *holds and H is of type $GU_2(2) \wr S_{n/2}$, then $\mathcal{G}_2(H) = \{\Omega_{\mathcal{D}^*}\}$ for some non-degenerate 1-space decomposition $\mathcal{D}^*$.*

Proof. By the remarks above, V_i contains precisely two non-degenerate 1-spaces W_{2i-1}, W_{2i}, and $V_i = W_{2i-1} \perp W_{2i}$. Consequently $H < \Omega_{\mathcal{D}^*}$, where $\mathcal{D}^*$ is the non-degenerate 1-decomposition of V given by $V = W_1 \perp \cdots \perp W_n$. Assume now that the group K appearing in (6.0.1) satisfies $K = \Omega_{\mathcal{D}_1}$, where $\mathcal{D}_1$ is an a-decomposition given by $V = U_1 \oplus \cdots \oplus U_b$, with $n = ab$. Since L is irreducible (Lemma 6.2.3.iii), $L \not\leq \Omega_{(\mathcal{D}_1)}$, and hence $J' \not\leq \Omega_{(\mathcal{D}_1)}$. But $\frac{n}{2} \geq 7$ in view of (6.0.3), and so J' is simple. Therefore $A_{n/2} \cong J' \preceq H^{\mathcal{D}_1} \preceq S_b$, forcing $b = \frac{n}{2}$ or n. If $b = \frac{n}{2}$, then $\mathcal{D}_1$ is also a non-degenerate 2-space decomposition, and so $\mathcal{D}$ and $\mathcal{D}_1$ are Ω-conjugate. But then H and K are Ω-conjugate, against (6.0.1). Therefore $b = n$. Thus H has an orbit on $V \backslash 0$ of size at most $3n$, namely the non-zero vectors in $U_1 \cup \cdots \cup U_n$. Now I_1 has two orbits on $V_1 \backslash 0$, of sizes 6 and 9 (the non-singular and singular vectors, respectively). Hence by Lemma 4.2.13, if $u \in U_1 \backslash 0$ and u has $\mathcal{D}$-length k, then $6^{k-1}\binom{n/2}{k} \leq |\mathrm{orb}_H(u)| \leq 3n$, which yields $k = 1$. Therefore U_1 is a non-degenerate 1-space in V_1. Similarly, every space U_i is a non-degenerate 1-space in some V_j, and hence $\mathcal{D}_1 = \mathcal{D}^*$, as required. ∎

In the next series of triples, H is of type $O_2^{\pm}(q) \wr S_{n/2}$, and they arise because the groups $O_2^{\pm}(q)$ are rather small. For example, we have already seen that $O_2^+(2)$ and $O_2^+(3)$ are reducible on the natural module; so in these cases we will see that H is contained in a member of $\mathcal{C}_1 \cup \mathcal{C}_2$. Also, we have seen in Proposition 2.10.9 that $O_2^+(4)$ is realized over a proper subfield on its natural module over $\mathbf{F}_4$ (in view of the isomorphism $O_2^-(2) \cong O_2^+(4)$); thus groups of type $O_2^+(4) \wr S_{n/2}$ are contained in subfield groups in $\mathcal{C}_5$. Now notice that the geometries of type $O_2^-(3)$ and $O_2^+(5)$ have discriminant a square (see Proposition 2.5.10), and so they contain subgroups in $\mathcal{C}_2$ of type $O_1(3) \wr S_2$ and $O_1(5) \wr S_2$. However $O_2^-(3)$, $O_2^+(5)$, $O_1(3) \wr S_2$ and $O_1(5) \wr S_2$ are all isomorphic to D_8, and hence $O_2^-(3)$ and $O_2^+(5)$ stabilize non-degenerate isometric 1-decompositions on their natural modules. Therefore if H is of type $O_2^-(3) \wr S_{n/2}$ or $O_2^+(5) \wr S_{n/2}$ then H is contained in a member of $\mathcal{C}_2$ of type $O_1(p) \wr S_n$, where p is 3 or 5.

Proposition 6.2.8. *Assume that case* $\mathbf{O}^+$ *holds and that H is of type $O_2^+(2) \wr S_{n/2}$.*

(i) *The set $\mathcal{G}_1(L) = \mathcal{G}_1(H)$ has size four. It consists of the stabilizer of a 1-space which is singular (respectively, non-singular) if $\frac{n}{2}$ is even (respectively, odd); the stabilizer of a totally singular $(\frac{n}{2} - 1)$-space; and the stabilizers of the two totally singular $\frac{n}{2}$-spaces containing this $(\frac{n}{2} - 1)$-space.*

(ii) *L fixes a unique non-zero vector in V.*

(iii) *If $\frac{n}{2}$ is even, then $\mathcal{G}_2(H) = \{\Omega_{\mathcal{D}^*}\}$, where $\mathcal{D}^*$ is a -2-space decomposition.*

Proof. Let e_i and f_i be the two non-zero singular vectors in V_i, and set $v_i = e_i + f_i$, so that v_i is the (unique) non-singular vector of V_i. Define $r_i = r_{v_i}$ and observe that $H = \Omega_{(\mathcal{D})}J$, where $\Omega_{(\mathcal{D})} = \langle r_ir_j \mid 1 \le i,j \le \frac{n}{2}\rangle \cong 2^{n/2-1}$ and $J \cong S_{n/2}$ acts naturally on the sets $\{e_1,\dots,e_{n/2}\}$ and $\{f_1,\dots,f_{n/2}\}$. Put $v = v_1 + \cdots + v_{n/2}$, so that $L \le H < N_\Omega(v)$. Now let W be an L-invariant subspace of V satisfying one of the conditions of Table 4.1.A. Suppose first that $W = \langle w\rangle$ for some vector $w \in V$, and write $w = x_1 + \cdots + x_{n/2}$ with $x_i \in V_i$. Since L contains all elements of the form r_ir_j (this follows easily from the argument in Lemma 6.2.3) it follows that $x_ir_i = x_i$ for all i, and hence $x_i \in \langle v_i\rangle$ for all i. Thus by the transitivity of $L^{\mathcal{D}}$ (Lemma 6.2.2) we conclude that $x_i = v_i$ for all i, and so $w = v$. Thus v is the unique non-zero L-invariant vector in V, proving (ii). Also note that $Q(v)$ is 0 or 1 according as $\frac{n}{2}$ is even or odd. Thus to complete the proof of (i) it suffices to consider the case where $\dim(W) \ge 2$ and W is either non-degenerate or totally singular.

We first argue

$$W \text{ is totally singular.} \tag{6.2.5}$$

For suppose that W is in fact non-degenerate. Then $V = W \perp W^\perp$, and so replacing W with $W^\perp$ if necessary, we can assume that $v \notin W$. Since L is transitive on the vectors v_i, we deduce that $v_i \notin W$ for all i (for otherwise W contains all the v_i and hence contains v). Now let w be a non-singular vector in W. Relabelling the indices if necessary, we may write $w = v_1 + \cdots + v_k + w_{k+1} + \cdots + w_{n/2}$, where $k < \frac{n}{2}$, k is odd and w_i is a singular vector in V_i. Suppose first that $w_i \ne 0$ for some i. Then as $r_1r_i \in L$, we know that W contains $w + wr_1r_i = v_i$, contrary to the fact that $v_i \notin W$ for all i. This leaves the case in which $w_i = 0$ for $k+1 \le i \le \frac{n}{2}$. As $v_1 \notin W$ and k is odd, we have $3 \le k \le \frac{n}{2} - 1$. Now let $h \in L$ induce the 3-cycle $(V_{k-1}, V_k, V_{k+1}) \in Alt(\mathcal{D})$. Then W contains $w + wh = v_{k-1} + v_{k+1}$. Thus by the 2-transitivity of $L^{\mathcal{D}}$, it follows that W contains all vectors of the form $v_i + v_j$, and so W contains $(v_1 + v_2) + \cdots + (v_{k-2} + v_{k-1}) + w = v_k$, which yields a contradiction again. Thus we have established (6.2.5).

Define $U = \langle v_i + v_j \mid 1 \le i,j \le n\rangle$, an H-invariant totally singular $(\frac{n}{2} - 1)$-space. Evidently U is the deleted permutation module for $L^{\mathcal{D}}$, as defined in (5.3.4). Thus if $W \le U$ then $W \in \{\langle v\rangle, U\}$, by Lemma 5.3.4. Next suppose that $W \not\le U$, so that W contains a vector $w = v_1 + \cdots + v_k + w_{k+1} + \cdots + w_{n/2}$, where k is even, w_i is a singular vector in V_i, and $w_j \ne 0$ for some j. If $k \ge 2$, then W contains $w + wr_1r_j = v_j$, which contradicts the fact that W is totally singular. Therefore $k = 0$. If $w_i = 0$ and $w_j \ne 0$ for some i,j, then W contains $w + wr_ir_j = v_j$, another contradiction. Therefore $w_i \ne 0$ for all i, and so W contains $w + wr_ir_j = v_i + v_j$ for all i,j. Thus 2-transitivity ensures that $U < W$, and hence by Lemma 2.5.8.iii, W is one of the two totally singular $\frac{n}{2}$-spaces containing U. The proof of (i) is now complete.

(iii). Now suppose that $H < K = \Omega_{\mathcal{D}^*}$, where $\mathcal{D}^*$ is an a-subspace decomposition of V given by $W_1 \oplus \cdots \oplus W_b$. First observe that (i) implies $L \not\le \Omega_{(\mathcal{D}^*)}$. So as in the proof of Proposition 6.2.7 we deduce $A_{n/2} \cong J' \preceq S_b$, which shows $b \ge \frac{n}{2}$. However a is even (since q is even), and so $b = \frac{n}{2}$. And because $H \ne K$ we know that $\mathcal{D}^*$ is a

-2-decomposition, and so $K \cong 3^{n/2}.2^{n/2-1}.S_{n/2}$ by Proposition 4.2.11(II). Recall from the beginning of this proof that $H \cong 2^{n/2-1}.S_{n/2}$, and thus it is clear that $H^{\mathcal{D}^*} \cong S_{n/2}$ and that $\Omega_{(\mathcal{D})} = O_2(H) \leq \Omega_{(\mathcal{D}^*)}$. Now take $g \in \Omega_{(\mathcal{D})}\backslash 1$ and choose j such that g acts non-trivially on W_j. Thus there is a vector $w \in W_j$ such that $w \neq wg \in W_j$, and so $W_j = \langle w, wg \rangle$. Now $N_H(W_j) \cong 2^{n/2-1}.S_{n/2-1}$, which has no subgroup of index 3. Therefore $N_H(W_j)$ has an orbit $\{w, wg\}$ in W_j, and so

$$|\mathrm{orb}_H(w)| = n. \tag{6.2.6}$$

It now follows from Lemma 4.2.13 that w has $\mathcal{D}$-length 1, $\frac{n}{2}-1$ or $\frac{n}{2}$. If w has $\mathcal{D}$-length 1 however, then $w \in V_i$ for some i, and since W_j is an $O_2^-(2)$-space, w must be non-singular. But then $w = v_i$, which forces $|\mathrm{orb}_H(w)| = \frac{n}{2}$, a contradiction. Next suppose that w has $\mathcal{D}$-length $\frac{n}{2}-1$. Then $w+v$ lies in the same orbit H-orbit as $e_1+\dots+e_s+v_{n/2}$ for some $s < \frac{n}{2}$ (possibly $s = 0$). Observe $|\mathrm{orb}_H(w+v)| = |\mathrm{orb}_H(w)| = n$, and so as before we cannot have $w+v = v_{n/2}$. Thus $s \geq 1$, and it is then easily checked that $|\mathrm{orb}_H(w+v)| \geq \binom{n/2}{s}(\frac{n}{2}-s)2^s > n$, a contradiction. We have therefore proved that w has $\mathcal{D}$-length $\frac{n}{2}$, and hence $w+v$ lies in the same H-orbit as $e_1+\dots+e_s$, where $1 \leq s \leq \frac{n}{2}-1$. Here $|\mathrm{orb}_H(w+v)| \geq \binom{n/2}{s}2^s$, which forces $s = 1$. Therefore $w+v$ lies in the same H-orbit as e_1, whence $w \in \{v+e_i, v+f_i\}$ for some i, which yields $W_j = \langle v+e_i, v+f_i \rangle$. With suitable relabelling, we obtain $W_i = \langle v+e_i, v+f_i \rangle$ for all i. It is rather easy to verify that for $\frac{n}{2}$ even, $V = \langle v+e_1, v+f_1 \rangle \perp \dots \perp \langle v+e_{n/2}, v+f_{n/2} \rangle$ is indeed an H-invariant -2-decomposition of V, and so the proof is complete. ■

Proposition 6.2.9. *Assume that case* $\mathbf{O}^+$ *holds with* H *of type* $O_2^+(3) \wr S_{n/2}$.

(i) *L fixes just two non-zero proper subspaces of V, which we call V_+ and V_-.*

(ii) *V_+ and V_- form a non-degenerate $\frac{n}{2}$-space decomposition $\mathcal{D}^*$ of V; they are $+(\frac{n}{2})$-spaces if $n \equiv 0 \pmod 8$, $-(\frac{n}{2})$-spaces if $n \equiv 4 \pmod 8$, and they are non-isometric yet similar if $\frac{n}{2}$ is odd.*

(iii) $\mathcal{G}_2(H) = \{\Omega_{\mathcal{D}^*}\}$.

Proof. Recall from the discussion just before Proposition 6.1.2 that V_i contains a unique ☐1-space V_i^+ and a unique ⊠1-space V_i^-. We put $V_\epsilon = V_1^\epsilon \perp \dots \perp V_{n/2}^\epsilon$ (where $\epsilon = \pm$), a non-degenerate $\frac{n}{2}$-space. Evidently $D(V_i^\epsilon) = \epsilon$, and so $D(V_\epsilon) = \epsilon^{n/2}$, and thus V_+ and V_- satisfy the description in (ii). Clearly H acts on the set $\{V_1^\epsilon, \dots, V_{n/2}^\epsilon\}$ and so $H < N_\Omega(V_\epsilon)$. Thus H fixes the non-degenerate $\frac{n}{2}$-space decomposition $\mathcal{D}^*$ given by $V = V_+ \perp V_-$, which proves $\Omega_{\mathcal{D}^*} \in \mathcal{G}_2(H)$. It remains to prove (i) and (iii).

(i). Let W be any non-zero proper L-invariant subspace of V. We seek to show that $W = V_\epsilon$. Replacing W by $W^\perp$ if need be, we may suppose that $\dim(W) \leq \frac{n}{2}$. Take $w \in W\backslash 0$ and write $w = v_1 + \dots + v_{n/2}$ with $v_i \in V_i$. Without loss, $v_1 \neq 0$. Now if $x_{i,j}$ is the element in $L_{(\mathcal{D})}$ which negates $V_i \perp V_j$ and centralizes $(V_i \perp V_j)^\perp$ (the proof of Lemma 6.2.3 shows that $L_{(\mathcal{D})}$ contains such an element), then W contains $v_1 = w - wx_{1,2} - wx_{1,3} + wx_{2,3}$. Suppose for the moment that v_1 is singular. Then Lemma 6.2.3.i and the fact that I_1 interchanges the two totally singular points in V_1

shows that $V_1 \leq W$. But then the transitivity of $L^{\mathcal{D}}$ forces $V_i \leq W$ for all i, which is to say $W = V$, a contradiction. Therefore $W \cap V_1 = \langle v_1 \rangle = V_1^{\epsilon}$, and so by the transitivity of $L^{\mathcal{D}}$ we see that $W \cap V_i = V_i^{\epsilon}$ for all i. Thus $V_\epsilon \leq W$, and since $\dim(W) \leq \frac{n}{2}$, we deduce $W = V_\epsilon$, as desired.

(iii). Suppose that $H < K = \Omega_{\mathcal{D}_1}$, where $\mathcal{D}_1$ is an a-subspace decomposition of V given by $W_1 \oplus \cdots \oplus W_b$. If $L \leq \Omega_{(\mathcal{D}_1)}$, then it follows directly from (i) that $\mathcal{D}_1 = \mathcal{D}^*$, as desired. So it may be assumed that $L \not\leq \Omega_{(\mathcal{D}_1)}$, and hence $J' \preceq S_b$. Thus b is $\frac{n}{2}$ or n.

Suppose first that $b = \frac{n}{2}$. Then as $H \neq K$, we see that $\mathcal{D}_1$ is a -2-decomposition. Thus if $w \in W_1 \backslash 0$, then $|\mathrm{orb}_H(w)| \leq 2n$, since each W_i has just four vectors of norm 1 and four of norm -1. But on the other hand, if w has $\mathcal{D}$-length k, then Lemma 4.2.13 shows that $2^{k-1}\binom{n/2}{k} \leq 2n$, which forces $k = 1$, and hence $w \in V_i$ for some i. This argument shows that $W_1 \cup \cdots \cup W_{n/2}$ is contained in $V_1 \cup \cdots \cup V_{n/2}$, which is a contradiction, for the former set has $2n$ vectors of norm 1, while that latter set has only n.

The case $b = n$ is treated in the same fashion. ■

Proposition 6.2.10. *If case* $\mathbf{O}^+$ *holds and* H *is of type* $O_2^+(4) \wr S_{n/2}$, *then* $|\mathcal{G}_5(H)| = 1$. *The overgroup of* H *in* $\mathcal{C}_5$ *is of type* $O_n^{\epsilon}(2)$, *where* $\epsilon = (-)^{n/2}$.

Proof. Let $(V_\sharp, \mathbf{F}_\sharp, Q_\sharp)$ be a subfield geometry as appearing in Table 4.5.A, with $|\mathbf{F}_\sharp| = 2$ and $\mathrm{sgn}(Q_\sharp) = (-)^{n/2}$. Put $\Omega_\sharp = \Omega(V_\sharp, \mathbf{F}_\sharp, Q_\sharp) \leq \Omega$. Then $V_\sharp$ admits a -2-space decomposition $\mathcal{D}_\sharp$ given by $V_\sharp = W_1 \perp \cdots \perp W_{n/2}$. Evidently $W_i\mathbf{F}$ (the $\mathbf{F}$-span of W) is a $+2$-space over $\mathbf{F}$, and hence $V = W_1\mathbf{F} \perp \cdots \perp W_t\mathbf{F}$ is a $+2$-space decomposition of V, and there is no loss in identifying it with $\mathcal{D}$. Now $\frac{1}{2}(O_2^-(2) \wr S_{n/2}) \cong (\Omega_\sharp)_{\mathcal{D}_\sharp} \leq H \cong \frac{1}{2}(O_2^+(4) \wr S_{n/2})$, and since $O_2^-(2) \cong O_4^+(2)$, we deduce $H = (\Omega_\sharp)_{\mathcal{D}_\sharp} < \Omega_\sharp \in \mathcal{G}_5(H)$. Now suppose that $H < K \in \mathcal{C}_5(\Omega)$. Then $K = \Omega_\flat = \Omega(V_\flat, \mathbf{F}_\flat, Q_\flat)$ for some subfield geometry, and by Proposition 4.5.11 there are involutions $\alpha_\sharp$ and $\alpha_\flat$ in $\Gamma\backslash\Delta$ which centralize $\Omega_\sharp$ and $\Omega_\flat$, respectively. Therefore $\alpha_\sharp\alpha_\flat \in C_\Delta(H)$, and since H is absolutely irreducible by Lemma 6.2.3.iii, we deduce $C_\Delta(H) = \mathbf{F}^*$. Consequently $\Omega_\sharp = C_\Omega(\alpha_\sharp) = C_\Omega(\alpha_\flat) = \Omega_\flat$, as desired. ■

Proposition 6.2.11. *Assume that case* $\mathbf{O}^\pm$ *holds and that* H *is of type* $O_2^-(3) \wr S_{n/2}$ *or* $O_2^+(5) \wr S_{n/2}$. *Then* $\mathcal{G}_2(H) = \{\Omega_{\mathcal{D}_\square}, \Omega_{\mathcal{D}_\boxtimes}\}$ *where* $\mathcal{D}_\square, \mathcal{D}_\boxtimes$ *is a* $\square 1$*-space,* $\boxtimes 1$*-space decomposition, respectively.*

Proof. As we remarked before Proposition 6.2.8, I_i preserves a $\square 1$-space decomposition $V_i = W_{2i-1}^\square \perp W_{2i}^\square$ and a $\boxtimes 1$-space decomposition $V_i = W_{2i-1}^\boxtimes \perp W_{2i}^\boxtimes$ of V_i. Consequently H fixes the $\square 1$-space decomposition $\mathcal{D}_\square$ given by $V = W_1^\square \perp \cdots \perp W_n^\square$, and also the $\boxtimes 1$-space decomposition $\mathcal{D}_\boxtimes$ defined similarly. Now suppose that $H < K = \Omega_{\mathcal{D}_1}$, where $\mathcal{D}_1$ is an a-space decomposition $W_1 \oplus \cdots \oplus W_b$. As usual we see that b is $\frac{n}{2}$ or n. However $|O_2^+(5)|$ does not divide $|O_2^-(5)|$, and $|O_2^-(3)|$ does not divide $|O_2^+(3)|$, and so it follows that $b \neq \frac{n}{2}$. Therefore $b = n$. But now the usual argument using Lemma 4.2.13 shows that $4^{k-1}\binom{n/2}{k} \leq |\mathrm{orb}_H(w)| \leq 2n$ for any $w \in W_i\backslash 0$, with w of $\mathcal{D}$-length k. Therefore $k = 1$, which means each W_i is contained in some V_j, and hence W_i is one of $W_{2j-1}^\square$, $W_{2j}^\square$, $W_{2j-1}^\boxtimes$ and $W_{2j}^\boxtimes$. Consequently $\mathcal{D}_1 = \mathcal{D}_\square$ or $\mathcal{D}_\boxtimes$, completing the proof. ■

Now consider the $O_3(3)$-geometry and take a non-degenerate isometric 1-decomposition therein, given by $\langle v_1 \rangle \perp \langle v_2 \rangle \perp \langle v_3 \rangle$, with $(v_i, v_i) = \lambda \in \{\pm 1\}$. The stabilizer of this decomposition is $O_1(3) \wr S_3 \cong 2^3{:}S_3$ which has order $2^4.3$. This is also the order of $O_3(3)$, which means $O_3(3)$ stabilizes this 1-decomposition. Moreover this is the unique non-degenerate isometric 1-decomposition stabilized by $O_3(3)$. For the centralizer of such a decomposition must be the unique normal 2^3 in $O_3(3)$, which equals $E = \langle r_{v_1}, r_{v_2}, r_{v_3} \rangle$. And using Proposition 2.10.11 we see that the spaces $\langle v_i \rangle$ are the only E-invariant 1-spaces. Now observe that since $O_3(3)$ stabilizes a non-degenerate isometric 1-decomposition, so do groups in $\mathcal{C}_2$ of type $O_3(3) \wr S_{n/3}$. This situation is described in the next Proposition.

Proposition 6.2.12. *If case* **O** *holds and* H *is of type* $O_3(3) \wr S_{n/3}$, *then* $\mathcal{G}_2(H) = \{\Omega_{\mathcal{D}^*}\}$ *for some non-degenerate isometric 1-decomposition* $\mathcal{D}^*$.

Proof. As we remarked above, I_i preserves a non-degenerate isometric 1-decomposition $V_i = W_{3i-2} \perp W_{3i-1} \perp W_{3i}$. Since the spaces V_i $(1 \le i \le \frac{n}{3})$ are all isometric, it follows that the spaces W_i $(1 \le i \le n)$ are all isometric, and thus $H \le \Omega_{\mathcal{D}^*}$, where $\mathcal{D}^*$ is the non-degenerate isometric 1-space decomposition of V given by $V = W_1 \perp \cdots \perp W_n$. Suppose now that $H < K = \Omega_{\mathcal{D}_1} \in \mathcal{G}_2(H)$ for some a-space decomposition $\mathcal{D}_1$ given by $V = U_1 \oplus \cdots \oplus U_b$, with $n = ab$. Since L is irreducible on V, it follows that $L \not\le \Omega_{(\mathcal{D}_1)}$, and so $J' \not\le \Omega_{(\mathcal{D}_1)}$. As $\frac{n}{3} \ge 5$, we know that $A_{n/3} \cong J' \preceq S_b$, and hence $a \le 3$. Since $H \ne K$, we have $a \ne 3$. Consequently $a \le 2$, and since $O_1(3)$ and $O_2^{\pm}(3)$ are 2-groups, $\Omega_{(\mathcal{D}_1)}$ is a 2-group. Now $\Omega_{(\mathcal{D})}$ contains an elementary abelian subgroup of order $3^{n/3}$, which necessarily intersects $\Omega_{(\mathcal{D}_1)}$ trivially. Therefore by Proposition 5.2.8.ii, $\frac{b}{3} \ge m_3(S_b) \ge \frac{n}{3}$, which forces $b = n$ and $a = 1$. Now pick $u \in U_1 \backslash 0$ with $\mathcal{D}$-length k. The orbits of I_1 on the non-zero vectors in V_1 have sizes 6, 8 and 12, and hence by Lemma 4.2.13, $6^{k-1}\binom{n/3}{k} \le |\mathrm{orb}_H(u)| \le 2n$, which forces $k = 1$. Therefore each U_i is contained in some V_j. However by our remarks before this Proposition, $W_{3j-2} \perp W_{3j-1} \perp W_{3j}$ is the unique isometric non-degenerate 1-space decomposition of V_j, and it now follows that $\mathcal{D}_1 = \mathcal{D}^*$. ■

The final triple occurs with $m = 4$. In the $O_4^+(2)$-geometry observe that the stabilizer of a -2-space has structure $O_2^-(2) \times O_2^-(2)$, which has order $2^2.3^2 = \frac{1}{2}|O_4^+(2)|$. Therefore there are just two -2-spaces in this geometry. Evidently, these two -2-spaces are orthogonal complements of one another, and so $O_4^+(2)$ stabilizes a unique non-degenerate isometric -2-space decomposition on its natural module.

Proposition 6.2.13. *If case* $\mathbf{O}^+$ *holds and* H *is of type* $O_4^+(2) \wr S_{n/4}$, *then* $\mathcal{G}_2(H) = \{\Omega_{\mathcal{D}^*}\}$ *for some* -2*-space decomposition* $\mathcal{D}^*$.

Proof. In view of the preceding discussion, V_i has precisely two -2-spaces W_{2i-1} and W_{2i}, and they satisfy $V_i = W_{2i-1} \perp W_{2i}$. Thus $H \le \Omega_{\mathcal{D}^*}$, where $\mathcal{D}^*$ is the -2-decomposition given by $V = W_1 \perp \cdots \perp W_{n/2}$. Suppose now that $H < K = \Omega_{\mathcal{D}_1} \in \mathcal{G}_2(H)$ for some a-space decomposition $\mathcal{D}_1$ given by $V = U_1 \oplus \cdots \oplus U_b$, with $n = ab$. As in the previous proofs, we see that $J' \not\leq \Omega_{(\mathcal{D}_1)}$, and so there is a non-trivial homomorphism from $A_{n/4}$ to A_b. Thus $\frac{n}{4} \leq b$, except possibly when $n = 16$; but $O^2(A_4) = A_4$, and so once again we must have $b \geq \frac{n}{4}$ here. Therefore $a \leq 4$. Moreover a is even (since q is even), and so $a \in \{2, 4\}$. Since $|O_4^+(2)|$ does not divide $|O_4^-(2)|$, and since $\mathcal{D}_1$ cannot also be a $+4$-decomposition, it follows that $a = 2$. Furthemore, because the stabilizer of a $+2$-decomposition in V is reducible, we know that $\mathcal{D}_1$ is a -2-decomposition. Therefore if $u \in U_1 \backslash 0$, then $|\mathrm{orb}_H(u)| \leq \frac{3n}{2}$. However the orbits of I_1 on V_1 have sizes 6 and 9, and so if u has $\mathcal{D}$-length k, we obtain $6^{k-1}\binom{n/4}{k} \leq \frac{3n}{2}$. This forces $k = 1$, and so by relabelling we may assume that $u \in U_1 \cap V_1$. Now take $v \in U_1 \backslash \langle u \rangle$, so that $v \in V_j$ for some j (using the same reasoning). If $j \neq 1$, then $(u, v) = 0$ since V_1 and V_j are orthogonal; but then $Q(u + v) = Q(u) + Q(v) = 1 + 1 = 0$, contrary to the fact that U_1 is a -2-space. Therefore $v \in V_1$, which means $U_1 = W_1$ or W_2. Similarly each U_i is equal to some W_j, which proves $\mathcal{D}_1 = \mathcal{D}^*$, as claimed. ∎

§6.3 The case $H \in C_4$

In this section, $H = \Omega_{\mathcal{D}}$, where $\mathcal{D}$ is a tensor product decomposition $(V, \kappa) = (V_1, \mathbf{f}_1) \otimes (V_2, \mathbf{f}_2)$ as described in §4.4 (see in particular the definition of C_4). Define $X_i = X(V_i, \mathbf{f}_i)$ as in §4.4 and define

$$L = \Omega_1 \otimes \Omega_2.$$

Also write $n_i = \dim(V_i)$. With no loss $n_1 \leq \sqrt{n} \leq n_2$, and so $n_2 \geq 4$. Thus Ω_2 is quasisimple, except when $\Omega_2 \approx \Omega_4^+(q)$. But in this case q is odd and Ω_1 is $Sp_4(q)$ or $\Omega_4^-(q)$, both of which are quasisimple. So by interchanging the indices we can assume that

$$\Omega_2 \text{ is quasisimple.} \qquad (6.3.1)$$

We also note that by Proposition 2.10.6

$$\Omega_2 \text{ is absolutely irreducible on } V_2. \qquad (6.3.2)$$

These first examples occur because $GL_2(2) \cong Sp_2(2) \cong S_3$, and so $GL_2(2)'$ and $Sp_2(2)'$ are not absolutely irreducible on the natural 2-dimensional module.

Proposition 6.3.1. *Suppose that $n_1 = q = 2$.*

(i) *If case* **L** *holds, so that H is of type $GL_2(2) \otimes GL_{n/2}(2)$, then the set $\mathcal{G}_3(H)$ is a singleton consisting of a group of type $GL_{n/2}(4)$.*

(ii) *If case* $\mathbf{O}^+$ *holds, so that H is of type $Sp_2(2) \otimes Sp_{n/2}(2)$, then the set $\mathcal{G}_3(H)$ is a singleton consisting of a group of type $GU_{n/2}(2)$.*

Proof. According to Propositions 4.4.10(II) and 4.4.12(II), we have $H = L = \Omega_1 \otimes \Omega_2$, which is equal to $GL(V_1) \otimes GL(V_2)$ in case **L** and $Sp(V_1) \otimes Sp(V_2)$ in case $\mathbf{O}^+$. Since $Sp(V_1) = GL(V_1) \cong S_3$, in both cases we therefore have $\Omega_1' \cong \mathbf{Z}_3$. Clearly Ω_1' is irreducible on V_1, and so $J = \Omega_1' \otimes \Omega_2$ is irreducible on V by Lemma 4.4.3.v. Thus $\mathbf{E} = \mathrm{End}_{\mathbf{F}J}(V)$ is a field extension of $\mathbf{F}$ by Schur's Lemma. Moreover $\Omega_1' \leq \mathbf{E}^*$, and hence $\mathbf{E}$ contains a field $\mathbf{F}_\sharp$ of order 4 (in fact $\Omega_1' = \mathbf{F}_\sharp^*$). Now according to Aschbacher's definition of the family $\mathcal{C}_3$ (see the Remark directly after Table 4.3.A), $N_\Omega(\mathbf{F}_\sharp)$ lies in $\mathcal{C}_3(\Omega)$. Thus by Proposition 4.3.3.iii, we have $H \leq N_\Omega(\mathbf{F}_\sharp) \in \mathcal{C}_3(\Omega)$. Note that as $\mathbf{F}_\sharp^* = \Omega_1' \leq \Omega$, it follows that $N_\Omega(\mathbf{F}_\sharp)$ is local; hence in case $\mathbf{O}^+$, $N_\Omega(\mathbf{F}_\sharp)$ is of type $GU_{n/2}(2)$ in view of Proposition 4.3.14(III). To prove uniqueness, suppose that $H < K \in \mathcal{C}_3$. Then by Proposition 4.3.3.ii, $H < K = N_\Omega(\mathbf{F}_\flat)$ for some field extension $\mathbf{F}_\flat$. Since $\Omega_1' \otimes \Omega_2$ is the only proper normal subgroup of H with cyclic quotient, $\Omega_1' \otimes \Omega_2 \leq C_\Omega(\mathbf{F}_\flat)$. Consequently $\mathbf{F}_\flat \subseteq \mathbf{E}$. Moreover $\Omega_2 \leq GL(V, \mathbf{F}_\flat)$, and since $R_2(Sp_{n/2}(2)) = R_2(L_{n/2}(2)) = \frac{n}{2}$ by Proposition 5.4.13, it follows that $|\mathbf{F}_\flat : \mathbf{F}| = 2$. Consequently $\mathbf{F}_\flat = \mathbf{F}_\sharp$, giving uniqueness. ∎

Assume here that $I_1 \approx O_3(3)$, so that either case $\mathbf{S}$ holds and H is of type $O_3(3) \otimes Sp_{n/3}(3)$, or case $\mathbf{O}^\epsilon$ holds and H is of type $O_3(3) \otimes O^\epsilon_{n/3}(3)$. Recall from §6.2 that I_1 stabilizes a unique non-degenerate isometric 1-decomposition $V_1 = Y_1 \perp Y_2 \perp Y_3$. Write $\mathcal{D}_1$ for this decomposition and write $\mathcal{D}_1 \otimes V_2$ for the $\frac{n}{3}$-decomposition $V = (Y_1 \otimes V_2) \oplus (Y_2 \otimes V_2) \oplus (Y_3 \otimes V_2)$. Since $\mathbf{f} = \mathbf{f}_1 \otimes \mathbf{f}_2$, it is clear that $\mathcal{D}_1 \otimes V_2$ is in fact non-degenerate and isometric, and hence $H \leq \Omega_{\mathcal{D}_1 \otimes V_2} \in \mathcal{C}_2(\Omega)$. This next Proposition shows that $\Omega_{\mathcal{D}_1 \otimes V_2}$ is in fact the unique element of $\mathcal{G}_2(H)$.

Proposition 6.3.2. *If case* $\mathbf{S}$ *or* $\mathbf{O}$ *holds and* H *is of type* $O_3(3) \otimes Sp_{n/3}(3)$ *or* $O_3(3) \otimes O^\epsilon_{n/3}(3)$, *then* $\mathcal{G}_2(H) = \{\Omega_{\mathcal{D}_1 \otimes V_2}\}$, *where* $\mathcal{D}_1$ *is the unique* I_1*-invariant non-degenerate isometric 1-decomposition of* V_1.

Proof. Assume that $H \leq \Omega_{\mathcal{D}^*} \in \mathcal{C}_2(\Omega)$, where $\mathcal{D}^*$ is a subspace decomposition $V = W_1 \oplus \cdots \oplus W_b$. As $P(\Omega_2) > n$ by Proposition 5.2.1 and Corollary 5.2.3.ii, we know that $\Omega_2 \leq \Omega_{(\mathcal{D}^*)}$. And since the irreducible constituents of Ω_2 all have dimension $\frac{n}{3}$, we have $b = 3$. Thus by Lemma 4.4.3.iv, $W_i = U_i \otimes V_2$ for some 1-space U_i in V_1. Moreover as $b = 3$, we know that $\mathcal{D}^*$ is isometric and non-degenerate, and thus because $\mathbf{f} = \mathbf{f}_1 \otimes \mathbf{f}_2$ it follows that the spaces U_i form an isometric, non-degenerate 1-decomposition of V_1. Since V_1 admits a unique such decomposition, we deduce $\mathcal{D}_1 = \{U_1, U_2, U_3\}$, and so $\mathcal{D}^* = \mathcal{D}_1 \otimes V_2$, as required. ∎

Just like $O_3(3)$, the group $GU_2(2)$ also fixes a unique non-degenerate isometric 1-decomposition on its natural module (see the discussion preceding Proposition 6.2.7), and so we obtain the following.

Proposition 6.3.3. *If case* $\mathbf{U}$ *holds and* H *is of type* $GU_2(2) \otimes GU_{n/2}(2)$, *then* $\mathcal{G}_2(H) = \{\Omega_{\mathcal{D}_1 \otimes V_2}\}$, *where* $\mathcal{D}_1$ *is the unique non-degenerate 1-decomposition of* V_1.

Proof. Argue as in the proof of Proposition 6.3.2. ∎

The final family of triples in this section arises because $O_4^+(q)$ preserves a tensor decomposition of type $Sp_2(q) \otimes Sp_2(q)$ on its natural module (see the proof of Proposition 2.9.1.iv).

Proposition 6.3.4. *Assume that case* $\mathbf{O}^+$ *holds and that* H *is of type* $O_4^+(q) \otimes O_{n/4}^{\epsilon}(q)$, *so that* q *is odd. If* $\frac{n}{4}$ *is even and* $D(\mathbf{f}_2)$ *is a square, then* $\mathcal{G}_4(H)$ *is empty. Otherwise,* $\mathcal{G}_4(H)$ *consists of just two groups of type* $Sp_2(q) \otimes Sp_{n/2}(q)$.

Proof. We first determine $\mathcal{G}_4(L)$. As mentioned before, I_1 preserves a symplectic tensor decomposition $(V_1, \mathbf{f}_1) = (V_3, \mathbf{f}_3) \otimes (V_4, \mathbf{f}_4)$. Moreover Ω_1 fixes each factor in this decomposition, and so L fixes both symplectic tensor decompositions $\mathcal{D}_i$ $(i = 3, 4)$ given by $V_i \otimes (V_{7-i} \otimes V_2)$. Note that $\Omega_{\mathcal{D}_i}$ is of type $Sp_2(q) \otimes Sp_{n/2}(q)$. We now aim to prove

$$\mathcal{G}_4(L) = \{H, \Omega_{\mathcal{D}_3}, \Omega_{\mathcal{D}_4}\}. \tag{6.3.3}$$

To prove (6.3.3), assume that $\Omega_{\mathcal{D}^*} \in \mathcal{G}_4(L)$ for some tensor decomposition $\mathcal{D}^*$ given by $V = (V_5, \mathbf{f}_5) \otimes (V_6, \mathbf{f}_6)$. Put $n_i = \dim(V_i)$ and $\Omega_i = \Omega(V_i, \mathbf{f}_i)$ for $i = 5, 6$. As in the beginning of this section, assume without loss that $n_5 \leq n_6$ and that Ω_6 is quasisimple.

Case $n_6 = 4$. Here $n_1 = n_2 = n_5 = n_6 = 4$. Since $\Omega_{\mathcal{D}^*} \notin \mathcal{C}_7$, we must have $\Omega_1 \approx \Omega_5 \approx \Omega_4^+(q)$ and $\Omega_2 \approx \Omega_6 \approx \Omega_4^-(q)$. Thus Ω_6 is the unique subgroup isomorphic to $L_2(q^2)$ in $\Omega_{\mathcal{D}^*}$. Therefore $\Omega_2 = \Omega_6$ and hence by Lemma 4.4.6, $\Omega_{\mathcal{D}^*} = N_\Omega(\Omega_6) = N_\Omega(\Omega_2) = H$, as desired.

Case $n_6 \geq 5$. Since $\frac{n}{4} = n_2 \geq \sqrt{n} \geq n_5$, our assumption $n_6 \geq 5$ ensures that $n_2 > n_5$. Suppose for the moment that the image of Ω_2 in $\overline{\Omega}_5$ is non-trivial. Then by Proposition 5.4.16.ii, we must have $\Omega_2 \approx \Omega_5(q)$ and $\Omega_5 \approx Sp_4(q)$. But this is impossible, for on the one hand $n = 4n_2 = 20$, yet on the other, $\mathbf{f}_i$ is symplectic for $i = 5, 6$, which means n_6 is even and so $8 \mid n$. This contradiction ensures that the image of Ω_2 in $\overline{\Omega}_5$ is trivial, which is to say $\Omega_2 \leq \Omega_6$. Thus as Ω_2 is homogeneous with irreducible constituents of dimension $\frac{n}{4}$, it follows that $n_6 = \frac{n}{4}$ or $\frac{n}{2}$. Suppose first that $n_6 = n_2 = \frac{n}{4}$. As Ω_2 and Ω_6 are both perfect orthogonal or symplectic groups of the same dimension $n_6 \geq 5$ over the same field $\mathbf{F}_q$, and since $\Omega_2 \leq \Omega_6$, it is clear that $\Omega_2 = \Omega_6$. Hence as in the previous paragraph we deduce $\Omega_{\mathcal{D}^*} = H$, as required. Next suppose that $n_6 = \frac{n}{2}$. Because of the restriction $n_i \geq 3$ given in the last row of Table 4.4.A, $\mathbf{f}_5$ and $\mathbf{f}_6$ are symplectic, and so $\Omega_5 \otimes \Omega_6 \approx Sp_2(q) \circ Sp_{n/2}(q)$.

Note that Ω_1 is perfect when $q \geq 5$, and when $q = 3$, we have $O^2(\Omega_1) = \Omega_1$. Thus for all q,

$$L = O^2(L) \leq O^2(\Omega_{\mathcal{D}^*}) \leq \Omega_5 \otimes \Omega_6. \tag{6.3.4}$$

Since L is irreducible, $\overline{L} \not\leq \overline{\Omega}_6$, and since we have already shown that $\Omega_2 \leq \Omega_6$, we have $\overline{\Omega}_1 \not\leq \overline{\Omega}_6$. On the other hand, $\overline{\Omega}_1 \cong L_2(q) \times L_2(q)$ does not embed in $\overline{\Omega}_5 \cong L_2(q)$ and therefore

$$1 < \overline{\Omega}_1 \cap \overline{\Omega}_6 < \overline{\Omega}_1. \tag{6.3.5}$$

Next we argue that

$$\overline{\Omega}_1 \cap \overline{\Omega}_6 = \overline{\Omega}_i \text{ for some } i \in \{3, 4\}. \tag{6.3.6}$$

When $q \geq 5$, the group $\overline{\Omega}_1$ is the product of the two simple groups $\overline{\Omega}_3, \overline{\Omega}_4$ and hence (6.3.6) is immediate from (6.3.5). Now take $q = 3$. Then $\overline{\Omega}_1 \cong L_2(3) \times L_2(3)$ and the irreducibility of L ensures that the image of $\overline{\Omega}_1$ in $\overline{\Omega}_5$ is an irreducible subgroup of $PSL(V_5) \cong L_2(3)$. Since any irreducible subgroup of $PSL(V_5)$ contains an involution, and since any non-trivial image of $L_2(3)$ has order divisible by 3, $\overline{\Omega}_1$ must project onto $PSL(V_5)$ with kernel isomorphic to $L_2(3)$. However the groups $\overline{\Omega}_3, \overline{\Omega}_4$ are the only normal subgroups of $\overline{\Omega}_1$ isomorphic to $L_2(3)$, and so (6.3.6) holds in this case as well.

It now follows that $\Omega_i \leq \Omega_6$, and hence $\Omega_i\Omega_2 \leq \Omega_6$. Since both $\Omega_i\Omega_2$ and Ω_6 act homogeneously on V, and with absolutely irreducible constituents of dimension $\frac{n}{2}$, it follows from Lemma 4.4.3.i that $C_{GL(V)}(\Omega_i\Omega_2) \cong C_{GL(V)}(\Omega_6) \cong GL_2(q)$. Thus the two centralizers are equal. And since $\Omega_{7-i} \cong \Omega_5 \cong SL_2(q)$, it follows that $\Omega_{7-i} = \Omega_5$. Therefore by Lemma 4.4.6, $\Omega_{\mathcal{D}^*} = N_\Omega(\Omega_5) = N_\Omega(\Omega_{7-i}) = \Omega_{\mathcal{D}_{7-i}}$, and so the proof of (6.3.3) is now finished.

We are now in a position to complete the proof. For (6.3.3) implies that

$$\mathcal{G}_4(H) = \begin{cases} \text{is empty} & \text{if } H \text{ interchanges } \Omega_3 \text{ and } \Omega_4 \\ \{\Omega_{\mathcal{D}_3}, \Omega_{\mathcal{D}_4}\} & \text{if } H \text{ does not interchange } \Omega_3 \text{ and } \Omega_4. \end{cases} \tag{6.3.7}$$

Thus the Proposition follows from (6.3.7) and Propositions 4.4.14(IV), 4.4.15(IV) and 4.4.17(IV). ■

Chapter 7
DETERMINING THE MAXIMALITY OF THE MEMBERS OF $\mathcal{C}$ PART I

The goal in this chapter is to classify all triples of the form (H, K, Ω), satisfying

$$H < K < \Omega, \tag{7.0.1}$$

with $H, K \in \mathcal{C}$, and where $\Omega = \Omega(V, \mathbf{F}, \kappa)$ is as in the previous chapter. We will in fact show that every such triple appears in Table 3.5.H. In §7.i below we will consider the case in which $H \in \mathcal{C}_i$. Recall the assumption $n \geq 13$ which we imposed at the beginning of Chapter 6 (see (6.0.3)). Furthermore, notice that as we are assuming that H is a proper subgroup of K, it must be the case that

$$H \text{ and } K \text{ are not of the same type.} \tag{7.0.2}$$

For otherwise H and K would be A-conjugate by Theorem 3.1.1, which would force H and K to be equal. The main result of this chapter may be stated as follows.

Theorem 7.0.1. *All triples (H, K, Ω) as in (7.0.1) with $H, K \in \mathcal{C}$ appear in Table 3.5.H.*

§7.1. The case $H \in \mathcal{C}_1$

Here we assume that $H \in \mathcal{C}_1$, so that $H = N_\Omega(W)$ or $N_\Omega(W, U)$, where W is an m-space in V and U is an $(n-m)$-space, as described in Table 4.1.A.

Proposition 7.1.1. *If H is of type P_m, so that $H = N_\Omega(W)$ and W is totally singular, then (H, K, Ω) appears in Table 3.5.H.*

Proof. Here H contains a Sylow p-subgroup P of Ω by Proposition 4.1.14, and it is easily checked that no member of $\mathcal{C}_2 \cup \cdots \cup \mathcal{C}_8$ contains such a Sylow subgroup. Therefore $K \in \mathcal{C}_1$. Indeed, either $K = N_\Omega(W_1)$ for some non-zero proper totally singular subspace W_1 of V, or case **L** holds and K is of type $P_{k,n-k}$ where $1 \leq k < \frac{n}{2}$. However the latter alternative is impossible, for in case **L** a group of type P_m has irreducible constituents of dimensions m and $n-m$, while a group of type $P_{k,n-k}$ has irreducible constituents of dimensions $k, k, n-2k$. Therefore $K = N_\Omega(W_1)$. Evidently P fixes a 1-space in both W and W_1. Yet Proposition 4.1.14 implies that P fixes a unique totally singular 1-space in V. Therefore $W \cap W_1 \neq 0$. According to Corollary 4.1.10, $SL(W) \leq H^W$, and in particular H^W is irreducible on W. It now follows that $W < W_1 < W^\perp$, which means H is reducible on $W^\perp/W$. Now let Y and X be as described in Lemma 4.1.12, so that $V = (W \oplus Y) \perp X$ and $W \perp X = W^\perp$. Evidently $\Omega(X, \kappa) \leq H$, and since $H^{W^\perp/W}$ is reducible, $\Omega(X, \kappa)$ is also reducible. Therefore Proposition 2.10.6 implies that case **O** holds and $\Omega(X, Q) \approx \Omega_2^\pm(q)$. In particular, $\dim(X) = 2$ and so $\dim(W) = \frac{n}{2} - 1$. Therefore $\dim(W_1) = \frac{n}{2}$ and hence H is of type $P_{n/2-1}$ in case $\mathbf{O}^+$. Thus (H, K, Ω) appears in Table 3.5.H in view of Proposition 6.1.1. ∎

Proposition 7.1.2. *If case* **L** *holds and H is of type $P_{m,n-m}$, then (H, K, Ω) appears in Table 3.5.H.*

Proof. Here $\Omega = SL(V)$ and $H = N_\Omega(U, W)$, where U and W are as described in Table 4.1.A. Now H contains a Sylow p-subgroup of Ω, and as above we deduce that $K \in \mathcal{C}_1$. We now appeal to Proposition 6.1.3. ∎

Proposition 7.1.3. *If H is of type $GL_m(q) \oplus GL_{n-m}(q)$ in case* **L**, *or if W is non-degenerate in cases* **U**, **S** *or* **O**, *then (H, K, Ω) appears in Table 3.5.H.*

Proof. Here H contains a subgroup $L = \Omega_1 \times \Omega_2$, where $\Omega_i = \Omega(W_i, \kappa)$, $W_1 = W$, and $W_2 = W^\perp$ in cases **U**, **S** and **O**, and $W_2 = U$ in case **L**. Putting $n_i = \dim(W_i)$, our convention is $n_2 \geq n_1$, and hence

$$n_2 \geq \left\lceil \frac{n}{2} \right\rceil \geq 7. \tag{7.1.1}$$

In particular, Ω_2 is quasisimple. First we prove

$$K \notin \mathcal{C}_2 \cup \mathcal{C}_4 \cup \mathcal{C}_7. \tag{7.1.2}$$

Otherwise, $K = \Omega_{\mathcal{D}}$, where $\mathcal{D}$ is an a-subspace decomposition $V_1 \oplus \cdots \oplus V_t$ with $at = n$ (see §4.2) or a tensor decomposition $V_1 \otimes \cdots \otimes V_t$ (see §4.4). Since $P(\Omega_2) \geq n_2^2 + 3 > n$ by Corollary 5.2.3 and Proposition 5.2.1.i, we see that $\Omega_2 \leq \Omega_{(\mathcal{D})}$, and hence we obtain a non-trivial projective representation of Ω_2 in $PGL(V_i)$ for some i. Therefore $\dim(V_i) \geq R_p(\Omega_2/Z(\Omega_2))$ by Corollary 5.3.3. However $R_p(\Omega_2/Z(\Omega_2)) = n_2$ by Proposition 5.4.15, and since $\dim(V_i)$ properly divides n, it follows from (7.1.1) that $n_2 = \dim(V_i) = \frac{n}{2}$. Since W is not similar to $W^\perp$, case $\mathbf{O}^-$ holds, $\frac{n}{2}$ is even, W is a $\pm\frac{n}{2}$-space and $W^\perp$ is a $\mp\frac{n}{2}$-space. But according to Table 4.2.A and Proposition 4.2.11(II), in case $\mathbf{O}^-$ there are no members of $\mathcal{C}_2$ with $t = 2$ and $\frac{n}{2}$ even. Obviously $K \notin \mathcal{C}_7$ (as $(\frac{n}{2})^t > n$) and also $K \notin \mathcal{C}_4$ in view of the restriction $\dim(V_1) \geq 3$ in the bottom row of Table 4.4.A. Thus we reach the desired contradiction.

Next we prove

$$K \notin \mathcal{C}_3. \tag{7.1.3}$$

Otherwise $\Omega_2 \leq \Omega_\sharp = \Omega(V_\sharp, \mathbf{F}_\sharp, \kappa_\sharp)$ where $(V_\sharp, \mathbf{F}_\sharp, \kappa_\sharp)$ is a field extension geometry as described in §4.3. Here $\mathbf{F}_\sharp$ is an extension field of degree r over $\mathbf{F}$. Therefore by Corollary 5.3.3, $n_2 = R_p(\Omega_2/Z(\Omega_2)) \leq \dim_{\mathbf{F}_\sharp}(V_\sharp) = \frac{n}{r}$, whence $n_2 = \frac{n}{r} = \frac{n}{2}$. So as before, case $\mathbf{O}^-$ holds, $\frac{n}{2}$ is even, and we obtain $\Omega^+_{n/2}(q) \times \Omega^-_{n/2}(q) \cong L \leq \Omega_\sharp \cong \Omega^-_{n/2}(q^2)$, which violates Proposition 5.5.7.

Furthermore, we have

$$K \notin \mathcal{C}_5. \tag{7.1.4}$$

For otherwise, $\Omega_2 \leq \Omega(V_\sharp, \mathbf{F}_\sharp, \kappa_\sharp)$ for some subfield geometry $(V_\sharp, \mathbf{F}_\sharp, \kappa_\sharp)$ as described in §4.5. Here $\mathbf{F}_\sharp$ is a proper subfield of $\mathbf{F}$. But then $\mathbf{F}_p(\Omega_2) \subseteq \mathbf{F}_\sharp$, contrary to Proposition 2.10.9.

Now we establish

$$K \notin \mathcal{C}_6. \tag{7.1.5}$$

Otherwise, $n = r^b$ for some prime r distinct from p, and so there is a non-trivial projective representation of Ω_2 to $PGL_{2b}(r)$. And since $n_2 \geq 7$, it follows from Corollary 5.3.10 that $R_{p'}(\Omega_2/Z(\Omega_2)) \geq 8$, and so $b \geq 4$. However $n_2 \geq \frac{r^b}{2} \geq 8$. Now if $n_2 = 8$, then it must be the case that $n = 16$, and as before this forces case $\mathbf{O}^-$ to hold. However $\mathcal{C}_6$ is void in case $\mathbf{O}^-$, and hence $n_2 > 8$. But then by Corollary 5.3.10.i, $2b \geq R_{p'}(\Omega_2/Z(\Omega_2)) > n_2^2$, which conflicts with the fact that $n_2 \geq \frac{1}{2}r^b$. This proves (7.1.5).

Our next claim is

$$K \notin \mathcal{C}_8. \tag{7.1.6}$$

If case $\mathbf{S}$ holds, then q is even and $K = N_\Omega(P)$ for some quadratic form P satisfying $\mathbf{f}_P = \mathbf{f}$. Since $(W, \mathbf{f})$ is non-degenerate, (W, P) is non-degenerate. But this forces $Sp_{n_2}(q) \cong \Omega_2 \leq I(W, P) \cong O^{\pm}_{n_2}(q)$, which is impossible. If case $\mathbf{L}$ holds, then $K = N_\Omega(\kappa)$ for some non-degenerate form κ. However the only $SL(W_2)$-invariant form on W_2 is the zero from, which means W_2 is totally singular with respect to κ. But (V, κ) contains no totally singular spaces of dimension greater than $\frac{n}{2}$, and this contradiction completes the proof of (7.1.6).

Obviously (7.1.2)-(7.1.6) imply that $K \in \mathcal{C}_1$, which means H fixes some non-zero proper subspace of V other than W_1 or W_2. Since Ω_2 is irreducible on W_2, it follows from Lemma 2.10.11 that H^{W_1} is reducible on W_1. By Lemma 4.1.1.v, $H^{W_1} = I(W_1, \kappa)$, and hence by Proposition 2.10.6 case $\mathbf{O}$ holds, W is a $+2$-space and $q \leq 3$. Thus (H, K, Ω) appears in Table 3.5.H in view of Proposition 6.1.2. ■

Lemma 7.1.4. *If W is a non-singular 1-space in case $\mathbf{O}^{\pm}$ with q even, then $\mathcal{G}_\mathcal{C}(H)$ is empty.*

Proof. Here $H \cong Sp_{n-2}(q)$ and the result is left as an exercise to the reader. The proof is in fact an easier version of that given in Proposition 7.1.3. ■

§7.2 The case $H \in \mathcal{C}_2$

In this section assume that $H \in \mathcal{C}_2(\Omega)$, so that $H = \Omega_\mathcal{D}$, where $\mathcal{D}$ is an m-space decomposition of V given by $V = V_1 \oplus \cdots \oplus V_t$, as described in Table 4.2.A. Thus $n = mt$ where $m = \dim(V_i)$.

We first consider the case in which m is small. Indeed, all but the last Proposition of this section is devoted to the case $m \leq 4$. For these small values of m, we know that H is of type $GL^{\pm}_m(q) \wr S_t$, $Sp_m(q) \wr S_t$ or $O^{\xi}_m(q) \wr S_t$. We retain the notation $X_i = X(V_i, \kappa)$ introduced in (4.2.8), so that $I_{(\mathcal{D})} = I_1 \times \cdots \times I_t$. Further, $H_I = I_{(\mathcal{D})}{:}J$, where J appears in Lemma 4.2.1 (see also Lemma 4.2.8), and as in (6.2.2) we set $L = (J')^H$. We begin with a preliminary Lemma which enables us to show that H is rarely contained in a member of $\mathcal{C}_5$.

Lemma 7.2.1. *Assume that $t \geq 4$ and that H is contained in a member of $\mathcal{C}_5$. Then H is of type $O_2^+(4) \wr S_{n/2}$.*

Proof. Since H is contained in a member of $\mathcal{C}_5$, we know that $L \leq \Omega_\sharp = \Omega(V_\sharp, \mathbf{F}_\sharp, \kappa_\sharp)$ for some subfield geometry as described in §4.5. Consequently, using the notation given in (2.10.4), $\mathbf{F}_p(L) \subseteq \mathbf{F}_\sharp$, which means $\mathbf{F}_p(L)$ is a proper subfield of $\mathbf{F}$. Now by the proof of Lemma 6.2.3, we see that L contains all elements of the form $x_g = (g, g^{-2}, g, 1, \ldots, 1)(234)$ for all $g \in I_1$. Evidently $\mathrm{tr}(x_g) = \mathrm{tr}_{V_1}(g) + m(t-4)$, and so $\mathbf{F}_p(I_1) \subseteq \mathbf{F}_p(L)$. The Lemma now follows from Proposition 2.10.9.ii (along with the fact that $q = p$ when H is of type $O_1(q) \wr S_n$ — this is the restriction given in row 4 of Table 4.2.A). ∎

Proposition 7.2.2. *If $m = 1$ then (H, K, Ω) appears in Table 3.5.H.*

Proof. Since $n \geq 13$, we have by Proposition 5.3.7.i

$$R(J') = n - 2 > \frac{n}{2}. \tag{7.2.1}$$

Now the non-abelian composition factors of the members of $\mathcal{C}_3 \cup \mathcal{C}_4 \cup \mathcal{C}_6 \cup \mathcal{C}_7$ are either classical groups of dimension at most $\frac{n}{2}$ or alternating groups of degree at most $\log_2(n)$. So in light of (7.2.1), J' embeds in none of these composition factors, which shows

$$K \notin \mathcal{C}_3 \cup \mathcal{C}_4 \cup \mathcal{C}_6 \cup \mathcal{C}_7. \tag{7.2.2}$$

Next we argue that

$$K \notin \mathcal{C}_2. \tag{7.2.3}$$

For otherwise, $K = \Omega_{\mathcal{D}_1}$ for some m_1-space decomposition $\mathcal{D}_1$ of V. By (7.2.1), $J' \not\leq \Omega_{(\mathcal{D}_1)}$ and hence $J'^{\mathcal{D}_1} \cong A_n$. But then $m_1 = 1$, which means H and K are of the same type, violating (7.0.2). This proves (7.2.3), and so in view of (7.2.2) and Lemma 7.2.1 we deduce $K \in \mathcal{C}_1 \cup \mathcal{C}_8$. Now if $K \in \mathcal{C}_1$, then Lemma 6.2.3.iii ensures that Case **L** holds and $q = 2$; hence (H, K, Ω) appears in Table 3.5.H by Proposition 6.2.4.

Therefore we are left with the case in which $K \in \mathcal{C}_8$. Since $\mathcal{C}_8$ is void in cases **O** and **U**, and since case **S** does not hold, we know that case **L** holds. Thus $L \leq K^\infty = \Omega_\sharp = \Omega(V, \mathbf{F}, \kappa_\sharp)$, where $(V, \mathbf{F}, \kappa_\sharp)$ is a geometry as described in §4.8. When $q = 2$ we may appeal to Proposition 6.2.4 and so for the rest of this proof we can take $q \geq 3$. Since $L_{(\mathcal{D})}$ contains elements of the form $(g, g^{-2}, g, 1, \ldots, 1)$ (see the proof of Lemma 6.2.3), it follows that $\overline{L}$ has a non-trivial abelian normal subgroup, and hence L is contained in a member of $\mathcal{C}(\Omega_\sharp)$ by Aschbacher's Theorem 1.2.1. Moreover L is absolutely irreducible on V by Lemma 6.2.3.iii, and hence L is not contained in a member of $\mathcal{C}_1(\Omega_\sharp) \cup \mathcal{C}_3(\Omega_\sharp)$. The proofs of (7.2.2) and (7.2.3) show that L is not contained in a member of $\mathcal{C}_4(\Omega_\sharp) \cup \mathcal{C}_6(\Omega_\sharp) \cup \mathcal{C}_7(\Omega_\sharp)$, and Lemma 7.2.1 implies that L is not contained in a member of $\mathcal{C}_5(\Omega_\sharp)$. Therefore L is contained in a member of $\mathcal{C}_2(\Omega_\sharp) \cup \mathcal{C}_8(\Omega_\sharp)$. Suppose for the moment that L is contained in a member of $\mathcal{C}_8(\Omega_\sharp)$. Since $\mathcal{C}_8$ is void for the unitary and orthogonal groups, and for the symplectic groups in odd characteristic, it follows that q is even, $\kappa_\sharp$ is symplectic, and L is contained in a member of $\mathcal{C}(\Omega(V,P))$ for some non-degenerate quadratic form P whose associated bilinear form is $\kappa_\sharp$. But then repeating the above argument yields $L \preceq \mathcal{C}_2(\Omega(V,P))$. Since q is even, $\mathcal{C}_2(\Omega(V,P))$ contains no 1-decompositions, and hence L must fix a k-decomposition for some $k > 1$. But this forces $A_n \preceq GL_k(q) \wr S_{n/k}$, which is impossible. Consequently L is contained in a member of $\mathcal{C}_2(\Omega_\sharp)$, and L must fix a non-degenerate 1-decomposition in $(V, \kappa_\sharp)$. Therefore either $\kappa_\sharp$ is unitary or q is odd and $\kappa_\sharp$ is orthogonal. Suppose first that q is odd and $\kappa_\sharp$ is orthogonal, so that $L \preceq 2^{n-1}.A_n$. Now by Lemma 6.2.3.ii, $(q-1)^{n-2}|A_n|$ divides $|L|$. Hence $(q-1)^{n-2} \mid 2^{n-1}$, which forces $q = 3$. Therefore (H,K,Ω) appears in Table 3.5.H by Proposition 6.2.5.i. Now take the case where $\kappa_\sharp$ is unitary. Then $L \preceq (\mathbf{Z}_{\sqrt{q}+1})^{n-1}.A_n$, and consequently $(q-1)^{n-2} \mid (\sqrt{q}+1)^{n-1}$. Thus $q = 4$ and the result follows from Proposition 6.2.5.ii. ∎

Proposition 7.2.3. *If $m = 2$ then (H,K,Ω) appears in Table 3.5.H.*

Proof. We begin by showing

$$K \notin \mathcal{C}_3. \tag{7.2.4}$$

For if false, then $L \leq GL(V, \mathbf{F}_\sharp)$ for some field extension $\mathbf{F}_\sharp$ of $\mathbf{F}$, where $|\mathbf{F}_\sharp : \mathbf{F}| = r$ is a prime divisor of n. Therefore $\frac{n}{r} \geq R(A_{n/2})$ and so it follows easily from Proposition 5.3.7 that $r = 2$. Now because L is not absolutely irreducible on V, Lemma 6.2.3 implies that case $\mathbf{O}^+$ holds and that V_i is a $+2$-space with $q \leq 3$. Suppose first that $q = 2$. Then L fixes the non-zero vector v, as described in the proof of Proposition 6.2.8. Therefore L also fixes the space $W = v\mathbf{F}_\sharp$, which has dimension 2 over $\mathbf{F}$. But L is perfect, and so L centralizes W. This contradicts Proposition 6.2.8.ii, which asserts that L fixes only one non-zero vector in V. This leaves the case $q = 3$. But here $I_1 \cong 2^2$, and so $2^{n-4} \preceq L$ by Lemma 6.2.3.ii; this yields a contradiction for $m_2(GL_{n/2}(9)) = \frac{n}{2} < n-4$ by Lemma 5.5.2.i.

Next we prove

$$K \notin \mathcal{C}_4 \cup \mathcal{C}_7. \tag{7.2.5}$$

Otherwise, write $K = \Omega_{\mathcal{D}_1}$, where $\mathcal{D}_1$ is the tensor decomposition $W_1 \otimes \cdots \otimes W_b$ of V, with $m_i = \dim(W_i)$ (see §§4.4, 4.7). Clearly $\frac{n}{2} > \log_2(n) \geq b$, and so $A_{n/2} \npreceq S_b$. Therefore $J' \leq \Omega_{(\mathcal{D}_1)} \leq GL(W_1) \otimes \cdots \otimes GL(W_b)$. Consequently $J' \preceq PGL(W_i)$ for some i, and with no loss we can take $i = 1$. Suppose first that $m_1 < \frac{n}{2}$. Then $m_1 \leq \frac{n}{3}$, and since $n \geq 14$ it follows from Proposition 5.3.7 that $(n, b, m_1, p) = (16, 2, 4, 2)$. Therefore either case $\mathbf{O}^+$ holds and $K \cong Sp_4(q) \wr 2$ (see Proposition 4.7.5(II)) or case $\mathbf{L}^\epsilon$ holds and $K \cong L_4^\epsilon(q) \wr 2$ (see Proposition 4.7.3(II)). In the latter case, $|H| = |L_2^\epsilon(q)^8|(q-\epsilon)^7|S_8|$, and so $H \npreceq K$ by Lagrange's Theorem. In the former case, we must have $A_8 \preceq Sp_4(q)$. But then the subgroup 7:3 of A_8 must act irreducibly on a 3-dimensional subspace X of the natural 4-dimensional module for $Sp_4(q)$. However X cannot be totally singular (since the maximal totally singular spaces in the $Sp_4(q)$-geometry have dimension 2) and X cannot be non-degenerate (since non-degenerate symplectic spaces have even dimension). Thus $A_8 \npreceq Sp_4(q)$, and we have therefore eliminated the case in which $m_1 < \frac{n}{2}$. This leaves the case in which $m_1 = \frac{n}{2}$, and hence $K \in \mathcal{C}_4$. Since $R(A_{n/2}) > 2$, we know that

$J' \leq GL(W_1)$, and since K normalizes $GL(W_1)$, we see that $L \leq GL(W_1)$. Consequently L is reducible on V, and so as before, case $\mathbf{O}^+$ holds, V_i is a $+2$-space and $q \leq 3$. If $q = 2$, then K is of type $Sp_{n/2}(2) \otimes Sp_2(2)$, and hence by Proposition 6.3.1, K is contained in a member of $\mathcal{C}_3$, which contradicts (7.2.4). And if $q = 3$, then L fixes the four $\frac{n}{2}$-spaces of the form $W_1 \otimes w_2$ $(w_2 \in W_2 \backslash 0)$, and this runs contrary to Propositions 6.2.9.i. This final contradiction completes the proof of (7.2.5).

Now we argue

$$K \notin \mathcal{C}_6. \tag{7.2.6}$$

Otherwise, $n = 2^b$ with $b \geq 4$ and K is described in §4.6. Writing $\overline{\overline{}}$ to denote reduction modulo $O_2(K)$, we have $A_{2^b-1} \cong J' \preceq \overline{\overline{K}} \preceq GL_{2b}(2)$. Therefore by Proposition 5.3.7, we must have $b = 4$. Thus if r is a prime divisor of $|\Omega_i|$, then $r^8 \mid |H|$, and hence $r^8 \mid |K|$. But $\overline{\overline{K}}$ is isomorphic to $Sp_8(2)$, $\Omega_8^{\pm}(2)$ or $O_8^{\pm}(2)$, and so $r = 2$. Therefore Ω_i is a 2-group and consequently case $\mathbf{O}^+$ holds, $\Omega_i \approx \Omega_2^{\xi}(q)$ and $|H| = \frac{1}{4}(|O_2^{\xi}(q) \wr S_8|) = 2^6(q-\xi)^8 8!$ and $|K|$ divides $2^9|O_8^+(2)| = 2^{22}.3^5.5^2.7$ (see Propositions 4.2.11 and 4.6.8). Therefore $q - \xi = 2$, which is to say $(q, \xi) = (3, +)$. We now see that $H \cong 2^{14}.S_8$, and $K \cong 2_+^{1+8}.\Omega_8^+(2)$. However $m_2(O_2(K)) = 5$, and hence $|\overline{\overline{O_2(H)}}| \geq 2^9$. But then $|\overline{\overline{H}}|_2 = |\overline{\overline{O_2(H)}}|_2 |\overline{\overline{J}}|_2 \geq 2^9.2^7 > |\overline{\overline{K}}|_2$.

Thus $K \in \mathcal{C}_1 \cup \mathcal{C}_2 \cup \mathcal{C}_5 \cup \mathcal{C}_8$. Note that when $K \in \mathcal{C}_5$ we may appeal to Lemma 7.2.1 and Proposition 6.2.10 to conclude that (H, K, Ω) appears in Table 3.5.H. We consider the remaining possibilites for K in turn.

Case $K \in \mathcal{C}_1$. Here L is reducible and so Lemma 6.2.3 ensures that case $\mathbf{O}^+$ holds, $\mathcal{D}$ is a $+2$-decomposition and $q \leq 3$. The result follows from Propositions 6.2.8 and 6.2.9.

Case $K \in \mathcal{C}_2$. Write $K = \Omega_{\mathcal{D}^*}$ for some a-space decomposition $\mathcal{D}^*$ given by $V = W_1 \oplus \cdots \oplus W_b$ (so that $n = ab$). Since $R(A_{n/2}) \geq \frac{n}{2} - 4$ by Proposition 5.3.7, either $a \geq \frac{n}{2} - 4$ or $b \geq \frac{n}{2}$. Assume first that $a \geq \frac{n}{2} - 4$. Then $a \geq \frac{n}{4}$ and hence $b \leq 4$. Therefore $J' \leq \Omega_{(\mathcal{D}^*)}$, and hence $L \leq \Omega_{(\mathcal{D}^*)}$. Thus L is reducible, and so by Lemma 6.2.3, case $\mathbf{O}^+$ holds, V_i is a $+2$-space and $q \leq 3$. The result now follows from Propositions 6.2.8 and 6.2.9. Next take $b = \frac{n}{2}$. Since $H \neq K$, the only possibility occurs in case $\mathbf{O}^+$, with $\mathcal{D}$ a $+2$-decomposition and $\mathcal{D}^*$ a -2-decomposition. Thus $2^{n/2-2}(q-1)^{n/2}(\frac{n}{2})! = |H|$ divides $2^{n/2-2}(q+1)^{n/2}(\frac{n}{2})! = |K|$, and so $q \leq 3$. The result now follows from Propositions 6.2.8 and 6.2.9, again. Finally consider the case $b = n$. Then case $\mathbf{S}$ does not hold. Suppose that case $\mathbf{L}^{\epsilon}$ holds. Then $SL_2(q)^{n/2} \preceq K \preceq \mathbf{Z}_{q-\epsilon} \wr S_n$. If $L_2(q)$ is simple, then the image of $SL_2(q)^{n/2}$ in S_n is a covering group of $L_2(q)^{n/2}$, and hence Proposition 5.2.7 implies that $n \geq \frac{n}{2}P(L_2(q))$, which is absurd. Therefore $L_2(q)$ is not simple, and hence $q \leq 3$. Since $3 \mid |SL_2(q)|$, we have $m_3(K) \geq \frac{n}{2}$, and since $m_3(S_n) < \frac{n}{2}$ (see Proposition 5.2.8), we have $3 \mid q - \epsilon$, and so $(q, \epsilon) = (2, -)$. We may now appeal to Proposition 6.2.7. This leaves case $\mathbf{O}$. Here $q = p$ is odd, H is of type $O_2^{\xi}(p) \wr S_{n/2}$, K is of type $O_1(p) \wr S_n$, and $I_{\mathcal{D}^*} \cong 2 \wr S_n$. By Proposition 5.2.8, $m_r(K) \leq \frac{n}{3}$ for all odd primes r, and so $q - \xi = 2^i$ for some i. Thus $|H| = 2^{(i+1)n/2-2}(\frac{n}{2})!$ divides $2^{n-1}n!$, and so $i \leq 2$. Therefore $(q, \xi) \in \{(3, +), (5, +), (3, -)\}$, and the result now follows from Propositions 6.2.9 and 6.2.11.

Case $K \in \mathcal{C}_8$. Here $K = N_\Omega(\kappa_\sharp)$ for a suitable form $\kappa_\sharp$ as described in §4.8. It follows from Lemma 6.2.3.i that $\Omega_{(\mathcal{D})}^{V_i} = I_i$, and hence $I_i \leq I(V_i, \kappa_\sharp)$. Since case **L** or **S** holds, I_i is irreducible on V_i, and hence $(V_i, \kappa_\sharp)$ is either totally singular or non-degenerate. However Ω_i acts non-trivially on V_i (indeed $\Omega_i = SL(V_i)$ or $Sp(V_i)$) yet Ω_i centralizes V/V_i. And Lemma 4.1.12.ii shows that any element which acts non-trivially on a totally singular space also acts non-trivially on the quotient space. Therefore $(V_i, \kappa_\sharp)$ must be non-degenerate. Suppose first that $\kappa_\sharp$ is quadratic. Then $SL_2(q) \cong \Omega_i \leq I(V_i, \kappa_\sharp) \cong D_{2(q\pm1)}$, which forces $q = 2$. Thus by the definition of $\mathcal{C}_8$ (see Table 4.8.A) we know that case **S** holds and the result follows from Proposition 6.2.6. Next assume that $\kappa_\sharp$ is symplectic, so that case **L** holds. Here $GL_2(q) \cong I_i \leq I(V_i, \kappa_\sharp) \cong Sp_2(q)$, and so once again $q = 2$. We appeal to Proposition 6.2.6 as before. Finally, assume that $\kappa_\sharp$ is unitary, so that case **L** holds. Here $GL_2(q)$ embeds in $GU_2(q^{1/2})$, which is impossible by Lagrange's Theorem. ∎

Proposition 7.2.4. *Assume that $m \geq 3$ and that $\Omega_{(\mathcal{D})}$ is soluble. Then (H, K, Ω) appears in Table 3.5.H.*

Proof. Note that according to Proposition 2.9.2, $\Omega_i \approx SU_3(2)$, $\Omega_3(3)$, $\Omega_4^+(2)$ or $\Omega_4^+(3)$. Now according to Lemma 6.2.3 L is absolutely irreducible on V, and hence

$$K \notin \mathcal{C}_1 \cup \mathcal{C}_3. \tag{7.2.7}$$

Next we prove

$$K \notin \mathcal{C}_4. \tag{7.2.8}$$

Assume false, so that $K = \Omega_{\mathcal{D}_1}$, where $\mathcal{D}_1$ is a tensor decomposition $V = W_1 \otimes W_2$ with $m_i = \dim(W_i)$ and $m_2 \geq m_1$.

Case $\Omega_i \approx \Omega_4^+(2)$. Here $\frac{1}{2}m_1m_2 = \frac{n}{2} \leq m_3(H) \leq m_3(K) = m_3(Sp_{m_1}(2) \otimes Sp_{m_2}(2)) = \frac{1}{2}(m_1 + m_2)$, which is absurd.

Case $\Omega_i \approx \Omega_4^+(3)$. If $4 \leq \frac{n}{4} \leq 6$, then K must be of type $O_4^+(3) \otimes O_4^-(3)$, $Sp_2(3) \otimes Sp_8(3)$, $Sp_2(3) \otimes Sp_{10}(3)$, $O_4^+(3) \otimes O_5(3)$, $Sp_2(3) \otimes Sp_{12}(3)$, $O_3(3) \otimes O_8^+(3)$, $Sp_4(3) \otimes Sp_6(3)$ or $O_4^{\epsilon_1}(3) \otimes O_6^{\epsilon_2}(3)$, and a contradiction is easily obtained from Lagrange's Theorem (note that the orders of $|H|$ and $|K|$ can be determined using Propositions 4.2.11, 4.4.12, 4.4.14, 4.4.15, 4.4.16 and 4.4.17). So we may take $\frac{n}{4} \geq 7$. Since L is irreducible, $L \not\leq 1 \otimes \Omega(W_2)$, and hence $J' \not\leq 1 \otimes \Omega(W_2)$. Therefore $A_{n/4} \preceq P\Omega(W_1)$. If $n = 28$, then $\sqrt{n} \geq m_1 \geq R_3(A_{n/4}) = 4$, and hence so $m_1 = 4$ and $m_2 = 7$. Thus K is of type $O_4^\pm(3) \otimes O_7(3)$, and this forces A_7 to embed in $P\Omega_4^\pm(3)$, which is false. Therefore $\frac{n}{4} \geq 8$, and hence $\sqrt{n} \geq m_1 \geq R_3(A_{n/4}) \geq \frac{n}{4} - 2$, which is impossible.

Case $\Omega_i \approx \Omega_3(3)$. As before, we may take $\frac{n}{3} \geq 7$ and we obtain $A_{n/3} \preceq PGL(W_1)$. But then we find $\sqrt{n} \geq \frac{n}{3} - 2$, which is false.

Case $\Omega_i \approx SU_3(2)$. Lagrange's Theorem eliminates the case $\frac{n}{3} \leq 8$. And if $\frac{n}{3} \geq 9$, then arguing as before we obtain $\sqrt{n} \geq R(A_{n/3}) \geq \frac{n}{3} - 2$, which is absurd. This completes the proof of (7.2.8).

Now we prove

$$K \notin \mathcal{C}_5. \tag{7.2.9}$$

The only possibility arising here is $\Omega_i \approx SU_3(2)$, and hence $K \cong Sp_n(2)$. But here $GU_3(2)^{n/3-2} \preceq H$ (Lemma 6.2.3.ii), and so $m_3(H) \geq (\frac{n}{3} - 2)m_3(GU_3(2)) = n - 6$. On the other hand, an easy application of Maschke's Theorem shows that any elementary 3-group contained in $Sp_n(2)$ has order at most $3^{n/2}$. Since $n - 6 > \frac{n}{2}$, we obtain the desired contradiction.

Next we establish

$$K \notin \mathcal{C}_6. \tag{7.2.10}$$

Here the only relevant cases are $\Omega_i \approx \Omega_4^+(3)$ and $\Omega_i \approx SU_3(2)$. In the first case we have $n = 2^r$ and $K \cong 2_+^{1+2r}.\Omega_{2r}^+(2)$ and $H \cong \Omega_4^+(3)^{\frac{n}{4}}.2^{n/2-2}.S_{n/4}$. We can eliminate the case $r \leq 5$ with Lagrange's Theorem. And when $r \geq 6$, then $A_{2^{r-2}} \preceq \Omega_{2r}^+(2)$, violating Proposition 5.3.7. In the second case, $n = 3^r$ with $r \geq 3$ and $K \cong 3^{1+2r}.Sp_{2r}(3)$. But then $A_{3^{r-1}} \preceq Sp_{2r}(3)$, contravening Proposition 5.3.7, again.

The next step is

$$K \notin \mathcal{C}_7. \tag{7.2.11}$$

Otherwise $K = \Omega_{\mathcal{D}_1}$, where $\mathcal{D}_1$ is a tensor decomposition $V = W_1 \otimes \cdots \otimes W_b$, with $a = \dim(W_i)$ and $n = a^b$. First suppose that $t = 4$. Then $\Omega_i \approx \Omega_4^+(2)$ or $\Omega_4^+(3)$, which means $\Omega \approx \Omega_{16}^+(2)$ or $\Omega_{16}^+(3)$. In the first case, $O^2(K) \cong Sp_4(2)' \times Sp_4(2)'$, and so $|K|$ is not divisible by 3^8 while $|H|$ is. Similarly, in the second case, $O^2(K) \cong \Omega_4^-(3) \times \Omega_4^-(3)$ or $Sp_4(3) \circ Sp_4(3)$, and so $|K|$ is not divisible by 3^9 while $|H|$ is. Therefore $t \geq 5$. As $n = a^b$ with $b \geq 2$, it actually follows that $t \geq 8$. And if $t = 8$, then $\Omega \approx \Omega_{32}^+(q)$ with $q \leq 3$; but this is impossible, for $\mathcal{C}_7(\Omega)$ is void as then $Sp_2(q)$ and $O_2^\pm(q)$ are not quasisimple. Thus $t \geq 9$, which means $R(A_t) \geq t - 2 \geq \frac{n}{4} - 2$, by Proposition 5.3.7. Moreover $t \geq \frac{n}{4} > \log_2(n) \geq b$, and so $J' \leq \Omega_{(\mathcal{D}_1)}$. Thus $A_t \preceq PGL(W_1)$, and so $\sqrt{n} \geq a \geq \frac{n}{4} - 2$, which forces $n < 32$. But then $n = 27$ and $\Omega \approx \Omega_{27}(3)$ or $SU_{27}(2)$. However, in both these cases $\mathcal{C}_7$ is void.

Since $\mathcal{C}_8$ is void in cases **U** and **O**, assertions (7.2.7)-(7.2.11) imply that $K \in \mathcal{C}_2$. Thus we may write $K = \Omega_{\mathcal{D}_1}$, where $\mathcal{D}_1$ is a a-space decomposition of V given by $W_1 \oplus \cdots \oplus W_b$ (with $n = ab$). If $\Omega_i \approx \Omega_3(3)$ or $\Omega_4^+(2)$ then the result follows from Propositions 6.2.12 and 6.2.13. So we can assume that $\Omega_i \approx \Omega_4^+(3)$ or $SU_3(2)$. First take $t = 4$. Here H is of type $O_4^+(3) \wr S_4$ in $\Omega_{16}^+(3)$, and using Lagrange's Theorem we can show that K must be of type $O_8^+(3) \wr S_2$. The subgroup $\Omega_4^+(3)^4 = O^2(H_{(\mathcal{D})})$ satisfies the conditions of Lemma 2.10.11, and hence $W_1 = V_1 + V_2$ and $W_2 = V_3 + V_4$ (with suitable relabelling). But then $H^{\mathcal{D}}$ preserves the pair of sets $\{V_1, V_2\}$ and $\{V_3, V_4\}$, which is impossible as $H^{\mathcal{D}}$ is the full symmetric group $Sym(\mathcal{D})$. Therefore $t \geq 5$. Since L is absolutely irreducible on V, it follows that $L \not\leq \Omega_{(\mathcal{D}_1)}$, and so $J' \not\leq \Omega_{(\mathcal{D}_1)}$. Thus J' acts faithfully on the b spaces W_i, which proves $t \leq b$. Moreover, because $|O_4^+(3)|$ does not divide $|O_4^-(3)|$, we actually have $t < b$, which means $b \in \{\frac{n}{3}, \frac{n}{2}, n\}$ in case $\mathbf{O}^+$, and $b \in \{\frac{n}{2}, n\}$ in case **U**. But now one checks that $|H|$ does not divide $|K|$ using Propositions 4.2.9, 4.2.11, 4.2.14

and 4.2.15. For example, if H is of type $O_4^+(3) \wr S_{n/4}$ and K is of type $O_3(3) \wr S_{n/3}$, then $|H|$ is divisible by $3^{n/2}$ while $|K|_3 = 3^{n/3}(\frac{n}{3}!)_3 < 3^{n/3}3^{n/6} = 3^{n/2}$. ∎

It remains now to consider the case where $m \geq 3$ and $\Omega_{(\mathcal{D})}$ is insoluble. In this case, putting $N = H^{\infty}_{(\mathcal{D})}$, we have

$$N = \begin{cases} \Omega'_1 \times \cdots \times \Omega'_t & \text{if } \mathcal{D} \text{ is non-degenerate or if case } \mathbf{L} \text{ holds} \\ H'_{(\mathcal{D})} \cong SL_{n/2}(q^u) & \text{if } \mathcal{D} \text{ is totally singular and case } \mathbf{U}, \mathbf{S} \text{ or } \mathbf{O} \text{ holds.} \end{cases}$$

(Recall the definition of u given in (2.1.17).) In the latter case, N acts in the way described in Lemma 4.2.3 (recall α is defined in (2.1.18)). Note that in both cases, we have $N/Z(N) \cong X^k$ for some non-abelian simple group X and some integer k. Evidently

$$X \cong \begin{cases} L_2(q) & \text{if } \Omega_i \approx \Omega_4^+(q), \text{ and in this} \\ & \text{case } q \geq 4 \text{ and } k = \frac{n}{2} \\ L_{n/2}(q^u) & \text{if } \mathcal{D} \text{ is totally singular, and in this} \\ & \text{case } k = 1 \\ P\Omega'_1 & \text{otherwise, and in these cases } k = t. \end{cases}$$

The group N is in fact a covering group of X^k, and according to Lemma 4.2.17, H is transitive on the components of N, except in case $\mathbf{O}$ where H is of type $O_{n/2}(q)^2$ (in this exceptional case, q and $\frac{n}{2}$ are odd and V_1 and V_2 are similar but not isometric).

Proposition 7.2.5. *If $m \geq 3$ and $H_{(\mathcal{D})}$ is insoluble then (H, K, Ω) appears in Table 3.5.H.*

Proof. First we prove

$$K \notin \mathcal{C}_2. \tag{7.2.12}$$

Otherwise, $K = \Omega_{\mathcal{D}_1}$, where $\mathcal{D}_1$ is an a-space decomposition of V given by $W_1 \oplus \cdots \oplus W_b$ (with $n = ab$). Suppose for the moment that $N \not\leq \Omega_{(\mathcal{D}_1)}$. Then there is a non-trivial homomorphism from some component of N to S_b, which means $P(X) \leq b \leq n$. If $t \leq 2$, then $X \cong C\ell_{n/2}(q)$ or $C\ell_{n/2}(q^2)$, and so by Corollary 5.2.3 $P(X) \geq \frac{n^2}{4} + 3 > n$, a contradiction. Therefore $t \geq 3$, which means H is transitive on the components of N. It follows that no component of N lies in $\Omega_{(\mathcal{D}_1)}$, and hence $N_{(\mathcal{D}_1)} \leq Z(N)$. Thus $N^{\mathcal{D}_1}$ is a covering group of X^k, and hence by Proposition 5.2.7 we have $n \geq b \geq k\,P(X)$. But Corollary 5.2.3 implies that $P(X) > \frac{n}{k}$, and so we reach a contradiction. Therefore $N \leq \Omega_{(\mathcal{D}_1)}$. Clearly N satisfies the conditions of Lemma 2.10.11, and so each W_j is a sum of $\frac{t}{b}$ of the spaces V_i. Consequently the spaces W_j give rise to a system of imprimitivity for the action of H on $\mathcal{D}$, against Corollary 4.2.2.iii.

Now we establish

$$K \notin \mathcal{C}_3. \tag{7.2.13}$$

Otherwise, L is not absolutely irreducible, and so Lemma 6.2.3 implies that $t = 2$. If $\mathcal{D}$ is non-degenerate or if case $\mathbf{L}$ holds, then we have $\Omega_1 \times \Omega_2 \preceq GL_{n/r}(q^r)$ for some prime divisor r of n, which violates Proposition 5.5.7. Therefore $\mathcal{D}$ is totally singular and case **U**, **S** or **O** holds. We easily eliminate case **U**, for here r must be odd, and hence $H^\infty \cong SL_{n/2}(q^2)$, which does not embed in $K^\infty \cong SU_{n/r}(q^r)$ by Proposition 5.4.15 and Corollary 5.3.3. In case **S**, $H^\infty \cong SL_{n/2}(q)$, and so using Proposition 5.4.15 and Corollary 5.3.3 again we see that $r = 2$ and so K is of type $Sp_{n/2}(q^2)$ or $GU_{n/2}(q)$. However Proposition 5.4.16.i shows that this cannot occur. Case **O** is discarded in the same way.

Next we argue

$$K \notin \mathcal{C}_4 \cup \mathcal{C}_7. \tag{7.2.14}$$

Otherwise $K = \Omega_{\mathcal{D}_1}$, where $\mathcal{D}_1$ is a tensor decomposition given by $W_1 \otimes \cdots \otimes W_b$ as described in §§4.4, 4.7. Put $m_i = \dim(W_i)$, and (without loss) assume that $m_1 \leq \sqrt{n}$. Arguing as in the proof of (7.2.12) we conclude $N \leq \Omega_{(\mathcal{D}_1)}$. Suppose for the moment that some component C of N is contained in $GL(W_2) \otimes \cdots \otimes GL(W_b)$. Since $GL(W_2) \otimes \cdots \otimes GL(W_b)$ acts homogeneously on V, it follows that C has at least m_1 non-trivial isomorphic irreducible constituents on V. The only possibility is for H to be of type $O_4^+(q) \wr S_{n/4}$ in $\Omega_n^+(q)$; here $C \cong SL_2(q)$, and C has just two non-trivial irreducible constituents on V. Thus $m_1 = 2$, and so either K is of type $Sp_2(q) \otimes Sp_{n/2}(q)$ in $\mathcal{C}_4$, or of type $Sp_2(q) \wr S_b$ in $\mathcal{C}_7$. In the first case, it is clear that at most one component of N can project non-trivially into $PGL(W_1)$, and so at least $\frac{n}{2} - 1$ components project non-trivially into $PGL(W_2)$. But then $PGL(W_2)$ contains a covering group of $L_2(q)^{n/2-1}$, which is impossible by Proposition 5.5.7. In the second case, $n = 2^b$ and so N is a covering group of $L_2(q)^{2^{b-1}}$, and so N cannot embed in $\Omega_{(\mathcal{D}_1)}^\infty$, which is a covering group of $L_2(q)^b$. Thus we have proved that no component of N is contained in $GL(W_2) \otimes \cdots \otimes GL(W_b)$, and so every component projects non-trivially in $PGL(W_1)$. Therefore, a covering group of X^k embeds in $PGL(W_1)$. But then by Propositions 5.5.7 and 5.4.15, $\sqrt{n} \geq m_1 \geq k\,R_p(X) \geq \frac{n}{2}$, a contradiction.

Our next result is

$$K \notin \mathcal{C}_5. \tag{7.2.15}$$

Otherwise, $N < GL(V_\sharp, \mathbf{F}_\sharp)$ for some proper subfield $\mathbf{F}_\sharp$ of $\mathbf{F}$. Clearly this contradicts Proposition 2.10.9.i if $\mathcal{D}$ is non-degenerate or if case **L** holds. Thus case **U**, **S** or **O**$^+$ holds and $\mathcal{D}$ is totally singular. Here $N^{V_1} = SL(V_1, \mathbf{F})$, and hence there exists $g \in N$ such that $\mathrm{tr}_{V_1}(g) = \mu$, where $\langle \mu \rangle = \mathbf{F}^*$. Then $\mathrm{tr}(g) = \mu + \mu^{-\alpha} \in \mathbf{F}_\sharp$. First suppose that case **S** or **O** holds, so that $\mu + \mu^{-1} \in \mathbf{F}_\sharp$. Arguing as in the proof of Proposition 2.10.9, with μ replacing μ^2, we see that $\mu + \mu^{-1}$ lies in a proper subfield only if $|\mathbf{F}| = 4$ and $|\mathbf{F}_\sharp| = 2$. In case **S** we have imposed the restriction q odd upon members of $\mathcal{C}_2$ stabilizing totally singular $\frac{n}{2}$-decompositions (see Table 4.2.A), and so we are left to consider case **O**. However $SL_{n/2}(4)$ does not embed in $\Omega_n^\pm(2)$ by Lagrange's Theorem, and we reach the desired contradiction. Now assume that case **U** holds, so that $N \approx SL_{n/2}(q^2)$ and $\mu + \mu^{-q} \in \mathbf{F}_\sharp$. Put $\lambda = \mu^{q+1}$, a generator for $\mathbf{F}_q$ and let $h \in N$ satisfy $\mathrm{tr}_{V_1}(h) = \lambda$. Then $\lambda + \lambda^{-1} \in \mathbf{F}_\sharp \cap \mathbf{F}_q$. If $\mathbf{F}_\sharp \cap \mathbf{F}_q \neq \mathbf{F}_q$, then as before we see that $q = 4$ and $\mathbf{F}_\sharp \cap \mathbf{F}_q = \mathbf{F}_2$. But this forces $\mathbf{F} = \mathbf{F}_{16}$ and $\mathbf{F}_\sharp = \mathbf{F}_2$, contrary to the fact that $|\mathbf{F} : \mathbf{F}_\sharp|$ is prime.

Therefore $\mathbf{F}_\sharp \cap \mathbf{F}_q = \mathbf{F}_q$, which implies $\mathbf{F}_\sharp = \mathbf{F}_q$. We therefore obtain an embedding of $SL_{n/2}(q^2)$ in $Sp_n(q)$ or $\Omega_n^\epsilon(q)$. Let W be the natural n-dimensional module over $\mathbf{F}_q$ for $Sp_n(q)$ or $\Omega_n^\epsilon(q)$. Clearly $SL_{n/2}(q^2)$ is irreducible on W (use Lagrange's Theorem), and since $n < (\frac{n}{2})^2$, it follows from Proposition 5.4.6 that it is not absolutely irreducible. Therefore by Proposition 4.3.3.iii, $SL_{n/2}(q^2)$ is contained in a member X of $\mathcal{C}_3(Sp_n(q))$ or $\mathcal{C}_3(\Omega_n^\epsilon(q))$. Since $R_p(L_{n/2}(q^2)) = \frac{n}{2}$, it follows that X must be of type $Sp_{n/2}(q^2)$, $GU_{n/2}(q)$ or $O_{n/2}^\chi(q^2)$. This forces $L_{n/2}(q^2)$ to be involved in $PSp_{n/2}(q^2)$, $U_{n/2}(q)$ or $P\Omega_{n/2}^\chi(q^2)$, which is not so. This contradiction establishes (7.2.15).

And now

$$K \notin \mathcal{C}_6. \tag{7.2.16}$$

Otherwise $n = r^b$ for some prime r and $K/O_r(K) \preceq GL_{2b}(r)$. We know that $N/(N \cap O_r(K))$ is a covering group of X^k and so by Proposition 5.5.7, $2\log_r(n) \geq 2b \geq k\,R_r(X)$. Now in case **O** or **S** q is odd, and so Ω_i is not $\Omega_4^-(2)$ or $Sp_4(2)$. Thus in all cases, it follows easily from Corollary 5.3.11 that $k\,R_r(X) > \frac{n}{2}$, which forces $2\log_r(n) > \frac{n}{2}$. This is impossible, and thus (7.2.16) holds.

And finally

$$K \notin \mathcal{C}_8. \tag{7.2.17}$$

Otherwise, $N \leq K^\infty = \Omega_\sharp$, as described in §4.8. Since N is irreducible on V_1, it follows that $(V_1, \kappa_\sharp)$ is either non-degenerate or totally singular. If case **L** holds, then $(V_1, \kappa_\sharp)$ cannot be non-degenerate, for $SL_m(q)$ does not embed in $\Omega_m^\pm(q)$, $Sp_m(q)$ or $SU_m(q^{1/2})$. Similarly $Sp_m(q)$ does not embed in $\Omega_m^\pm(q)$, and so $(V_1, \kappa_\sharp)$ cannot be non-degenerate in case **S** either. Therefore $(V_1, \kappa_\sharp)$ is totally singular, and hence Ω_1 has at least two irreducible constituents of degree m, namely V_1 and $V/V_1^{\perp_\sharp}$ (here $\perp_\sharp$ denotes the perp in the $(V, \mathbf{F}, \kappa_\sharp)$-geometry). Consequently case **S** holds, q is even and $\mathcal{D}$ is a singular $\frac{n}{2}$-space decomposition. But this case has been excluded from $\mathcal{C}$ (see Table 4.2.A). This establishes (7.2.17).

Assertions (7.2.12)-(7.2.17) imply that $K \in \mathcal{C}_1$. Therefore H is reducible on V, and so Lemma 6.2.3.iv ensures that $t = 2$. Now V_1 and V_2 are non-isomorphic as N-modules, and so by Lemma 2.10.11 we have $K = N_\Omega(V_i)$ for some i, and without loss $i = 1$. If V_1 is non-degenerate, then V_1 and V_2 are similar. However, according to the definition of $\mathcal{C}_1$, there are no members of $\mathcal{C}_1$ which are the stabilizers of non-degenerate subspaces which are similar to their orthogonal complements (see the Remark after Table 4.1.A). Thus V_1 must be totally singular. Furthermore, H cannot interchange V_1 and V_2, and so the only possibility is for case $\mathbf{O}^+$ to hold with $\frac{n}{2}$ odd. The result now follows from Proposition 6.2.1. ■

§7.3 The case $H \in \mathcal{C}_3$

Here $H = N_\Omega(V_\sharp, \mathbf{F}_\sharp, \kappa_\sharp)$ as described in Table 4.3.A. We have $n = mr$, where $r = |\mathbf{F}_\sharp : \mathbf{F}|$ is prime, and as usual we define $X_\sharp = X(V_\sharp, \mathbf{F}_\sharp, \kappa_\sharp)$ as X ranges over the symbols in (2.1.15). In the following lemmas we record some pertinent information about the structure of H, most of which follows directly from §4.3. Recall the restriction $m \geq 3$ appearing in case $\mathbf{O}^\epsilon$ in Table 4.3.A.

Lemma 7.3.1. *Exactly one of the following holds:*

(i) $m \geq 2$ *and* $\Omega_\sharp$ *is quasisimple;*

(ii) *case* $\mathbf{O}^+$ *holds,* $m = 4$ *and* $\Omega_\sharp \approx \Omega_4^+(q^r) \cong SL_2(q^r) \circ SL_2(q^r)$;

(iii) *case* $\mathbf{L}^\epsilon$ *holds,* $m = 1$ *and* $H \cong \mathbf{Z}_{(q^n-\epsilon)/(q-\epsilon)}.\mathbf{Z}_n$.

Proof. See Proposition 2.9.2 and Proposition 4.3.6(II). ∎

Now define the subgroup L of H as follows:

$$L = \begin{cases} \Omega_\sharp & \text{if (i) or (ii) occurs in Lemma 7.3.1} \\ \text{a group of order } p_{\epsilon fn} & \text{if (iii) occurs in Lemma 7.3.1,} \end{cases} \tag{7.3.1}$$

where $p_{\epsilon fn}$ is defined at the end of §5.2. Thus in all cases, L is either simple or quasisimple.

Lemma 7.3.2.

(i) $|\overline{H}| \geq q^{n-2}$.

(ii) *The group* $\overline{H}$ *contains an element of order at least 122.*

(iii) $P(L) > n$.

(iv) $P_f(\overline{L}) > n + 2$.

(v) *Let* L_1 *be a non-trivial quotient of* L. *If* $m \geq 2$, *then* $\zeta_p(L_1) = fk$, *where* k *is given in Lemma 4.3.1.ii. If* $m = 1$, *then* $\zeta_p(L_1) = ufn$.

(vi) *We have* $\zeta_p(\overline{L}) \geq \frac{ufn}{2}$, *with equality if and only if* H *is of type* $O_4^+(q^{n/4})$ *and* $\Omega \approx \Omega_n^+(q)$ *or of type* $GU_2(q^{n/2})$ *and* $\Omega \approx SU_n(q)$.

(vii) L *is irreducible on* $(V, \mathbf{F})$.

Proof. Part (i) follows from the structure results appearing in §4.3. Part (vii) follows from Lemma 4.3.2 when $m \geq 2$, and when $m = 1$ it is easy to check that $q_{\epsilon n}$ does not divide the order of any reducible subgroup of $\Omega \approx SL_n^\epsilon(q)$. Part (v) is immediate from Lemma 4.3.1.ii, and (vi) is a direct consequence of (v) (coupled with the fact that $\zeta_p(\overline{L}) = ufn$ when $m = 1$). We now prove (iii) and (iv). Note that when $m = 1$, we have $P(L) = P_f(\overline{L}) = |L| = p_{\epsilon nf} \geq 2n + 1$ (see Proposition 5.2.15 and observe that $n + 1 \neq p_n$ since $n + 1$ is even). Now assume that $m \geq 2$. Here $P(L) \geq P(L/Z(L))$ by Proposition 5.2.5.iii. When $L \approx \Omega_4^+(q^r)$ (with $r \geq 5$), we therefore have $P(L) \geq P(L_2(q^r)) \geq q^r + 1 > 4r + 2 = n + 2$. When L is quasisimple, use Theorem 5.2.2.

It remains to prove (ii). Suppose first that case $\mathbf{O}^\pm$ or $\mathbf{S}$ holds and that H is of type $GU_{n/2}(q)$. Then $U_{n/2}(q)$ is involved in $\overline{H}$, and hence so is $U_7(q)$. Now $U_7(q)$ has an element of order $(q^7+1)/(q+1)(q+1,7)$, which is greater than 122 provided $q \geq 3$. So take $q = 2$. In this case Proposition 4.3.18(II) implies that $GU_{n/2}(q) \preceq \overline{H}$. Hence $\overline{H}$ contains $GU_7(2)$, which has an element of order $2^7 + 1$. Now suppose that case $\mathbf{L}^\pm$ holds. Then H contains $\frac{1}{q-\epsilon} GL_m^\epsilon(q^r)$. Thus if $\epsilon = +$ or if n is odd, then H contains

$\frac{1}{q-\epsilon}GL_1^\epsilon(q^n)$, which is cyclic of order $(q^n-\epsilon)/(q-\epsilon)$; thus $\overline{H}$ contains an element of order $(q^n-\epsilon)/(q-\epsilon)(q-\epsilon,n) \geq 122$. And if $\epsilon = -$ and n is even, then m is also even and H contains $\frac{1}{q+1}GL_{m/2}(q^{2r})$, and hence contains $\frac{1}{q+1}GL_1(q^n)$, which is cyclic of order $(q^n-1)/(q+1)$; here $\overline{H}$ contains an element of order $(q^n-1)/(q+1)(q+1,n) \geq 122$. Next take case **S** with H of type $Sp_m(q^r)$. Then $PSp_2(q^{n/2}) \preceq PSp_m(q^r) \preceq \overline{H}$, and so $\overline{H}$ contains an element of order $(q^{n/2}+1)/(q-1,2) \geq 122$. Now take case $\mathbf{O}^\epsilon$, and assume that H is of type $O_m^\epsilon(q^r)$ with m even; then $\overline{H}$ contains $P\Omega_2^\epsilon(q^{n/2})$, which is cyclic of order $(q^{n/2}-\epsilon)/(4,q-\epsilon) \geq 122$. Finally take case **O** with qm odd, so that $\Omega_m(q^r) \preceq \overline{H}$; here $\Omega_2^-(q^{(n-r)/2}) \preceq \Omega_{m-1}^-(q^r) \preceq \Omega_m(q^r)$, and so $\overline{H}$ contains an element of order $\frac{1}{2}(q^{(n-r)/2}+1) \geq 122$, with equality when $(q,n,r) = (3,15,5)$. ∎

Lemma 7.3.3. *Assume that L_1 is a non-trivial quotient of L and that $|L_1|$ divides $|GL_k(q^u)|$ for some k. Then*
(i) $k \geq \frac{n}{2}$;
(ii) *if equality holds in* (i), *then H is of type $O_4^+(q^{n/4})$ with $\Omega \approx \Omega_n^+(q)$ or of type $GU_2(q^{n/2})$ with $\Omega \approx SU_n(q)$;*
(iii) *in case* **L**, $k \geq n$.

Proof. This is immediate from Lemma 7.3.2.v, since $\zeta_p(GL_k(q^u)) \leq ufk$. ∎

We are now ready for the main result of this section.

Proposition 7.3.4. *If $H \in \mathcal{C}_3$, then H is maximal among members of $\mathcal{C}(\overline{\Omega})$.*

Proof. As usual, we proceed in several steps. First of all, in view of Lemma 7.3.2.vii we have

$$K \notin \mathcal{C}_1. \tag{7.3.2}$$

Next we prove

$$K \notin \mathcal{C}_2 \cup \mathcal{C}_4 \cup \mathcal{C}_7. \tag{7.3.3}$$

Otherwise $K = \Omega_{\mathcal{D}}$, where $\mathcal{D}$ is some subspace or tensor decomposition of V given by $V = V_1 \oplus \cdots \oplus V_t$ or $V = V_1 \otimes \cdots \otimes V_t$. By Lemma 7.3.2.iii, $L \leq \Omega_{(\mathcal{D})}$. And since L is irreducible by Lemma 7.3.2.vii, it must be the case that $\mathcal{D}$ is a tensor decomposition. Clearly the image of L in $PGL(V_i)$ is non-trivial for some i (without loss $i = 1$) and hence $\dim(V_1) = \frac{n}{2}$ by Lemma 7.3.3.i. Thus $t = \dim(V_2) = 2$, and so Lemma 7.3.3.i ensures that the image of L in $PGL(V_2)$ is trivial. But then $L \leq GL(V_1) \otimes 1$, which is a contradiction as L is irreducible.

And now we show

$$K \notin \mathcal{C}_3. \tag{7.3.4}$$

Otherwise $K \leq N_\Omega(\mathbf{F}_\flat)$ for some field extension $\mathbf{F}_\flat$ of $\mathbf{F}$ as described in Table 4.3.A. The group L has no non-trivial cyclic quotient of order less than n (see Lemma 7.3.2.iii) and hence $L \leq C_\Omega(\mathbf{F}_\flat)$. Therefore $\mathbf{F}_\flat \subseteq \mathrm{End}_L(V)$. But also $\mathbf{F}_\sharp \subseteq \mathrm{End}_L(V)$, and as $\mathbf{F}_\sharp \neq \mathbf{F}_\flat$ (since $H \neq K$), we know that $\mathbf{F}_\sharp \neq \mathrm{End}_L(V)$. In other words, L is not absolutely irreducible on $(V_\sharp, \mathbf{F}_\sharp)$. But this is a contradiction, for $\Omega_\sharp \not\approx \Omega_2^\pm(q^r)$ (see Proposition 2.10.6.i).

The next step is

$$K \notin \mathcal{C}_5. \tag{7.3.5}$$

Otherwise, $\overline{K} \preceq PGL_n(p^e)$, where $e \mid uf$ and $e < uf$. Therefore $\zeta_p(\overline{K}) \leq \frac{nuf}{2}$. On the other hand, $\frac{nuf}{2} \leq \zeta_p(\overline{L}) \leq \zeta_p(\overline{K})$, and hence equality holds in Lemma 7.3.2.vi and $e = \frac{uf}{2}$. But an easy calculation shows that $|\Omega_4^+(q^{n/4})|$ does not divide $|\Omega_n^{\pm}(q^{1/2})|$ and $(q^{n/2}+1)|U_2(q^{n/2})|/(q+1)$ divides neither $2|PSp_n(q)|$ nor $2|PSO_n^{\pm}(q)|$ (see Propositions 4.3.6(II), 4.5.5(II) and 4.5.6(II)).

Now we show

$$K \notin \mathcal{C}_6. \tag{7.3.6}$$

Otherwise, $n = r^b$ for some integer b and $m = r^{b-1}$. If $b = 1$, then case $\mathbf{L}^{\pm}$ holds and $|\overline{L}| = q_{\pm r} \geq 2r+1$. But then $|\overline{L}|$ does not divide $r^3(r^2-1) = |\overline{K}|$, a contradiction. This leaves the case $b \geq 2$. Clearly $m \geq 5$ since $n \geq 13$, and so $L = \Omega_\sharp$ is quasisimple. Moreover $L \preceq r^{1+2b}.GL_{2b}(r)$, and so a covering group of L embeds in $PGL_{2b}(r)$, which implies $R_r(L/Z(L)) \leq 2b$. On the other hand, Corollary 5.3.11.iii implies $R_r(L/Z(L)) \geq m$ (note $\Omega_\sharp \not\approx \Omega_5(3)$ or $\Omega_6^-(2)$ since $\mathbf{F}_\sharp \neq \mathbf{F}_p$). Therefore $R_r(L/Z(L)) \geq r^{b-1}$, whence $2b \geq r^{b-1}$. Since $n \geq 13$, the only possibility is $n = 16$. Therefore $L \approx SL_8(q^2)$, $SU_8(q)$, $Sp_8(q^2)$ or $\Omega_8^{\pm}(q^2)$, none of which has a non-trivial r-modular representation of degree 8 by Corollary 5.3.10.ii.

We complete the proof of the Proposition by establishing

$$K \notin \mathcal{C}_8. \tag{7.3.7}$$

Otherwise $H \leq \Delta(V, \mathbf{F}, \kappa_\flat)$ for a suitable form $\kappa_\flat$. First assume that Lemma 7.3.1.iii holds (with $\epsilon = +$). Then n is odd and so K is of type $O_n(q)$ or $U_n(q^{1/2})$. However $\zeta_p(O_n(q)) = f(n-1) < \zeta_p(L)$, so $\kappa_\flat$ cannot be quadratic; moreover $(q^n-1)/(q-1)$ does not divide $|SU_n(q^{1/2})|(q-1)$, and so $\kappa_\flat$ cannot be unitary either. Therefore part (i) or (ii) of Lemma 7.3.1 holds. And since $\mathcal{C}_8$ is void in case $\mathbf{O}$, we are left with case in which Lemma 7.3.1.i holds. Thus $L = \Omega_\sharp$ is quasisimple, and hence $\Omega_\sharp \leq \Omega_\flat = \Omega(V, \mathbf{F}, \kappa_\flat)$. Note that L is irreducible on $(V, \mathbf{F})$ by Lemma 7.3.2.vii, and hence according to Proposition 4.3.12, L is contained a group $J \in (\mathcal{C}_2 \cup \mathcal{C}_3 \cup \mathcal{C}_8)(\Omega_\flat)$. The proof of (7.3.3) shows that L cannot be contained in a member of $\mathcal{C}_2(\Omega_\flat)$. Suppose for the moment that $J \in \mathcal{C}_3(\Omega_\flat)$. The proof of (7.3.4) shows that L normalizes no fields in $\mathrm{End}_{\mathbf{F}}(V)$ other than $\mathbf{F}_\sharp$ and $\mathbf{F}$ itself. Consequently $J = N_{\Omega_\flat}(\mathbf{F}_\sharp)$, and hence J^∞ is a classical group of dimension m. In case $\mathbf{L}$, the inclusion $L \leq J^\infty$ leads to an embedding $SL_m(q^r) \preceq Sp_m(q^r)$, $\Omega_m^\epsilon(q^r)$ or $SU_m(q^{r/2})$. This is possible only when $L \approx SL_2(q^r)$ and $J^\infty \approx Sp_2(q^r)$. But in this case, $|H| = |SL_2(q^r)|(q^r-1)r/(q-1)$, which does not divide $|\Delta(V, \mathbf{F}, \kappa_\flat)| = |Sp_{2r}(q)|(q-1)$. In case $\mathbf{S}$, since we have excluded groups of type $GU_{n/2}(q)$ in $Sp_n(q)$ when q is even, we obtain an embedding $Sp_m(q^r) \preceq \Omega_m^{\pm}(q^r)$ or $Sp_{n/2}(q^2) \preceq SU_{n/2}(q)$, both of which are impossible. It remains to consider the situation in which $J \in \mathcal{C}_8(\Omega_\flat)$. Then it must be the case that $\Omega \approx SL_n(q)$ with q even, $\Omega_\flat \approx Sp_n(q)$, and $J^\infty \approx \Omega_n^{\pm}(q)$. As before, L is contained in a member of $(\mathcal{C}_2 \cup \mathcal{C}_3 \cup \mathcal{C}_8)(J^\infty)$. We eliminate $\mathcal{C}_2$ is before, and since

$\mathcal{C}_8(J^\infty)$ is void, L is contained in a member of $\mathcal{C}_3(J^\infty)$. Using the previous argument, this forces an embedding $SL_m(q^r) \preceq \Omega_m^\pm(q^r)$ or $SL_{n/2}(q^2) \preceq SU_{n/2}(q)$, both of which are impossible. ∎

§7.4 The case $H \in \mathcal{C}_4$

Here $H = \Omega_\mathcal{D}$, where $\mathcal{D}$ is a tensor product decomposition $(V_1, \mathbf{f}_1) \otimes (V_2, \mathbf{f}_2)$ as described in §4.4. As in §4.4 we set $\Omega_i = \Omega(V_i, \mathbf{f}_i)$, $n_i = \dim(V_i)$, and we put $L = \Omega_1 \otimes \Omega_2 \leq H$. With no loss $n_1 \leq \sqrt{n} \leq n_2$. Thus $n_2 \geq 4$, and so Ω_2 is quasisimple, except when $\Omega_1 \approx \Omega_4^+(q)$. But in this case, q is odd and $\Omega_1 \approx Sp_4(q)$ or $\Omega_4^-(q)$, both of which are quasisimple. So by interchanging the indices we can assume that

$$\Omega_2 \text{ is quasisimple.} \tag{7.4.1}$$

As a convenience, we also record the following restrictions on Ω_2 :

$$\begin{gathered}\Omega_2 \text{ is not an orthogonal group in even characteristic,}\\ \Omega_2 \not\approx Sp_4(q) \text{ when } q \text{ is even,}\\ \text{and } \Omega_2 \not\approx SL_3^\pm(q) \text{ or } SL_4^\pm(q).\end{gathered} \tag{7.4.2}$$

This fact follows directly from the definition of $\mathcal{C}_4$ and the condition $n \geq 13$. For example, if $\Omega_2 \approx SL_4^\pm(q)$, then case **L** holds and so $\Omega_1 \approx SL_m^\pm(q)$ with $m < 4$, which forces $n \leq 12$, a contradiction.

Note that Ω_i is absolutely irreducible on V_i in view of Proposition 2.10.6.i and the restriction $\Omega_1 \not\approx \Omega_2^\pm(q)$ given in Table 4.4.A. Thus we obtain the following useful Lemma.

Lemma 7.4.1. *L is absolutely irreducible on V.*

We now come to the main result of this section.

Proposition 7.4.2. *If $H \in \mathcal{C}_4(\Omega)$ then (H, K, Ω) appears in Table 3.5.H.*

Proof. First we establish

$$K \notin \mathcal{C}_5. \tag{7.4.3}$$

Assume otherwise, so that by (4.5.5), $K = \Delta_\sharp \mathbf{F}^* \cap \Omega$ for some subfield geometry $\Delta_\sharp = \Delta(V_\sharp, \mathbf{F}_\sharp, \kappa_\sharp)$, as described in §4.5. If L is perfect, then $L \leq K' \leq \Delta_\sharp \leq GL(V_\sharp, \mathbf{F}_\sharp)$. If on the other hand L is not perfect, then neither is Ω_1. In cases **L**, **S** and **O** we have $q \geq 4$, and so by Proposition 2.9.2, the only way in which Ω_1 can fail to be perfect is for case **U** to hold and $\Omega_1 \approx SU_2(2)$, $SU_2(3)$ or $SU_3(2)$. In this situation, $|\mathbf{F}^*| = 3$, 8 or 3, repsectively, and so there is no non-trivial homomorphism from Ω_1 to $\mathbf{F}^*$. So once again we conclude $L \leq \Delta_\sharp \leq GL(V_\sharp, \mathbf{F}_\sharp)$. However, it follows from Proposition 2.10.9.iii that Ω_1 contains an element g_1 such that $\mathrm{tr}_{V_1}(g_1) \in \mathbf{F}_p^*$. Moreover, for all $g_2 \in \Omega_2$ we have $\mathrm{tr}_V(g_1 g_2) = \mathrm{tr}_{V_1}(g_1)\mathrm{tr}_{V_2}(g_2)$. It follows that $\{\mathrm{tr}_V(g) \mid g \in L\}$ generates the same field over $\mathbf{F}_p$ as $\{\mathrm{tr}_{V_2}(g_2) \mid g_2 \in \Omega_2\}$. In other words, $\mathbf{F}_p(L^V) = \mathbf{F}_p(\Omega_2^{V_2})$. However, $\mathbf{F}_p(\Omega_2^{V_2}) = \mathbf{F}$ by Proposition 2.10.9.i, and this contradicts the fact $L \leq GL(V_\sharp, \mathbf{F}_\sharp)$.

Next we prove

$$K \notin \mathcal{C}_6. \tag{7.4.4}$$

Otherwise $n = r^m$ for some prime r, with $m \geq 2$, $n_2 \geq r^{m/2}$ and $\overline{\Omega}_2 \preceq K/O_r(K) \preceq GL_{2m}(r)$. Consequently

$$R_{p'}(\overline{\Omega}_2) \leq 2m \leq 4\log_r(n_2).$$

As $n_2 \geq 4$, n_2 is a power of r, and Ω_2 is quasisimple, the above inequality and Theorem 5.3.9 imply that $\Omega_2 \approx SL_4^{\pm}(2)$, $Sp_4(2)$, $\Omega_4^-(2)$, $\Omega_8^+(2)$, $SU_4(3)$, $Sp_4(3)$, $\Omega_4^-(3)$ or $\Omega_5(3)$. However $p \neq r$, which eliminates the first four possibilities, and (7.4.2) eliminates the fifth. Also we observe that $\Omega_2 \not\approx \Omega_5(3)$, since $\mathcal{C}_6$ is void in case $\mathbf{O}^\circ$. This leaves the case in which $\Omega_2 \approx Sp_4(3)$ or $\Omega_4^-(3)$, which implies H is of type $O_4^+(3) \otimes O_4^-(3)$ in $O_{16}^+(3)$ or $O_4^{\pm}(3) \otimes Sp_4(3)$ in $Sp_{16}(3)$. Now if H is of type $O_4^{\pm}(3) \otimes Sp_4(3)$, then $3^6 \mid |H|$, while 3^6 does not divide the order of $K \cong 2_-^{1+8}.\Omega_8^-(2)$. Finally take the case where H is of type $O_4^+(3) \otimes O_4^-(3)$ in $O_{16}^+(3)$. Here $\overline{L}$ contains $\mathbf{Z}_3 \times \mathbf{Z}_3 \times A_6$, which does not embed in $K/O_2(K) \cong \Omega_8^+(2)$ (see [At, p. 85]). This final contradiction finishes the proof.

Our next claim is

$$K \notin \mathcal{C}_7. \tag{7.4.5}$$

Otherwise $K = \Omega_{\mathcal{D}_1}$ where $\mathcal{D}_1$ is a tensor decomposition given by $V = (V_3, \mathbf{f}_3) \otimes \cdots \otimes (V_{t+2}, \mathbf{f}_{t+2})$, where $m = \dim(V_i)$ for $3 \leq i \leq t+2$ and $n = m^t$. We have $t \leq \log_2(n) \leq \log_2(n_2^2) \leq n_2$, and Corollary 5.2.3.i ensures that $P(\overline{\Omega}_2) > t$. Consequently $\Omega_2 \leq \Omega_{(\mathcal{D}_1)}$. Putting $X_i = X(V_i, \mathbf{f}_i)$ (where X is one of the symbols in (2.1.15)), we argue

$$\Omega_2 \leq \Omega_i \text{ for some } i \in \{3, \ldots, t+2\}. \tag{7.4.6}$$

Assume for a contradiction that this fails. Then (7.4.1) ensures that $\overline{\Omega}_2 \preceq \overline{\Omega}_i$ for at least two values of $i \geq 3$, and without loss we can take $i = 3, 4$. Thus by Proposition 5.4.16.ii, $m \geq n_2 - 1$ for $i = 3, 4$ (observe that if $\Omega_2 \approx \Omega_6^{\pm}(q)$, then case $\mathbf{S}$ or $\mathbf{O}$ holds, and so $\Omega_i \not\approx SL_m^{\pm}(q)$ for $3 \leq i \leq t$). And since $n_2 \geq \sqrt{n} \geq m$, it follows that $t = 2$ and $n_1 = n_2 = m = \sqrt{n}$. Moreover, $C_{\overline{\Delta}_i}(\overline{\Omega}_2) = 1$ for $i = 3, 4$. But then $\overline{\Omega}_1 \leq C_{\overline{\Delta}_3 \times \overline{\Delta}_4}(\overline{\Omega}_2) = 1$, a contradiction. This establishes (7.4.6), and as Ω_2 and Ω_i act homogeneously on V, we have $n_2 \mid m$. However $m = \sqrt[t]{n}$, and hence $n_1 = n_2 = m = \sqrt{n}$. Therefore, because $H \notin \mathcal{C}_7$, we know that H is of type $O_m^+(q) \otimes O_m^-(q)$ or $O_m^{\pm}(q) \otimes Sp_m(q)$, with q odd. However, since $t = 2$ case $\mathbf{S}$ does not hold, and hence H is of type $O_m^+(q) \otimes O_m^-(q)$. Also K is of type $Sp_m(q) \wr S_2$ or $O_m^{\pm}(q) \wr S_2$. But $|H|$ does not divide the order of groups of type $O_m^{\pm}(q) \wr S_2$, and this forces K to be of type $Sp_m(q) \wr S_2$. However $\Omega_m^-(q)$ does not embed in $Sp_m(q)$ when q is odd, by Proposition 5.4.16.iii,iv, and this provides the desired contradiction.

Now we argue

$$K \notin \mathcal{C}_8. \tag{7.4.7}$$

Otherwise, $K = N_\Omega(\kappa_\sharp)$, as described in §4.8. As $\mathcal{C}_4$ is void in case $\mathbf{S}$ with q even, case $\mathbf{L}$ holds and so $\Omega_2 \approx SL_{n_2}(q)$. Let W be an irreducible Ω_2-submodule of V. Since Ω_2 acts homogeneously on V, we have $\dim(W) = n_2$ and $\Omega_2^W = SL(W)$. Consequently $(W, \kappa_\sharp)$ is totally singular. But then Ω_2 acts on the n_2-space $V/W^\perp$ in the dual representation (or the Frobenius dual in case **U**), which contradicts the fact that Ω_2 is homogeneous.

Clearly Lemma 7.4.1 implies $K \notin \mathcal{C}_1$, and hence in view of the preceding argument we deduce $K \in \mathcal{C}_2 \cup \mathcal{C}_3 \cup \mathcal{C}_4$. We consider these cases in turn.

Case $K \in \mathcal{C}_2$. Write $K = \Omega_{\mathcal{D}_2}$, where $\mathcal{D}_2$ is an a-space decomposition $V = W_1 \oplus \cdots \oplus W_t$, with $n = ta$. First we prove

$$\Omega_2 \leq \Omega_{(\mathcal{D}_2)}. \tag{7.4.8}$$

For if this fails, then $P(\overline{\Omega}_2) \leq t \leq n \leq n_2^2$, and so Ω_2 is one of the groups appearing in Corollary 5.2.3.ii. Using (7.4.2) and the fact that $P(\Omega_5(3)) = 27 > 5^2$, we are left with the case in which $\Omega_2 \approx \Omega_4^-(3) \cong A_6$. Here $n = 16$, $t \in \{8, 16\}$ and H is of type $O_4^+(3) \otimes O_4^-(3)$ or $O_4^-(3) \otimes Sp_4(3)$. Lemma 7.4.1 implies that $L^{\mathcal{D}_2}$ is transitive on $\mathcal{D}_2$. However, the index of any proper subgroup of A_6 is either 6, 10, 15, or is greater than 16. So it is easy to check that neither S_8 nor S_{16} contains a transitive subgroup which normalizes an A_6. This establishes (7.4.8). Since Ω_2 acts homogeneously on V with irreducible constituents of dimension n_2, it follows from (7.4.8) that $n_2 \mid a$ and $t \mid n_1$. And since $L^{\mathcal{D}_2}$ is transitive, (7.4.8) ensures that $\Omega_1^{\mathcal{D}_2}$ is transitive. Therefore $P(\overline{\Omega}_1) \leq n_1$, and hence Corollary 5.2.3.i ensures that Ω_1 is not quasisimple. Thus $\Omega_1 \approx Sp_4(2)$, $\Omega_4^+(q)$, or Ω_1 appears in Proposition 2.9.2.i. According to our definition of $\mathcal{C}_4$, we know that $\Omega_1 \not\approx \Omega_2^\pm(q)$ or $\Omega_4^+(2)$, and if $\Omega_1 \approx \Omega_4^+(q)$ with $q \geq 4$ then $P(\overline{\Omega}_1) > 4 = n_1$. If $\Omega_1 \approx SL_2^\pm(3)$ or $Sp_2(3)$, then $P(\overline{\Omega}_1) = 3 > n_1$. And if $\Omega_1 \approx SU_3(2)$, then $t = 3$, whereas Ω_1 has no transitive permutation representation of degree 3. Therefore one of the following holds:

(a) $t = 2$ and $\Omega_1 \approx SL_2(2)$ or $Sp_2(2)$
(b) $t = 3$ and $\Omega_1 \approx \Omega_3(3)$
(c) $t = 2$ and $\Omega_1 \approx SU_2(2)$
(d) $t = 2$ and $\Omega_1 \approx Sp_4(2)$
(e) $t = 4$ and $\Omega_1 \approx \Omega_4^+(3)$.

Cases (b) and (c) are treated in Propositions 6.3.2 and 6.3.3. In cases (a) and (d), observe that $\Omega_1' \leq \Omega_{(\mathcal{D}_2)}$, and hence $\Omega_1' \otimes \Omega_2$ is reducible. But this is impossible by Lemma 4.4.3.v, for Ω_2 is absolutely irreducible on V_2 and Ω_1' is irreducible on V_1. Finally, consider (e). We may write $\Omega_1 = SL(U_1) \otimes SL(U_2)$, where $V_1 = U_1 \otimes U_2$ is a tensor decomposition of V_1. Now $\overline{\Omega}_1 \cong L_2(3) \times L_2(3) \cong A_4 \times A_4$, and since $\Omega_1^{\mathcal{D}_2}$ is transitive, it is clear that one of the factors $SL(U_i)$ is contained in $\Omega_{(\mathcal{D}_2)}$. Therefore $SL(U_i) \otimes \Omega_2 \leq \Omega_{(\mathcal{D}_2)}$. Now on the one hand, $SL(U_i) \otimes \Omega_2$ acts irreducibly on $U_i \otimes V_2$, which has dimension $\frac{n}{2}$. On the other hand, $t = n_1 = 4$, which means $a = n_2 = \frac{n}{4}$, which implies that all irreducible constituents of $\Omega_{(\mathcal{D}_2)}$ have dimension at most $\frac{n}{4}$. Clearly this is a contradiction, and so (e) cannot occur.

Case $K \in \mathcal{C}_3$. Write $K = N_\Omega(\mathbf{F}_\sharp)$ as in §4.3, and observe that $L' \leq K' \leq GL(V, \mathbf{F}_\sharp)$, and so L' is not absolutely irreducible. Thus Lemma 4.4.3.vi implies that Ω_1' is not absolutely irreducible on V_1, and so Ω_1 is soluble by Corollary 2.10.7. Therefore Ω_1 appears in Proposition 2.9.2.i, and as before we note that $\Omega_1 \not\approx \Omega_2^{\pm}(q)$ or $\Omega_4^{+}(2)$. Moreover, $SL_2^{\pm}(3)'$, $Sp_2(3)'$, $SU_3(2)'$ and $\Omega_4^{+}(3)'$ are absolutely irreducible on their natural modules, and hence $\Omega_1 \approx SL_2(2)$, $Sp_2(2)$, $SU_2(2)$ or $\Omega_3(3)$. The first two cases are treated in Proposition 6.3.1. If $\Omega_1 \approx SU_2(2)$, then case **U** holds and $|\mathbf{F}_\sharp : \mathbf{F}| = r$ is an odd prime. But then $O^r(\Omega_1) = \Omega_1$, whence $O^r(L)$ is absolutely irreducible on V, while $O^r(K)$ is not. Finally, assume that $\Omega_1 \approx \Omega_3(3) \cong A_4$. Since $O^r(\Omega_1)$ is absolutely irreducible on V_1 for all primes r other than 3, the argument in the preceding example shows that $|\mathbf{F}_\sharp : \mathbf{F}| = 3$. However, if H is of type $O_3(3) \otimes Sp_{n/3}(3)$ in $Sp_n(3)$ or of type $O_3(3) \otimes O^{\pm}_{n/3}(3)$ in $O^{\pm}_n(3)$ with n even, then it follows from Propositions 4.4.11 and 4.4.17 that $S(V_1, \mathbf{f}_1) \otimes \Omega_2 \leq H$; and since $S(V_1, \mathbf{f}_1) \approx SO_3(3) \cong S_4$, it follows that $O^3(H)$ is absolutely irreducible while $O^3(K)$ is not. If H is of type $O_3(3) \otimes O_{n/3}(3)$ in $O_n(3)$ with n odd, then $H = (\Omega_1 \otimes \Omega_2)\langle g_1 \otimes g_2 \rangle$, where $g_i \in S(V_i, \mathbf{f}_i) \backslash \Omega_i$. Thus the image of H in $PGL(V_1)$ is $SO_3(3)$, and so once again $O^3(H) = H$, which is absolutely irreducible. Thus $H \not\leq K$.

Case $K \in \mathcal{C}_4$. Write $K = \Omega_{\mathcal{D}_3}$, where $\mathcal{D}_3$ is a tensor decomposition $V = (V_3, \mathbf{f}_3) \otimes (V_4, \mathbf{f}_4)$. Put $\Omega_i = \Omega(V_i, \mathbf{f}_i)$, $n_i = \dim(V_i)$, and assume that $n_4 \geq \sqrt{n} \geq n_3$. Arguing as in the proof of (7.4.6) we find that $\Omega_2 \leq \Omega_i$ for some $i \in \{3, 4\}$ and hence $n_2 \mid n_i$ (as Ω_2 is homogeneous). Assume first that $n_2 = n_i$. Then it follows from Proposition 5.4.16.iii,iv that $\Omega_2 \approx \Omega^{\pm}_{\sqrt{n}}(q)$ and $\Omega_i \approx Sp_{\sqrt{n}}(q)$ with q even. But this violates (7.4.2). Therefore $n_2 < n_i$, and hence $i = 4$. Moreover $n_3 \mid n_1$ and $n_3 < n_1$, and so $n_1 \geq 4$. If $\Omega_1 \approx \Omega_4^{+}(3)$ then the result follows from Proposition 6.3.4. So we can assume that $\Omega_1 \not\approx \Omega_4^{+}(3)$, and hence Ω_1 is perfect. Moreover $\overline{\Omega}_1$ does not embed in $\overline{\Omega}_3$ by Proposition 5.4.16.ii. Therefore $\overline{\Omega}_1 \cap \overline{\Omega}_4 \neq 1$. Furthermore $\overline{\Omega}_1 \not\leq \overline{\Omega}_4$ since L is irreducible while Ω_4 is reducible. Consequently $1 < \overline{\Omega}_1 \cap \overline{\Omega}_4 \trianglelefteq \overline{\Omega}_1$, which means $\overline{\Omega}_1$ is not simple. And as $n_1 \geq 4$, we deduce $\Omega_1 \approx \Omega_4^{+}(q)$, and the result follows again from Proposition 6.3.4. ■

§7.5 The case $H \in \mathcal{C}_5$

In this section assume that $H \in \mathcal{C}_5$, so that $H^\infty = \Omega(V_\sharp, \mathbf{F}_\sharp, \kappa_\sharp)$ for some subfield geometry $(V_\sharp, \mathbf{F}_\sharp, \kappa_\sharp)$, as described in §4.5. Clearly H^∞ is quasisimple as $n \geq 13$ and $\overline{H}^\infty$ is simple.

Proposition 7.5.1. *H is maximal among $\mathcal{C}$-groups.*

Proof. First of all, H^∞ is absolutely irreducible on V, and thus $K \notin \mathcal{C}_1 \cup \mathcal{C}_3$. Moreover $R_{p'}(\overline{H}^\infty) > R_p(\overline{H}^\infty) = n$ by Corollary 5.3.10.i and $P(\overline{H}^\infty) > n$ by Corollary 5.2.3.i, and so it follows easily that $K \notin \mathcal{C}_2 \cup \mathcal{C}_4 \cup \mathcal{C}_6 \cup \mathcal{C}_7$. Now if case **L** holds, then H is of type $GL_n(q_o)$ and $H^\infty \cong SL_n(q_o)$ does not fix a bilinear or unitary form on V, whence $K \notin \mathcal{C}_8$. Similarly, if case **S** holds, then H is of type $Sp_n(q_o)$, and $H^\infty \approx Sp_n(q_o)$, which does not fix a quadratic form, and again $K \notin \mathcal{C}_8$. Finally, if $K \in \mathcal{C}_5$, then Lagrange's Theorem and Zsigmondy's Theorem 5.2.14 eliminate all possibilities for K, except when case **U** holds, H is of type $O_n^{\pm}(q)$ and K is of type $Sp_n(q)$. But here q is odd, and we can apply Proposition 5.4.16.iii. ■

§7.6 The case $H \in \mathcal{C}_6$

In this section we assume that $H \in \mathcal{C}_6(\Omega)$, so that $n = r^m$ for some prime $r \neq p$ and H is the normalizer of a suitable symplectic type r-group, as described in §4.6. We let

$$L = H^{\infty} \cong \begin{cases} r^{1+2m}.Sp_{2m}(r) \text{ or } (4 \circ 2^{1+2m}).Sp_{2m}(2) & \text{in case } \mathbf{L}^{\pm} \\ 2_{\pm}^{1+2m}.\Omega_{2m}^{\pm}(2) & \text{in cases } \mathbf{S} \text{ and } \mathbf{O}^{+}. \end{cases} \tag{7.6.1}$$

Also we put $R = O_r(L)$, so that $\overline{R} = \mathrm{soc}(\overline{L})$.

Lemma 7.6.1. *We have*
(i) $P(\overline{L}) > n$.
(ii) $P_f(\overline{L}) \geq n(n-1) - 2$.
(iii) $R_{r'}(\overline{L}) = n$.

Proof. Let X be any proper subgroup of $\overline{L}$. If $\overline{R}X \neq \overline{L}$, then $|\overline{L} : X| \geq |\overline{L} : \overline{R}X| \geq P(L/R)$. Using Theorem 5.2.2 it is easy to check that $P(L/R) > r^m = n$. On the other hand, if $\overline{R}X = \overline{L}$, then as $\overline{R} \cap X$ is normal in both $\overline{R}$ and X, it is normal in $\overline{L}$. And since $\overline{R}$ is a minimal normal subgroup of $\overline{L}$, we have $\overline{R} \cap X = 1$. Therefore $|\overline{L} : X| = |\overline{R}| = r^{2m} = n^2 > n$, and assertion (i) follows.

For (ii), assume that X is a subgroup of $\overline{L}$ which contains no non-trivial normal subgroup of $\overline{L}$. If $X \cap \overline{R} = 1$, then $|\overline{L} : X| \geq |\overline{R}| = n^2$, as before. So we can assume that $1 < X \cap \overline{R} < \overline{R}$. It follows that X acts reducibly on $\overline{R}$, regarded as a $2m$-dimensional space over $\mathbf{F}_r$. When $L/R \cong Sp_{2m}(r)$, it follows, using Theorem 5.2.2 (together with a straightforward calculation with reducible subgroups of $Sp_{2m}(2)$), that $|\overline{L} : X\overline{R}| \geq (r^{2m} - 1)/(r - 1) \geq r^{2m-1}$, and hence $|\overline{L} : X| = |\overline{L} : X\overline{R}||\overline{R} : \overline{R} \cap X| \geq r^{2m} = n^2$. If $L/R \cong \Omega_{2m}^{\pm}(2)$, then by Theorem 5.2.2, $|\overline{L} : X\overline{R}| \geq (2^m + 1)(2^{m-1} - 1)$, and so we deduce as before that $|\overline{L} : X| \geq (2^m + 1)(2^m - 2) \geq n(n-1) - 2$.

To prove (iii), we apply Lemma 5.5.3. First suppose that $L \cong 2_{\pm}^{1+2m}.\Omega_{2m}^{\pm}(2)$. Then $\overline{L} \cong 2^{2m}.\Omega_{2m}^{\pm}(2)$, and so it is clear that Lemma 5.5.3 applies (with $K = \overline{L}$, $N = \overline{R}$ and $r = 2$). Thus we deduce $R_{r'}(\overline{L}) \geq \min\{P(\Omega_{2m}^{\pm}(2)), 2^m\} = 2^m = n$. The same proof works when $\overline{L} \cong r^{2m}.Sp_{2m}(r)$ or $\overline{L} \cong 2^{2m}.Sp_{2m}(2)$. ■

The main result of this section is

Proposition 7.6.2. *H is maximal among $\mathcal{C}$-groups.*

Proof. Since L is perfect and absolutely irreducible on V, and since the conditions on $\mathbf{F}$ in the definition of $\mathcal{C}_6$ ensure that the representation of L can be written over no proper subfield of $\mathbf{F}$, we have

$$K \notin \mathcal{C}_1 \cup \mathcal{C}_3 \cup \mathcal{C}_5. \tag{7.6.2}$$

Next we prove

$$K \notin \mathcal{C}_2 \cup \mathcal{C}_4 \cup \mathcal{C}_7. \tag{7.6.3}$$

Otherwise $K = \Omega_{\mathcal{D}}$ where $\mathcal{D}$ is either a subspace or tensor decomposition of V given by $V_1 \oplus \cdots \oplus V_t$ or $V_1 \otimes \cdots \otimes V_t$. Obviously Lemma 7.6.1.i implies $L \leq \Omega_{(\mathcal{D})}$. As L is irreducible, $\mathcal{D}$ must be a tensor decomposition, and thus $\overline{L} \preceq PGL(V_1) \times \cdots \times PGL(V_t)$. Clearly the image of $\overline{R}$ in $PGL(V_i)$ is non-trivial for some i, and hence $\overline{L} \preceq PGL(V_i)$ (since $\overline{R}$ is the unique minimal normal subgroup of $\overline{L}$). But then Lemma 7.6.1.iii forces $\dim(V_i) = n$, a contradiction.

In view of (7.6.2) and (7.6.3) and the fact that $K \notin \mathcal{C}_6$ (by (7.0.2)), it remains to prove

$$K \notin \mathcal{C}_8. \tag{7.6.4}$$

Otherwise $K = N_\Omega(\kappa_\sharp)$ as described in §4.8, and $L \leq \Omega(V, \kappa_\sharp)$. Now $\mathcal{C}_8$ is void in cases **O** and **U** and also in case **S** with q odd. Thus case **L** holds. According to the definition of $\mathcal{C}_6$ in case **L**, we know that $f = \log_p(q)$ is odd, hence $\kappa_\sharp$ is orthogonal or symplectic. But then $\mathbf{Z}_{r(r,2)} \cong Z(L) \leq Z(\Omega(V, \kappa_\sharp)) \preceq \mathbf{Z}_2$, which is impossible. This contradiction finishes the proof. ∎

§7.7 The case $H \in \mathcal{C}_7$

In this section H stabilizes a tensor product decomposition $\mathcal{D}$ given by $(V_1, \mathbf{f}_1) \otimes \cdots \otimes (V_t, \mathbf{f}_t)$, as described in §4.4 and §4.7 (see (4.4.5)). We use the notation of §4.7, so that $\dim(V_i) = m$, $n = m^t$, and $X_i = X(V_i, \mathbf{f}_i)$ where X ranges over the symbols in (2.1.15). Also define

$$L = \Omega_1' \otimes \cdots \otimes \Omega_t'.$$

Thus L is a central product of t quasisimple groups. In the following Lemma we collect some more facts about L.

Lemma 7.7.1.
(i) $C_{\overline{\Omega}}(\overline{L}) = 1$.
(ii) *If* $t \geq 3$, *then* H *is transitive on the components of* L.
(iii) $P_f^t(\overline{L}) \geq (m+1)^t$.
(iv) $\mathbf{F}_p(L) = \mathbf{F}$.
(v) $\overline{L}$ *normalizes no non-trivial soluble subgroup of* $\overline{\Omega}$.

Proof. Since Ω_i' is absolutely irreducible on V_i (Corollary 2.10.7), L is absolutely irreducible on V (Lemma 4.4.3.vi). Thus assertion (i) follows from Lemma 4.0.5.ii. Recall from (4.7.1) that $I_{\mathcal{D}}$ has a subgroup $J \cong S_t$ naturally permuting the tensor factors V_i. Thus $A_t \cong J' \leq \Omega_{\mathcal{D}} = H$. Evidently J also naturally permutes the components of L, and hence (ii) is now clear (for A_t is transitive when $t \geq 3$). Part (iii) is immediate from Corollary 5.2.3 and Proposition 5.2.7. To prove (iv), we use induction on t. First of all, we claim that the action of $N := \Omega_2' \otimes \cdots \otimes \Omega_t'$ on $W := V_2 \otimes \cdots \otimes V_t$ satisfies $\mathbf{F}_p(N) = \mathbf{F}$.

This holds when $t = 2$ by Proposition 2.10.9.i, and holds by induction when $t \geq 3$. (Note that $\Omega_1 = \Omega_1'$, except when $\Omega_1 \approx Sp_4(2)$; in this case $\mathbf{F} = \mathbf{F}_2$, and so there is nothing to prove.) Thus $\mathbf{F}_p[\mathrm{tr}_W(x) \mid x \in N] = \mathbf{F}$. Now according to Proposition 2.10.9, there exists $g_1 \in \Omega_1$ such that $\mathrm{tr}_{V_1}(g_1) \in \mathbf{F}_p^*$. Since $\mathrm{tr}_V(g_1 \otimes x) = \mathrm{tr}_{V_1}(g_1)\mathrm{tr}_W(x)$ for all $x \in N$, it follows that $\mathbf{F}_p(L)$ contains $\mathbf{F}_p[\mathrm{tr}_W(x) \mid x \in N] = \mathbf{F}$, as desired.

To prove (v), assume for a contradiction that $\overline{L}$ normalizes a non-trivial soluble subgroup of $\overline{\Omega}$. Then there is a prime r such that $\overline{L}$ normalizes some non-trivial elementary abelian r-group R, which is irreducible when regarded as an $\mathbf{F}_r\overline{L}$-module. Clearly $r \neq p$ as L is irreducible on V. Note that $\overline{L}$ acts non-trivially on R by (i). Relabel the indices so that $\overline{\Omega}_i'$ acts non-trivially on R if and only if $1 \leq i \leq s$, and define $L_o = \Omega_1' \otimes \cdots \otimes \Omega_s'$. Let $\widetilde{R}$ be a minimal preimage of R in Ω. The minimality of $\widetilde{R}$ implies that $[\widetilde{R}, L_o] = \widetilde{R}$, whence $\widetilde{R}$ is L_o-invariant. Since R centralizes $\overline{\Omega}_{s+1}' \times \cdots \times \overline{\Omega}_t'$, Lemma 4.0.5 implies that $\widetilde{R} \leq C_\Omega(\Omega_{s+1}' \otimes \cdots \otimes \Omega_t')$, whence $\widetilde{R} \leq GL(V_o)$, where $V_o = V_1 \otimes \cdots \otimes V_s$. Thus we may regard $R\overline{L}_o$ as a subgroup of $PGL(V_o)$. Now write $|R| = r^a$, and observe that Lemma 5.5.5 shows that $a \geq d^s$, where d is the smallest integer for which $\overline{\Omega}_1' \preceq GL_d(\overline{\mathbf{F}}_r)$, where $\overline{\mathbf{F}}_r$ is the algebraic closure of $\mathbf{F}_r$. On the other hand, Lemma 5.5.2.ii implies that $a \leq \dim(V_o) = m^s$, and so we deduce $d \leq m$. Obviously $d \geq R_{p'}(\overline{\Omega}_1')$, and so it follows from Corollary 5.3.11 that $\Omega_1 \approx SL_3(2)$, $SU_4(2)$, $Sp_2(4)$, $Sp_2(5)$, $Sp_4(2)$, $Sp_4(3)$, $\Omega_3(5)$, $\Omega_3(7)$, $\Omega_3(9)$, $\Omega_4^-(3)$, $\Omega_5(3)$ or $\Omega_6^-(3)$. Suppose for the moment that $\widetilde{R}$ is abelian. We have $\widetilde{R}{:}L_o \leq GL(V_o)$ and L_o acts on the homogeneous components of $\widetilde{R}$ in V_o. More specifically, $\overline{L}_o$ acts faithfully on these components. For if Ω_i' fixes each component for some $i \in \{1, \ldots, s\}$, then Ω_i' centralizes $\widetilde{R}$ (since $\widetilde{R}$ acts as scalars on each of its homogeneous components). But then $\overline{\Omega}_i'$ centralizes R, a contradiction. It follows that there are at least $P_f^t(\overline{L}_o)$ such components, and this forces

$$m^s = \dim(V_o) \geq P_f^t(\overline{L}_o) = P(\overline{\Omega}_1')^s.$$

This of course violates Corollary 5.2.3.i. Consequently $\widetilde{R}$ is non-abelian and in particular $r \mid |Z(\Omega)|$. Thus the only possibilities are $r = |Z(\Omega)| = 2$ and $\Omega_1 \approx Sp_2(5)$, $Sp_4(3)$, $\Omega_4^-(3)$ or $\Omega_6^-(3)$. Moreover, $Z(\Omega) \leq \widetilde{R}$, and since L_o acts irreducibly on $\widetilde{R}/Z(\Omega)$, it must be the case that $\widetilde{R}$ is an extraspecial 2-group. It follows from Proposition 4.6.3 that

$$m^s = \dim(V_o) \geq 2^{a/2}. \tag{7.7.1}$$

Recall that $a \geq d^s$, and hence

$$m^s \geq 2^{d^s/2}. \tag{7.7.2}$$

For $\Omega_1 \approx Sp_2(5)$, $Sp_4(3)$, $\Omega_4^-(3)$, $\Omega_6^-(3)$, we have $m = 2, 4, 4, 6$ and $d \geq 2, 4, 3, 6$, respectively. When $\Omega_1 \approx Sp_2(5)$, (7.7.2) implies $s \leq 2$. Since $|PSp_2(5)|$ does not divide $|L_3(2)|$, we must have $a \geq 4$, and so (7.7.1) implies $s = 2$ and $a = 4$; however $PSp_2(5) \times PSp_2(5)$ does not embed in $GL_4(2)$, and this rules out $Sp_2(5)$. In the remaining three cases, (7.7.2) implies that $s = 1$ and $\Omega_1 \approx Sp_4(3)$ or $\Omega_4^-(3)$. If $\Omega_1 \approx Sp_4(3)$, then $a \geq 6$ as $|PSp_4(3)|$ does does not divide $|L_5(2)|$. But then (7.7.1) fails. If $\Omega_1 \approx \Omega_4^-(3)$, then in view of

(7.7.1) we must have $a = 4$. But the extraspecial groups $2_{\pm}^{1+4}$ have automorphism groups $2^4.O_4^{\pm}(2)$, and so $\Omega_4^-(3)$ cannot act faithfully on the Frattini factor $\widetilde{R}/Z(\Omega) \cong 2^4$. This final contradiction completes the proof of the Lemma. ∎

Lemma 7.7.2. *Let X be a direct product of k copies of $L_2(q)$ with q odd, $k \geq 2$ and $q \geq 5$, and assume that X is contained irreducibly in $PGL_{2^k}(q)$. Then X embeds in $\overline{\Omega} = PSp_{2^k}(q)$ if k is odd and in $\overline{\Omega} = P\Omega_{2^k}^+(q)$ if k is even. Moreover, $N_{\overline{\Omega}}(X) \in \mathcal{C}_7(\overline{\Omega})$.*

Proof. Let $X_1, \ldots, X_k$ be the components of a minimal preimage $\widetilde{X}$ of X in $GL_{2^k}(q)$. Write V for the associated vector space of dimension 2^k over $\mathbf{F}_q$. As in the proof of Lemma 5.5.5, we see that $\widetilde{X}$ preserves a tensor decomposition $V_1 \otimes \cdots \otimes V_k$ of V, with $X_i \leq 1 \otimes \cdots \otimes GL(V_i) \otimes \cdots \otimes 1$. Since $\dim(V) = 2^k$, it follows that $\dim(V_i) = 2$ for all i, and so V_i is the natural module for $X_i \cong SL_2(q)$. In particular, X_i fixes a non-degenerate symplectic form $\mathbf{f}_i$ on V_i, and so $\widetilde{X}$ preserves $\mathbf{f} = \mathbf{f}_1 \otimes \cdots \otimes \mathbf{f}_k$ on V. By (4.4.2), $\mathbf{f}$ is symplectic or orthogonal according as k is odd or even. And when k is even, we see that $\langle v_1 \rangle \otimes V_2 \otimes \cdots \otimes V_k$ is a totally singular 2^{k-1}-space, and so the corresponding quadratic form has maximal Witt index. The final assertion is clear. ∎

The main result of this section is

Proposition 7.7.3. *H is maximal among $\mathcal{C}$-groups.*

Proof. As we observed in the proof of Lemma 7.7.1.i, L is absolutely irreducible on V. This fact, along with Lemma 7.7.1.iv,v, yields

$$K \notin \mathcal{C}_1 \cup \mathcal{C}_3 \cup \mathcal{C}_5 \cup \mathcal{C}_6. \qquad (7.7.3)$$

Now we show that

$$K \notin \mathcal{C}_2. \qquad (7.7.4)$$

Otherwise $K = \Omega_{\mathcal{D}_1}$, for some subspace decomposition $\mathcal{D}_1$ given by $W_1 \oplus \cdots \oplus W_b$. Since L is irreducible on V, we know that $L^{\mathcal{D}_1}$ is transitive. However $\overline{L}$ does not permute $\mathcal{D}_1$ faithfully in view of Lemma 7.7.1.iii. Therefore H is not transitive on the components of L, and so by Lemma 7.7.1.ii, $t = 2$. Indeed, we may write $\Omega_1' \leq \Omega_{(\mathcal{D}_1)}$ and $\Omega_2' \not\leq \Omega_{(\mathcal{D}_2)}$. Since Ω_1' acts homogeneously on V, it follows that $\dim(W_i) \geq m = \sqrt{n}$, and hence $b \leq m$. But then $P(\Omega_2') \leq m$, against Corollary 5.2.3.i.

Next we claim

$$K \notin \mathcal{C}_4 \cup \mathcal{C}_7. \qquad (7.7.5)$$

Otherwise $K = \Omega_{\mathcal{D}_2}$, where $\mathcal{D}_2$ is a tensor decomposition of V given by $(U_1, \mathbf{g}_1) \otimes \cdots \otimes (U_b, \mathbf{g}_b)$. Evidently $L^{\mathcal{D}_2} \cong (\overline{\Omega}_1')^k$, where $0 \leq k \leq t$. Thus by Proposition 5.2.7, $kP(\overline{\Omega}_1') \leq b \leq \log_2(n) = t\log_2(m)$. Obviously $P(\overline{\Omega}_1') > \log_2(m)$, and so we must have $k < t$. Thus some component of L lies in $\Omega_{(\mathcal{D}_2)}$. If H is transitive on the components of L, then clearly $L \leq \Omega_{(\mathcal{D}_2)}$. If on the other hand H is intransitive, then as before $t = 2$. But then $P(\overline{\Omega}_i') > 2\log_2(m) \geq b$, and so once again we have $L \leq \Omega_{(\mathcal{D}_2)}$. If the image of L in $PGL(U_1)$ fixes a subspace U_1' of U_1, then L fixes the subspace $U_1' \otimes U_2 \otimes \cdots \otimes U_b$ of V.

Therefore the image of L in $P\Omega(U_1, \mathbf{g}_1)$ is irreducible. In particular, the image of some Ω_i' in $P\Omega(U_1, \mathbf{g}_1)$ is non-trivial. Therefore by Proposition 5.4.16.ii $m_1 := \dim(U_1) \geq m-1$. But $m_1 \mid m^t$, and so $m_1 \geq m$, and similarly $m_i := \dim(U_i) \geq m$ for all i. Consider the case $t = 2$; here case $\mathbf{L}^{\pm}$ or $\mathbf{O}$ holds, $b = 2$ and $m_1 = m_2 = m = \sqrt{n}$. Since H and K are not of the same type, we know that case $\mathbf{O}^{+}$ holds with q odd, H is of type $O_m^{\pm}(q) \wr S_2$ or $Sp_m(q) \wr S_2$, and K is of type $O_m^{\pm}(q) \wr S_2$, $Sp_m(q) \wr S_2$ or $O_m^{+}(q) \otimes O_m^{-}(q)$. Using Proposition 5.4.16.iii we find that the only possibility is for H to be of type $O_4^{-}(q) \wr S_2$ and K to be of type $Sp_4(q) \wr S_2$, with q odd. Taking preimages, this forces $\Omega_4^{-}(q) \precsim Sp_4(q)$, which contradicts Proposition 5.4.16.iv. This leaves the case in which $t \geq 3$. Recalling the notation of §4.7, we see that $I_{\mathcal{D}}$ contains a subgroup $J \cong S_t$ acting naturally on the t components of L, and so H contains $J' \cong A_t$. We claim that

$$J_{(\mathcal{D}_2)} = 1. \tag{7.7.6}$$

Assume for a contradiction that $J_{(\mathcal{D}_2)} \neq 1$. Since any non-trivial normal subgroup of A_t is transitive on t letters, $J_{(\mathcal{D}_2)}$ is transitive on the components of L, and since Ω_i' has non-trivial image in $PGL(U_1)$ it follows that every Ω_j' has a non-trivial image in $PGL(U_1)$. In other words, $\overline{L} \precsim PGL(U_1)$, and hence by Proposition 5.5.7.ii, $m_1 \geq R_p^i(\overline{L}) = R_p(\overline{\Omega_1'})^t$. Moreover, Proposition 5.4.15 guarantees that $R_p(\overline{\Omega_1'}) \geq \sqrt{m}$, and we deduce $m_1 \geq \sqrt{m}^t = \sqrt{n}$. Similarly $m_2 \geq \sqrt{n}$, and this forces equality in the previous two expressions. In particular, $R_p(\overline{\Omega_1'}) = \sqrt{m}$, which shows $\Omega_i \approx \Omega_4^{-}(q)$ (see Proposition 5.4.15). Thus $\dim(U_1) = 2^t$, and so by Corollary 5.5.6 the image of L in $PGL(U_1)$ preserves a tensor decomposition $U_1 = Y_1 \otimes \cdots \otimes Y_t$ with $\dim(Y_i) = 2$ and $\overline{\Omega_i'} \leq PGL(Y_1)$. This is of course absurd, as $\Omega_4^{-}(q)$ does not embed in $PGL_2(q)$. This contradiction establishes (7.7.6). Thus $A_t \precsim S_b$, and hence $3 \leq t \leq b$. In particular, $K \in \mathcal{C}_7$, and so we may put $m_i = a$ for all i. And as $m^t = a^b$, we deduce $a \mid m$. However we already saw that $\overline{\Omega}_1 \precsim PGL(U_1)$, and since H and K are not of the same type, it follows from Proposition 5.4.16 that q is odd, H is of type $O_4^{-}(q) \wr S_t$ and K is of type $Sp_4(q) \wr S_t$. Since $\Omega_4^{-}(q)^2$ does not embed in $PSp_4(q)$, it follows that each component of H lies in a component of K, which means $\Omega_4^{-}(q) \precsim Sp_4(q)$. However this is a contradiction, as before.

Finally we prove,

$$K \notin \mathcal{C}_8. \tag{7.7.7}$$

For a contradiction assume that $K \in \mathcal{C}_8$. Then $L \leq \Omega_\sharp = \Omega(V, \mathbf{F}_\sharp, \kappa_\sharp)$ as described in §4.8. Now $\mathcal{C}_7$ is void in case $\mathbf{S}$ with q even, and so case $\mathbf{L}$ holds. In view of the restriction $m \geq 3$, we know that $SL_m(q)$ supports neither a non-degenerate bilinear form nor a non-degenerate unitary form on its natural module, and so the proof of (7.4.7) shows that Ω_1' cannot fix $\kappa_\sharp$ on V. This final contradiction completes the proof. ■

§7.8 The case $H \in \mathcal{C}_8$

In this section assume that $H \in \mathcal{C}_8$, so that $H = N_\Omega(\kappa_\sharp)$ for some non-degenerate form $\kappa_\sharp$, as described in §4.8. Here $H^\infty = \Omega(V, \mathbf{F}, \kappa_\sharp)$ is quasisimple and $\overline{H}^\infty$ is simple.

Proposition 7.8.1. *H is maximal among $\mathcal{C}$-groups.*

Proof. First of all, H^∞ is absolutely irreducible on V, and $\mathbf{F}_p(H^\infty) = \mathbf{F}$ (see Proposition 2.10.9.i). Thus $K \notin \mathcal{C}_1 \cup \mathcal{C}_3 \cup \mathcal{C}_5$. Moreover $R_{p'}(\overline{H}^\infty) > R_p(\overline{H}^\infty) = n$ by Proposition 5.4.15 and Corollary 5.3.10, and $P(\overline{H}^\infty) > n$ by Corollary 5.2.3.i. Consequently $K \notin \mathcal{C}_2 \cup \mathcal{C}_4 \cup \mathcal{C}_6 \cup \mathcal{C}_7$. Finally, if $K \in \mathcal{C}_8$, then as H and K are not of the same type, case **L** holds and H and K are of types $Sp_n(q)$, $O_n^\epsilon(q)$ or $GU_n(q^{1/2})$. One checks that the only in way in which $|H|$ can divide $|K|$ is for H to be of type $O_n^\pm(q)$ and K to be of type $Sp_n(q)$. According to the definition $\mathcal{C}_8$, groups of type $O_n^\pm(q)$ occur only when q is odd, and in this case $\Omega_n^\pm(q)$ is not contained in $Sp_n(q)$ by Proposition 5.4.16.iii. ■

Chapter 8
DETERMINING THE MAXIMALITY OF THE MEMBERS OF $\mathcal{C}$
Part II

§8.1 Introduction

In this chapter we complete the proof of the major result of this book, the Main Theorem stated in §3.1, by determining precisely when a subgroup in $\mathcal{C}$ is contained in a subgroup in $\mathcal{S}$. Throughout the chapter, we adopt the following hypotheses. We have $\Omega = \Omega(V, \mathbf{F}, \kappa)$, where $\Omega(V, \mathbf{F}, \kappa)$ is a classical geometry of dimension $n \geq 13$, and where $\mathbf{F}$ is a finite field of characteristic p. In particular, $\mathbf{F} = \mathbf{F}_{q^u} = \mathbf{F}_{p^{fu}}$, as described in (2.1.17). Furthermore

$$H < K < \Omega, \text{ where } H \in \mathcal{C}(\Omega) \text{ and } \overline{K} \in \mathcal{S}. \tag{8.1.1}$$

Define

$$K_o = \operatorname{soc}(\overline{K}),$$

so that K_o is a non-abelian simple group satisfying the conditions of the definition of $\mathcal{S}$ in §1.2. Then K^∞ (the last term of the derived series of K) is quasisimple and is a covering group of K_o. The main result of this chapter is:

Theorem 8.1.1. *The triple (H, K, Ω) appears in Table 3.5.I. Conversely, each triple in Table 3.5.I occurs.*

We first prove the second statement in Theorem 8.1.1.

Proposition 8.1.2.
(i) *If $\Omega = \Omega_{2^t}^+(q)$ with q even and $q \geq 4$, and H is of type $Sp_2(q) \wr S_t$ in $\mathcal{C}_7(\Omega)$, then there exists $K \in \mathcal{S}$ with $K \cong Sp_{2t}(q)$ such that $H < K < \Omega$.*
(ii) *If $\Omega = \Omega_{4^t}^+(q)$ with q even and H is of type $Sp_4(q) \wr S_t$ in $\mathcal{C}_7(\Omega)$, then there exists $K \in \mathcal{S}$ with $K \cong Sp_{4t}(q)$ such that $H < K < \Omega$.*
(iii) *If $\Omega = \Omega_{2^t}^+(q)$ with q even, $q \geq 4$ and t odd, and H is of type $Sp_2(q) \wr S_t$ in $\mathcal{C}_7(\Omega)$, then there exists $M \in \mathcal{S}$ with $M \cong \Omega_{2t+2}^+(q)$ such that $H < M < \Omega$.*
(iv) *In each case, V is a spin module for K and M, and $N_\Omega(K) = K$, $N_\Omega(M) = M$.*

Proof. (i). To prove this, we take a spin module U for $K = Sp_{2t}(q)$ and show that a subgroup $Sp_2(q) \wr S_t$ in $\mathcal{C}_2(K)$ acts irreducibly on U and preserves a decomposition of U as a tensor product of t 2-spaces; we then show that K preserves a quadratic form of type $\Omega_{2^t}^+(q)$ on U; so identifying U with V, we have $Sp_2(q) \wr S_t \in \mathcal{C}_7(\Omega)$. Since $I(V, \mathbf{F}, \kappa)$ is transitive on the members of $\mathcal{C}_7(\Omega)$ of type $Sp_2(q) \wr S_t$, this is enough to establish (i).

We define the spin module as in the section on spin modules in §5.4. Thus let (W, Q') be a $(2t+2)$-dimensional space of type $\Omega_{2t+2}^+(q)$, and let $\{e_1, \ldots e_{t+1}, f_1, \ldots, f_{t+1}\}$ be a standard basis of W, as in Proposition 2.5.3.i. Define $v = e_{t+1} + f_{t+1}$, a non-singular vector in W, and let K be the stabilizer of v in $\Omega(W, Q')$. Then $K \cong Sp_{2t}(q)$ by Proposition 4.1.7.

As in §5.4, we let $E = \langle e_1, \ldots, e_{t+1} \rangle$ and identify the spin module with $C_+(E)$. Write $U = C_+(E)$, so that U has a basis

$$\{e_S \mid S \subseteq \{1,\ldots,t+1\},\ |S| \text{ even}\},$$

where $e_S = \prod_{s \in S} e_s$.

Now let H_1 be a subgroup $Sp_2(q) \wr S_t$ in $\mathcal{C}_2(K)$, and let $B = B_1 \times \cdots \times B_t$ be the normal subgroup $Sp_2(q)^t$, where $B_i \cong Sp_2(q)$. We may take B_i to act naturally as $Sp_2(q)$ on $\langle e_i, f_i, v \rangle / \langle v \rangle$ and to centralize $\langle e_j, f_j, v \rangle / \langle v \rangle$ for $j \neq i, t+1$. Thus B_i is generated by the elements

$$r_{\alpha e_i + v} r_v, \quad r_{\beta f_i + v} r_v, \quad (\alpha, \beta \in \mathbf{F}_q).$$

We now calculate the actions of these elements on the spin module U. Identify K with the stabilizer of v in the special Clifford group X^+ described in §5.4, and let ρ_+ be the spin representation of K on U. Then if $f = f_1 \ldots f_{t+1}$, we have

$$(\rho_+(c).u)f = cuf \qquad (c \in K,\ u \in U = C_+(E)).$$

Put $c_i = c_i(\alpha) = (\alpha e_i + v)v$ and $d_i = d_i(\beta) = (\beta f_i + v)v$ $(1 \leq i \leq t)$, so that $B_i = \langle c_i(\alpha), d_i(\beta) \mid \alpha, \beta \in \mathbf{F}_q \rangle$. We calculate that

$$\rho_+(c_i) \text{ sends } e_S \mapsto \begin{cases} e_S & \text{if } i \in S \\ e_S + \alpha e_{S \Delta \{i,t+1\}} & \text{if } i \notin S \end{cases}$$

(where Δ denotes symmetric difference of sets), and

$$\rho_+(d_i) \text{ sends } e_S \mapsto \begin{cases} e_S & \text{if } i \notin S \\ e_S + \beta e_{S \Delta \{i,t+1\}} & \text{if } i \in S. \end{cases}$$

Now each B_i acts on its 2-dimensional spin module $U_i = \langle 1, e_i e_{t+1} \rangle$. We define a map $\phi : U_1 \otimes \cdots \otimes U_t \to U$ by

$$\phi(e_{S_1} \otimes \cdots \otimes e_{S_t}) = e_{S_1 \Delta S_2 \Delta \cdots \Delta S_t} \quad (\text{where } S_i = \emptyset \text{ or } \{i, t+1\}),$$

and extending linearly. It is straightforward to check that

$$\phi((1 \otimes \cdots \otimes c_i \otimes \cdots \otimes 1)(e_{S_1} \otimes \cdots \otimes e_{S_t})) = \rho_+(c_i).e_{S_1 \Delta \cdots \Delta S_t},$$
$$\phi((1 \otimes \cdots \otimes d_i \otimes \cdots \otimes 1)(e_{S_1} \otimes \cdots \otimes e_{S_t})) = \rho_+(d_i).e_{S_1 \Delta \cdots \Delta S_t}.$$

Consequently, if we identify U with $U_1 \otimes \cdots \otimes U_t$ via ϕ, then the group $B = B_1 \times \cdots \times B_t$ acts irreducibly on U, preserving a tensor decomposition. Moreover, as B_i preserves a non-degenerate symplectic form $\mathbf{f}_i$ on U_i, the group B preserves a non-degenerate quadratic form $Q = Q(\mathbf{f}_1, \ldots, \mathbf{f}_t)$ on U, as we discussed in §4.4. We know in fact that $\mathrm{sgn}(Q) = +$ (see (4.4.3)).

To prove that $H_1 = B.S_t$ lies in $\mathcal{C}_7(\Omega(U,Q))$ here, we must check that a complement S_t of B in H_1 permutes the factors $U_1, \ldots, U_t$ in the required fashion. For $i, j \in \{1, \ldots, t\}$ define

$$t_{ij} = (e_i + e_j + v)(f_i + f_j + v)(e_i + f_i + e_j + f_j + v)v \in K.$$

Then, in its action on the natural module W, t_{ij} interchanges e_i with e_j and f_i with f_j, while fixing v and fixing e_k, f_k for $k \le t$ and $k \ne i, j$. Hence $t_{ij} \in H_1$. It is enough to check that t_{ij} permutes the tensor factors naturally; via the identification ϕ, this amounts to showing that

$$\rho_+(t_{ij}).e_{S_1 \Delta \cdots \Delta S_t} = e_{(S_1 \Delta \cdots \Delta S_t)(ij)},$$

where (ij) is a 2-cycle in the symmetric group on $\{1, \ldots, t+1\}$. This can be checked by straightforward calculation. Thus we have shown that H_1 lies in $\mathcal{C}_7(\Omega(U,Q))$.

To complete the proof of (i), it remains to show that K preserves a quadratic form of type $\Omega^+_{2^t}(q)$ on the spin module U. Since B is absolutely irreducible on U by Lemma 4.4.3.vi, we know by Lemma 2.10.3 that Q is (up to scalar multiplication) the unique non-degenerate B-invariant quadratic form on U. Now $t \ge 4$ (as $n \ge 13$), so there are subgroups $A_1 = Sp_2(q) \times Sp_{2t-2}(q)$ and $A_2 = Sp_4(q) \times Sp_{2t-4}(q)$ of K such that $B \le A_1 \cap A_2$. Similar calculations to those above show that A_1 and A_2 preserve decompositions of U as a tensor product of the spin modules for the direct factors of A_1 and A_2. Hence by (4.4.3), A_1 and A_2 both preserve quadratic forms on U. But $B \le A_1 \cap A_2$, so by the uniqueness of Q, both A_1 and A_2 preserve Q. Since $\langle A_1, A_2 \rangle = K$, we deduce that K preserves Q. This completes the proof of (i).

To prove (iii), let $M = \Omega^+_{2t+2}(q)$ and let U be the spin module for M of dimension 2^t. We see exactly as in the proof of (i) that if v is a non-singular vector in the natural $(2t+2)$-dimensional module for M, then the subgroup $H_1 = Sp_2(q) \wr S_t$ of M_v preserves a decomposition of U as a tensor product of t 2-spaces. By Proposition 5.4.9.ii, since t is odd, M preserves a quadratic form Q on U of type $\Omega^+_{2^t}(q)$, and so again H_1 lies in $\mathcal{C}_7(\Omega(U,Q))$, as required.

Finally, we prove part (ii). This is similar to the proof of (i), so we give only a sketch. Let (W, Q') be a space of type $\Omega^+_{4t+2}(q)$ with standard basis $\{e_1, \ldots, e_{2t+1}, f_1, \ldots, f_{2t+1}\}$, and let $v = e_{2t+1} + f_{2t+1}$. Then if K is the stabilizer in $\Omega(W, Q')$ of v, we have $K \cong Sp_{4t}(q)$. Let $E = \langle e_1, \ldots, e_{2t+1} \rangle$ and $U = C_+(E)$, the spin module for K. Choose a subgroup $H_1 = Sp_4(q) \wr S_t$ in $\mathcal{C}_2(K)$, with base group $B = B_1 \times \cdots \times B_t$, where $B_i \cong Sp_4(q)$ acts as $Sp_4(q)$ on $\langle e_{2i-1}, e_{2i}, f_{2i-1}, f_{2i}, v \rangle / \langle v \rangle$ and centralizes $\langle e_j, f_j, v \rangle / \langle v \rangle$ for $j \ne 2i-1, 2i, 2t+1$. As before we identify K with the stabilizer of v in the special Clifford group, and let ρ_+ be the spin representation of K on U. Each B_i acts on its 4-dimensional spin module $U_i = \langle 1, e_{2i-1}e_{2i}, e_{2i-1}e_{2t+1}, e_{2i}e_{2t+1} \rangle$. If we again define $\phi : U_1 \otimes \cdots \otimes U_t \to U$ by

$$\phi(e_{S_1} \otimes \cdots \otimes e_{S_t}) = e_{S_1 \Delta \cdots \Delta S_t}$$

(where $S_i \subseteq \{2i-1, 2i, 2t+1\}$ and $|S_i|$ is even), extending linearly, we find that for

$b_i \in B_i$,

$$\phi((1 \otimes \cdots \otimes b_i \otimes \cdots \otimes 1)(e_{S_1} \otimes \cdots \otimes e_{S_t})) = \rho_+(b_i).e_{S_1 \Delta \cdots \Delta S_t}.$$

Hence, identifying U with $U_1 \otimes \cdots \otimes U_t$ via ϕ, we see (as before) that B preserves the tensor decomposition $(U_1, \mathbf{f}_1) \otimes \cdots \otimes (U_t, \mathbf{f}_t)$ of (U, Q), where $\mathbf{f}_i$ is a non-degenerate symplectic form and $Q = Q(\mathbf{f}_1, \ldots, \mathbf{f}_t)$. We now check as before that H_1 also preserves this tensor decomposition. Since by the proof of (i), K preserves a quadratic form of type $\Omega^+_{4^t}(q)$ on U, we have $H_1 \in \mathcal{C}_7(\Omega(U, Q))$. This proves (ii).

Finally, note that $N_\Omega(K) = K$ and $N_\Omega(M) = M$ in (i), (ii) and (iii), since field and graph automorphisms of K and M do not fix the spin module, by Proposition 5.4.2.i,ii. ∎

We now embark on the proof of the first statement of Theorem 8.1.1. The most difficult case is usually that in which the group K_o lies in $Lie(p)$, with p the characteristic of $\mathbf{F}$. The next three lemmas give some preliminary information in this situation. The first follows immediately from Proposition 5.4.6 and Remark 5.4.7. Recall from (2.1.17) that u is defined by $\mathbf{F} = \mathbf{F}_{q^u}$, so that $u = 2$ in case $\mathbf{U}$ and $u = 1$ in the other three cases.

Lemma 8.1.3. *Suppose that K_o is of Lie type over $\mathbf{F}_{p^a}$. Assume also that $n < R_p(K_o)^2$. Then one of the following holds:*

(i) *K_o is untwisted or of type ${}^2B_2$, ${}^2G_2$ or ${}^2F_4$, and $q^u = p^a$;*

(ii) *K_o is of type ${}^2A_\ell$, ${}^2D_\ell$, ${}^2E_6$ or ${}^3D_4$, $V \not\cong V^{\tau_o}$ as K^∞-modules (notation of Proposition 5.4.6), and $q^u = p^{2a}$ ($q^u = p^{3a}$ if K_o is of type ${}^3D_4$);*

(iii) *K_o is of type ${}^2A_\ell$, ${}^2D_\ell$, ${}^2E_6$ or ${}^3D_4$, $V \cong V^{\tau_o}$ as K^∞-modules and $q^u = p^a$.*

Lemma 8.1.4. *Suppose that $K_o = C\ell_d(s)$ for some prime power s.*

(i) *$d < n$.*

(ii) *If $p \mid s$, then $n \geq \min\{\frac{1}{2}d(d-1) - 2,\ 2^{[(d-1)/2]}\}$.*

Proof. (i) Suppose that $d \geq n$. Since $R_p(K_o) \leq n$ it is immediate from Corollary 5.3.10 that $s = p^a$ for some a. Then $d = n$ by Proposition 5.4.11 and V is a minimal module for K_o. Moreover, by Lemma 8.1.3, either $q^u = p^a$ or $K_o = U_n(p^a)$ and $q^u = p^{2a}$. Since every minimal module for K_o is quasiequivalent to the natural module (Proposition 5.4.11), we deduce from Lemma 2.10.14 that K is contained in a member of $\mathcal{C}_8$. However, we pointed out in the Remark before Theorem 1.2.1, by Lemma 2.10.15 the definition of $\mathcal{S}$ prevents members of $\mathcal{S}$ from being contained in members of $\mathcal{C}_8$, and this provides the desired contradiction.

(ii) This follows from (i) and Proposition 5.4.11. ∎

This next Lemma is a consequence of Proposition 5.4.9 and parts (d), (e) and (f) of the definition of $\mathcal{S}$ given in §1.2.

Lemma 8.1.5. *Suppose that K_o is a symplectic or orthogonal group over $\mathbf{F}_{p^a}$ and that V is a spin module for K.*

(i) *If $K_o \cong Sp_{2\ell}(2^a)$, then $q = 2^a$ and $\Omega \approx \Omega^+_{2\ell}(q)$.*
(ii) *If $K_o \cong P\Omega^+_{2\ell}(p^a)$ with ℓ even, then $\mathbf{F} = \mathbf{F}_{p^a}$ and $\Omega \approx \Omega^+_{2^\ell-1}(q)$ or $Sp_{2^\ell-1}(q)$, according to the conditions on q and ℓ given in Proposition 5.4.9.ii. If ℓ is odd, then $\Omega \approx SL_{2^{\ell-1}}(q)$.*
(iii) *If $K_o \cong P\Omega^-_{2\ell}(p^a)$, then $\mathbf{F} = \mathbf{F}_{p^{2a}}$ and $\Omega \approx SU_{2^{\ell-1}}(q)$, $\Omega^+_{2^\ell-1}(q^2)$ or $Sp_{2^\ell-1}(q^2)$, according to the conditions on q and ℓ given in Proposition 5.4.9.iii.*
(iv) *If $K_o = \Omega_{2\ell+1}(p^a)$ with p odd, then $\mathbf{F} = \mathbf{F}_{p^a}$ and $\Omega \approx \Omega^+_{2^\ell}(q)$ or $Sp_{2^\ell}(q)$ according to the conditions given in Proposition 5.4.9.i.*

We can now exclude some possibilities for H.

Lemma 8.1.6. *The group H does not lie in $\mathcal{C}_1 \cup \mathcal{C}_5 \cup \mathcal{C}_8$.*

Proof. Assume first that $H \in \mathcal{C}_1 \cup \mathcal{C}_8$. Then provided G_o is not $\Omega_{13}(q)$, we check using the results in §§4.1,4.8 that $|H| > \max\{(n+2)!, q^{3un}\}$, which contradicts Theorem 5.2.4. When $G_o \cong \Omega_{13}(q)$, we have $H \in \mathcal{C}_1$ (since $\mathcal{C}_8$ is void), and we check that $|H| > \max\{15!, q^{30}\}$. However [Lie$_1$, Th. 4.2] determines all $\mathcal{S}$-groups of order greater than q^{2n+4}, and there are no such for $\Omega_{13}(q)$.

Now let $H \in \mathcal{C}_5$. Then $\overline{H}^\infty$ is a classical simple group of dimension n over some subfield $\mathbf{F}_{q_o}$ of $\mathbf{F}_q$. Since $n \geq 13$, Proposition 5.4.15 implies that $n = R_p(\overline{H}^\infty)$, and hence $n = R_p(K_o)$. Moreover, $R_{p'}(\overline{H}^\infty) > n^2$ by Corollary 5.3.10, so K_o cannot be a group in $Lie(p')$ or an alternating group. Further, K_o is not a sporadic group by Proposition 5.2.9. Thus $K_o \in Lie(p)$. If $K_o = C\ell_d(p^a)$, then $d = n$ by Proposition 5.4.15, contrary to Lemma 8.1.4. Finally, if K_o is exceptional of Lie type, then the fact that $C\ell_n(q_o) \preceq K_o$ and $n = R_p(K_o)$ gives a contradiction, using Proposition 5.4.13 and Lemma 5.2.12.i (e.g., $C\ell_{25}(p) \npreceq F_4(p^a)$ and $C\ell_{248}(p) \npreceq E_8(p^a)$). ∎

For the final result in this section, we recall the definition of $\mathcal{G}_\mathcal{S}(\overline{H})$ given in (3.4.2).

Lemma 8.1.7. *Assume that $\overline{K}$ is minimal member of $\mathcal{G}_\mathcal{S}(\overline{H})$. Then either*
(i) *$\overline{H}$ is maximal in $\overline{K}$ or*
(ii) *H is one of the non-maximal $\mathcal{C}$-groups listed in Table 3.5.H.*

Proof. Suppose that $\overline{H}$ is non-maximal in $\overline{K}$, say $\overline{H} < M < \overline{K}$. By the minimality of K we know that $M \notin \mathcal{S}$, and hence by Theorem 1.2.1, M is contained in a member of $\mathcal{C}(\overline{\Omega})$. Hence (ii) holds by Theorem 7.0.1. ∎

§8.2 The case $H \in \mathcal{C}_2$

Here $H = \Omega_\mathcal{D}$ for some m-space decomposition $\mathcal{D}$ given by $V = V_1 \oplus \cdots \oplus V_t$, with $\dim(V_i) = m$ for all i, as described in §4.2. The possible structures of H are given explicitly throughout §4.2. As in (4.2.8), when H is of type $GL^\pm_m(q) \wr S_t$, $Sp_m(q) \wr S_t$, $O^\xi_m(q) \wr S_t$ or $O_{n/2}(q)^2$ we write $X_i = X(V_i, \kappa)$ as X ranges over the symbols Ω, S and

I. Thus one of the following holds:

$$\Omega_1 \wr A_t \leq H \tag{8.2.1}$$

$$SL_{n/2}(q^u) \leq H \tag{8.2.2}$$

Observe that if Ω_1 is insoluble, then $H^{\infty}_{(\mathcal{D})}$ contains $\prod_{i=1}^t \Omega_i'$, a central product of t quasisimple groups (or $2t$ quasisimple groups when $\Omega \approx \Omega_4^+(q) \cong SL_2(q) \circ SL_2(q)$).

The goal of this section is to prove that $\overline{H}$ is contained in no member of $\mathcal{S}$, and hence to obtain a contradiction we can assume that $\overline{K}$ is minimal member of $\mathcal{G}_{\mathcal{S}}(\overline{H})$ so that Lemma 8.1.7 applies. The first Lemma uses the notation introduced in §5.3 (see (5.3.1), in particular).

Lemma 8.2.1. *We have* $R(\overline{H}) \geq \frac{n}{3}$.

Proof. If (8.2.2) holds, we have $R(\overline{H}) \geq R(L_{n/2}(q^u))$ by Corollary 5.3.3, and so the result is immediate from Proposition 5.4.15 and Corollary 5.3.10. So assume that (8.2.1) holds. Suppose first that Ω_1 is quasisimple. Then by Proposition 5.4.15 and Corollary 5.3.11.iii we have $R(\Omega_1/Z(\Omega_1)) \geq \frac{m}{2}$, and hence by Proposition 5.5.7.i

$$R(\overline{H}) \geq t\,R(\Omega_1/Z(\Omega_1)) \geq \frac{n}{2},$$

as desired. The same argument applies if $\Omega_1 \approx Sp_4(2)$, and if $\Omega_1 \approx \Omega_4^+(q)$ (with $q \geq 4$), then $\overline{H}$ contains a covering group of $L_2(q)^{2t}$, and since $R(L_2(q)) = 2$, we have $R(\overline{H}) \geq 2t\,R(L_2(q)) = 4t = n$.

Now assume that Ω_1 is soluble. If $m \geq 3$, then by Proposition 2.9.2, $\Omega_1 \approx SU_3(2)$, $\Omega_3(3)$, $\Omega_4^+(2)$ or $\Omega_4^+(3)$. If $\Omega_1 \approx SU_3(2) \cong 3^{1+2}.Q_8$, then $m_3(\Omega_1) = 2$ and $m_2(\Omega_1) = 1$. Thus $m_3(H) \geq 2t$ and $m_2(H) \geq t$. Since $\Omega \approx SU_n(2)$, we know that $|Z(\Omega)| \mid 3$, and so $m_3(\overline{H}) \geq 2t - 1 > t$ and $m_2(\overline{H}) \geq t$. Thus Lemma 5.5.2.ii shows that $R(\overline{H}) \geq t = \frac{n}{3}$. The same reasoning works when $\Omega_1 \approx \Omega_3(3) \cong A_4$, for here $m_2(H) \geq 2t$, $m_3(H) \geq t$, and $|Z(\Omega)| \mid 2$. In the last two cases, $\Omega_1 \approx \Omega_4^+(q)$ with $q = 2$ or 3, and $\overline{H}$ contains both $2^{n/2}$ and $3^{n/2}$, whence $R(\overline{H}) \geq \frac{n}{2}$ once again.

Finally, assume that $m \leq 2$. If $m = 1$, then $\overline{H}$ contains A_n, and so $R(\overline{H}) \geq R(A_n) \geq \frac{n}{3}$ by Proposition 5.3.7. Now consider $m = 2$ and $t = \frac{n}{2}$. Then $\overline{H}$ contains a subgroup $2^{t-k}.A_t$, with $k \in \{1,2\}$ (here, the normal 2^{t-k} is the deleted permutation module or the fully deleted permutation module for the quotient A_t). If $p \neq 2$, then $R_p(\overline{H}) \geq t > \frac{n}{3}$ by Corollary 5.5.4.i. So let $p = 2$ and define $a = R_2(\overline{H})$. If $n \geq 18$, then $a \geq R_2(A_t) \geq t - 2 \geq \frac{n}{3}$ by Proposition 5.3.7. This leaves the cases $n = 14, 16$. Now $2^{t-k}.A_t$ is contained in $L_a(2^b)$ for some b, and it must act reducibly on the natural a-dimensional projective module for $L_a(2^b)$ (see [Is, 15.37], for example). Hence when $n = 14$, we have $a \geq 1 + R_2(A_7) = 5 \geq \frac{n}{3}$. If $n = 16$, we have $2^{8-k}.A_8 \cong 2^{8-k}.L_4(2) \leq L_a(2^b)$. If $a = 5$, the only reducible maximal subgroup of $L_a(2^b)$ which can contain $L_4(2)$ has structure $2^{4b}.GL_4(2^b)$. But in $2^{4b}.GL_4(2^b)$, any $L_4(2)$ acts homogeneously on the elementary abelian 2^{4b} with constituents of dimension 4, and hence there is no subgroup $2^6.L_4(2)$ or $2^7.L_4(2)$. This contradiction proves $a \geq 6 \geq \frac{n}{3}$, completing the proof. ■

Lemma 8.2.2. *K_o is not an alternating group.*

Proof. Assume for a contradiction that $K_o \cong A_k$. Since $n \geq 13$ we have $k \leq n+2$ by Proposition 5.3.7.i. It is trivial to see that $Aut(A_6)$ cannot contain a $\mathcal{C}_2$-group when $n \geq 13$, and hence $k \geq 7$. Therefore $\overline{K} \preceq S_k$ by Theorem 5.1.3, and so in the notation of §5.2, $k \geq P_f(\overline{H})$. Since $P(SL_{n/2}(q^u)) > n+2$ (see Proposition 5.2.1.i and Corollary 5.2.3.ii), it follows that (8.2.2) cannot hold, and so for the remainder of this proof we can assume that (8.2.1) holds.

Suppose here that Ω_1 is insoluble. If $\Omega_1 \not\approx \Omega_4^+(q)$, then Ω_1' is quasisimple and $P(\Omega_1'/Z(\Omega_1')) \geq m+1$ by Corollary 5.2.3. Therefore, by Proposition 5.2.7.i, $P_f(\overline{H}) \geq t\,P(\Omega_1'/Z(\Omega_1')) \geq t(m+1) = n+t$, which forces $t=2$. But then $m = \frac{n}{2} \geq 6$, and so we obtain the desired contradiction using Corollary 5.2.3.ii. Similarly, if $\Omega_1 \approx \Omega_4^+(q)$ with $q \geq 4$, then $\overline{H}$ contains a covering group of $L_2(q)^{2t}$, and hence $P_f(\overline{H}) \geq 5 \times 2t > n+2$, another contradiction.

We are left with the case in which Ω_1 is soluble. We now argue that $P_f(\overline{H}) \geq n$. We see from §4.2 that $\overline{H}$ contains one of the groups $3^{n/2}.A_{n/4}$ (if $\Omega_1 \approx \Omega_4^+(q)$ with $q \leq 3$), $3^{n/3}.A_{n/3}$ (if $\Omega_1 \approx \Omega_3(3)$ or $SU_3(2)$), $2^{n/2-\ell}.A_{n/2}$ with $\ell \leq 2$ (if $m=2$), and A_n (if $m=1$). Obviously $P_f(\overline{H}) \geq n$ if $A_n \leq \overline{H}$, and since $m_3(S_{n-1}) < \frac{n}{3}$ by Proposition 5.2.8.ii we have $P_f(\overline{H}) \geq n$ in the first two cases. As for the third case, the normal $2^{n/2-\ell}$ is either the fully deleted permutation module or the deleted permutation module for the quotient group $A_{n/2}$ and so once again $P_f(\overline{H}) \geq n$ by Proposition 5.2.8.iii. Thus we have

$$n+2 \geq k \geq P_f(\overline{H}) \geq n. \tag{8.2.3}$$

Since $n \geq 13$, it follows from Proposition 5.3.5 that $k = n+b$ with $b \leq 2$, and V is the fully deleted permutation module for K_o. Moreover by the discussion of forms on the fully deleted permutation module given in §5.3, we see that case **S** or **O** holds. Hence H is either of type $Sp_2(2) \wr S_{n/2}$ in $Sp_n(2)$, or of type $O_m^\epsilon(p) \wr S_{n/m}$ with $m \leq 4$. In particular, observe that $\overline{H}$ is local (see §4.2).

Suppose in this paragraph that $\overline{H}$ is maximal in $\overline{K}$. Since $\overline{H}$ is local, the O'Nan-Scott Theorem 1.3.1 implies that $\overline{H}$ is the intersection of $\overline{K}$ with one of the groups $S_a \times S_{k-a}$ $(a \leq 4)$, $S_a \wr S_{k/a}$ $(a \leq 4)$ and $AGL_d(r)$ (r prime, $k = r^d$). Assume for the moment that $\overline{H} = AGL_d(r) \cap \overline{K}$. Now the only possible non-abelian composition factor of H is A_t with $t \geq 5$, and hence $A_t \cong L_d(r)$. Appealing to Theorem 5.1.2 and Proposition 2.9.1, we see that $t \in \{3,4,5,6,8\}$. Since $m \leq 4$ and $n \geq 13$, we cannot have $t=3$. If $t=4$, then $m=4$ and hence $n=16$. But then $AGL_d(r) \cong \mathbf{Z}_{17}{:}\mathbf{Z}_{16}$, which does not involve A_4. If $t=8$, then $n = 16$, 24 or 32; but then $k = 17$ or 25, and neither $\mathbf{Z}_{17}{:}\mathbf{Z}_{16}$ nor $5^2{:}GL_2(5)$ involve A_8. The cases $t = 5, 6$ are also easy to eliminate, and thus we conclude $\overline{H} \neq AGL_d(r) \cap \overline{K}$. Moreover, if $\overline{H} = (S_a \times S_{k-a}) \cap \overline{K}$, then $k-a \geq n-3 > \frac{n}{2}$, and so the non-abelian composition factor of $\overline{H}$ must be A_n. Here $m=1$ and so by the remark at the end of the previous paragraph, H is of type $O_1(p) \wr S_n$. However it is clear from Proposition 4.2.15(II) that $\overline{H}$ is not isomorphic to $(S_a \times S_{k-a}) \cap \overline{K}$. This leaves the case in which $\overline{H} = (S_a \wr S_{k/a}) \cap \overline{K}$. Now if $S_a \wr S_{k/a}$ is soluble, then $k = 16$ and $a = 4$; but then $n = 14$ or 15 and so $t = 7$ or 5, which means $\overline{H}$ is insoluble, a contradiction. Therefore $S_a \wr S_{k/a}$ is insoluble, and hence so is $\overline{H}$. But as Ω_1 and S_a are soluble, it follows that A_t and $A_{k/a}$ are respectively the unique non-abelian composition factors of $\overline{H}$ and $S_a \wr S_{k/a}$. Consequently $t = \frac{k}{a} \geq 5$. But then $\frac{k}{a}$ divides both k and n, forcing $\frac{k}{a} = 2$, which is absurd.

Thus $\overline{H}$ is non-maximal in $\overline{K}$, so by Lemma 8.1.7 H belongs to the list of non-maximal $\mathcal{C}$-groups given by Table 3.5.H. If $m \geq 3$, then $\Omega_1 \approx \Omega_3(3)$ or $\Omega_4^+(2)$, so $\overline{H}$ contains $2^{2n/3}$ or $3^{n/2}$ and so by Proposition 5.2.8.ii $P_f(\overline{H})$ is at least $\frac{4n}{3}$ or $\frac{3n}{2}$, respectively, contravening (8.2.3). Thus $m \leq 2$ and since case **S** or **O** holds, Ω_1 is either $Sp_2(2)$ or $\Omega_2^{\pm}(q)$. In all cases except for $\Omega_2^+(2)$, we see using Propositions 4.2.10(II) and 4.2.11(II) that $P_f(\overline{H}) > n+2$ by Proposition 5.2.8.ii,iii.

Thus we are left to consider the case

$$\Omega \cong \Omega_n^+(2), \quad H \text{ of type } O_2^+(2) \wr S_t \ (n = 2t), \quad K_o \cong A_k \ (k = n+b,\ 1 \leq b \leq 2).$$

The module V and the quadratic form on V preserved by K are described in §5.3.

Here $Z(\Omega) = 1$, and so we may drop the symbol $\overline{}$ for the rest of this discussion. Let K act naturally on $X = \{1, \ldots, 2t+b\}$. We write elements of V as $(2t+b)$-tuples $(a_1, \ldots, a_{2t+b})$, where $a_i \in \mathbf{F}_2$ and $a_1 + \cdots + a_{2t+b} = 0$, and when $b = 2$, two $(2t+2)$-tuples are identified if their sum is $(1, \ldots, 1)$. Furthermore, $Q(a_1, \ldots, a_{2t+b})$ is as defined in (5.3.7). Now by Proposition 4.2.11(II), $H \cong 2^{t-1}.S_t$ and the derived group $H' \cong 2^{t-1}.A_t$ is perfect, and it is easy to check that $O_2(H)$ is the deleted permutation module for $H'/O_2(H)$. Furthermore, $H'/O_2(H) \cong A_t$ acts on the $O_2(H)$-orbits on X. Suppose $O_2(H)$ has an orbit Y with $|Y| \geq 4$. Then H' must fix Y as $4t > 2t+2$. Since $O_2(H)$ is the deleted permutation module for $H'/O_2(H)$, it follows from Lemma 5.3.4 that any proper subgroup of $O_2(H)$ which is normal in H' has order at most 2. Therefore $|O_2(H)^Y| \geq 2^{t-2}$. Since an abelian transitive group is regular, this forces $2t+2 \geq |Y| \geq 2^{t-2}$, which is impossible. Thus all orbits of $O_2(H)$ have size at most 2. Since the orbits of size 2 are permuted non-trivially, there must be t of them and b orbits of size 1. Without loss of generality these orbits are $\{1,2\}, \{3,4\}, \ldots, \{2t-1, 2t\}, \{2t+1\}, \ldots, \{2t+b\}$. The stabilizer in S_{2t+b} of this collection of $O_2(H)$-orbits is isomorphic to $(2^t{:}S_t) \times b$, and its derived group is $2^{t-1}{:}A_t$. Hence H' is the derived subgroup of the stabilizer in S_{2t+b} of this collection of orbits. Thus $g = (13)(24)(57)(68) = [(135)(246), (17)(28)(35)(46)] \in H'$, and consider the commutator space $[V, g] = \langle vg + v \mid v \in V \rangle$. Now on the one hand,

$$\begin{aligned} [V,g] = \langle &(1,0,1,0,\ldots,0),\ (0,1,0,1,0,\ldots,0), \\ &(0,0,0,0,1,0,1,0,\ldots,0),\ (0,0,0,0,0,1,0,1,0,\ldots,0) \rangle. \end{aligned} \tag{8.2.4}$$

On the other hand, consider the image $g^{\mathcal{D}}$ of g in $Sym(\mathcal{D})$. Recall that $\mathcal{D}$ is the $+2$-decomposition given by $V = V_1 \perp \cdots \perp V_t$. Since $g^{\mathcal{D}}$ is a non-trivial involution in $Alt(\mathcal{D})$, we may write (without loss)

$$g^{\mathcal{D}} = (V_1, V_2)(V_3, V_4) \cdots (V_{4k-3}, V_{4k-2})(V_{4k-1}, V_{4k}),$$

with $k \geq 1$. Then $\dim([V,g]) \geq 4k$, and since we have already seen in (8.2.4) that $\dim([V,g]) = 4$, we must have $k = 1$. Thus if $\{e_i, f_i\}$ is a standard basis for V_i, then (with suitable labelling)

$$[V,g] = \langle e_1 + e_2, f_1 + f_2, e_3 + e_4, f_3 + f_4 \rangle. \tag{8.2.5}$$

Evidently (8.2.4) and (8.2.5) are incompatible, for the space in (8.2.4) contains non-singular vectors, while the space in (8.2.5) is totally singular. This final contradiction completes the proof. ∎

Lemma 8.2.3. *K_o does not lie in $Lie(p')$.*

Proof. Suppose false. First let $K_o \cong C\ell_d(r^a)$, where $(p,r) = 1$ and with d chosen minimally. By Lemma 8.2.1, we have $d \geq \lceil \frac{n}{3} \rceil \geq 5$. Furthermore,

$$R_p(C\ell_d(r^a)) \leq n \leq 3d.$$

It now follows from Theorem 5.3.9 that K_o is one of the groups $L_5(2)$, $U_5(2)$, $Sp_6(2)$, $PSp_6(3)$ and $\Omega_8^{\pm}(2)$. Consider the case $K_o \cong PSp_6(3)$, so that $|\overline{K}| \mid 2^{10}.3^9.5.7.13$. Evidently (8.2.2) fails as $|L_7(q)|$ does not divide $|\overline{K}|$, and so (8.2.1) must hold. In addition, $|\Omega_1/Z(\Omega_1)|$ must be a $\{2,3\}$-group. In particular, Ω_1 is soluble, and so by Proposition 2.9.2, $\Omega_1 \approx SL_2^{\pm}(2)$, $Sp_2(2)$, $\Omega_2^{\pm}(q)$, $SU_3(2)$ or $\Omega_4^{+}(2)$. If $\Omega_1 \approx SL_2(2)$, then $t \geq 7$ as $n \geq 13$, and so $\overline{H}$ contains $SL_2(2) \wr S_7$. But then $2^{11} \mid |\overline{H}|$, a contradiction. Similarly, Lagrange's Theorem may be used to eliminate all possibilities for Ω_1, except for $\Omega_2^{\pm}(2)$ and $\Omega_2^{+}(4)$, with $t = 7$. However, in these instances $\overline{H}'$ contains $2^6.A_7$, by Proposition 4.2.11(II). But then $R_3(H) \geq 7$ by Corollary 5.5.4, a contradiction. The other choices for K_o are easier to eliminate, and we leave these to the reader.

Now suppose that K_o is an exceptional group in characteristic r. Again using Lemma 8.2.1 we obtain

$$R_r(K_o) \geq \frac{n}{3}, \quad R_p(K_o) \leq n.$$

Thus $R_{r'}(K_o) \leq 3R_r(K_o)$. Lower bounds for $R_{r'}(K_o)$ are given by Theorem 5.3.9, and the values of $R_r(K_o)$ by Proposition 5.4.13. From these results we deduce that K_o is one of the groups

$${}^2B_2(8),\ {}^3D_4(2),\ {}^2F_4(2)',\ G_2(3),\ G_2(4),\ F_4(2),$$

and that n is at least 13, 24, 16, 14, 13, 44 in the respective cases. Once again we see using Lagrange's Theorem that $\overline{K}$ cannot contain $\overline{H}$. For example, take $K_o \cong G_2(3)$, so that $|\overline{K}| \mid 3^6.2^7.7.13$ and $14 \leq n \leq 21$. Since 5 does not divide $|\overline{K}|$, we have $t \leq 4$ and so $m \geq 4$. Moreover no classical group of dimension 7 or more is contained in $G_2(3)$, and so $t \geq 3$. Then $|\Omega_1/Z(\Omega_1)|$ must divide $2^2 3^2$, and as $m \geq 4$, the only possibility is $\Omega_1 \approx \Omega_4^{+}(2)$. But in this case, $\overline{H}$ contains $\Omega_4^{+}(2) \wr 3$, which has order divisible by 3^7. ∎

Lemma 8.2.4. *K_o does not lie in $Lie(p)$.*

Proof. Suppose false. First let $K_o = C\ell_d(p^a)$, with d chosen minimally as in the previous proof. As before, we have $d \geq \lceil \frac{n}{3} \rceil \geq 5$. Observe that n and d satisfy the inequalities given in Proposition 5.4.11, and hence the representation of K_o on V appears in Table 5.4.A. Thus from Lemmas 8.1.3, 8.1.4 and 8.1.5 we conclude that one of the following holds:

(i) $n = \frac{1}{2}d(d-1) - b$ with $0 \leq b \leq 2$ and $q = p^a$

(ii) $n = 15$ and $K_o = L_5^\epsilon(q)$

(iii) $n = 14$ and $K_o = PSp_6(q)$ with q odd

(iv) $n = 2^\ell$, V is a spin module for K_o, and $q = p^a$ (or $q = p^{2a}$ when $K_o = P\Omega_d^-(p^a)$ with $4 \mid d$), where

$$\ell = \begin{cases} d/2 & \text{if } K_o = Sp_d(q) \text{ with } q \text{ and } d \text{ even} \\ (d-1)/2 & \text{if } K_o = \Omega_d(q) \text{ with } q \text{ and } d \text{ odd} \\ (d-2)/2 & \text{if } K_o = P\Omega_d^\epsilon(q) \text{ or } P\Omega_d^-(q^{1/2}) \text{ with } d \text{ even.} \end{cases}$$

In case (i), the fact that $n \leq 3d$ forces $d \leq 7$. Thus in cases (i), (ii) and (iii) we see using Propositions 5.4.2 and 5.4.3 that either

$$(K_o, \overline{\Omega}) \in \{(L_5^\epsilon(q), L_{15}^\epsilon(q)), (L_6^\epsilon(q), L_{15}^\epsilon(q)), (L_7^\epsilon(q), L_{21}^\epsilon(q)), (\Omega_7(q), \Omega_{21}(q))\};$$

or

$$K_o = PSp_6(q) \text{ with } 13 \leq n \leq 14.$$

If $(K_o, \overline{\Omega}) = (L_5^\epsilon(q), L_{15}^\epsilon(q))$, then H cannot be of type $GL_1^\epsilon(q) \wr S_{15}$ as $R(A_{15}) > 5$ (see Proposition 5.3.7). But on the other hand, H cannot be of type $GL_3^\epsilon(q) \wr S_5$ or $GL_5^\epsilon(q) \wr S_3$ by Lagrange's Theorem. This eliminates the first pair, and the following three pairs are eliminated in exactly the same way. When $K_o = PSp_6(q)$, case **S** or **O** holds by Proposition 5.4.3. As $R(A_{13}) > 6$ and since $PSp_6(q)$ does not contain a classical group of dimension 7, we must have $n = 14$ and $(m,t) = (2,7)$. Lagrange's Theorem eliminates all possibilities except for

$$K = K_o = Sp_6(2), \quad \Omega \cong \Omega_{14}^+(2), \quad \text{and } H \text{ of type } O_2^+(2) \wr S_7.$$

Because $H \cong 2^6.S_7$ is a 2-local subgroup of K, we know that H is reducible on the natural 6-dimensional module for K. This means H is contained in a member of $\mathcal{C}_1(K)$; but according to Propositions 4.1.3(II) and 4.1.19(II), no member of $\mathcal{C}_1(K)$ involves A_7.

Finally consider (iv). Here the fact that $n \leq 3d$ implies that $d \leq 12$ and n is 16 or 32. Using Proposition 5.3.7 and Lagrange's Theorem as above, we see that K cannot contain a $\mathcal{C}_2$-group if $n = 32$, and so by Lemma 8.1.5, we are left with the case in which

$$(K_o, \overline{\Omega}) \in \{(\Omega_9(q), P\Omega_{16}^+(q)), (Sp_8(q), P\Omega_{16}^+(q)), (P\Omega_{10}^\pm(q), L_{16}^\pm(q))\}.$$

Again, we check using Lagrange's Theorem that the only possibility is

$$K = K_o = Sp_8(2), \quad \Omega = \Omega_{16}^+(2), \quad \text{and } H \text{ of type } O_2^+(2) \wr S_8.$$

Now write $K = I(W, \mathbf{F}_2, \mathbf{f})$, where W is an 8-dimensional vector space over $\mathbf{F}_2$ and $\mathbf{f}$ is a non-degenerate symplectic form on W. We have $H \cong 2^7.S_8$, which is 2-local, and so as before we see that H is contained in a member of $\mathcal{C}_1(K)$. According to Propositions 4.1.3(II) and 4.1.19(II), the only member of $\mathcal{C}_1(K)$ which involves S_8 is a group P of type P_1, which is isomorphic to $2^7.Sp_6(2)$. Thus if $\{e_1, \ldots, f_4\}$ is a symplectic basis for $(W, \mathbf{F}_2, \mathbf{f})$, then we may assume that $P = N_K(e_1)$. Now let Q be the unique quadratic form on W which satisfies $\mathbf{f}_Q = \mathbf{f}$ and $Q(e_i) = Q(f_i) = 0$ for all i. Thus $O_8^+(2) \cong I(W, \mathbf{F}_2, Q) \leq K$ and K has a subgroup $N \cong O_6^+(2)$ which is the centralizer in $I(W, \mathbf{F}_2, Q)$ of the non-degenerate 2-space $\langle e_1, f_1 \rangle$. Here N acts faithfully on the 6-space $\langle e_2, e_3, e_3, f_1, f_2, f_3 \rangle$, and it contains a subgroup $O_2^+(2) \wr S_3$ permuting these 6 vectors; it follows from the subsection on spin modules in §5.4 that this $O_2^+(2) \wr S_3$ has two orbits of size 8 on the 16 weights of V. Therefore the irreducible constituents of N on V have dimension at least 8. Observe that N fixes e_1, and hence $N < P$. Consequently $O_2(P)N \cong 2^7.O_6^+(2)$ and the irreducible constituents of $O_2(P)N$ on V have dimension at least 8. We are now in a position to obtain the desired contradiction. For observe that $Sp_6(2)$ has a unique conjugacy class of subgroups $O_6^+(2)$. Consequently P has a unique conjugacy class of subgroups $2^7.O_6^+(2)$, and so in view of the isomorphism $O_6^+(2) \cong S_8$ (see the remarks after Proposition 2.9.1), we know that $O_2(P)N$ and H must be conjugate in P. However H fixes a non-zero vector in V (see Proposition 6.2.8) while the irreducible constituents of $O_2(P)N$ have dimension at least 8. This contradiction completes the analysis when $K_o = C\ell_d(p^a)$.

Now take K_o to be an exceptional group in characteristic p. By Lemma 8.2.1 we have $n \leq 3R_p(K_o)$. The values of $R_p(K_o)$ are given by Proposition 5.4.13. Hence we deduce from Proposition 5.4.12 that one of the following holds:

type of K_o	n
G_2^ϵ	$n = 14$ or $n \geq 18$
F_4^ϵ	$n = 25, 26$ or 52
E_6^ϵ	$n = 27, 77$ or 78
E_7	$n = 56, 132$ or 133
E_8	$n = 248$

Since $n < R_p(K_o)^2$, Lemma 8.1.3 implies that q^u is a power of p^a. In all cases it is easy to check using Proposition 5.2.13 that K cannot contain a $\mathcal{C}_2$-group. For example, consider $K_o \cong E_6^\epsilon(p^a)$ with $n = 27$. Then $\Omega \approx SL_{27}^\epsilon(q)$ (see Proposition 5.4.16). Since by Proposition 5.2.13 for any prime $s \notin \{2, p\}$ we have $m_s(K_o) \leq 6$, H must be of type $GL_9^\epsilon(q) \wr S_3$. But then $\overline{H}$ does not embed in $\overline{K}$ by Lagrange's Theorem. ■

Lemma 8.2.5. *K_o is not a sporadic group.*

Proof. Suppose false. Since by Proposition 5.2.9, no sporadic group involves $L_7(q)$, (8.2.1) holds.

Suppose first that $t = 2$. Then $m \geq 7$, and so by Proposition 5.2.9, K_o is one of Co_1,

Fi_{22}, Fi_{23}, Fi'_{24}, BM and M. But then $n \geq 24$ by Proposition 5.3.8, so in fact $m \geq 12$, violating Proposition 5.2.9.

This leaves the case $t \geq 3$. Now if $m \geq 9$, then by Proposition 5.2.9 $\Omega_1 \approx \Omega_{10}^{\pm}(2)$. But $|\Omega_{10}^{\pm}(2)|^2$ does not divide $|K_o|$. Therefore $m \leq 8$; moreover $t \leq 12$ by Proposition 5.2.9. Hence $n \leq 96$, and so by Proposition 5.3.8, K_o is not J_4, Ly, Fi_{23}, Fi'_{24}, BM or M. If $m \geq 5$ then the fact that $|\Omega_1/Z(\Omega_1)|^3$ divides $|K|$ gives a contradiction.

Hence $m \leq 4$, which implies $t \geq 4$. If there exists a prime $r \geq 5$ dividing $|\Omega_1|$, then r^4 dvides $|K_o|$, forcing $K_o = Co_1$ or HN and $r = 5$. But then by Proposition 5.3.8 $n \geq 24$ or 56 in the respective cases, and hence $t \geq 6$ or 14. This is impossible as 5^6 and 5^{14} do not divide $|Co_1|$, $|HN|$, respectively. Consequently there is no such prime r, and so Ω_1 is soluble. Thus Ω_1 appears in Proposition 2.9.2.i. First take the case where $m \leq 2$. Now $7 \leq t \leq 12$ by Proposition 5.2.9, and so $14 \leq n \leq 24$. If $t \geq 10$, then K_o involves A_{10}, so by Proposition 5.2.9, K_o is one of Ly, Fi_{22}, Fi_{23}, Fi'_{24}, BM, M and HN; but then $n > 24$ by Proposition 5.3.8, a contradiction. Hence $t \leq 9$, and so $n \leq 18$. Since K_o involves $A_{n/2}$, we deduce from Propositions 5.2.9 and 5.3.8 that K_o is M_{22}, M_{23}, M_{24} or Suz. However we check from the lists of the maximal subgroups of these groups in [At] that none of them contains $2^6.A_7$.

Thus m is 3 or 4 and Ω_1 is $SU_3(2)$, $\Omega_3(3)$ or $\Omega_4^+(q)$ with $q \leq 3$. If Ω_1 is not $\Omega_3(3)$, then $3^{\lceil n/2 \rceil + 1}$ divides $|\overline{K}|$ with $n \geq R(K_o)$ and $n \geq 13$, which is impossible by Proposition 5.3.8. And if $\Omega_1 \approx \Omega_3(3)$, then $12^{n/3}$ divides $|\overline{K}|$, which by Proposition 5.3.8 implies that K_o is Suz, Co_1, Co_2 or Fi_{22}. But we check from lists of their maximal subgroups (see [At,K-W$_2$]) that none of these groups contains $\Omega_3(3) \wr A_{n/3}$ with $n \geq R(K_o)$. ∎

§8.3 The case $H \in \mathcal{C}_3$

Here $H = N_\Omega(V_\sharp, \mathbf{F}_\sharp, \kappa_\sharp)$ as described in §4.3 and we write $X_\sharp = X(V_\sharp, \mathbf{F}_\sharp, \kappa_\sharp)$, as usual. Recall the subgroup L of H defined in (7.3.1), and also the notation $\zeta_p(X)$ introduced in (5.2.4).

Lemma 8.3.1. *K_o is not an alternating group.*

Proof. Suppose that $K_o \cong A_t$ for some $t \geq 5$. Then $t \leq n+2$ by Proposition 5.3.7. Now obviously $t \geq 7$ by Lemma 7.3.2.ii, and hence $A_t \leq \overline{K} \leq S_t$. Consequently $P_f(\overline{L}) \leq t \leq n+2$, and this runs contrary to Lemma 7.3.2.iv. ∎

Lemma 8.3.2. *K_o does not lie in $Lie(p')$.*

Proof. Suppose false. It follows from Lemma 7.3.2.i and the fact that $n \geq 13$ that

$$|\overline{K}| \geq \max\{q^{11}, q^{R_p(K_o)-2}\}. \tag{8.3.1}$$

Now Theorem 5.3.9 can be used to obtain the list of possible groups K satisfying this inequality. For example, if $K_o = PSp_{2d}(s)$ with s odd and $d \geq 2$, then (8.3.1), Theorem 5.3.9 and Table 5.1.A imply $s^{2d^2+d}\log(s) \geq |\overline{K}| \geq p^{(s^d-1)/2-2}$, and so (d, s) is one of $(2,3),(2,5),(2,7),(3,2),(3,3),(4,3)$. Using this reasoning for the other groups of Lie type we find that K_o is one of the following:

$$\begin{gathered} L_2(s),\ s \le 37, \\ L_3(3),\ L_3(4),\ L_4(3),\ L_5(2), \\ U_3(3),\ U_3(4),\ U_3(5),\ U_4(3),\ U_5(2),\ U_6(2), \\ PSp_4(3),\ PSp_4(5),\ PSp_4(7),\ Sp_6(2),\ PSp_6(3),\ PSp_8(3), \\ \Omega_7(3),\ \Omega_8^+(2), \\ G_2(3),\ G_2(4),\ {}^2F_4(2)'. \end{gathered} \tag{8.3.2}$$

Now $\overline{K}$ must contain an element of order at least 122 by Lemma 7.3.2.ii. With just three exceptions, the groups appearing in (8.3.2) and their automorphism groups occur in [At], from which we check that none has an element of the required order. The three exceptions are $L_2(37)$, $PSp_4(7)$ and $PSp_8(3)$. Take $K_o = PSp_8(3)$, so that $|Aut(K_o)| = 2^{15}.3^{16}.5^2.7.13.41$ and $R_p(K_o) \ge 40$. Thus by (8.3.1), $|\overline{K}| \ge q^{38}$ and so $q = 2$. Evidently $\zeta_2(\overline{K}) = 20$. But $\zeta_2(\overline{K}) \ge \zeta_2(\overline{L}) \ge \frac{nuf}{2} \ge 20$. Therefore $n = 40$ and $u = 1$. However Lemma 7.3.2.vi implies that H must be of type $O_4^+(q^{n/4})$ or of type $GU_2(q^{n/2})$, which is clearly absurd as $\frac{n}{4}$ and $\frac{n}{2}$ are not prime. The other two possibilities for K_o are dealt with similarly. Alternatively, it is easy to establish that neither $Aut(L_2(37))$ nor $Aut(PSp_4(7))$ contains an element of order at least 122. ■

Lemma 8.3.3. *K_o does not lie in $Lie(p)$.*

Proof. First suppose that $K_o = C\ell_d(p^e)$ (with d chosen minimally), and that K_o is untwisted (that is, $K_o \not\cong U_d(p^e)$ or $P\Omega_d^-(p^e)$). Then $\zeta_p(\overline{K}) = \zeta_p(K_o) \le ed$ by Propositions 5.2.16 and 5.2.17. Now according to Proposition 5.4.6.i, $uf \mid e$ and $n \ge d^{e/uf}$. Thus by Lemma 7.3.2.vi, we have

$$ed \ge \zeta_p(\overline{K}) \ge \zeta_p(\overline{H}) \ge \frac{ufn}{2} \ge \frac{ufd^{e/uf}}{2}. \tag{8.3.3}$$

Therefore $2ed \ge ufd^{e/uf}$. Suppose first that $e \ge 2uf$. Then as n is an $(e/uf)^{\text{th}}$-power by Proposition 5.4.6.i, we must have $n = 16$ and $d \le 4$. But when $n = 16$, the group $\overline{\Omega}_\sharp$ is a classical quasisimple group of dimension 8, and hence cannot be contained in a classical simple group of dimension at most 4. Therefore $e = uf$, which means $d \ge \lceil \frac{n}{2} \rceil \ge 7$, and hence the representation of K on V is given by Proposition 5.4.11. As $d < n$ by Lemma 8.1.4, and since $\frac{1}{2}d(d-1) - 2 > 2d$ for $d \ge 7$, we deduce that V is a spin module for K_o. Therefore by Lemma 8.1.5, $K_o \cong B_4(q)$ or $D_5^\epsilon(q)$ and $\Omega \approx \Omega_{16}^+(q)$ or $SL_{16}^\epsilon(q)$, respectively. But here L has a non-abelian composition factor $P\Omega_8^\epsilon(q^2)$, $U_8(q)$ or $L_8(q^2)$, which is clearly not involved in K_o.

Next suppose that $K_o = U_d(p^e)$ with $d \ge 3$. Here $\zeta_p(\overline{K}) = \zeta_p(K_o) \le 2ed$ (see Proposition 5.2.16), and hence

$$2ed \ge \frac{ufn}{2}. \tag{8.3.4}$$

We now apply Proposition 5.4.6.ii. If Proposition 5.4.6.ii(b) holds, then $n \ge d^{2e/uf}$ and the argument in the previous paragraph goes through — one need only replace e by $2e$.

Assume therefore that Proposition 5.4.6.ii(a) holds. Then $uf \mid e$ and $n = x^{e/uf}$, where $x = \dim(M)$ and $M \cong M^{\tau_o}$. By Proposition 5.4.8 we have $x \geq \frac{1}{2}d(d-1)$. Thus from (8.3.4) we obtain

$$\frac{4ed}{uf} \geq \left(\tfrac{1}{2}d(d-1)\right)^{e/uf}. \tag{8.3.5}$$

Thus if $e > uf$, then $K_o = U_3(q^{2u})$ or $U_3(q^{3u})$. If $K_o = U_3(q^{2u})$, then $n \leq 24$ (by (8.3.4)) and also n is a square; this forces $n = 16$, and we obtain a contradiction as in the previous paragraph. If $K_o = U_3(q^{3u})$, then $n \leq 36$ and n is a cube; therefore $n = 27$ and again we reach a contradiction since no simple classical group of dimension 9 can embed in $U_3(q^{3u})$. Now consider the case $e = uf$, so that $K_o = U_d(q^u)$. If $d \geq 7$, then it follows from Proposition 5.4.8 that $n \geq d(d-1)$. But then $\zeta_p(K_o) \geq \frac{1}{2}ufd(d-1)$ by Lemma 7.3.2.vi, which is a contradiction because $\zeta_p(U_d(q^u)) \leq 2ufd$. Therefore $d \leq 6$, and since $\zeta_p(K_o) \geq \frac{ufn}{2} \geq \frac{13uf}{2}$ we must have $d \geq 5$. Thus d is 5 or 6, whence $\zeta_p(K_o) = 10uf$, which shows $n \leq 20$. On the other hand, Proposition 5.4.8 implies that $n \geq 20$. Thus $n = 20$, and we now reach a contradiction as neither $C\ell_{10}(q)$ nor $C\ell_4(q^5)$ can be involved in $U_5(q^u)$ or $U_6(q^u)$.

Next assume that $K_o = P\Omega_d^-(p^e)$ with $d \geq 8$. Again we apply Proposition 5.4.6.ii. If Proposition 5.4.6.ii(b) occurs, then we are led to $ed \geq \frac{1}{2}ufd^{2e/uf}$ using Lemma 7.3.2.vi, which in turn forces $uf = 2e$ and $n \leq d$, against Lemma 8.1.4. And if Proposition 5.4.6.ii(a) holds, we obtain $uf = e$, which forces $n \leq 2d$. Thus by Proposition 5.4.11, $d = 10$ and V is a spin module of dimension 16. But then by Lemma 8.1.5 $\Omega \approx SU_{16}(q)$, and so $\mathcal{C}_3$ is void.

Finally, consider the case where K_o is an exceptional group of Lie type. If $K_o = E_8(p^e)$, then by Proposition 5.4.6.i we have $uf \mid e$ and $n \geq 248^{e/uf}$. On the other hand, $\zeta_p(\overline{K}) = \zeta_p(K_o) = 30e$, and thus we are led to the absurdity $30e \geq \frac{1}{2}uf248^{e/uf}$. Similarly we show that K_o cannot be $E_6(p^e)$, $E_7(p^e)$ or $F_4(p^e)$. If $K_o = G_2(p^e)$, then the same argument shows that $6e \geq \frac{1}{2}uf6^{e/uf}$, which forces $e = uf$; but then $\zeta_p(\overline{H}) \geq \frac{1}{2}ufn > \frac{13}{2}uf > 6uf = \zeta_p(K_o) = \zeta_p(\overline{K})$, a contradiction. If $K_o = {}^2E_6(p^e)$ then $\zeta_p(\overline{K}) = 18e$ and $R_p(K_o) = 27$. Thus if Proposition 5.4.6.ii(b) holds, then we obtain $18e \geq \frac{1}{2}uf27^{2e/uf}$, which is impossible. And if Proposition 5.4.6.ii(a) occurs, it follows from Proposition 5.4.8 that $18e \geq \frac{1}{2}uf72^{e/uf}$, which is again impossible. If $K_o = {}^3D_4(p^e)$, then $\zeta_p(\overline{K}) = 12e$ and we now apply Remark 5.4.7.a. If $uf \mid e$, then we obtain $12e \geq \frac{1}{2}uf8^{e/uf}$, which forces $e = uf$. Moreover $n \geq 24$, by Proposition 5.4.8, and hence $\zeta_p(\overline{H}) \geq \frac{ufn}{2} \geq 12uf = \zeta_p(K_o)$. So in fact $n = 24$. But this yields a contradiction as neither $C\ell_{12}(q^2)$ nor $C\ell_8(q^3)$ is contained in ${}^3D_4(q^u)$. If uf does not divide e and $uf \mid 3e$, then $n \geq 8^{3e/uf}$, and hence $uf = 3e$; but this forces $n \leq 8$, violating hypothesis (6.0.3). The remaining three families ${}^2B_2$, ${}^2G_2$ and ${}^2F_4$ are handled using Remark 5.4.7.b. If $K_o = {}^2G_2(3^e)$, then by Remark 5.4.7.b we have $uf \mid e$ and $n \geq 7^{e/uf}$. Thus $6e = \zeta_3(\overline{K}) \geq \frac{nuf}{2} \geq \frac{1}{2}uf7^{e/uf}$, which forces $e = uf$. But then $n \leq 12$, against (6.0.3). The same reasoning applies to ${}^2B_2$ and ${}^2F_4$. ∎

Lemma 8.3.4. *K_o is not sporadic.*

Proof. This is easy, for $\overline{H}$ contains an element of order at least 122 by Lemma 7.3.2.ii, while no automorphism group of a sporadic group has an element of such an order (see [At]). ■

§8.4 The case $H \in \mathcal{C}_4$

In this section we use the notation of §7.4, so that $H = \Omega_{\mathcal{D}}$ for some tensor decomposition $\mathcal{D}$ given by $(V_1, \mathbf{f}_1) \otimes (V_2, \mathbf{f}_2)$. We write $\Omega_i = \Omega(V_i, \mathbf{f}_i)$, $n_i = \dim(V_i)$, and we set $L = \Omega_1 \otimes \Omega_2 \leq H$. We arrange matters so that $n_1 \leq \sqrt{n} \leq n_2$ and Ω_2 is quasisimple. Thus both (7.4.1) and (7.4.2) hold here.

Lemma 8.4.1. *K_o is not alternating.*

Proof. Suppose that $K_o = A_k$, so that $k \leq n+2$ by Proposition 5.3.7. Then $P(\overline{\Omega}_2) \leq n+2 \leq n_2^2+2$, and so in view of Corollary 5.2.3.ii and (7.4.2), $\Omega_2 \approx \Omega_4^-(3)$ or $\Omega_5(3)$. If $\Omega_2 \approx \Omega_5(3)$, then case **O** holds and it must be the case that $n_1 \leq 4$, since $H \notin \mathcal{C}_7$. Therefore $n \leq 20$ and $k \leq 22$, which yields a contradiction as $P(\Omega_5(3)) = 27$. This leaves the case where $\Omega_2 \approx \Omega_4^-(3)$, and hence $\overline{L} \approx P\Omega_4^+(3) \times P\Omega_4^-(3) \cong A_4 \times A_4 \times A_6$. Note further that 17 does not divide $|P\Omega_{16}^+(3)|$, and so $k \leq 16$. Write $X = \{1, \ldots, k\}$ for the set permuted naturally by K_o, and define $J \leq K_o$ as the alternating group on the set of fixed points of $\overline{\Omega}_2$ in X. Now the non-trivial transitive permutation representations of A_6 of degree at most 16 have degrees 6, 10 and 15, and all these representations are primitive. Since the centralizer of a non-abelian primitive group is trivial (see [Wie, 4.5′]), it follows that $C_{K_o}(\overline{\Omega}_2)$ is either J or a group twice as big as J in the event that $\overline{\Omega}_2$ has two orbits of size 6. As $O^2(\overline{\Omega}_1) = \overline{\Omega}_1$, it now follows that $\overline{\Omega}_1 \leq J$, and hence J has degree $j \geq 8$. Hence $\widetilde{J}$ is a covering group of A_j for $j \geq 8$, where $\widetilde{J}$ is the last term of the derived series of the preimage of J in $GL(V)$. Now by Lemma 4.0.5.i $\widetilde{J} \leq C_\Omega(\Omega_2)$, and hence by Lemma 4.4.3.i $\widetilde{J} \leq GL(V_1) \cong GL_4(3)$. This, however, violates Proposition 5.3.7. ■

Lemma 8.4.2. *K_o does not lie in $Lie(p')$.*

Proof. First assume that $K_o = C\ell_d(r^e)$, where $(p, r) = 1$ and with d chosen minimally. Then $R_{p'}(\overline{\Omega}_2) \leq d$. Suppose first that $d \geq 9$. Then Corollary 5.3.10.i implies $n \geq R_p(K_o) > d^2$, and hence $n_2 > d$. But then Corollary 5.3.10.i implies that $R_{p'}(\overline{\Omega}_2) > d$, and we have reached a contradiction. Therefore

$$d \leq 8. \tag{8.4.1}$$

Thus it follows from Theorem 5.3.9 and (7.4.2) that

$$\Omega_2 \approx Sp_4(3),\ \Omega_4^-(3),\ \Omega_5(3),\ \Omega_6^-(3) \text{ or } Sp_6(2). \tag{8.4.2}$$

Suppose for the moment that $K_o = \Omega_8^+(2)$. Then $q = 3$, and since 3^7 does not divide $|Aut(\Omega_8^+(2))|$ and 3^6 does not divide $|\Omega_8^+(2)|$, it follows that H must be of type $O_4^+(3) \otimes O_5(3)$, $O_3(3) \otimes O_5(3)$, $O_4^+(3) \otimes Sp_4(3)$ or $O_4^+(3) \otimes O_4^-(3)$. In the first three cases, $\overline{H} \geq \overline{\Omega}_1 \times \overline{\Omega}_2$, which contains $A_4 \times \Omega_5(3)$. Since $Out(\Omega_8^+(2)) \cong S_3$, it follows that $2 \times \Omega_5(3)$ embeds in $\Omega_8^+(2)$. But then by Proposition 5.2.10, $\Omega_5(3)$ must embed in a maximal parabolic subgroup of $\Omega_8^+(2)$. This forces $\Omega_5(3)$ to embed in $\Omega_6^+(2) \cong L_4(2)$, which is impossible. In the fourth case, $\overline{H}$ contains $A_4 \times A_4 \times A_6$, and so reasoning as before we see that $3 \times A_6$ embeds in some maximal parabolic of $\Omega_8^+(2)$. This forces $3 \times A_6$ to embed in $L_4(2)$, which is again impossible. Thus we have proved

$$K_o \neq \Omega_8^+(2). \tag{8.4.3}$$

Now consider the case in which $d \geq 7$. It now follows from Corollary 5.3.10.iii and (8.4.3) that $n \geq 27$, and hence by (8.4.2) H is of type $O_5(3) \otimes O_6^-(3)$, $O_6^+(3) \otimes O_6^-(3)$ or $Sp_6(3) \otimes O_6^-(3)$. But then $R_{3'}(\overline{\Omega}_1) \geq 4$, and so by Proposition 5.5.7, $d \geq R_{3'}(\overline{L}) \geq 4+6$, violating (8.4.1). Therefore

$$d \leq 6. \tag{8.4.4}$$

And since $R_{2'}(Sp_6(2)) = 7$, we have $\Omega_2 \not\approx Sp_6(2)$, and hence $q = 3$ in view of (8.4.2). In particular,

$$r \neq 3. \tag{8.4.5}$$

Suppose for the moment that $\Omega_2 \approx \Omega_6^-(3)$. Then $R_{3'}(\overline{\Omega}_2) = 6$, and hence $d = 6$. On the other hand, $R_3(K_o) \leq n \leq n_2^2 = 36$. Thus the only possibilities for K_o are $L_6^\pm(2)$ and $Sp_6(2)$. Of these, only $U_6(2)$ contains $P\Omega_6^-(3)$. However, $P\Omega_6^-(3)$ does not embed in a parabolic subgroup of $U_6(2)$ and $Out(U_6(2)) \cong S_3$. Thus the argument used to prove (8.4.3) also shows that $K_o \neq U_6(2)$. This establishes $\Omega_2 \not\approx \Omega_6^-(3)$, and so $n_2 \leq 5$. And since $H \notin \mathcal{C}_7$, we cannot have $n_1 = 5$. Therefore

$$n \leq 20. \tag{8.4.6}$$

Combining facts (8.4.4), (8.4.5) and (8.4.6), along with the fact that $3^4 \mid |\overline{K}|$ (since $3^4 \mid |\overline{H}|$), we check from Theorem 5.3.9 that the only possibilities for K_o are $L_3(4)$, $U_3(5)$, $U_4(2)$, $U_5(2)$ and $Sp_6(2)$. Each of these easy to eliminate, using the elementary ideas in the proof of (8.4.3).

Now assume that K_o is an exceptional group of Lie type in characteristic r, with $r \neq p$. By Theorem 5.3.9, either $R_p(K_o) \geq 44$ or K_o is one of the groups

$$G_2(3),\ G_2(4),\ {}^3D_4(2),\ {}^2F_4(2)',\ Sz(8).$$

If $R_p(K_o) \geq 44$, then $n \geq 44$ and so $n_2 \geq \lceil\sqrt{n}\,\rceil \geq 7$. Moreover, $\Omega_2 \not\approx \Omega_8^+(2)$ by (7.4.2), and so it follows from Corollary 5.3.10 that $R_r(K_o) \geq R_{p'}(\overline{\Omega}_2) \geq 27$. Thus by Proposition 5.4.13, K_o is of type $E_6^\pm$, E_7 or E_8. Consequently $n \geq 1,536$ by Theorem 5.3.9, and so $n_2 \geq 40$; but then $R_r(\overline{\Omega}_2) \geq 1,600$ by Corollary 5.3.10.i, contrary to the fact that $R_r(K_o) \leq 248$. Hence K_o is one of the five groups above. Now $p^4 \mid |\overline{L}|$, and this fact forces $p = 2$ if $K_o = G_2(3)$, forces $p = 3$ if $K_o = {}^3D_4(2)$, and eliminates the remaining three groups. If $K_o = G_2(3)$, then (7.4.2) implies that either $L_5^\mp(2)$ or $Sp_6(2)$ is involved in $\overline{\Omega}_2$, yet the orders of these groups do not divide $|G_2(3)|$. And if $K_o = {}^3D_4(2)$, then $n \geq 24$ by Theorem 5.3.9, and hence $n_2 \geq 5$. But none of the groups $L_5^{\pm}(3)$, $\Omega_5(3)$ and $PSp_6(3)$ are involved in ${}^3D_4(2)$ by Lagrange's Theorem, and we reach the desired contradiction. ■

Lemma 8.4.3. *K_o is not a classical group in characteristic p.*

Proof. Assume for a contradiction that $K_o = C\ell_d(p^e)$, with d minimal. Let $(V_\sharp, \mathbf{F}_\sharp, \kappa_\sharp)$ be the classical geometry associated with K_o. From the results on automorphisms of classical groups in Chapter 2, we see that $\overline{K}' \leq P\Delta(V_\sharp, \mathbf{F}_\sharp, \kappa_\sharp)$, except possibly when $K_o = P\Omega_8^+(p^e)$; in this exceptional case, $O^3(\overline{K}') \leq P\Delta(V_\sharp, \mathbf{F}_\sharp, \kappa_\sharp)$. Thus in all cases we obtain an embedding

$$O^3(\overline{H}') \leq P\Delta(V_\sharp, \mathbf{F}_\sharp, \kappa_\sharp), \tag{8.4.7}$$

where $(V_\sharp, \mathbf{F}_\sharp, \kappa_\sharp)$ is the classical geometry associated to K_o. We proceed by considering various possibilities for n_1.

Case $n_1 = 2$. Here $n_2 \geq 7$ by (6.0.3), and so $d \geq R_p(\overline{\Omega}_2) \geq n_2$ by Proposition 5.4.15. Consequently $n < \frac{1}{2}d^2$, and so it follows from Lemma 8.1.4 and Proposition 5.4.11 that the representation of K_o on V appears in Table 5.4.A. Since $\frac{1}{2}d(d-1)-2 > 2d$ (as $d \geq 7$), and since $2d \geq 2n_2 = n$, we see that V must be a spin module. Thus $2d \geq n \geq 2^{[(d-1)/2]}$, which forces $n = 16$ and $8 \leq d \leq 10$. If $d = 8$, then $K_o = Sp_8(q)$ with q even and by Lemma 8.1.5 we have $\Omega = \Omega_{16}^+(q)$. But then H is of type $Sp_2(q) \otimes Sp_8(q)$, and so $\overline{H} \cong Sp_2(q) \times Sp_8(q)$ (see Proposition 4.4.12), which does not embed in $Aut(Sp_8(q))$. If $d = 9$, then $K_o = \Omega_9(q)$ with q odd, and $\Omega = \Omega_{16}^+(q)$. Therefore H is of type $Sp_2(q) \otimes Sp_8(q)$, which means $\overline{\Omega}_2 \cong PSp_8(q)$, which does not embed in $Aut(\Omega_9(q))$. Finally, if $d = 10$, then $K_o = P\Omega_{10}^{\pm}(q)$ and $\Omega = SL_{16}^{\pm}(q)$. But then H is of type $GL_2^{\pm}(q) \otimes GL_8^{\pm}(q)$, which yields a contradiction as $|L_8^{\pm}(q)|$ does not divide $|P\Omega_{10}^{\pm}(q)|$.

Case $n_1 = 3$. We first argue that $d \geq R_p(\overline{\Omega}_2) + 2$. If Ω_1 is insoluble, then this is immediate from Proposition 5.5.7. If Ω_1 is soluble, then $\Omega_1 \approx \Omega_3(3)$ or $SU_3(2)$. In these two cases, $\overline{\Omega}_1 \cong A_4$ or $3^2{:}Q_8$, respectively, and so $O^3(\overline{H}')$ contains $2^2 \times \overline{\Omega}_2$ or $3^2.2 \times \overline{\Omega}_2$, respectively. Now apply Lemma 5.5.11. Assume for now that $n_2 \geq 7$. Then $d \geq R_p(\overline{\Omega}_2) + 2 = n_2 + 2 \geq 9$, and hence $n < \frac{1}{2}d^2$ (since $n = 3n_2$). Thus as before, the representation of K_o on V appears in Table 5.4.A. Since $d \geq \frac{n}{3} + 2$, we have $n < \frac{1}{2}d(d-1) - 2$, and hence V must be a spin module. But this is absurd, since $3 \mid n$. This contradiction shows that $5 \leq n_2 \leq 6$, and hence $n = 15$ or 18. Moreover, by Proposition 5.4.15 $R_p(\overline{\Omega}_2) \geq 4$, and so $d \geq 6$. If $d \geq 7$, then the representation of K_o on V appears in Table 5.4.A (since $n \leq 18 < \frac{1}{2}d^2$). If $d = 6$, then by the minimality of d, it must be the case that $K_o = L_6^{\pm}(p^e)$ or $PSp_6(p^e)$, and so once again the representation appears in Table 5.4.A. But dimension 15 or 18 occurs in Table 5.4.A only for the exterior square of the natural module for $L_6^{\pm}(p^e)$. Since these modules are not self-dual (see Propositions 5.4.2 and 5.4.3), $\Omega \approx SL_{15}^{\pm}(q)$, which means $\overline{L} \cong L_3^{\pm}(q) \times L_5^{\pm}(q)$. But then $R_p(\overline{\Omega}_2) = 5$, which forces $d \geq 7$ by the remark above.

Case $n_1 = 4$. As in the previous case we have $d \geq R_p(\overline{\Omega}_2) + 2$. Assume first that $n_2 \geq 7$. Then $d \geq n_2 + 2 = \frac{n}{4} + 2 \geq 9$, and so the representation of K_o on V appears in Table 5.4.A. As $4(d-2) < \frac{1}{2}d(d-1) - 2$, it must be the case that V is a spin module, whence $4(d-2) \geq n \geq 2^{[(d-1)/2]}$, and so $n \leq 32$ and $d \leq 12$. On the other hand, we are assuming $n_2 \geq 7$, and this forces $n_2 = 8$, $n = 32$ and $10 \leq d \leq 12$. If $d = 10$, then $K_o \cong Sp_{10}(q)$ with q even, $\Omega \approx \Omega^+_{32}(q)$ and H is of type $Sp_4(q) \otimes Sp_8(q)$. But then by Propositions 5.4.15 and 5.5.7, we have $R_p(\overline{L}') \geq 3 + 8 > d$, a contradiction. If $d = 11$, then $K_o \cong \Omega_{11}(q)$ with q odd, $\Omega \approx Sp_{32}(q)$ and H is of type $O^\pm_4(q) \otimes Sp_8(q)$ or $Sp_4(q) \otimes O^\pm_8(q)$. Since $R_p(PSp_4(q) \times P\Omega^\pm_8(q)) = 12 > d$, it must be the case that H is of type $O^\epsilon_4(q) \otimes Sp_8(q)$. If $q \geq 5$ or if $\epsilon = -$, then either $L_2(q) \times L_2(q) \times PSp_8(q)$ or $L_2(q^2) \times PSp_8(q)$ embeds in $\Omega_{11}(q)$ — but this is impossible by Lagrange's Theorem. And if $q = 3$ and $\epsilon = +$, then we can again appeal to Lagrange's Theorem coupled with the fact that $\overline{H} \cong (PO^+_4(3) \times PSp_8(3)).2 \cong (L_2(3) \times L_2(3) \times PSp_8(3)).[8]$ (see Proposition 4.4.11). Finally, take $d = 12$, so that $K_o = P\Omega^+_{12}(q)$ or $P\Omega^-_{12}(q^{1/2})$. If q is even, then $\Omega = \Omega^+_{32}(q)$ and H is of type $Sp_4(q) \otimes Sp_8(q)$; but this is impossible by Lagrange's Theorem. If q is odd, then $\Omega = Sp_{32}(q)$ and H is of type $O^\pm_4(q) \otimes Sp_8(q)$ or $Sp_4(q) \otimes O^\pm_8(q)$. If H is of type $Sp_4(q) \otimes O^\pm_8(q)$, then an application of Corollary 5.5.9 shows that $\overline{L} \cong PSp_4(q) \times P\Omega^+_8(q)$ must fix a subspace decomposition of type $O^\pm_4(q) \perp O^\pm_8(q)$ or $O^\pm_4(q^{1/2}) \perp O^\pm_8(q^{1/2})$, which is impossible. The case in which H is of type $O^-_4(q) \otimes Sp_8(q)$ can be ruled out using Lagrange's Theorem. Finally suppose that H is of type $O^+_4(q) \otimes Sp_8(q)$. Then $L_2(q) \times L_2(q) \times PSp_8(q) \preceq P\Omega^+_{12}(q)$, and so arguing as in the proof of the parabolic argument Proposition 5.2.11, we see that $PSp_8(q)$ is contained in a non-maximal parabolic subgroup P of $P\Omega^+_{12}(q)$; however the only possible non-abelian composition factors of P are of the form $L_m(q)$ $(2 \leq m \leq 5)$ and $P\Omega^+_8(q)$, none of which contains $PSp_8(q)$. This completes the analysis when $n_2 \geq 7$. Now take $n_2 = 6$. Since 24 does not appear as the dimension of a module in Proposition 5.4.11, it must be the case that $d \leq 6$. On the other hand, $R_p(\overline{\Omega}_2) \geq 4$, and so $d \geq 6$. It now follows that $d = 6$ and $R_p(\overline{\Omega}_2) = 4$, forcing $\Omega_2 \approx \Omega^\pm_6(q)$. Therefore $\Omega_1 \approx \Omega^\pm_4(q)$ and $\Omega \approx \Omega^+_{24}(q)$, with q odd. Also, the minimality of d and Proposition 5.4.6 ensure that $K_o \cong L^\pm_6(q)$ or $PSp_6(q)$. Observe that $m_2(\overline{\Omega}_2) \geq m_2(\Omega_5(q)) \geq 4$, and $m_2(P\Omega^+_4(q)) = 4$. On the other hand, $m_2(K_o) \leq 6$ by Lemma 5.5.2.ii, whence H is of type $O^\pm_6(q) \otimes O^-_4(q)$. But then $(q^2+1)^2$ divides $|\overline{L}|$, yet it does not divide $|K_o|$. Next take $n_2 = 5$. Here H is of type $L^\pm_4(q) \otimes L^\pm_5(q)$, $Sp_4(q) \otimes O_5(q)$ or $O^\pm_4(q) \otimes O_5(q)$. In the first two instances, $d \geq 8$ by Proposition 5.5.7, and this violates Proposition 5.4.11. In the third instance, we see as before that $d \geq 6$. Quoting Propositions 5.4.6 and 5.4.11 again, we find that $d = 6$ and $K_o \cong L^\pm_6(q)$. Thus the 2-rank argument above ensures that H is of type $O^-_4(q) \otimes O_5(q)$, and we reach a contradiction as before, since $(q^2+1)^2$ does not divide $|K_o|$. Finally, take $n_2 = 4$. Here q is odd and H is of type $O^+_4(q) \otimes O^-_4(q)$ or $O^\pm_4(q) \otimes Sp_4(q)$. Therefore $m_2(O^3(\overline{L}')) \geq 6$, and so by (8.4.7) and Lemma 5.5.2.ii we have $d \geq 6$. Then Proposition 5.4.11 implies that $K_o \cong \Omega_9(q)$. In view of Proposition 5.4.9, case **O** holds, and so H is of type $O^+_4(q) \otimes O^-_4(q)$. Write $V_\sharp$ for the natural 9-dimensional $\mathbf{F}_q K_o$-

module. If $q \geq 5$, we have $L_2(q) \times L_2(q) \times L_2(q^2) \cong \overline{L} \leq K_o$, and if $q = 3$, then $A_4 \times A_4 \times L_2(q^2) \cong O^2(\overline{L}) \leq O^2(\overline{K}) = K_o$. So in either case, $\overline{L} \cap K_o$ contains a subgroup $N \cong A_4 \times A_4 \times L_2(q^2)$. Observe that N is irreducible on V, and in particular, N does not lie in a parabolic subgroup of $\overline{\Omega}$. Hence N does not fix a non-zero totally singular space in $(V_\sharp, \kappa_\sharp)$. Thus we may write $V_\sharp = W_1 \perp \cdots \perp W_k$, where each W_i is a non-degenerate and is an irreducible N-submodule of $V_\sharp$. If $\dim(W_i) = 1$ for some i, then N is contained in a subgroup $\Omega_8^{\pm}(q)$ of K_o. But this is impossible, since such a subgroup cannot be irreducible on V by Proposition 5.4.11. Hence $\dim(W_i) \geq 2$ for each i. Arrange the indices so that $\overline{\Omega}_2 \cong L_2(q^2)$ acts non-trivially on W_1. Since $L_2(q^2)$ does not embed in $GL_3(q)$, it follows from Clifford's Theorem that $\dim(W_1) \geq 4$ and that W_1 is an irreducible $\overline{\Omega}_2$-module. Therefore $A_4 \times A_4$ acts trivially on W_1 and hence faithfully on $W_2 \perp \cdots \perp W_k$. But it is easy to see that any faithful $(A_4 \times A_4)$-module in odd characteristic has dimension at least 6, and we have therefore shown that N does not embed in K_o. This is the desired contradiction.

Case $n_1 = 5$. Here $n_2 \geq 6$ (since $H \notin \mathcal{C}_7$), and $d \geq 4 + R_p(\overline{\Omega}_2)$, by Propositions 5.4.15 and 5.5.7. If $R_p(\overline{\Omega}_2) \geq n_2$, then $d \geq 4 + n_2 = 4 + \frac{n}{5} \geq 10$, and so $n < \frac{1}{2}d^2$. Thus the representation of K_o on V is given in Table 5.4.A. Since $n_1 = 5$, V is not a spin module, and hence $n \geq \frac{1}{2}d(d-1) - 2$. But this cannot occur because $5(d-4) \geq n$ and $d \geq 10$. We have therefore shown that $R_p(\overline{\Omega}_2) < n_2$, and since $n_2 \geq 6$, it follows from Proposition 5.4.15 that $n_2 = 6$, and $\Omega_2 \approx \Omega_6^{\pm}(q)$ with q odd. Therefore $n = 30$ while $d \geq 8$, and so the representation of K_o on V must appear in Table 5.4.A. But there is no module in Table 5.4.A with dimension 30.

Case $n_1 = 6$. As in the previous case, $d \geq 4 + R_p(\overline{\Omega}_2)$. Now if $n_2 = 6$, then $n = 36$ and H is of type $O_6^+(q) \otimes O_6^-(q)$ or $O_6^{\pm}(q) \otimes Sp_6(q)$ with q odd. Observe, however, that Proposition 5.4.11 ensures that $d \leq 9$. Therefore by Proposition 5.5.7 we know that H cannot be of type $O_6^{\pm}(q) \otimes Sp_6(q)$. Thus H is of type $O_6^+(q) \otimes O_6^-(q)$, $\Omega = \Omega_{36}^+(q)$ and $8 = R_p(\overline{\Omega}_1) + R_p(\overline{\Omega}_2) \leq d \leq 9$. If $d = 9$, then it follows from Proposition 5.4.11 and Lemma 8.1.3 that $K_o = L_9^{\pm}(q)$, $U_9(q^{1/2})$ or $\Omega_9(q)$. However the 36-dimensional modules for $L_9^{\pm}$ are not self-dual, which leaves $K_o = \Omega_9(q)$. But $\overline{L}$ contains $\Omega_4^+(q) \times P\Omega_6^-(q) \cong (SL_2(q) \circ SL_2(q)) \times P\Omega_6^-(q)$, and thus an application of the parabolic argument (Proposition 5.2.11) forces $P\Omega_6^-(q)$ to embed in a non-maximal parabolic of $\Omega_9(q)$. This is not possible, as the non-abelian composition factors of such a parabolic are among $L_2(q)$, $L_3(q)$ and $\Omega_5(q)$. If $d = 8$, then K_o is $L_8^{\pm}(q)$, $U_8(q^{1/2})$, $PSp_8(q)$, $P\Omega_8^{\pm}(q)$ or $P\Omega_8^-(q^{1/2})$. We eliminate $L_8^{\pm}$ as before, and the remaining groups can be handled using Corollary 5.5.9. The argument so far has taken care of the case where $n_2 = 6$. We can assume therefore that $n_2 \geq 7$, and hence $d \geq 4 + R_p(\overline{\Omega}_2) = 4 + n_2 = 4 + \frac{n}{6} \geq 11$. As usual, we deduce that $n < \frac{1}{2}d^2$, and so the representation of K_o on V appears in Table 5.4.A. Since $n_1 = 6$, V is not a spin module, and so we reach the desired contradiction for $n \leq 6(d-4) < \frac{1}{2}d(d-1) - 2$.

Case $n_1 \geq 7$. Here $d \geq n_1 + n_2 \geq 14$. But then it is easy to show that Lemma 8.1.4.ii is violated, and the proof is now complete. ∎

Lemma 8.4.4. *K_o is not an exceptional group of Lie type in characteristic p.*

Proof. Suppose false, and let K_o be an exceptional group over $\mathbf{F}_{p^e}$. Also let ℓ, ℓ_1 ℓ_2 be the Lie ranks of K_o, $\overline{\Omega}_1$ and $\overline{\Omega}_2$, respectively. Note that when $\overline{\Omega}_i$ is simple we use the definition of Lie rank given in §5.1, and when $\overline{\Omega}_1$ is soluble, we use the definition given in (5.2.3). Then by Proposition 5.2.12,

$$\ell_1 + \ell_2 \leq \ell. \tag{8.4.8}$$

Because $\ell_i \geq \frac{1}{2}(n_i - 2)$ (this follows from the definition of ℓ_i), we have

$$n_1 + n_2 \leq 2\ell + 4. \tag{8.4.9}$$

Moreover

$$n = n_1 n_2 \leq \tfrac{1}{4}(n_1 + n_2)^2 \leq (\ell + 2)^2, \tag{8.4.10}$$

so the representation of K on V is determined by Propositions 5.4.11 and 5.4.12 (recall from Theorem 5.4.1 and the sentence which follows that the irreducible representations of ${}^2B_2(2^e)$ and ${}^3D_4(p^e)$ extend to irreducible representations of the corresponding untwisted groups $B_2(2^e)$ and $D_4(p^{3e})$). In particular, observe that $K \neq E_8(p^e)$ for $R_p(E_8(p^e)) = 248 > (8+2)^2$.

If $K_o = E_7(p^e)$, then $n \leq 81$ by (8.4.10), and so by Propositions 5.4.12 and 5.4.18, $\Omega \approx Sp_{56}(q)$ with q odd or $\Omega^{\pm}_{56}(q)$ with q even. However $\ell_1 + \ell_2 \leq 7$, and so the only possibility is $\overline{L} \cong \Omega_7(q) \times PSp_8(q)$, with q odd. But then the parabolic argument Lemma 5.2.11 forces $\Omega_7(q)$ to embed in a Levi factor Y of K_o, where $O^{p'}(Y)$ has Lie rank 3. The only possible non-abelian composition factors of Y are $L_m(p^e)$ with $2 \leq m \leq 4$, and we reach a contradiction as $R_p(\Omega_7(q)) = 7$. If $K_o = E^{\pm}_6(p^e)$, then $n \leq 64$ and so $\Omega \approx SL^{\pm}_{27}(q)$ by Propositions 5.4.12 and 5.4.17. But then $\overline{L} \cong L^{\pm}_3(q) \times L^{\pm}_9(q)$, and so $\ell_1 + \ell_2 > \ell$, a contradiction. If $K_o = F^{\pm}_4(p^e)'$, then n is 26 (if $p \neq 3$) or 25 (if $p = 3$). But $\mathcal{C}_4$ is void when $n = 25$, and hence $n = 26$ and $n_2 = 13$; this forces $\ell_2 \geq 6 > \ell$, against (8.4.8). If K_o is $G_2(p^e)$ or ${}^3D_4(p^e)$ then $n \leq 16$ and so by Propositions 5.4.11 and 5.4.12, $n = 14$ and $K_o = G_2(p^e)$. But then $n_2 = 7$ and so $\ell_2 \geq 3 > \ell$, against (8.4.8). Finally, as $n \geq 13$, we see that $\ell \geq 2$, and so K_o is not of type ${}^2G_2$ or ${}^2B_2$. The proof is now complete. ∎

Lemma 8.4.5. *K_o is not a sporadic group.*

Proof. Suppose false. It follows from Proposition 5.2.9 and (7.4.2) that $n_2 \leq 8$ (since $\Omega_2 \not\approx \Omega^{\pm}_{10}(2)$), and so $n \leq 64$. Thus by Proposition 5.3.8 we eliminate J_4, Fi_{23}, Fi'_{24}, Ly, BM and M. Furthermore, Proposition 5.2.9 shows that if $K_o = HN$ or Th, then K_o contains no simple classical groups of dimension 7 or more, and so $n_2 \leq 6$. But then $n \leq 36$, contrary to Proposition 5.3.8. For the rest of this proof we shall assume that K_o is one of the remaining 18 sporadic groups. We refer to [G-L, pp.40-70] for information concerning centralizers of elements in these sporadic groups.

Suppose that $p = 2$. Then by (7.4.2), Ω_2 contains either $Sp_6(2)$ or $L_5^{\pm}(2)$, and also $2^{13} \mid |\overline{L}|$, so $2^{13} \mid |Aut(K_o)|$. So using Table 5.1.C and Proposition 5.2.9, we deduce that K_o is Suz, Co_1, Co_2 or Fi_{22}. If $K_o = Suz$, then Proposition 5.2.9 forces $\Omega_2 \approx SU_5(2)$, and so H is of type $GU_{n_1}(2) \otimes GU_5(2)$, with $3 \leq n_1 \leq 4$. But this is impossible because any subgroup of order 11 in Suz is self-centralizing. If $K_o = Co_1$ or Co_2, then $n \geq 22$ by Proposition 5.3.8, and so by Proposition 5.2.9, $\Omega_2 \approx SU_6(2)$ or $Sp_6(2)$. Therefore H is of type $GU_{n_1}(2) \otimes GU_6(2)$ $(4 \leq n_1 \leq 5)$ or $Sp_4(2) \otimes Sp_6(2)$. The former is impossible because groups of order 11 in K_o cannot be centralized by $U_4(2)$ (the 11-centralizer in Co_1 has order 66). The latter is impossible because $Sp_6(2)$ does embed in the centralizer of an element of order 5 in Co_1 (the non-abelian composition factors in the centralizer of a 5-element in Co_1 are A_5 or J_2). If $K_o = Fi_{22}$, then $n \geq 27$, and since K_o contains neither $Sp_8(2)$ nor $L_7^{\pm}(2)$, it must be the case that H is of type $L_5^{\pm}(2) \otimes L_6^{\pm}(2)$. But then $2^{25} \mid |\overline{L}|$, a contradiction.

Now suppose that $p = 3$. Since $3^4 \mid |\overline{L}|$, we know that K_o is one of J_3, Suz, McL, $O'N$, Co_1, Co_2, Co_3 and Fi_{22}. Using Lagrange's Theorem (note that the power of 3 dividing $|\overline{L}|$ is the same as that dividing $|I_1||I_2|$, and so it can be read off directly from Table 2.1.C) and Proposition 5.3.8, we see that the only cases to consider are as follows:

(a) $K_o = J_3$ with H of type $O_4^+(3) \otimes O_4^-(3)$ or $O_3(3) \otimes O_5(3)$

(b) $K_o = Suz$ with H of type $O_4^{\pm}(3) \otimes Sp_4(3)$, $O_3(3) \otimes O_m^{\epsilon}(3)$ with $5 \leq m \leq 6$, or $O_4^{\pm}(3) \otimes O_m^{\epsilon}(3)$ with $4 \leq m \leq 5$,

(c) $K_o = Co_1$ and H of type $O_4^{\epsilon}(3) \otimes O_6^{\delta}(3)$.

Now (a) cannot occur, for neither $\Omega_4^-(3) \cong A_6$ nor $\Omega_5(3)$ is contained in an involution centralizer of J_3 (the only non-abelian composition factor of an involution centralizer in J_3 is A_5). We now treat (b). First assume that H is of type $O_4^+(3) \otimes O_4^-(3)$. Then $O^2(\overline{H})$ contains $A_4 \times A_4 \times A_6$. But this is impossible, for the centralizers of elements of order 5 in Suz have structures $5 \times A_5$ and $5 \times A_6$, neither of which contains $5 \times A_4 \times A_4$. In the remaining cases $O^2(\overline{H})$ contains $A_4 \times PSp_4(3)$, and hence Suz must contain $2^2 \times PSp_4(3)$. Now the non-abelian composition factors of the involution centralizers in Suz are $U_4(2)$, $L_3(4)$ and J_2. Only the first contains $PSp_4(3)$, and so we obtain an embedding $2^2 \times PSp_4(3) \leq C := 2_-^{1+6}.\Omega_6^-(2)$ (see [G-L, p. 56]). However this cannot occur, for $PSp_4(3) \cong \Omega_6^-(2)$ and $C/O_2(C) \cong \Omega_6^-(2)$ acts irreducibly on $O_2(C)/Z(O_2(C)) \cong 2^6$; therefore the subgroup $PSp_4(3)$ cannot normalize a 2^2 in C. As for (c), the non-abelian composition factors in the centralizers of the involutions of Co_1 are $\Omega_8^+(2)$, $G_2(4)$ and M_{12}, none of which contains $P\Omega_6^{\pm}(3) \cong L_4^{\pm}(3)$.

Finally, suppose that $p \geq 5$. Again $p^4 \mid |\overline{K}|$, and so the only possibility is $K_o = Co_1$ with $p = 5$. But as 5^5 does not divide $|\overline{K}|$, it must be the case that H is of type $O_4^+(5) \otimes O_4^-(5)$. But then $n = 16$, and this violates Proposition 5.3.8. ∎

§8.5 The case $H \in \mathcal{C}_6$

Here $n = r^m$ for some prime r and H is the normalizer of a suitable r-group of symplectic type, as described in §4.6. As in §7.6 we set $L = H^{\infty}$, so that the structure of L is given in (7.6.1).

Lemma 8.5.1. *The prime r divides the order of the multiplier $M(K_o)$.*

Proof. This is clear as $\mathbf{Z}_r \trianglelefteq Z(L) \leq Z(K^\infty)$ and K^∞ is a covering group of K_o. ■

Lemma 8.5.2. *K_o is not alternating.*

Proof. This is immediate from Proposition 5.3.7 and Lemma 7.6.1.ii. ■

Lemma 8.5.3. *K_o is not a classical group.*

Proof. Otherwise, $K_o = C\ell_d(s^a)$ for some prime s. If $s \neq r$, then $d \geq n$ by Lemma 7.6.1.iii, contradicting Lemma 8.1.4. Therefore $s = r$. Since $r \mid |M(K_o)|$ by Lemma 8.5.1, it follows that K_o must appear in Table 5.1.D, and so $r \leq 3$. But if $r = 2$, then $m \geq 4$ by (6.0.3), and so $\overline{L}$ contains $2^8.\Omega_8^\pm(2)$, which does not embed in $L_2(4)$, $L_3(2)$, $L_4^\pm(2)$, $Sp_6(2)$, $Sp_4(2)'$, $L_3(4)$, $U_6(2)$ or $\Omega_8^+(2)$. And if $r = 3$, then $m \geq 3$, and so $\overline{L}$ contains $3^6.Sp_6(3)$, which does not embed in $L_2(9)$, $\Omega_7(3)$ or $U_4(3)$. ■

Lemma 8.5.4. *K_o is not an exceptional group.*

Proof. Suppose that K_o is an exceptional group in characteristic s for some prime s. First assume that $s \neq r$. Then by Lemma 7.6.1.iii we have

$$R_p(K_o) \leq n \leq R_s(\overline{L}) \leq R_s(K_o).$$

Since $M(^2F_4(2)') = 1$, we know that $K_o \neq {}^2F_4(2)'$ by Lemma 8.5.1, and hence Corollary 5.4.14 implies that $s = p$. Thus $n = R_p(K_o)$. Consequently by Proposition 5.4.13, either $n = 25$, $p = 3$ and $K_o = F_4(3^e)$, or $n = 27$ and $K_o = E_6^\pm(p^e)$. The first case does not occur because 5 does not divide $|M(F_4(3^e))|$ (see Lemma 8.5.1). In the second case we have $3^6.Sp_6(3) \trianglelefteq K_o = E_6^\pm(p^e)$. Now K_o has a 78-dimensional representation on the Lie algebra of type E_6 in characteristic p (see Table 5.4.B), and this yields an inclusion $K_o \leq GL_{78}(\overline{\mathbf{F}}_p)$ (here $\overline{\mathbf{F}}_p$ denotes the algebraic closure of $\mathbf{F}_p$). Let $\mathcal{L}$ be the 78-dimensional Lie algebra over $\overline{\mathbf{F}}_p$, and let W be an irreducible composition factor in $\mathcal{L}$ for $3^6.Sp_6(3)$ such that the normal 3^6 acts non-trivially. By Clifford's Theorem, the factor group $Sp_6(3)$ acts transitively on the homogeneous components of W regarded as a module for the 3^6. However, $Sp_6(3)$ also acts transitively on the $3^6 - 1$ non-trivial irreducible representations of 3^6, and this forces $\dim(W) \geq 3^6 - 1$, a contradiction.

To complete the proof, assume that $s = r$. Then as in the proof of Lemma 8.5.3, we see that K_o must appear in Table 5.1.D. If $r = 3$, then K_o must be $G_2(3)$; however $\overline{L}$ contains $3^6.Sp_6(3)$, which does not embed in $G_2(3)$. This leaves the case where $r = 2$ and K_o is one of $G_2(4)$, $F_4(2)$, $^2E_6(2)$ and $^2B_2(8)$. However $\overline{L}$ contains $2^8.\Omega_8^\pm(2)$, which does not embed in $G_2(4)$ or $^2B_2(8)$. If K_o is $F_4(2)$, then $n \geq 44$ by Theorem 5.3.9 and so $m \geq 6$; but then $\overline{L}$ contains $2^{12}.\Omega_{12}^\pm(2)$, which does not embed in $F_4(2)$. And if $K_o = {}^2E_6(2)$, then $n \geq 1,536$ and so $m \geq 11$; again $\overline{L}$ does not embed in K_o. ■

Lemma 8.5.5. *K_o is not sporadic.*

Proof. Suppose false. From Theorem 5.1.4 and Lemma 8.5.1 we see that $r \leq 3$. If $r = 3$, then $m \geq 3$, which yields a contradiction, for by Proposition 5.2.9, $PSp_6(3)$ is involved in none of the sporadic groups. This leaves the case $r = 2$, and here m is 4 or 5, by Proposition 5.2.9. If $m = 5$, then 2^{30} divides $|K_o|$ and $R(K_o) \leq 2^5$, which is impossible by Table 5.1.C and Proposition 5.3.8. Similarly, if $m = 4$ then 2^{20} divides $|K_o|$ and $R(K_o) \leq 16$, which is again not possible. ■

§8.6 The case $H \in \mathcal{C}_7$

Here $H = \Omega_{\mathcal{D}}$ for some tensor decomposition $\mathcal{D}$ of V as described in §4.7. Using the notation of §7.7, $n = m^t$ and H has a normal subgroup $L = \Omega_1' \otimes \cdots \otimes \Omega_t'$, where the Ω_i' are isomorphic quasisimple groups. Recall

$$C_{\overline{\Omega}}(\overline{L}) = 1 \tag{8.6.1}$$

according to Lemma 7.7.1.i.

Lemma 8.6.1. *K_o is not alternating.*

Proof. Suppose that $K_o \cong A_k$, so that $k \leq n+2$ by Proposition 5.3.7. Without loss, we can choose k to be minimal. Write $X = \{1, \ldots, k\}$ for the set upon which K_o acts naturally. Obviously $k \geq 7$, and so $A_k \leq \overline{K} \leq S_k$ (see Theorem 5.1.3), hence $\overline{K}$ also acts naturally on X. Now let $X_1, \ldots, X_r$ be the non-trivial orbits of $\overline{L}$ on X, and let X_o be the set of fixed points of $\overline{L}$. Thus $X = X_o \cup \cdots \cup X_r$. Let B_i be the full alternating group on X_i, so that $\overline{L} \leq B_o \times \cdots \times B_r \leq K_o$. Clearly B_o centralizes $\overline{L}$, and hence by (8.6.1) $B_o = 1$, whence $|X_o| \leq 2$. Obviously $\overline{H}$ acts on X_o, and so if $|X_o| = 1$, then $\overline{H} \leq N_{\overline{K}}(X_o) \cong A_{k-1}$ or S_{k-1}, contrary to the minimality of k. If $|X_o| = 2$, then it must be the case that $\overline{K} = K_o$ (otherwise $\overline{L}$ is centralized by a transposition); but then $\overline{H} \leq N_{\overline{K}}(X_o) \cong S_{k-2}$, another contradiction. Therefore

$$\overline{L} \text{ has no fixed points.} \tag{8.6.2}$$

For $1 \leq i \leq r$, define h_i as the number of components of $\overline{L}$ which act non-trivially on X_i. Quoting Proposition 5.2.7.iii we obtain

$$|X_i| \geq s^{h_i}, \tag{8.6.3}$$

where $s = P(\overline{\Omega}_1')$. Since $(m+1)^t > m^t + 2 = n + 2 \geq k$, it follows from (8.6.3) and Corollary 5.2.3.i that

$$h_i < t \text{ for all } i. \tag{8.6.4}$$

We now claim that

$$h_i = 1 \text{ for all } i. \tag{8.6.5}$$

Otherwise, $h := h_i \geq 2$ for some i. In view of (8.6.4), $t \geq 3$, and hence the action of $\overline{H}$ on the components of $\overline{L}$ contains the full alternating group A_t. Since A_t is transitive on h-sets, there are at least $\binom{t}{h}$ indices j for which $h = h_j$. Now according to Corollary 5.5.6, the group $B_1 \times \cdots \times B_r$ preserves a tensor decomposition

$$V = U_1 \otimes \cdots \otimes U_r, \tag{8.6.6}$$

where each U_i is a non-trivial projective B_i-module. Furthermore, for these $\binom{t}{h}$ indices j we have $|X_j| \geq s^h \geq 5^2$, so by Proposition 5.3.7 $\dim(U_j) \geq R(A_{s^h}) \geq s^h - 2$. Consequently

$$m^t = n \geq (s^h - 2)^{\binom{t}{h}}.$$

On the other hand, $s \geq m + 1$, whence $s^h \geq m^h + 2$, and therefore

$$(s^h - 2)^{\binom{t}{h}} \geq m^{h\binom{t}{h}}.$$

Since $2 \leq h \leq t - 1$, these last two inequalities are incompatible, and this contradiction establishes (8.6.5).

Since each component of $\overline{L}$ projects non-trivially into some B_i, it follows from (8.6.5) that

$$r \geq t.$$

Moreover

$$m^t = \prod_{i=1}^{r} \dim(U_i) \geq R_p(A_s)^r \geq R_p(A_s)^t. \tag{8.6.7}$$

In particular,

$$m \geq R_p(A_s). \tag{8.6.8}$$

Now when $m \geq 6$, it follows from Corollary 5.2.3 that $s \geq m^2 + 3$, and hence Proposition 5.3.7 forces $R_p(A_s) \geq m^2 + 1$, violating (8.6.8). Therefore $m \leq 5$, and so (8.6.7) and Proposition 5.3.7 now yield $s \leq 8$. Consequently Theorem 5.2.2 shows that $\Omega_1 \approx SL_3(2)$, $SL_4(2)$, $Sp_2(4)$, $Sp_2(5)$, $Sp_2(7)$, $Sp_2(9)$, $Sp_4(2)$, $\Omega_3(5)$, $\Omega_3(7)$, $\Omega_3(9)$ or $\Omega_4^-(3)$. However (8.6.8) eliminates $SL_3(2)$, $Sp_2(7)$ and $\Omega_3(7)$.

We now argue that

$$r = t. \tag{8.6.9}$$

First note that if $m = R_p(A_s)$, then equality holds throughout (8.6.7), and hence (8.6.9) holds. This is the case when $\Omega_1 \approx SL_4(2)$, $Sp_2(4)$, $Sp_2(5)$ and $Sp_2(9)$. If $\Omega_1 \approx \Omega_3(5)$, then $n = 3^t$, so $\dim(U_i)$ must be a power of 3 for each i; in particular, $\dim(U_i) \geq 3$ and so (8.6.9) once again follows from (8.6.7). The same reasoning works when $\Omega_1 \approx \Omega_3(9)$ or $Sp_4(2)$. Finally, when $\Omega_1 \approx \Omega_4^-(3)$, we see that $\dim(U_i) \geq 4$ since A_6 does not embed in $PGL_3(3)$. Thus (8.6.9) holds in this case, too.

In view of (8.6.9), with suitable relabelling we may write $\overline{\Omega_i'} \leq B_i$ for each i. Evidently $B_1 \leq C_{\overline{\Omega}}(\overline{N})$, where $N = \Omega_2' \otimes \cdots \otimes \Omega_t'$. Hence by Lemma 4.0.5.i, $\widetilde{B}_1 \leq C_\Omega(N)^\infty$, where $\widetilde{B}_1$ denotes the last term of the derived series of the preimage of B_1 in Ω. Now N is absolutely irreducible on $V_2 \otimes \cdots \otimes V_t$ and thus by Lemma 4.4.3.i, $C_{GL(V)}(N) = GL(V_1) \otimes 1 \otimes \cdots \otimes 1$. A straightforward adaptation of the proof of Lemma 4.4.6 shows that $(GL(V_1) \otimes 1 \otimes \cdots \otimes 1) \cap \Delta = \Delta_1 \otimes 1 \otimes \cdots \otimes 1 = \Delta_1$. Thus $C_\Delta(N) = \Delta_1$, whence $C_\Omega(N)^\infty = C_\Delta(N)^\infty = \Delta_1^\infty = \Omega_1'$. We conclude $\Omega_1' = \widetilde{B}_1$. Consequently $\overline{\Omega_1'} \cong B_1$, and thus by Theorem 5.1.2 and Proposition 2.9.1, the degree $b = |X_1|$ of B_1 is 5, 6, or 8. Similarly $\overline{\Omega_i'} = B_i$ for all i, and so we have

$$\overline{L} = \overline{\Omega_1'} \times \cdots \times \overline{\Omega_1'} = B_1 \times \cdots \times B_t < K_o,$$

with $B_i \cong A_b$ and $K_o = A_{bt}$ (recall (8.6.2)).

If $b = 8$, then $\Omega_1 \approx SL_4(2)$ and $\Omega \approx SL_{4^t}(2)$. Consequently Lemma 4.4.3.ii implies that $N_\Omega(\Omega_1)/C_\Omega(\Omega_1) \cong \Omega_1$. On the other hand, $K_o \cong A_{8t}$ contains an element inducing an outer automorphism on B_1, and this fact provides the desired contradiction.

Therefore $b = 5$ or 6. Define $R = \Omega_3' \otimes \cdots \otimes \Omega_t'$, and observe that $(\overline{\Omega_1'} \times \overline{\Omega_2'}) \times \overline{R} \leq A_{2b} \times A_{(t-2)b} \leq K_o$, and we let J be the factor A_{2b}. Thus $\overline{\Omega_1'} \times \overline{\Omega_2'} \leq J$ and J centralizes $\overline{R}$. Writing $\widetilde{J}$ for the last term of the derived series of the preimage of J in Ω, we see that $\widetilde{J}$ is a covering group of A_{2b} and by Lemma 4.0.5.i, $\widetilde{J}$ centralizes R. Thus by Lemma 4.4.3, $\widetilde{J} \leq GL(V_1 \otimes V_2)$. Furthermore, $\Omega_1' \otimes \Omega_2' \leq \widetilde{J}$, and consequently $\widetilde{J}$ is absolutely irreducible on the m^2-dimensional space $V_1 \otimes V_2$.

First consider $b = 5$. Then $\Omega_1 \approx Sp_2(4)$, $Sp_2(5)$ or $\Omega_3(5)$ by Proposition 2.9.1. Since $R(A_{10}) \geq 8$, we must have $m \geq 3$, and so $\Omega_1 \approx \Omega_3(5)$. Thus $\widetilde{J}$ is absolutely irreducible in $GL_9(5)$. Since $Z(\widetilde{J})$ must act as scalars, it follows from Theorem 5.1.4.i that $\widetilde{J} \cong A_{10}$. But this contradicts Proposition 5.3.5, since the fully deleted permutation module for A_{10} in characteristic 5 has dimension 8.

Finally, let $b = 6$. Here $J \cong A_{12}$, and since $R(A_{12}) > 9$ we must have $m \geq 4$. Therefore by Proposition 2.9.1, $\Omega_1 \approx Sp_4(2)$ or $\Omega_4^-(3)$ and $\widetilde{J}$ is absolutely irreducible in $GL_{16}(q)$, with $q = 2$ or 3, respectively. First take $q = 2$. Evidently $C_K(R)$ is either $Alt(X_1 \cup X_2) \cong A_{12}$ or $Sym(X_1 \cup X_2) \cong S_{12}$, according as $K \cong A_k$ or S_k. Since $C_K(R) \geq \Omega_1 \times \Omega_2 \cong S_6 \times S_6$, we deduce that $K = S_k$ and that $C_K(R) \cong S_{12}$. Therefore $C_K(R)$ acts absolutely irreducibly in $GL(V_1 \otimes V_2)$. But according to [Ja$_2$, p.141], S_{12} has no irreducible representation in $GL_{16}(2)$. Next take $q = 3$. Now $N_{\overline{K}}(B_1 \times \cdots \times B_t)$ is clearly $(S_6 \wr S_t) \cap \overline{K}$, which has structure $A_6^t.2^t.S_t$ or $A_6^t.2^{t-1}.S_t$. However, when $t \geq 3$, Proposition 4.7.7 implies that $\overline{H} \cong PSO_4^-(3)^t.[2^{2t-1}].S_t \cong A_6^t.[2^{2t-1}].S_t$; because $2t - 1 > t$, we see that $\overline{H}$ does not embed in $\overline{K}$. So we are left to treat $t = 2$. Then $\overline{H} \cong \Omega_4^-(3)^2.[8] \cong A_6^2.[8]$, and thus it must be the case that $\overline{K} \cong S_{12}$. But according to [Ja$_2$, pp.148-149] and Proposition 5.3.6, S_{12} has no absolutely irreducible projective representation in $PGL_{16}(3)$, and the proof is now finished. ∎

Lemma 8.6.2. *K_o is not an exceptional group of Lie type.*

Proof. Assume first that K_o is an exceptional group over $\mathbf{F}_{p^e}$. Write ℓ for the Lie rank of K_o and ℓ_1 for that of Ω_1. Here $\ell_1 \geq \lceil \frac{1}{2}(m-2) \rceil$ and so as in the proof of Lemma 8.4.4,

$$\lceil \tfrac{1}{2}(m-2) \rceil t \leq \ell_1 t \leq \ell. \tag{8.6.10}$$

Clearly $\ell_1 \geq 1$ and so $\ell \geq 2$. Thus K_o is not of type ${}^2G_2$ or ${}^2B_2$. Suppose for the moment that $\ell = 2$, so that K_o is of type G_2, ${}^3D_4$ or ${}^2F_4$. Obviously $t = 2$ and $\ell_1 = 1$, which means H is of type $O_4^-(q) \wr S_2$ with $\Omega \approx \Omega_{16}^+(q)$. However this is impossible, for according to Propositions 5.4.11 and 5.4.12, K_o has no absolutely irreducible projective representation of degree 16. We are left with the case in which $\ell \geq 3$, so that K_o is of type E_8, E_7, $E_6^\pm$ or F_4. Note also that if K_o is untwisted, then $p^e = q^a$ for some a by Proposition 5.4.6.

First let $K_o = E_8(q^a)$. It follows from Proposition 5.4.12 and the fact that $248 \neq m^t$ that $n \geq 1024$. Thus as $t \leq 8$ we deduce $m \geq 3$. Also, from (8.6.10) we see that $\ell_1 \leq \frac{8}{3}$, and hence $\ell_1 \leq 2$; therefore $m \leq 6$. Thus $3 \leq m \leq 6$. If $m = 3$, then $t \geq 7$, $\ell_1 = 1$, and so H contains $\overline{\Omega}_1 \wr A_7$, where $\Omega_1 \approx \Omega_3(q)$ or $SU_3(q)$. Applying the parabolic argument (Lemma 5.2.11) four times shows that $R = \overline{\Omega}_1 \wr 3$ lies in a rank 4 Levi factor of K_o. Since $\overline{\Omega}_1^3$ is the unique minimal normal subgroup of R, it follows that $\overline{\Omega}_1^3$ embeds in some component C of the Levi factor, and the Lie rank of C must be at least 3. Therefore C is of type $A_3(q^a)$, $A_4(q^a)$ or $D_4(q^a)$ (using the Lie notation). Thus $\overline{\Omega}_1^3$ embeds in $L_4(q^a)$, $L_5(q^a)$ or $P\Omega_8^+(q^a)$ (using the classical notation). In particular, $R_p(\overline{\Omega}_1^3) \leq 8$. Since $R_p(U_3(q)) = 3$, it follows from Proposition 5.5.7 that $\Omega_1 \not\approx SU_3(q)$, and so $\Omega_1 \approx \Omega_3(q)$ with q odd. Proposition 5.5.7 also implies that $R_p(\overline{\Omega}_1^3) \geq 6$, and so it must be the case that C is of type $D_4(q^a)$. Since the Levi factor induces only inner and diagonal automorphisms on C (see [Ca$_1$, Chs. 8,12]), and since $D_4(q^a)$ has no outer diagonal automorphisms of order 3, it follows that all of R embeds in $P\Omega_8^+(q^a)$. Let $(W, \mathbf{F}_{q^a}, P)$ be the natural 8-dimensional geometry associated to C (here P is the quadratic form). Using the reasoning employed so far, we see that R cannot be contained in a parabolic subgroup of C, and so R fixes no non-zero totally singular subspace of $(W, \mathbf{F}_{q^a}, P)$. Thus if R is reducible on W, then it fixes a decomposition $W = W_1 \perp W_2$. Thus R embeds in $P\Omega_s^\pm(q^a)$ for some $s \leq 7$. As $R_p(R) \geq 6$ and $R_p(P\Omega_6^\pm(q^a)) = 4$, the only possibility is $s = 7$. Thus R must be irreducible in $PGL_7(p^e)$, and so Clifford's Theorem shows that $\overline{\Omega}_1^3$ is also irreducible. But this is impossible in view of Proposition 5.5.7.ii. We conclude that R must be irreducible in $PGL(W)$. But this violates Lemma 7.7.2, and we have now eliminated the case $m = 3$. Next take $m = 4$ or 5, so that $t \geq 5$ (as $n = m^t \geq 1024$). Clearly (8.6.10) implies that $\ell_1 = 1$, and hence $\Omega_1 \approx \Omega_4^-(q)$. Since $|\Omega_4^-(q)|^5$ does not divide $|E_8(q)|$, it follows that $a \geq 2$, and hence $n \geq 248^a$ by Proposition 5.4.6. But then $t > 8$, against (8.6.10). It remains to consider $m = 6$. Here $t \geq 4$, and so $\ell_1 \leq 2$. Therefore $\Omega_1 \approx \Omega_6^-(q)$ and $\overline{L} \cong P\Omega_6^-(q)^4$. Thus applying the parabolic argument six times shows that $P\Omega_6^-(q)$ embeds in a Levi factor of rank 2. But any such Levi factor is of type L_3 or of type $L_2 \times L_2$, and so we reach a contradiction as $R_p(P\Omega_6^-(q)) = 4$.

Next let $K_o = E_7(q^a)$. Since $m^t \neq 56$, 132 or 133, we see from Proposition 5.4.12 that $n \geq 244$. As in the previous analysis we deduce $3 \leq m \leq 6$. If $m = 3$, then $t \geq 6$ and so $\ell_1 = 1$. Thus $\Omega_1 \approx \Omega_3(q)$ or $SU_3(q)$, and $\overline{H}$ contains $\overline{\Omega}_1 \wr A_6$. Applying the parabolic argument three times shows that $\overline{\Omega}_1 \wr 3$ is contained in a Levi factor of rank 4, and we can invoke the same argument which was used to treat E_8. If $m = 4$, then $t \geq 4$, and so $\ell_1 = 1$. Therefore $\Omega_1 \approx \Omega_4^-(q)$. As $|\Omega_4^-(q)|^4$ does not divide $|E_7(q)|$, it follows that $a \geq 2$.

Then using Propositions 5.4.6.i and 5.4.12 (and observing that $4^t \neq 56^a, 132^a, 133^a$), we see that $4^t = n \geq 244^2$, which forces $t \geq 8$, against (8.6.10). If $m = 5$ or 6, then $t \geq 4$ as $n > 243$; but this violates (8.6.10).

Now take $K_o = E_6^{\pm}(p^e)$. Suppose first that $n \geq 82$. Since $t \leq 6$, we have as before $3 \leq m \leq 6$. If $m = 3$, then $t \geq 5$ and so $K_o = E_6(q^a)$ (since the Lie rank of $E_6^-(p^e)$ is only 4). Arguing as for E_8, we deduce $\Omega_1 \approx \Omega_3(q)$ or $SU_3(q)$, and that $\overline{\Omega}_1 \wr 3$ embeds in a rank 4 parabolic of K_o; this situation leads to a contradiction as before. If $m = 4$, then $t \geq 4$. If $K_o = E_6^-(p^e)$, then $\ell = 4$, and hence $t = 4$ and $\ell_1 = 1$. Thus $\Omega_1 \approx \Omega_4^-(q)$ with q odd, and $\overline{\Omega}_1 \wr 3$ embeds in a rank 3 Levi factor of K_o. The only such Levi factor which can contain $\overline{\Omega}_1 \wr 3$ is that with a unique component of type ${}^2D_4$. At this point we may appeal to the argument used already. If $K_o = E_6^+(q^a)$, then as $|\Omega_4^-(q)|^4$ does not divide $|E_6(q)|$, we must have $a \geq 2$, and hence $4^t = n > 81^2$. This forces $t \geq 7$, against (8.6.10). The previous reasoning easily eliminates the case where m is 5 or 6, and thus we have reduced to the case where $n \leq 81$. Thus quoting Proposition 5.4.12, we see that $n = 27$. By Proposition 5.4.17, $\Omega \approx SL_{27}^{\epsilon}(q)$ and $K_o = E_6^{\epsilon}(q)$ (where $\epsilon = \pm$). Thus H is of type $GL_3^{\epsilon}(q) \wr S_3$. Using the parabolic argument, $\overline{H}$ contains a subgroup $L_3^{\epsilon}(q) \times L_3^{\epsilon}(q)$ which lies in parabolic subgroup of $\overline{K}_o$. On the one hand, this $L_3^{\epsilon}(q) \times L_3^{\epsilon}(q)$ has three irreducible constituents on V, each of dimension 9. On the other hand, it is known (see [As$_3$, §3] for example) that a parabolic subgroup in $E_6^{\epsilon}(q)$ stabilizes a subspace in V of dimension 1, 2, 3, 5, 6 or 10. These two facts are incompatible, so we reach the desired contradiction.

Finally, assume that $K_o = F_4(q^a)$. The usual arguments eliminate the case in which $n \geq 96$, and so by Proposition 5.4.12 we have $n = 25$. Thus $m = 5$ and $t = 2$, which means $\ell_1 = 2$ and $\Omega_1 \approx \Omega_5(q)$ or $SU_5(q)$. Moreover, $K_o = F_4(q)$ (that is, $a = 1$) and $p = 3$. Clearly $|U_5(q)|^2$ does not divide K_o, and so H is of type $O_5(q) \wr S_2$. Using the parabolic argument twice, we obtain an embedding of $\Omega_5(q)$ in a rank 2 Levi factor. However, the component of such a Levi factor is the full covering group of either $B_2(q)$ or $C_2(q)$, neither of which is simple as q is odd. In other words, neither $B_2(q)$ nor $C_2(q)$ contains the simple group $\Omega_5(q)$, and this final contradiction completes the analysis when K_o is of Lie type in characteristic p.

Now assume that K_o is an exceptional group of Lie type over $\mathbf{F}_{r^e}$, where r is prime and $r \neq p$. Choose an odd prime divisor s of $|\overline{\Omega}_1'|$ such that $s \neq r$. Then $t \leq m_s(\overline{L}) \leq m_s(K_o)$. Moreover by Proposition 5.2.13, $m_s(K_o) \leq \ell$, where ℓ is the Lie rank of K_o if $K_o \neq {}^2E_6(r^e)$, and $\ell = 6$ if $K_o = {}^2E_6(r^e)$. In particular, $\ell \geq 2$, and so K_o is not of type ${}^2G_2$ or ${}^2B_2$.

Suppose in this paragraph that $r^e \neq 2$ and that K_o is not $G_2(3)$ or $G_2(4)$. Now $n = m^t \geq R_p(K_o)$, and so

$$m \geq R_p(K_o)^{1/t} \geq R_p(K_o)^{1/\ell}. \tag{8.6.11}$$

This implies by Theorem 5.3.9 that $m \geq 8$, and hence $R_{p'}(\overline{\Omega}_1') \geq 28$ (see Corollary 5.3.10). Thus by Proposition 5.5.7, $R_r(K_o) \geq 28t \geq 56$, and so by Proposition 5.4.13, $K_o =$

$E_7(r^e)$ or $E_8(r^e)$. If $K_o = E_7(r^e)$, then $R_r(K_o) = 56$ which means $t = 2$ and $R_r(\overline{\Omega}_1') = 28$. Thus by Corollary 5.3.10, $\Omega_1 \approx Sp_8(2)$ and $\Omega \approx \Omega_{64}^+(2)$. But this is impossible, for by Theorem 5.3.9, $R_2(E_7(r^e)) > 64$. If $K_o = E_8(r^e)$, then $R_p(K_o) \geq r^{27e}(r^{2e} - 1)$, and so (8.6.11) shows $m \geq 16$. But then by Corollary 5.3.10, $R_r(\overline{\Omega}_1') > 16^2 > 248$, and we reach the desired contradiction once again.

We are left with the case in which either $r^e = 2$ or K_o is $G_2(3)$ or $G_2(4)$. Since $|\overline{\Omega}_1'|$ is divisible by at least two odd primes p_1, p_2, we know that $|K_o|$ is divisible by $(p_1p_2)^t$. In particular, $K_o \neq G_2(3)$. If $K_o = E_8(2)$, the fact that $(p_1p_2)^t$ divides $|K_o|$ implies that $t \leq 5$. Thus by Theorem 5.3.9 and (8.6.11) we obtain $m \geq (2^{27}.3)^{1/5} > 16$, and so $R_2(\overline{\Omega}_1') > 248$ by Corollary 5.3.10. This yields a contradiction. Similarly, if $K_o = E_7(2)$, then $t \leq 3$, and so $m \geq (2^{15}.3)^{1/3} > 9$, which means $R_2(\overline{\Omega}_1') > 56$. The same argument eliminates $E_6^{\pm}(2)$. If $K_o = F_4(2)$, then $t \leq 2$ and $R_{2'}(K_o) \geq 44$, and so $m \geq 7$; but then $R_2(K_o) \geq 27$ by Corollary 5.3.10.iii, a contradiction. The argument used at the end of the proof of Lemma 8.4.2 eliminates the remaining groups $G_2(4)$, ${}^2F_4(2)'$ and ${}^3D_4(2)$. ∎

Lemma 8.6.3. *K_o is not a sporadic group.*

Proof. First of all, $|K_o|$ must be divisible by s^t, where s is the largest prime divisor of $\overline{\Omega}_1'$. Obviously $s \geq 5$, and this observation eliminates the Mathieu groups, J_1 and J_3 (see Table 5.1.C). Our treatment of the case where K_o is the Monster M will require a few extra arguments, so let us assume for the rest of this paragraph that $K_o \neq M$. Thus we see from Table 5.1.C that $s \leq 11$. We use [G-L, pp.40-70] for information about centralizers of elements of prime order in the sporadic groups. Suppose for the moment that $s = 11$, so that $K_o = J_4$. This means $R := \overline{\Omega}_2' \times \cdots \times \overline{\Omega}_t'$ centralizes an element of order 11 in J_4. But this is impossible, for the centralizer of such an element is a soluble group (see [G-L, p.48]). Next suppose that $s = 7$. Since the centralizer of an element of order 7 in K_o has a non-abelian composition factor with order divisible by 7, it follows that K_o is Co_1, He, Fi_{24}', Th or BM. Apart from BM, there is a unique such non-abelian composition factor, and it is either $L_2(7)$ or A_7. So in these cases, $t = 2$ and $\overline{\Omega}_1' \cong L_2(7)$. But as $t = 2$, we must have $m \geq 4$ (as $n \geq 13$), and this yields a contradiction, for no classical group of dimension at least 4 is isomorphic to $L_2(7)$. To handle the case $s = 7$ and $K_o = BM$, we see that $t = 2$ as 7^3 does not divide $|BM|$, and that $m \leq 10$ by Proposition 5.2.9. But then $n \leq 100$, which violates Proposition 5.3.8. Now we treat the case $s = 5$. Here the centralizer of an element of order 5 must have a non-abelian composition factor with order divisible by 5. If there is a unique such composition factor and it happens to be A_5, then we are forced to conclude $t = 2$; and since $\Omega_1 \not\approx \Omega_4^-(2)$, we have $m \leq 3$. But this violates our assumption $n \geq 13$. Therefore it cannot be the case that the unique non-abelian composition factor of an element of order 5 is A_5. This observation eliminates all possibilities for K_o apart from Suz, Co_1, Ly, Fi_{23}, Fi_{24}', HN and BM. Note that as $\overline{\Omega}_1'$ is a $\{2,3,5\}$-group, we have

$$m \leq 5.$$

Take $K_o = Co_1$. From [G-L, p.49] we see that R embeds in either $A_5 \times A_5$ or in J_2. Since any proper simple subgroup of J_2 with order divisible by 5 is isomorphic to A_5, we are forced to conclude that $t = 3$ and $\overline{\Omega}_1' \cong A_5$. The only possibility here is $\Omega_1 \approx \Omega_3(5)$, and hence $\Omega \approx \Omega_{27}(5)$. However, the simple group Co_1 has no non-trivial 27-dimensional representation over $\mathbf{F}_5$. For let W be a non-trivial irreducible $\mathbf{F}_5 Co_1$-module, and consider the restriction of W to the subgroup $2^{11}.M_{24}$. The orbits of the factor M_{24} on the non-trivial linear characters of the normal 2^{11} have sizes 276 and 1,771 (see [At, p.183]), and so we conclude from Clifford's Theorem that $\dim(W) \geq 276$. If $K_o = Ly$, then [G-L, p.59] shows that $R \leq SL_2(9)$. However, although $SL_2(9)$ contains the quasisimple subgroup $SL_2(5)$, it contains no simple subgroups. Hence the desired contradiction. Now assume that $K_o = HN$. The centralizers of elements of order 5 appear in [G-L, p.66], and since $SL_2(5)$ contains no non-abelian simple groups, we see that R embeds in $U_3(5)$. Therefore $t = 2$, and so by Proposition 5.3.8 $m \geq 8$, contrary to the fact that $m \leq 5$. Next take $K_o = Fi_{23}$. Because $n \geq 234$ by Proposition 5.3.8, we have $t \geq 4$. But this is impossible as 5^4 does not divide $|Fi_{23}|$. The same reasoning handles Fi_{24}'. Consider $K_o = BM$. We have already remarked that R does not embed in A_5, and hence [G-L, p.68] implies that R embeds in HS. Since HS has non-abelian Sylow 5-subgroups of order 125, it follows that $t \leq 3$. Therefore Proposition 5.3.8 implies that $m \geq 7$, which contradicts the earlier result $m \leq 5$. If $K_o = Suz$, then as $R \not\leq A_5$ we deduce from [G-L, p.56] that $R \cong A_6$. However, it is shown in [Wi$_4$, §3] that $A_6 \times A_6$ does not embed in Suz.

In this paragraph we treat the case where K_o is the Monster M. Here $n \geq 729$ by Proposition 5.3.8. Since $\Omega_1 \not\approx \Omega_{10}^{\pm}(2)$, it follows from Proposition 5.2.9 that $m \leq 8$, and hence $t \geq 4$. Thus $|\overline{\Omega}_1|^4 \mid |M|$, and so using Table 5.1.C we deduce $|\overline{\Omega}_1'| \mid 2^{11}.3^5.5^2.7$. If $m \geq 6$, then the only possibility is $\Omega_1 \approx Sp_6(2)$. However it follows from [G-L, p.69] that the centralizer of an element of order 7 in M does not contain $Sp_6(2)^3$, and hence this situation does not arise. Therefore $m \leq 5$, and so $t \geq 5$. Since $m_7(M) < 5$, it follows that $|\overline{\Omega}_1'|$ is not divisible by 7, and hence it divides $2^9.3^4.5$. It follows that the centralizer of an element of order 5 in M contains $R \cong \overline{\Omega}_1'^4$, and so using [G-L], we deduce that R embeds in either J_2 or HN. However $|J_2|$ is not divisible by 5^4, and so R embeds in HN. However the argument used in the previous paragraph (where the case $s = 5$ and $K_o = HN$ was eliminated) shows that HN cannot contain R. This completes the proof of the Lemma. ∎

Lemma 8.6.4. $\overline{\Omega}_i' = C_{\overline{\Omega}}(\prod_{j \neq i} \overline{\Omega}_j')^\infty$.

Proof. Define $C = C_{\overline{\Omega}}(\prod_{j \neq i} \overline{\Omega}_j')^\infty$, and write $\widetilde{C}$ for a minimal preimage of C in Ω. Then by Lemmas 4.0.5.i and 4.4.3.i, $\widetilde{C} \leq GL(V_i)$. Arguing as in the proof of Lemma 4.4.6, we deduce $\widetilde{C} \leq \Delta_1$, and since $\widetilde{C}$ is perfect, it lies in Ω_1'. ∎

Proposition 8.6.5. *If K_o is a classical group, then one of the following holds.*

(i) $\Omega \approx \Omega_{m^t}^+(q)$ *with q even, $m = 2$ or 4, the group H is of type $Sp_m(q) \wr S_t$ and $K_o \cong Sp_{m^t}(q)$ or $\Omega_{m^t+2}(q)$. Moreover,*

(a) *if* $K_o \cong Sp_{mt}(q)$, *then* H *is of type* $Sp_m(q) \wr S_t$ *in* $\mathcal{C}_2(K_o)$;

(b) *if* $K_o \cong \Omega_{mt+2}(q)$, *then* $H < N < K_o$, *where* N *is of type* $Sp_{mt}(q)$ *in* $\mathcal{C}_1(K_o)$, *and* H *is of type* $Sp_m(q) \wr S_t$ *in* $\mathcal{C}_2(N)$.

(ii) $\Omega \approx \Omega^+_{4^t}(q)$ *with* q *odd*, H *is of type* $O^-_4(q) \wr S_t$, $K_o \cong PSp_{2^t}(q^2)$ *or* $P\Omega^+_{2^t}(q^2)$, *and* $\overline{L}$ *acts absolutely irreducibly on the natural* 2^t*-dimensional projective module for* K_o.

Proof. This proof will occupy the next ten pages or so. Write

$$K_o \cong C\ell_d(r^e), \tag{8.6.12}$$

with r prime and d chosen minimally. Also write $(V_\sharp, \mathbf{F}_\sharp, \kappa_\sharp)$ for the classical geometry associated with K_o, and let $X_\sharp = X(V_\sharp, \mathbf{F}_\sharp, \kappa_\sharp)$ for the appropriate symbols X appearing in (2.1.15). As usual, $PX_\sharp$ denotes the corresponding projective group; in particular, $P\Omega_\sharp = K_o$.

Obviously $\overline{L} \le K_o = P\Omega_\sharp$, and we let $L_\sharp$ be a minimal preimage of $\overline{L}$ in $\Omega_\sharp$. Thus $L_\sharp$ is a covering group of $\overline{L}$. Furthermore, we have

$$C_{\Omega_\sharp}(L_\sharp) \le \mathbf{F}^*_\sharp. \tag{8.6.13}$$

For if $L_\sharp$ centralizes $g \in \Omega_\sharp$, then the image of g in $P\Omega_\sharp = K_o$ must be trivial by Lemma 7.7.1.i and so $g \in \mathbf{F}_\sharp$. Notice that if $\overline{L}$ fixes a totally singular non-zero proper subspace of $(V_\sharp, \mathbf{F}_\sharp, \kappa_\sharp)$, then $\overline{L}$ normalizes a non-trivial r-subgroup of K_o (as $\overline{L}$ is contained in a parabolic subgroup of K_o), contrary to Lemma 7.7.1.v. Consequently

$$\overline{L} \text{ fixes no non-zero proper totally singular subspace of } (V_\sharp, \mathbf{F}_\sharp, \kappa_\sharp). \tag{8.6.14}$$

Therefore exactly one of three situations arises:

(A) $K_o \cong L_d(r^e)$ and L is irreducible on $V_\sharp$;

(B) $\kappa_\sharp$ is non-degenerate and all $L_\sharp$-invariant subspaces of $V_\sharp$ are non-degenerate;

(C) $(V_\sharp, \mathbf{F}_\sharp, \kappa_\sharp)$ is an orthogonal geometry in even characteristic and $L_\sharp$ fixes a non-singular vector.

Here we consider (A) and (B) together. Case (C) is postponed until near the end of the proof. We may write

$$V_\sharp = W_o \perp W_1 \perp \cdots \perp W_k, \tag{8.6.15}$$

where $W_o = C_{V_\sharp}(L_\sharp)$ and for $i \ge 1$, the W_i are the non-trivial Wedderburn components of $L_\sharp$. Note that (8.6.14) ensures that either every W_i is non-degenerate, or else $K_o = L_d(r^e)$, $W_o = 0$ and $k = 1$. Moreover, $L_\sharp \le \Omega(W_1)' \times \cdots \times \Omega(W_k)' \le \Omega_\sharp$, where $\Omega(W_i)$ is as described in (4.1.1). Hence $\overline{L} \le T \le K_o$, where T is isomorphic to a central product $T_1 \circ \cdots \circ T_k$, with T_i the image of $\Omega(W_i)'$ in $P\Omega_\sharp = K_o$. We record a few facts concerning T in the following Step.

Step 1.
(i) *T is a direct product of at least k non-abelian simple groups.*
(ii) *If $V_\sharp \neq W_1$, then $T_i \cong \Omega(W_i)$ for all $i = 1, \ldots, k$.*
(iii) *If $V_\sharp \neq W_1$, then $Z(\Omega(W_i)) = 1$.*

Proof. Since $L_\sharp$ acts non-trivially on W_i for $i = 1, \ldots, k$, the image of $\overline{L}$ in T_i is always non-trivial. Therefore each T_i is insoluble, and hence T is actually a central product of quasisimple groups (see Proposition 2.9.2). Thus part (i) holds in view of Lemma 7.7.1.i. If $V_\sharp \neq W_1$, then $\Omega(W_i) \cap \mathbf{F}_\sharp^* = 1$, and so (ii) holds. Assertion (iii) is immediate from (i) and (ii). This finishes Step 1.

Let $\widetilde{T}$ be a minimal preimage of T in Ω. Then $\widetilde{T}$ is a central product of quasisimple groups. It now follows from Lemma 5.5.5 and the fact that $\widetilde{T}$ is absolutely irreducible that $\widetilde{T}$ preserves a tensor decomposition

$$V = U_1 \otimes \cdots \otimes U_k,$$

where U_i is an absolutely irreducible module for a covering group of T_i. (N.b., it may be the case that $T_i \cong L_2(r^e) \times L_2(r^e)$ in the event that $\Omega(W_i) \approx \Omega_4^+(r^e)$ — in this case U_i itself is tensor decomposable as a $\widetilde{T}_i$-module, yet the tensor decomposition above still remains accurate.)

Put

$$d_i = \dim_{\mathbf{F}_\sharp}(W_i)$$

for $0 \leq i \leq k$, and arrange the indices so that $d_1 \leq d_2 \leq \cdots \leq d_k$. Observe that

$$d_o \leq 1. \tag{8.6.16}$$

For if $d_o \geq 2$, then either W_o contains a non-zero singular vector, or $d_o = 2$ and W_o is an O_2^--space in orthogonal geometry. The first possibility contradicts (8.6.14); as for the second, note that $\Omega_2^-(r^e) \cong \mathbf{Z}_{(r^e+1)/(2,r-1)}$, and so $L_\sharp$ will centralize a non-scalar in $\Omega_\sharp$, contrary to (8.6.13). Thus (8.6.16) holds.

For $i \geq 1$, write h_i for the number of components of L projecting non-trivially in T_i, so that $h_i \geq 1$ for $1 \leq i \leq k$. Also write L_i for the subgroup of L generated by these h_i components (so that $\overline{L}_i$ is a direct product of h_i copies of $\overline{\Omega}_1'$). As in the proof of (7.7.5), the image of $\overline{L}$ in $PGL(U_1)$ is irreducible. The image of Ω_j' in T_i is non-trivial if and only if its image in $PGL(U_i)$ is non-trivial, and so precisely h_i components of $\overline{L}$ project non-trivially in $PGL(U_i)$. Hence by Proposition 5.5.7.ii,

$$\dim_{\mathbf{F}}(U_i) \geq R_p(\overline{\Omega}_1')^{h_i}. \tag{8.6.17}$$

Furthermore, a minimal preimage of $\overline{L}_i$ in $L_\sharp$ acts faithfully and homogeneously on W_i (since W_i is a Wedderburn component), and hence Proposition 5.5.7.ii again yields

$$d_i \geq R_r(\overline{\Omega}_1')^{h_i}. \tag{8.6.18}$$

Three further observations are in order. First, each component of L projects non-trivially in some T_j, and hence

$$\sum_{i=1}^{k} h_i \geq t. \tag{8.6.19}$$

Second,

$$\text{if } h_i = 1, \text{ then } T_i = \overline{\Omega}'_j \text{ for some } j. \tag{8.6.20}$$

To prove (8.6.20), choose j such that Ω'_j projects non-trivially into T_i. Since $h_i = 1$, we know that Ω'_ℓ projects trivially into T_i for all $\ell \neq j$. Hence by Lemma 8.6.4, $T_i \leq C_{\overline{\Omega}}(\prod_{\ell\neq j} \overline{\Omega}'_\ell)^\infty = \overline{\Omega}'_j$. Therefore $T_i \trianglelefteq \overline{\Omega}'_j$ (since $T_i \trianglelefteq T$), and hence $T_i = \overline{\Omega}'_j$ (by the simplicity of $\overline{\Omega}'_j$). Third, we see that (8.6.20) extends to the following fact:

$$\begin{gathered}\text{if } h_i = 1 \text{ for all } i \geq 1, \text{ then } k = t\\ \text{and } \overline{\Omega}'_i = T_{\rho(i)} \text{ for some permutation } \rho \text{ of } \{1, \ldots, t\}.\end{gathered} \tag{8.6.21}$$

We now make a few reductions, the first of which is

Step 2. *We have* $t \geq 3$.

Proof. Suppose for a contradiction that $t = 2$. Here $m \geq 4$ by (6.0.3). Now if $h_i = 2$ for some i, then $d \geq d_i \geq R_r(\overline{\Omega}'_1)^2$ by (8.6.18). And if $h_i = 1$ for all i, then $k = 2$ by (8.6.21), whence $d \geq \sum_{i=1}^{k} R_r(\overline{\Omega}'_i) \geq 2\,R_r(\overline{\Omega}'_1)$. Thus in either case, we have

$$d \geq 2\,R_r(\overline{\Omega}'_1). \tag{8.6.22}$$

If $K_o \cong \Omega_8^+(2)$ of order $2^{12}.3^5.5^2.7$, then $|\overline{\Omega}'_1| \mid 2^6.3^2.5$, and the only possibilities (since $m \geq 4$ and Ω'_1 is quasisimple) are $\Omega_1 \approx Sp_4(2)$ or $\Omega_4^-(3)$. However it is easy to show that $A_6 \times A_6$ does not embed in $\Omega_8^+(2)$; to see this, assume that $A_6 \times A_6$ is contained in $\Omega_8^+(2)$. Since $R_2(A_6) = 3$, it follows from Proposition 5.5.7.ii that the $A_6 \times A_6$ must be reducible on $V_\sharp$. Note that a Sylow 5-subgroup of $\Omega_8^+(2)$ is contained in a subgroup $\Omega_4^-(2) \times \Omega_4^-(2)$, acting irreducibly on two non-degenerate 4-spaces. These two 4-spaces are in fact the only non-zero proper subspaces of $V_\sharp$ which are invariant under this Sylow 5-subgroup, and this forces $A_6 \times A_6 \preceq O_4^-(2) \times O_4^-(2)$, which is impossible.

Consider the case in which $r \neq p$. If $m \geq 6$, then $R_r(\overline{\Omega}'_1) \geq m$ by Corollary 5.3.11.iii (since $\Omega_1 \not\approx \Omega_6^-(2)$), and so $d \geq 12$. But then by Corollary 5.3.10, $n > d^2$, which forces $m^2 > d^2 \geq (2m)^2$, which is absurd. Therefore $m \leq 5$, and so $n \leq 25$. But then by Corollary 5.3.10.iii and the fact that $K_o \not\cong \Omega_8^+(2)$, we have $d \leq 6$, which forces $R_r(\overline{\Omega}'_1) \leq 3$, and so $\Omega_1 \approx \Omega_4^-(3)$ or $Sp_4(2)$, by Corollary 5.3.11.iii. In particular, $m = 4$ and $n = 16$. If $\Omega_1 \approx \Omega_4^-(3)$, then $R_{3'}(\overline{\Omega}'_1) = 3$, which forces $d \geq 6$. Therefore $d = 6$, and since K_o is not of type $P\Omega_6^\pm$ (by the minimality of d), the condition $R_{r'}(K_o) \leq 16$ forces $K_o \cong Sp_6(2)$ by Theorem 5.3.9; but $|\Omega_4^-(3)|^2$ does not divide $|Sp_6(2)|$. If $\Omega_1 \approx Sp_4(2)$, then $4 \leq d \leq 6$ and r is odd. Thus K_o is $U_4(3)$, $PSp_4(3)$, $PSp_4(5)$ or $PSp_6(3)$, none of which has order divisible by $|Sp_4(2)'|^2$.

Now consider the case where $r = p$. If $m \geq 7$, then $R_p(\overline{\Omega}_1') = m$ by Proposition 5.4.15, and so $d \geq 2m = 2\sqrt{n} \geq 14$. But this violates Lemma 8.1.4.ii. This leaves the case $m \leq 6$. If $m = 5$, then $R_p(\overline{\Omega}_1') \geq 4$ (see Proposition 5.4.15), and so $d \geq 8$ and $n = 25$. Therefore $n < \frac{1}{2}d^2$, and we reach a contradiction here since no module in Table 5.4.A has dimension 25. Thus m is either 4 or 6. Take $m = 6$. Then $R_p(\overline{\Omega}_1') \geq 4$, which forces $d \geq 8$. If $d \geq 10$, then a contradiction is easily obtained from Proposition 5.4.11. Therefore $d \leq 9$, and so $R_p(\overline{\Omega}_1') \leq 4$. Hence $R_p(\overline{\Omega}_1') = 4$, forcing $\Omega_1 \approx \Omega_6^{\pm}(q)$, with q odd. In view of (8.6.18) and (8.6.19), we must have $k = 2$ and $h_1 = h_2 = 1$. Moreover $(d_o, d_1, d_2) = (0,4,4), (0,4,5)$ or $(1,4,4)$. Thus (8.6.21) applies, and so $P\Omega_6^{\pm}(q) \cong T_1 \cong \Omega(W_1)$, where $\Omega(W_1) \approx \Omega(W_1, \kappa_\sharp)$ is a 4-dimensional classical group. The only possibility is $\Omega(W_1, \kappa_\sharp) \approx SL_4^{\pm}(q)$; but this is impossible by Step1.iii, for $SL_4^{\pm}(q)$ has a non-trivial center, as q is odd. Finally, take $m = 4$, so that $n = 16$. Then it follows from Proposition 5.4.11 that $d \in \{4,5,8,9,10\}$, and we consider these possibilities in turn. First assume that $d = 4$. Then by (8.6.22) $R_p(\overline{\Omega}_1') = 2$, and so $\Omega_1 \approx \Omega_4^-(q)$ with q odd (see Proposition 5.4.15) and $\Omega \approx \Omega_{16}^+(q)$. Moreover, the minimality of d ensures that $K_o \cong L_4^{\pm}(p^e)$ or $PSp_4(p^e)$. Suppose for the moment that $k \geq 2$. Then it follows from (8.6.18) that $k = 2$ and $(d_o, d_1, d_2) = (0,2,2)$; but then $K_o \cong PSp_4(p^e)$ and $T_i \cong Sp_2(p^e)$, which has a non-trivial centre as p is odd. This is a contradiction, and so we are left with the case $k = 1$, with $L_\sharp$ irreducible on $V_\sharp$. It follows from Lemma 7.7.2 that $K_o \not\cong PSp_4(p^e)$, whence $K_o \cong L_4^{\pm}(p^e)$. According to Proposition 5.4.6, $K_o \cong L_4^{\pm}(q)$, $L_4^{\pm}(q^2)$ or $U_4(q^{1/2})$. On the other hand, K_o contains $\overline{L} \cong L_2(q^2) \times L_2(q^2)$, and hence $K_o \cong L_4^{\pm}(q^2)$. Now Proposition 5.4.6.ii implies $K_o \cong L_4(q^2)$. Moreover, Proposition 5.4.6 shows that as $SL_4(q^2)$-modules, $V \otimes \mathbf{F}_{q^2} \cong V_\sharp \otimes V_\sharp^{\psi}$, where ψ generates Gal($\mathbf{F}_{q^2}$:$\mathbf{F}$); but this module is never self dual, a contradiction (recall $\Omega \approx \Omega_{16}^+(q)$ here). Next take $d = 5$, so that $K_o \cong L_5^{\pm}(p^e)$. As before, $R_p(\overline{\Omega}_1') = 2$ and so $\Omega_1 \approx \Omega_4^-(q)$ with q odd. Clearly Corollary 5.5.6 guarantees that $\overline{L}$ is reducible on $V_\sharp$, and hence $(d_o, \dots, d_k) = (1,2,2), (1,4)$ or $(0,2,3)$ (recall (8.6.16)). Thus $K_o \not\cong L_5(p^e)$ by (8.6.14), and we can eliminate $U_5(p^e)$ with Step 1.iii, since $\Omega(W_1) \cong SU_2(p^e)$ or $SU_4(p^e)$, both of which have a non-trivial centre. Next take $d = 8$. Then by Proposition 5.4.11, $K_o \cong Sp_8(2^e)$ and V is a spin module for K_o. Moreover, Lemma 8.1.5 shows that $K \cong Sp_8(q)$ and that $\Omega \approx \Omega_{16}^+(q)$. Therefore H must be of type $Sp_4(q) \wr S_2$. In addition, $R_2(\overline{\Omega}_1') = R_2(Sp_4(q)') \geq 3$, and so $h_i = 1$ for all $i \geq 1$ by (8.6.18), and so it follows from (8.6.21) that $V_\sharp = W_1 \perp W_2$, and this leads to conclusion (i)(a) of the Proposition. Now assume that $d = 9$. Arguing as before, we see that $K_o \cong \Omega_9(q)$ with q odd, and $\Omega \approx \Omega_{16}^+(q)$. Moreover, H is of type $Sp_4(q) \wr S_2$ or $O_4^-(q) \wr S_2$, and so $\Omega_1 \approx \Omega_4^-(q)$ or $Sp_4(q)$. Now it must be the case that $d_o = 0$; for otherwise $\overline{L}$ is contained in a group $P\Omega_8^{\pm}(q)$, yet it follows from Proposition 5.4.11 that the full covering group of $P\Omega_8^{\pm}(q)$ has no absolutely irreducible p-modular representation of degree 16. Furthermore, $\overline{\Omega}_1'$ is not contained in $\Omega_3(q)$, and so $d_i \geq 4$ for $i \geq 1$. Therefore $(d_o, \dots, d_k) = (0,9)$ or $(0,4,5)$. Suppose for the moment $(0,9)$ occurs. Then $L_\sharp$ acts faithfully on an irreducible $L_\sharp$-submodule of $V_\sharp$, and such a submodule has dimension 3 or 9. Since $L_\sharp$ does not embed in $GL_3(q)$, it follows that $L_\sharp$ is in fact irreducible on $V_\sharp$. Consequently $\overline{L}$ must stabilize a tensor product decomposition of $V_\sharp$ by Corollary 5.5.6, and this forces $\overline{\Omega_1}'$ to be involved in $GL_3(q)$, which is false. Therefore $(0,4,5)$ occurs, and so $T_1 \cong \Omega_4^\pm(q)$, $T_2 \cong \Omega_5(q)$. But then $\overline{L}$ embeds in neither T_1 nor T_2, and hence $h_1 = h_2 = 1$. We reach the desired contradiction by appealing to (8.6.21). Finally take $d = 10$. Here $K_o \cong P\Omega_{10}^\pm(q)$ and $\Omega \approx SL_{16}^\pm(q)$. Thus $\Omega_1 \approx SL_4^\pm(q)$, so $R_p(\overline{\Omega_1}') = 4$, and hence $k = 2$ and $h_1 = h_2 = 1$. This forces $4 \le d_1 \le 5$, which means $T_1 \approx \Omega_4^\pm(q)$ or $\Omega_5(q)$. On the other hand, (8.6.21) shows that $T_1 \cong \overline{\Omega_1}'$, and so we have reached yet another contradiction. This completes the proof of Step 2.

Next we establish

Step 3. *We have $K_o \neq P\Omega_8^+(r^e)$.*

Proof. Assume otherwise. Then by Proposition 5.5.7, $8 \ge t\,R_r(\overline{\Omega_1}')$, and so $t \le 4$. Also $t \ge 3$ by Step 2, whence $R_r(\overline{\Omega_1}') = 2$. First suppose that $r \neq p$. Then by Corollary 5.3.11, $\Omega_1 \approx Sp_2(4)$, $Sp_2(5)$, $L_3(2)$, $\Omega_3(5)$ or $Sp_4(2)$. In all cases, $|\overline{\Omega_1}'|^3$ does not divide $|\Omega_8^+(2)|$, and so $r^e \ge 3$. It therefore follows from Theorem 5.3.9 that $n \ge 234$. Since $t \le 4$, we deduce $m \ge 4$, which means $\Omega_1 \approx Sp_4(2)$. As $|Sp_4(2)'|^4$ does not divide $|P\Omega_8^+(3)|$, we must have $r^e \ge 5$. But then Theorem 5.3.9 shows $n \ge 3,100 > 4^4$, a contradiction.

Next suppose that $r = p$. Thus $\Omega_1 \approx Sp_2(q)$, $\Omega_3(q)$ or $\Omega_4^-(q)$, by Proposition 5.4.15. If $\Omega_1 \approx Sp_2(q)$, then $t = 4$ and $n = 16$. But this is impossible, for according to Proposition 5.4.11, the full covering group of $P\Omega_8^+(p^e)$ does not have an absolutely irreducible p-modular representation of degree 16. Finally, assume that $\Omega_1 \approx \Omega_3(q)$ or $\Omega_4^-(q)$, so that q is odd. The case $k = 1$ may be eliminated using the fact that $t \ge 3$, coupled with Lemma 7.7.2. Therefore $k \ge 2$. Observe that $d_i \ge 3$ for all $i = 1, \dots, k$ (as $O_2^\pm(p^e)$ is soluble), and hence $k \le 2$. Therefore $k = 2$ and $(d_o, d_1, d_2) = (0,3,5)$, $(0,4,4)$ or $(1,3,4)$. Now as $t \ge 3$, we know that $h_i = 2$ for some i. Corollary 5.5.6 ensures that $d_i \neq 3, 5$, and hence $d_i = 4$. Indeed, W_i must be a $+4$-space (since $\Omega_4^-(p^e) \cong L_2(p^{2e})$ does not contain $L_2(q) \times L_2(q)$). But then T_i has non-trivial centre, which violates Step 1.iii. The proof of Step 3 is now finished.

In view of Step 3 and Theorem 2.1.4, we know that $PA_\sharp = Aut(K_o)$, except if $K_o \cong Sp_4(2^e)$. Consequently, from the structure of $PA_\sharp / P\Omega_\sharp$ given by the results in Chapter 2 (Propositions 2.2.3, 2.3.5, 2.4.4, 2.6.3, 2.7.3 and 2.8.2), we see that in any case, $P\Delta_\sharp \trianglelefteq Aut(K_o)$ and $Aut(K_o)/P\Delta_\sharp$ is abelian. Therefore $\overline{H}' \le \overline{K}' \le P\Delta_\sharp$. Our strategy now is to investigate the action of $\overline{H}'$ on the $(V_\sharp, \mathbf{F}_\sharp, \kappa_\sharp)$-geometry. The following Step will be quite useful for this investigation.

Step 4. *The action of $\overline{H}'$ on the t components of $\overline{L}$ is the full alternating group A_t.*

Proof. Now according to the results in §4.7, when $t \ge 4$ we know that H acts as the full symmetric group on the t components of L, and so Step 4 holds in this case. When $t = 3$ the same is true, except in case $\mathbf{O}^+$ with $\Omega_1 \approx \Omega_m^\pm(q)$, q odd, $m \equiv 2 \pmod 4$ (see Propositions 4.7.6, 4.7.7). Thus to complete the proof of Step 4 we must consider this exceptional case. So take H to be of type $O_m^\pm(q) \wr S_3$ with $m \equiv 2 \pmod 4$.

Suppose that $m \geq 10$. If $r \neq p$, then applying Corollary 5.3.10.i twice shows $d > m^2$ and $n > d^2 > m^4$, contrary to the fact that $n = m^3$. Therefore $r = p$, and hence $R_p(\overline{\Omega}_1') = m$. If $h_i \geq 2$ for some i, then $d \geq m^2$ by (8.6.18); but this leads to a contradiction in light of Lemma 8.1.4.ii. Therefore $h_i = 1$ for all i, and so (8.6.21) applies. Therefore by Theorem 5.1.2, we must have $T_i \approx \Omega_m^\epsilon(q)$ for all i. That is, K_o is an orthogonal group and W_i is an ϵm-space for $i = 1,2,3$. In particular, W_2 and W_3 are isometric, so by Propositions 2.5.11 and 2.5.13, the involution x negating $W_2 \perp W_3$ and centralizing $W_o \perp W_1$ lies in $\Omega_\sharp$ (recall q is odd). But clearly $x \in C_{\Omega_\sharp}(L_\sharp)$, and this violates (8.6.13) as $x \notin \mathbf{F}_\sharp^*$.

To complete the proof, assume that $m = 6$, so that $n = 316$. If $r \neq p$, then Corollary 5.3.10 implies $d \leq 17$. But then $R_{p'}(\overline{\Omega}_1') \leq \lfloor \frac{d}{3} \rfloor \leq 5$, which violates Corollary 5.3.11. Therefore $r = p$. Since 316 is not one of the dimensions appearing in Table 5.4.A, it must be the case that $n \geq \frac{1}{2}d^2$, which means $d \leq 25$. Since $R_p(\overline{\Omega}_1') = 4$, it follows from (8.6.18) that $h_i \leq 2$ for all i, and at most one of the h_i is equal to 2. Thus $h_i = 1$ for some i, and without loss $i = 1$. Therefore $T_1 = \overline{\Omega}_j'$ for some j by (8.6.20), and we can take $j = 1$. Now $Z(T_1) = 1$ by Step 1.iii. Therefore by Theorem 5.1.2 and the fact that $Z(SL_4^\pm(q)) \neq 1$ as q is odd, it must be the case that $T_1 \approx \Omega_6^\pm(q)$. Therefore K_o is an orthogonal group and W_1 is a ± 6-space. Arguing as in the previous paragraph, none of the other spaces W_i can be a ± 6-space isometric to W_1, and hence $h_j \geq 2$ for $j = 2, \ldots, k$. Therefore $d_j \geq 16$ for $j \geq 2$ by (8.6.18), and since $d \leq 25$ we have $k = 2$ and $d_2 \leq 19$. At this stage we have the situation $L_4^\pm(q) \times L_4^\pm(q) \cong \overline{\Omega}_2' \times \overline{\Omega}_3' \leq T_2 \cong \Omega(W_2)' \preceq GL(W_2)$, and here the $L_4^\pm(q) \times L_4^\pm(q)$ acts irreducibly on W_2. Since $\dim(W_2) \leq 19$ and since $R_p(L_4^\pm(q)) = 4$, it follows from Lemma 5.5.5 that $d_2 = 16$ and $L_4^\pm(q) \times L_4^\pm(q)$ preserves a tensor product decompositon of W_2 with two tensor factors of dimension 4. But this is impossible, for $L_4^\pm(q)$ does not embed in $GL_4(p^e)$ when p is odd. This final contradiction completes the proof of Step 4.

As we remarked before the proof of Step 4, $\overline{H}' \leq P\Delta_\sharp$, and we define

$$H_1 = \overline{H}' \cap PI_\sharp.$$

Thus we obtain an inclusion

$$H_\sharp \leq I_\sharp,$$

where $H_\sharp$ is a minimal preimage of H_1 in $I_\sharp$. Since A_t has no subgroup of index 2, and since $|P\Delta_\sharp : PI_\sharp| \leq 2$ (see Table 2.1.D), we deduce from Step 4 that

$$H_1 \text{ acts as } A_t \text{ on the } t \text{ components of } \overline{L}. \tag{8.6.23}$$

Evidently $L_\sharp$ is the group generated by the components of $H_\sharp$, and $H_\sharp$ acts as A_t on the t components in $L_\sharp$. Also $H_\sharp$ acts on the groups T_i.

We now establish

$$h_i < t \text{ for all } i. \tag{8.6.24}$$

Assume false, so that by (8.6.18), $d \geq R_r(\overline{\Omega_1'})^t$. Since $d < n$ by Lemma 8.1.4, we conclude that $R_r(\overline{\Omega_1'}) < m$, and so by Corollary 5.3.11.iii and Proposition 5.4.15, one of the following holds:

(a) $\Omega_1 \approx L_3(2)$;
(b) $\Omega_1 \approx Sp_4(2)$;
(c) $\Omega_1 \approx \Omega_3(q)$, with q odd;
(d) $\Omega_1 \approx \Omega_4^-(q)$, with q odd;
(e) $\Omega_1 \approx \Omega_5(q)$, with q odd;
(f) $\Omega_1 \approx \Omega_6^{\pm}(q)$, with q odd.

We consider these cases in turn. Note that since $t \geq 3$, we have $d \geq R_r(\overline{\Omega_1'})^t \geq 8$.

Case (a). By considering the subgroup 7:3 in $L_3(2)$, we see that $R_{7'}(L_3(2)) = 3$, and hence $r = 7$. Since $d \geq 8$, it follows from Corollary 5.3.10.ii that $R_{r'}(K_o) > d^2$. But this is clearly impossible, for $d^2 \geq (R_r(L_3(2))^t)^2 = 4^t$ while $R_{r'}(K_o) \leq n = 3^t$.

Case (b). If $r \neq 2$, then we reach the same contradiction as in case (a), unless $K_o \cong PSp_8(3)$ with $t = 3$. But $|Sp_4(2)'|^3$ does not divide $|PSp_8(3)|$, and so this situation cannot arise. Thus we are left to consider $r = p = 2$. Here $R_2(\overline{\Omega_1'}) = 3$, and so $d \geq 3^t$. But this violates Lemma 8.1.4.ii, since $n = 4^t$.

Case (c). As before, if $r \neq p$, then Corollary 5.3.10.ii implies that $K_o \cong \Omega_8^-(2)$, $PSp_8(2)$ or $PSp_8(3)$. Here $t = 3$, and we eliminate $\Omega_8^-(2)$ for its order is not divisible by the cube of any prime larger than 3. In addition, $R_{r'}(PSp_8(r)) > 27$ for $r = 2, 3$. Therefore $r = p$, and as $d \geq 2^t$ and $n = 3^t$, an application of Lemma 8.1.4.ii shows that $t = 3$. Thus $\Omega \approx \Omega_{27}(q)$ and Table 5.4.A implies that $K_o = PSp_8(p^e)$. According to Lemma 8.1.3.i, $p^e = q$, and using Step 1.iii we see that $L_\sharp$ is irreducible on $V_\sharp$ (recall q is odd). Now as an $Sp_8(q)$-module, the composition factors of $\Lambda^2(V_\sharp)$ are precisely V and a 1-dimensional module. On the other hand, Corollary 5.5.6 implies that $L_\sharp$ preserves a decomposition of $V_\sharp$ as a tensor product of three 2-spaces. Consequently the preimage of $\overline{\Omega_1'}$ in $L_\sharp$ acts homogeneously on $V_\sharp$ with irreducible constituents of dimension 2, and so this preimage fixes two 2-spaces $\langle v_1, v_2\rangle$ and $\langle v_3, v_4\rangle$, where the v_i span a 4-space in $V_\sharp$. But then $\overline{\Omega_1'}$ fixes the two linearly independent vectors $v_1 \wedge v_2$ and $v_3 \wedge v_4$ in $\Lambda^2(V_\sharp)$, and so $\overline{\Omega_1'}$ must fix at least one 1-space in V. This is clearly impossible, as all irreducible constituents of Ω_1' on V have dimension 3.

Case (d). If $r \neq p$, then $R_r(\overline{\Omega_1'}) \geq 3$, and so we reach a contradiction using the same arguments appearing in (a)-(c). This leaves the case $r = p$. Now as $h_i = t$, all components act non-trivially on W_i, and so $L_\sharp$ acts faithfully on W_i. Moreover, as W_i is a Wedderburn component of $V_\sharp$, it follows that $L_\sharp$ acts faithfully and irreducibly on some submodule W of W_i. Now if $L_\sharp$ fails to be absolutely irreducible on W, then $d \geq \dim_{\mathbf{F}_\sharp}(W) \geq 2\,R_p^i(\overline{L}) = 2 \times 2^t = 2\sqrt{n}$. But this impossible in view of Lemma 8.1.4 (note $m = 4$, $t \geq 3$, so $n \geq 64$). Therefore $L_\sharp$ must be absolutely irreducible on W, and so by Corollary 5.5.6, $\dim_{\mathbf{F}_\sharp}(W) \geq R_{\mathbf{F}_\sharp}(\overline{\Omega_1'})^t$. Now if $\mathbf{F}_\sharp$ does not contain $\mathbf{F}_{q^2}$, then $R_{\mathbf{F}_\sharp}(\overline{\Omega_1'}) = R_{\mathbf{F}_\sharp}(L_2(q^2)) \geq 3$, and so $d \geq 3^t$, against Lemma 8.1.4. Therefore $\mathbf{F}_\sharp$ must contain $\mathbf{F}_{q^2}$. On the other hand, as $d \geq 2^t = \sqrt{n}$ we know from Propositions 5.4.6 and

5.4.8 that $\mathbf{F}_\sharp$ cannot be strictly larger than $\mathbf{F}_{q^2}$. Therefore K_o is one of $L_d(q^2)$, $U_d(q)$, $PSp_d(q^2)$, $\Omega_d(q^2)$ and $P\Omega_d^\pm(q^2)$. If K_o is untwisted, then it follows from Proposition 5.4.6 that $d = 2^t$ and $L_\sharp$ is absolutely irreducible on $V_\sharp$. In particular, $K_o \neq \Omega_d(q^2)$ with d odd, and if $K_o = PSp_d(q^2)$ or $P\Omega_d^+(q^2)$, then conclusion (ii) of the Proposition holds. If $K_o = L_d(q^2)$, then using Proposition 5.4.6 we see that as $SL_d(q^2)$-modules, $V \otimes \mathbf{F}_{q^2} \cong V_\sharp \otimes V_\sharp^\psi$ (where $\langle\psi\rangle = \mathrm{Gal}(\mathbf{F}_{q^2}{:}\mathbf{F}_q)$), but this module is not self-dual. We now turn to the twisted groups. If $K_o \cong U_d(q)$, then Proposition 5.4.6 implies that $V \cong V^{\tau_o}$. Since $n \leq d^2$, it follows from [Lie$_2$, Th. 2.2] that $n = d^2 - 1$ or $d^2 - 2$. But neither $d^2 - 1$ nor $d^2 - 2$ can be of the form 4^t. Finally note that $K_o \neq P\Omega_d^-(q^2)$ by Lemma 7.7.2.

Cases (e) *and* (f). Here $R(\overline{\Omega}_1') \geq 4$, and so $d \geq 4^t \geq 64$. When $r \neq p$, we see from Corollary 5.3.10 that $R_p(K_o) > d^2 \geq 16^t > n$, a contradiction. And when $r = p$, we obtain a contradiction using Lemma 8.1.4. We have now proved (8.6.24).

We will now argue that

$$h_i = 1 \text{ for all } i. \tag{8.6.25}$$

Assume for a contradiction that $h := h_i \geq 2$. Thus by (8.6.24) we have $2 \leq h \leq t-1$. Without loss we can assume that $h_1 = h$. Since $L_\sharp \trianglelefteq H_\sharp \leq I_\sharp$, it is clear that $H_\sharp$ acts on the spaces W_i and hence on the groups T_i and the spaces U_i (the U_i are described after Step 1). Relabel the indices so that $\{W_1, \ldots, W_s\}$, $\{T_1, \ldots, T_s\}$ and $\{U_1, \ldots, U_s\}$ are $H_\sharp$-orbits. Since A_t acts transitively on h-sets, and since $H_\sharp$ acts as A_t on the set of t components in $L_\sharp$, it follows that $s \geq \binom{t}{h}$. In particular, $h_j = h$ for $j \leq \binom{t}{h}$. It now follows from (8.6.17) that $\dim_{\mathbf{F}}(U_j) \geq R_p(\overline{\Omega}_1')^h$ for $1 \leq j \leq \binom{t}{h}$, whence

$$m^t = n \geq R_p(\overline{\Omega}_1')^{h\binom{t}{h}} \geq \left(\max\{2, m-2\}\right)^{h\binom{t}{h}}. \tag{8.6.26}$$

As $2 \leq h \leq t-1$, we have $h\binom{t}{h} \geq t(t-1)$, and so (8.6.26) implies

$$m^t \geq \left(\max\{2, m-2\}\right)^{t(t-1)}.$$

And since $t \geq 3$, the only possibility is $m = 4$, $t = 3$ and $R_p(\overline{\Omega}_1') = 2$. Here $\Omega_1 \approx \Omega_4^-(q)$ with q odd, $h = 2$, $\binom{t}{h} = 3$, and hence $d_j \geq 4$ for $j = 1, 2, 3$ by (8.6.18). Therefore $d \geq 12$, and so $r = p$ by Corollary 5.3.10.i. Moreover, Proposition 5.4.11 implies that $K_o \cong PSp_{12}(p^e)$, $\Omega_{13}(p^e)$ or $P\Omega_{14}^\pm(p^e)$. Since $H_\sharp$ contains an element taking W_1 to W_2, these two spaces are isometric. Hence $W_1 \perp W_2$ is a non-degenerate 8-space, and if K_o is orthogonal, $W_1 \perp W_2$ is in fact an $O_8^+(p^e)$-space. Thus in all cases, $\Omega_\sharp$ contains an element negating $W_1 \perp W_2$ and centralizing $(W_1 \perp W_2)^\perp$, violating (8.6.13). This establishes (8.6.25).

We may now appeal to (8.6.21) to deduce

$$k = t \text{ and } \overline{\Omega}_i' = T_i \text{ for } 1 \leq i \leq t, \tag{8.6.27}$$

(with suitable relabelling). It follows immediately from (8.6.14) that $K_o \not\cong L_d(r^e)$. Also observe that $H_\sharp$ permutes the W_i naturally as A_t (since it does so to the T_i), and hence they are all isometric (as $H_\sharp \le I_\sharp$). Thus if r is odd, then $L_\sharp$ centralizes the involution negating $W_1 \perp W_2$, against (8.6.13) (n.b., we have already seen this argument in the proof of Step 4). And if K_o is unitary, then $L_\sharp$ centralizes the element which acts as λ on W_1 an λ^{-1} on W_2, where λ is a primitive $(r+1)^{\text{th}}$ root of unity, contrary to (8.6.13) once again. We have therefore established

$$r = 2 \text{ and } K_o \cong Sp_d(2^e) \text{ or } \Omega_d^{\mp}(2^e). \tag{8.6.28}$$

In addition,

$$T_i \approx Sp_{d_1}(2^e)' \text{ or } \Omega_{d_1}^{\pm}(2^e) \text{ for all } i. \tag{8.6.29}$$

We have already seen in (8.6.16) that $d_o \le 1$. If $K_o \cong Sp_d(2^e)$, then (8.6.14) implies that $d_o = 0$, and if $K_o \cong \Omega_d^{\mp}(2^e)$ then our initial assumption (B) (just after (8.6.14)) again implies that $d_o = 0$. Therefore

$$d = td_1.$$

Next we show

$$p = 2. \tag{8.6.30}$$

For assume that p is odd. Then by Theorem 5.1.2, (8.6.27) and (8.6.29), $T_i \approx Sp_2(4)$, $Sp_4(2)'$, $\Omega_4^-(2)$ or $\Omega_6^-(2)$. First suppose that $T_i \approx Sp_2(4)$, so that $K_o \approx Sp_{2t}(4)$. Then according to Theorem 5.3.9, $n \ge \frac{3}{2}4^{t-1}(4^{t-1}-1)$. On the other hand, $\Omega_1 \approx Sp_2(5)$ or $\Omega_3(5)$ (since $\overline{\Omega}_1'$ must be a classical group in odd characteristic isomorphic to $Sp_2(4)$), and so $n \le 3^t$. These last two inequalities are incompatible, and so this case cannot arise. The same argument handles the other possibilities for T_i. This establishes (8.6.30).

The next step is

$$\Omega \text{ is an orthogonal group.} \tag{8.6.31}$$

Assume false. Since $\mathcal{C}_7$ is void for the symplectic groups in characteristic 2, we are left to consider case $\mathbf{L}^{\pm}$. Here $\Omega_1 \approx SL_m^{\pm}(q)$ with $m \ge 3$. By Theorem 5.1.2, (8.6.27) and (8.6.29), $\Omega_i \approx SL_4^{\pm}(q)$ and $T_i \approx \Omega_6^{\pm}(q)$. We have $\Omega_1 \otimes \Omega_2 = T_1 \times T_2 \le Y := \Omega(W_1 \perp W_2) \approx \Omega_{12}^+(q)$, and observe that $Y \le C_\Omega(\Omega_3 \otimes \cdots \otimes \Omega_t)$. Therefore by Lemma 4.4.3, $\Omega_1 \otimes \Omega_2 \le Y \le GL(V_1 \otimes V_2)$. However this is impossible, for $\Omega_{12}^+(q)$ has no irreducible representation of degree 16 over $\mathbf{F}_q$ (see Proposition 5.4.11).

In view of (8.6.31) we know that H is of type $Sp_m(q) \wr S_t$ in $\Omega_{m^t}^+(q)$. In particular, $\Omega_1 \approx Sp_m(q)$. We now argue that

$$K_o \cong Sp_{mt}(q). \tag{8.6.32}$$

Quoting Theorem 5.1.2 and using (8.6.27) and (8.6.29), we see that the only other possibility is $T_i \approx \Omega_4^-(q^{1/2})$ and $\Omega_1 \approx Sp_2(q)$. Here $K_o \cong \Omega_{4t}^{\pm}(q^{1/2})$. But then arguing as in the proof of (8.6.31), we obtain an irreducible representation of $\Omega_8^+(q^{1/2})$ in $GL(V_1 \otimes V_2)$, which is impossible as $\dim(V_1 \otimes V_2) = 4$. Therefore (8.6.32) holds.

We now see that $\Omega_i \approx Sp_m(q) \approx T_i$, and the argument used in the proof of (8.6.31) shows that $Sp_{2m}(q)$ has an absolutely irreducible representation in $GL(V_1 \otimes V_2) \approx$

$GL_{m^2}(q)$. Obviously $m^2 < \frac{1}{2}(2m)^2$, and so this representation appears in Proposition 5.4.11. We deduce from Table 5.4.A that $m = 2$ or 4, and this leads to conclusion (i)(a) of the Proposition. This completes the analysis of cases (A) and (B), stated just below (8.6.14).

Now we analyse case (C). Here $L \leq N$, where N is the stabilizer in K_o of a non-singular vector $v \in V_\sharp$. Write $V_\flat$ for the symplectic space $v^\perp/\langle v \rangle$, so that $N = Sp(V_\flat) \cong Sp_{d-2}(2^e)$. Clearly Lemma 7.7.1.v implies that (8.6.14) holds with $\flat$ replacing $\sharp$, and so we may write $V_\flat = W_1 \perp \cdots \perp W_k$, just as we did in cases (A) and (B) above (note $C_{V_\flat}(\overline{L}) = 0$ in view of the '$\flat$' version of (8.6.14) and the fact that every vector in $V_\flat$ is singular). Carrying over the notation T_i, d_i, h_i we see that the proof of Step 2 goes through here. It is now easy to see that $K_o \neq \Omega_8^+(2^e)$. For otherwise, $N \cong Sp_6(2^e)$, and hence $R_2(\overline{L}) \leq 6$. Thus $t = 3$ and $R_2(\overline{\Omega}_1') = 2$ by Proposition 5.5.7. Therefore if q is odd, $\Omega_1 \approx Sp_2(5)$ or $\Omega_3(5)$ by Corollary 5.3.11, and if q is even, then $\Omega_1 \approx Sp_2(q)$ by Proposition 5.4.15. However $n \geq 13$ and $t = 3$, whence $m \geq 3$, and so $\Omega_1 \approx \Omega_3(5)$. Here $\Omega \approx \Omega_{27}(5)$, and since $R_5(\Omega_8^+(4)) > 27$, it follows that $K_o = \Omega_8^+(2)$. But 5^3 does not divide $|K_o|$, and we reach a contradiction. Therefore $K_o \neq \Omega_8^+(2^e)$, as claimed. Consequently $\overline{H} \leq P\Gamma_\sharp$, and so $\overline{H}' \leq P\Delta_\sharp$. Of course, $P\Delta_\sharp$ can be identified with $I_\sharp$ due to (2.7.2) and (2.8.3). Thus we may write $\overline{H}' \leq I_\sharp$.

We now claim that $\overline{H}'$ must fix v. For if not, $\overline{L}$ centralizes a 2-space spanned by v and some other non-singular vector $w \in V_\sharp$. In view of (8.6.14), $\langle v, w \rangle$ has no non-zero singular vectors, and so it is an anisotropic 2-space. But then $\overline{L}$ nis centralized by a subgroup $\Omega_2^-(2^e) \cong \mathbf{Z}_{2^e+1}$ in K_o, a contradiction. It follows that $\overline{H}' \leq N$. The proof of Step 4 goes through here, and so putting $H_1 = \overline{H}'$, we see that (8.6.23) holds. We may now argue as in case (B) above (starting at (8.6.24)), using the $(d-2)$-dimensional symplectic geometry for N instead of $(V_\sharp, \mathbf{F}_\sharp, \kappa_\sharp)$. We conclude from this that case (i)(b) of the Proposition holds.

This completes the proof of the Proposition. ∎

Proposition 8.6.6. *The situation described in* (ii) *of Proposition 8.6.5 does not occur.*

Proof. Assume for a contradiction that $\overline{H} < \overline{K} < \overline{\Omega}$, where $\Omega \approx \Omega_{4^t}^+(q)$ with q odd, H is of type $O_4^-(q) \wr S_t$ in $\mathcal{C}_7(\Omega)$ and $K_o = \mathrm{soc}(\overline{K}) \cong PSp_{2^t}(q^2)$ or $P\Omega_{2^t}^+(q^2)$. Let $(V_\sharp, \mathbf{F}_\sharp, \kappa_\sharp)$ be the natural 2^t-dimensional classical geometry over $\mathbf{F}_\sharp = \mathbf{F}_{q^2}$ associated with K_o, so that $K_o = P\Omega_\sharp$ in the usual notation. ¿From Proposition 8.6.5.ii we may also assume that a minimal preimage $L_\sharp$ of $\overline{L}$ in $\Omega_\sharp$ acts absolutely irreducibly on $V_\sharp$. It follows from Lemma 7.7.2 that $K_o \cong PSp_{2^t}(q^2)$ if t is odd and $K_o \cong P\Omega_{2^t}^+(q^2)$ if t is even. In particular, $t \geq 3$ since $K_o \neq P\Omega_4^+(q^2)$. Lemma 7.7.2 also implies that $N_{\Gamma_\sharp}(L_\sharp) \in \mathcal{C}_7(\Gamma_\sharp)$ and that $N_{\Gamma_\sharp}(L_\sharp)$ is of type $Sp_2(q^2) \wr S_t$ in $\mathcal{C}_7(\Gamma_\sharp)$. ¿From Proposition 5.4.6 we see that as an $\mathbf{F}_q K_o$-module, V is the module $V_\sharp \otimes V_\sharp^\sigma$ realized over $\mathbf{F}_q$, where σ is a field automorphism of K_o induced by the involutory field automorphism of $\mathbf{F}_\sharp$. It follows that $\overline{K} \leq P\Delta_\sharp\langle\sigma\rangle$. Thus if t is odd, $|\overline{K} : K_o|$ divides 4, and so Proposition 4.7.4 shows that

$$\overline{H} \preceq PSp_2(q^2)^t.2^{t-1}.S_t.[4]. \qquad (8.6.33)$$

And if t is even, it follows from parts (I) and (II) of Proposition 4.7.5 that a group of type $Sp_2(q^2) \wr S_t$ in $\mathcal{C}_7(P\Delta_\sharp)$ is isomorphic to $PSp_2(q^2)^t.2^{t-1}.S_t.2$ (this uses the facts that $|P\Delta_\sharp : P\Omega_\sharp| = 8$ and that $c = 4$ — see Lemma 4.0.3.ii,ix and Proposition 4.7.2). So (8.6.33) holds once again. On the other hand, we see from Proposition 4.7.7 that $\overline{H}$ has structure $\Omega_4^-(q)^t.[2^{2t-1}].S_t$, and this contradicts (8.6.33) and the fact that $t \geq 3$. ∎

The only task remaining is to prove uniqueness in situation (i) of Proposition 8.6.5.

Proposition 8.6.7. *Using the notation H, K, M, Ω appearing in Proposition 8.1.2, we have the following.*

(i) *If H is of type $Sp_2(q) \wr S_t$ in $\mathcal{C}_7(\Omega)$ with t odd, then $\mathcal{G}_S(H) = \{K, M\}$.*

(ii) *If H is of type $Sp_4(q) \wr S_t$ in $\mathcal{C}_7(\Omega)$, or if H is of type $Sp_2(q) \wr S_t$ in $\mathcal{C}_7(\Omega)$ with t even, then $\mathcal{G}_S(H) = \{K\}$.*

Proof. Let $X \in \mathcal{G}_S(H)$ with X simple. By Lemmas 8.6.1, 8.6.2, 8.6.3 and Propositions 8.6.5 and 8.6.6, X is as in conclusion (i) of Proposition 8.6.5. Write $H \cong Sp_{2b}(q) \wr S_t$ with $b \in \{1, 2\}$, q even, and $\Omega \approx \Omega^+_{2^{bt}}(q)$. Then $X \cong Sp_{2bt}(q)$ or $\Omega^+_{2bt+2}(q)$. Let U be the natural module for X of dimension $2bt$ or $2bt + 2$, respectively.

Step 1. *V is a spin module for X.*

First consider $X \cong Sp_{2bt}(q)$. Here by Proposition 8.6.5, H is of type $Sp_{2b}(q) \wr S_t$ in $\mathcal{C}_2(X)$, stabilizing a decomposition $U = U_1 \perp \cdots \perp U_t$. By Theorem 5.4.1, the representation of X on V lifts to an embedding

$$\widehat{X} = Sp_{2bt}(\widehat{\mathbf{F}}) < GL(\widehat{V}) = GL_{2^{bt}}(\widehat{\mathbf{F}}),$$

where $\widehat{\mathbf{F}}$ is the algebraic closure of $\mathbf{F} = \mathbf{F}_q$ and $\widehat{V} = V \otimes \widehat{\mathbf{F}}$. The natural module for $\widehat{X}$ is $U \otimes \widehat{\mathbf{F}}$, the symplectic form being the unique extension of the X-invariant symplectic form on U. Let L be the normal subgroup $Sp_{2b}(q)^t$ of H, so that L acts trivially on the set $\{U_1, \ldots, U_t\}$. Then $L < \widehat{L} = C_{\widehat{X}}(U_1 \otimes \widehat{\mathbf{F}}, \ldots, U_t \otimes \widehat{\mathbf{F}}) \cong Sp_{2b}(\widehat{\mathbf{F}})^t$. Now $H \in \mathcal{C}_7(\Omega)$, so L acts absolutely irreducibly on V, stabilizing a decomposition of V as a tensor product of t $2b$-spaces. Hence the same is true of the action of L on $V \otimes \widehat{\mathbf{F}}$. Let

$$V \otimes \widehat{\mathbf{F}} = \widehat{V}_1 \otimes \cdots \otimes \widehat{V}_t \tag{8.6.34}$$

be a such a tensor decomposition of $V \otimes \widehat{\mathbf{F}}$. Now the subgroup $Sp(U_1) \times \cdots \times Sp(U_{t-1}) \cong Sp_{2b}(q)^{t-1}$ of L acts absolutely irreducibly on $\widehat{V}_1 \otimes \cdots \otimes \widehat{V}_{t-1}$, and hence by Lemma 4.4.3.i, $Sp_{2b}(\widehat{\mathbf{F}}) \cong Sp(U_t \otimes \widehat{\mathbf{F}}) \leq C_{GL(\widehat{V})}(Sp_{2b}(q)^{t-1}) = GL(\widehat{V}_t)$. Similarly $Sp(\widehat{U}_i) \leq GL(\widehat{V}_i)$ for all i, and so $\widehat{L}$ preserves the tensor decomposition given in (8.6.34). For $1 \leq i \leq t$, $\widehat{V}_i$ is a symplectic $2b$-space for the ith factor $Sp_{2b}(\widehat{\mathbf{F}})$ of $\widehat{L}$. Let $\beta_i = \{e_i, f_i\}$ (if $b = 1$), $\beta_i = \{e_{2i-1}, e_{2i}, f_{2i-1}, f_{2i}\}$ (if $b = 2$) be a standard basis for $\widehat{V}_i$. For $\lambda_1, \ldots, \lambda_{bt} \in \widehat{\mathbf{F}}^*$, define $t_{\lambda_1,\ldots,\lambda_{bt}} \in \widehat{L}$ by

$$t_{\lambda_1,\ldots,\lambda_{bt}} : e_i \mapsto \lambda_i e_i,\ f_i \mapsto \lambda_i^{-1} f_i \ \ (1 \leq i \leq bt).$$

Also define $w_i \in \widehat{L}$ by

$$w_i : e_i \mapsto f_i,\ f_i \mapsto e_i,$$
$$e_j \mapsto e_j,\ f_j \mapsto f_j\ (j \neq i),$$

and when $b = 2$ define

$$x_i : e_{2i-1} \mapsto e_{2i},\ e_{2i} \mapsto e_{2i-1},$$
$$f_{2i-1} \mapsto f_{2i},\ f_{2i} \mapsto f_{2i-1},$$
$$e_j \mapsto e_j,\ f_j \mapsto f_j\ (j \neq 2i-1, 2i).$$

Set

$$W_o = \begin{cases} \langle w_i \mid 1 \leq i \leq t \rangle & \text{if } b = 1 \\ \langle w_{2i-1}w_{2i}, x_i \mid 1 \leq i \leq t \rangle & \text{if } b = 2. \end{cases}$$

Then $T = \{t_{\lambda_1,\dots,\lambda_{bt}} \mid \lambda_i \in \widehat{\mathbf{F}}^*\}$ is a maximal torus of $\widehat{X}$, normalized by $W_o \cong (\mathbf{Z}_2)^{bt}$. The weight-spaces of T on $\widehat{V}$ are the 2^{bt} 1-spaces $\langle v_1 \otimes \cdots \otimes v_t \rangle$, where $v_i \in \beta_i$. Clearly the subgroup W_o of $N(T)$ permutes these 1-spaces transitively, and each 1-space affords a different weight on T. Hence we conclude from Lemma 5.4.10 that $\widehat{V}$ is a spin module for $\widehat{X}$, which proves Step 1 for the case $X \cong Sp_{2bt}(q)$.

Now let $X \cong \Omega^+_{2bt+2}(q)$. Choose a standard basis a_i, b_i ($1 \leq i \leq bt+1$) for the natural module U, as in Proposition 2.5.3. In this case, by Proposition 8.6.5.i, X has a subgroup $Y \cong Sp_{2bt}(q)$ such that $H \in \mathcal{C}_2(Y)$, and we may take $Y = X_{\langle v \rangle}$, with $v = a_1 + b_1$. As above, using Theorem 5.4.1 we obtain an embedding

$$\widehat{Y} < \widehat{X} < GL(\widehat{V}),$$

where $\widehat{X} = \Omega_{2b+2}(\widehat{\mathbf{F}})$ and $\widehat{Y} = \widehat{X}_{\langle v \rangle} \cong Sp_{2bt}(\widehat{\mathbf{F}})$. Define

$$u_{\lambda_1,\dots,\lambda_{bt+1}} : a_i \mapsto \lambda_i a_i,\ b_i \mapsto \lambda_i^{-1} b_i\ \ (\lambda_i \in \widehat{\mathbf{F}}^*,\ 1 \leq i \leq bt+1)$$
$$v_i : a_i \mapsto b_i,\ b_i \mapsto a_i,\ a_j \mapsto a_j,\ b_j \mapsto b_j\ (j \neq i),$$

and set

$$T_1 = \{u_{\lambda_1,\dots,\lambda_{bt+1}} \mid \lambda_i \in \widehat{\mathbf{F}}^*\},\ T_2 = \{u_{1,\lambda_2,\dots,\lambda_{bt+1}} \mid \lambda_i \in \widehat{\mathbf{F}}^*\}$$

and

$$W_1 = \langle v_i v_j \mid \text{all } i, j \rangle.$$

Then T_1, T_2 are maximal tori of $\widehat{X}, \widehat{Y}$, respectively, and both T_1 and T_2 are normalized by $W_1 \cong (\mathbf{Z}_2)^{bt}$, a subgroup of $\widehat{Y}$. As we saw above, W_1 is transitive on the 2^{bt} weight-spaces for T_2 on $\widehat{V}$, and hence on the 2^{bt} weight-spaces for T_1. Hence as before, we conclude from Lemma 5.4.10 that $\widehat{V}$ is a spin module for $\widehat{X}$. This completes the proof of Step 1.

Step 2. *If $X \cong \Omega^+_{2bt+2}(q)$, then $b = 1$ and t is odd.*

Suppose that $X \cong \Omega^+_{2bt+2}(q)$. By Step 1, V is a spin module for X. As $\Omega \approx \Omega^+_{2bt}(q)$ we deduce from Proposition 5.4.9.ii that $bt + 1$ is even, and Step 2 follows.

Now recall the fixed overgroups $K \cong Sp_{2bt}(q)$, $M \cong \Omega_{2t+2}(q)$ in $\mathcal{G}_S(H)$, given by Proposition 8.1.2. By Step 2 we may choose $N \in \{K, M\}$ such that $N \cong X$. Let $\phi : X \to N$ be an isomorphism. Since all spin modules for X are quasiequivalent to each other (see Proposition 5.4.2), Step 1 shows that the two representations of X in $GL(V)$ given by the identity map and by ϕ, are quasiequivalent. Hence by Lemma 2.10.14, there exists $g \in GL(V, \mathbf{F})$ such that $N^g = X$. Since both X and N lie in $\Omega = \Omega(V, \mathbf{F}, Q)$, it follows that X fixes both the quadratic forms Q and Q^g on V (where $Q^g(v) = Q(vg)$ for $v \in V$). Consequently $Q^g = \lambda Q$ for some $\lambda \in \mathbf{F}$ by Lemma 2.10.3, and hence $g \in \Delta$. Since $\Delta = I \times \mathbf{F}^*$, we can take g to lie in I. Moreover both H and H^g are contained in X.

Step 3. *There exists $x \in X$ such that $gx \in N_I(H)$.*

First consider $X \cong Sp_{2bt}(q)$. According to Proposition 8.6.5, both H and H^g are of type $Sp_b(q) \wr S_t$ in $\mathcal{C}_2(X)$. Since by Proposition 4.2.10, X is transitive on such subgroups, H and H^g are conjugate in X, and the assertion of Step 3 follows. Now let $X \cong \Omega_{2t+2}(q)$ (with $b = 1$ and t odd). By Proposition 8.6.5.i, both H and H^g lie in subgroup of type $Sp_{2t}(q)$ in $\mathcal{C}_1(X)$. As X is transitive on such subgroups (Proposition 4.1.7), there exists $x_1 \in X$ such that H and H^{gx_1} both lie in a subgroup Y of type $Sp_{2t}(q)$. As before, H and H^{gx_1} are conjugate in Y, and Step 3 follows.

We can now complete the proof. It follows from Proposition 4.7.2 that $N_I(H) = H_I$, and from (4.7.6) that $H_I \cong Sp_{2b}(q) \wr S_t \cong H$. Therefore $N_I(H) = H$, and hence Step 3 shows that $gx \in H$ for some $x \in X$. As $H < X$, it follows that $g \in X$, and hence $X = N$. Since $N_\Omega(X) = X$ by Proposition 8.1.2, the Proposition follows. ∎

We have now completed the proof of Theorem 8.1.1. This was the last major result needed to complete the proof of the Main Theorem in Chapter 3, and so our work is finished at last.

References

[Ar] E. Artin, *Geometric Algebra*, Interscience, New York, 1957.

[As$_1$] M. Aschbacher, 'On the maximal subgroups of the finite classical groups', *Invent. Math.* **76** (1984), 469-514.

[As$_2$] M. Aschbacher, 'Chevalley groups of type G_2 as the group of a trilinear form', *J. Algebra* **109** (1987), 193-259

[As$_3$] M. Aschbacher , 'The 27-dimensional module for E_6. I', *Invent. Math.* **89** (1987), 159-195.

[As$_4$] M. Aschbacher , 'The 27-dimensional module for E_6. II', *J. London Math. Soc.* (2) **37** (1988), 275-293.

[As$_5$] M. Aschbacher , 'The 27-dimensional module for E_6. III', to appear in *Trans. Amer. Math. Soc.*

[As$_6$] M. Aschbacher , 'The 27-dimensional module for E_6. IV', to appear in *J. Algebra.*

[As$_7$] M. Aschbacher, 'The maximal subgroups of E_6', preprint.

[As$_8$] M. Aschbacher, *Finite Group Theory,* Cambridge University Press, 1986.

[As$_9$] M. Aschbacher , 'Subgroup structure of finite groups', in *Proceedings of the Rutgers Group Theory Year, 1983-1984*, M. Aschbacher, D. Gorenstein, et al., eds., Cambridge University Press (1984), 35-44.

[A-S] M. Aschbacher and L.L. Scott, 'Maximal subgroups of finite groups', *J. Algebra* **92** (1985), 44-80.

[At] J.H. Conway, R.T. Curtis, S.P. Norton, R.A. Parker and R.A. Wilson, *An ATLAS of Finite Groups*, Oxford University Press, 1985.

[Bo] A. Borel, 'Properties and linear representations of Chevalley groups', in *Seminar on Algebraic Groups and Related Finite Groups* (Borel et al.), *Springer Lecture Notes* **131**, Springer-Verlag, Berlin-Heidelberg-New York, 1970.

[Bor] A.V. Borovik, 'The structure of finite subgroups of simple algebraic groups', *Algebra i Logica* 28 (1989), 249-279 (in Russian).

[B-T] A. Borel and J. Tits, 'Éléments unipotents et sous-groupes paraboliques de groupes réductifs, I', *Invent. Math.* **12** (1971), 95-104.

[Bou] N. Bourbaki, *Groupes et algèbres de Lie*, Act. Sci. Ind. Hermann, Paris, 1969.

[B-W] N. Burgoyne and C. Williamson, 'On a theorem of Borel and Tits for finite Chevalley groups', *Arch. Math.* **27** (1976), 489-491.

[B-G-L] N. Burgoyne, R. Griess and R. Lyons, 'Maximal subgroups and automorphisms of Chevalley groups', *Pac. J. Math.* **71** (1977), 365-403.

[Bu] G. Butler, 'The maximal subgroups of the Chevalley group $G_2(4)$', in *Proceedings of Groups — St. Andrews, 1981* C.M. Campbell and L. Robertson, eds., London Mathematical Society Lecture Note Series **71**, Cambridge University Press, Cambridge New York (1982), 186-200.

[Ca_1] R.W. Carter, *Simple Groups of Lie Type*, Wiley-Interscience, 1972.

[Ca_2] R.W. Carter, *Finite Groups of Lie Type: conjugacy classes and complex characters* Wiley-Interscience, 1985.

[Cau] A. Cauchy, 'Mémoire sur le nombre des valeurs qu'une fonction peut acquérir lorsqu'on y permute de toutes les manieres possibles les quantités qu'elle renferme' *J. de l'Ecole Polytechnique* **10** (1815), 1-28.

[Ch_1] C. Chevalley, *Séminaire Chevalley Vols. I,II: Classifications des groupes de Li algébriques,* Paris, 1956-8.

[Ch_2] C. Chevalley, *The Algebraic Theory of Spinors,* Columbia University Press, New York, 1954.

[C-L-S-S] A.M. Cohen, M.W. Liebeck, J. Saxl and G.M. Seitz, 'The local maximal subgroups of the exceptional groups of Lie type', preprint.

[Col] M.J. Collins, ed. *Finite Simple Groups II*, Academic Press, 1980.

[Co_1] B.N. Cooperstein, 'Minimal degree for a permutation representation of a classical group', *Israel J. Math.* **30** (1978), 213-235.

[Co_2] B. N. Cooperstein, 'Maximal subgroups of $G_2(2^n)$', *J. Algebra* **70** (1981), 23-36.

[C-R] C.W. Curtis and I. Reiner, *Representation Theory of Finite Groups and Associative Algebras,* Wiley-Interscience, New York, 1962.

[Dic_1] L.E. Dickson, *Linear Groups, with an Exposition of the Galois Field Theory*, Dover Publications, New York, 1958.

[Dic_2] L.E. Dickson, 'Representations of the general symmetric groups as linear groups in finite and infinite fields', *Trans. Amer. Math. Soc.* **9** (1908), 121-148.

[Dic_3] L.E. Dickson, 'On the cyclotomic function', *Amer. Math. Monthly* **12** (1905), 86-89

[Di_1] J. Dieudonné, *Le géometrie des groupes classiques*, Ergebnisse der Mathematik und ihrer Grenzgebiete, 5 (1955), Springer.

[Di_2] J. Dieudonné, 'On the automorphisms of the classical groups', *Mem. Amer. Math Soc.* **2** (1951).

[Di_3] J. Dieudonné, 'Les isomorphismes exceptionels entre les groupes classiques finis' *Can. J. Math* **6** (1954), 305-315.

[Dy_1] R.H. Dye, 'Maximal subgroups of $GL_{2n}(K)$, $SL_{2n}(K)$, $PGL_{2n}(K)$ and $PSL_{2n}(K)$ associated with symplectic polarities', *J. Algebra* **66** (1980), 1-11.

[Dy_2] R.H. Dye, 'On the maximality of the orthogonal groups in the symplectic groups in characteristic 2', *Math. Zeit.* **172** (1980), 203-212.

[Dy_3] R.H. Dye, 'Maximal subgroups of finite orthogonal and unitary groups stabilizing anisotropic subspaces', *Math. Zeit.* **189** (1985), 111-129.

[Dy₄] R.H. Dye, 'Maximal subgroups of symplectic groups stabilizing spreads II', to appear in *J. London Math. Soc.*

[Fl₁] D.E. Flesner, 'The geometry of subgroups of $PSp_4(2^n)$', *Illinois J. Math.* **19** (1975), 48-70.

[Fl₂] D.E. Flesner, 'Maximal subgroups of $PSp_4(2^n)$ containing central elations or non-centered skew elations', *Illinois J. Math.* **19** (1975), 247-268.

[Ga₁] E. Galois, 'Oeuvres Mathématique: lettre a M. Auguste Chevalier', *J. Math. Pures Appl.* (Liouville) (1846), 400-415.

[Ga₂] E. Galois, 'Oeuvres Mathématique', *J. de Math.* **11** (1846), 381-444.

[G-S] P.B. Gilkey and G.M. Seitz, 'Some representations of exceptional Lie algebras', *Geom. Ded.* **25** (1988), 407-416.

[Go] D. Gorenstein, *Finite Simple Groups, an Introduction to their Classification*, Plenum Press, New York (1982).

[G-L] D. Gorenstein and R. Lyons, 'The local structure of finite groups of characteristic 2 type', *Mem. Amer. Math. Soc.* **42**, No. 276 (1983).

[Hi] G. Hiss, 'The modular characters of the Tits simple group and its automorphism group', *Comm. Alg.* **14** (1986), 125-154.

[Hu₁] J.E. Humphreys, *Linear Algebraic Groups,* Graduate Texts in Mathematics **21**, Springer, Berlin, 1975.

[Hu₂] J.E. Humphreys, *Ordinary and modular representations of Chevalley groups,* Springer Lecture Notes in Mathematics **528**, Berlin, 1976.

[Is] I.M. Isaacs, *Character Theory of Finite Groups*, Academic Press, 1976.

[Ja₁] G.D. James, 'On the minimal dimensions of irreducible representations of symmetric groups', *Math. Proc. Camb. Phil. Soc.* **94** (1983), 417-424.

[Ja₂] G.D. James, *The Representation Theory of the Symmetric Groups,* Springer Lecture Notes **682**, Berlin, 1978.

[Jo] C. Jordan, 'Note sur les equations modulaires', *C.R. Acad. Sci. Paris* **66** (1868), 308-312.

[Ka₁] W.M. Kantor, 'Permutation representations of the finite classical groups of small degree or rank', *J. Algebra* **60** (1979), 158-168.

[Ka₂] W.M. Kantor, 'Classical groups from a non-classical viewpoint', Lecture Notes, Mathematical Institute, Oxford, 1979.

[Ke₁] J.D. Key, 'Some maximal subgroups of $PSL_n(q)$, $n \geq 3$, $q = 2^r$', *Geom. Ded.* **4** (1975), 377-386.

[Ke₂] J.D. Key, 'Some maximal subgroups of certain projective unimodular groups', *J. London Math. Soc.* (2) **19** (1979), 291-300.

[K-L-S] W. Kimmerle, R. Lyons and R. Sandling, 'Composition factors from the group ring and Artin's theorem on the orders of simple groups', to appear in *Proc. London Math. Soc.*

[Ki_1] O.H. King, 'Some maximal subgroups of the classical groups', *J. Algebra* **68** (1981), 109-120.

[Ki_2] O.H. King, 'Maximal subgroups of the classical groups associated with non-isotropic subspaces of a vector space', *J. Algebra* **73** (1981), 350-375.

[Ki_3] O.H. King, 'Imprimitive maximal subgroups of the general linear, special linear, symplectic and general symplectic groups', *J. London Math. Soc.* (2) **25** (1982), 416-424.

[Ki_4] O.H. King, 'Imprimitive maximal subgroups of the orthogonal, special orthogonal, unitary and special unitary groups', *Math. Zeit.* **182** (1983), 193-203.

[Ki_5] O.H. King, 'Imprimitive maximal subgroups of the symplectic, orthogonal and unitary groups', *Geom. Ded.* **15** (1984), 339-353.

[Ki_6] O.H. King, 'On subgroups of the special linear group containing the special unitary group', *Geom. Ded.* **19** (1985), 297-310.

[Ki_7] O.H. King, 'On subgroups of the special linear group containing the special orthogonal group', *J. Algebra* **96** (1985), 178-193.

[Kl_1] P.B. Kleidman, 'The maximal subgroups of the finite 8-dimensional orthogonal groups $P\Omega_8^+(q)$ and of their automorphism groups', *J. Algebra* **110** (1987), 173-242.

[Kl_2] P.B. Kleidman, 'The low-dimensional finite classical groups and their subgroups', to appear in Longman Research Notes.

[Kl_3] P.B. Kleidman, 'The maximal subgroups of the Chevalley groups $G_2(q)$ with q odd, of the Ree groups ${}^2G_2(q)$, and of their automorphism groups', *J. Algebra* **117** (1988), 30-71.

[Kl_4] P.B. Kleidman, 'The maximal subgroups of the Steinberg triality groups ${}^3D_4(q)$ and of their automorphism groups', *J. Algebra* **115** (1988), 182-199.

[Kl_5] P.B. Kleidman, 'On the maximal subgroups of groups of type E_6 and ${}^2E_6$', in preparation.

[K-L] P.B. Kleidman and M.W. Liebeck, 'A survey of the maximal subgroups of the finite simple groups', *Geom. Ded.* **25** (1988), 375-389.

[K-W_1] P.B. Kleidman and R.A. Wilson, 'The maximal subgroups of J_4', *Proc. London Math. Soc.* (3) **56** (1988), 484-510.

[K-W_2] P.B. Kleidman and R.A. Wilson, 'The maximal subgroups of Fi_{22}', *Math. Proc. Cam. Phil. Soc.* **102** (1987), 17-23.

[K-W_3] P.B. Kleidman and R.A. Wilson, 'The maximal subgroups of $E_6(2)$ and $Aut(E_6(2))$', to appear in *Proc. London Math. Soc.*

[K-P-W] P.B. Kleidman, R.A. Parker and R.A. Wilson, 'The maximal subgroups of the Fischer group Fi_{23}', *J. London Math. Soc.* (2) **39** (1989), 89-101.

[L-Se] V. Landazuri and G.M. Seitz, 'On the minimal degrees of projective representations of the finite Chevalley groups', *J. Algebra* **32** (1974), 418-443.

[La] S. Lang, *Algebra*, Addison-Wesley, 1965.

[L-N] V.M. Levchuk and Ya. N. Nuzhin, 'Structure of Ree groups', *Algebra i Logika* **24** (1985), 26-41.

[Li_1] S. Li, 'Overgroups in $GL(nr, F)$ of certain subgroups of $SL(n, K)$', to appear in *J. Algebra.*

[Li_2] S. Li, 'Overgroups in $GL(U \otimes W)$ of certain subgroups of $GL(U) \otimes GL(W)$, I', preprint.

[Lie_1] M.W. Liebeck, 'On the orders of maximal subgroups of the finite classical groups', *Proc. London Math. Soc.* (3) **50** (1985), 426-446.

[Lie_2] M.W. Liebeck, 'The affine permutation groups of rank three', *Proc. London Math. Soc.* (3) **54** (1987), 477-516.

[Lie_3] M.W. Liebeck, 'The local maximal subgroups of the finite simple groups', *Proc. Symp. Pure Math.* **47** (1987), 455-461.

[L-P-S_1] M.W. Liebeck, C.E. Praeger and J. Saxl, 'A classification of the maximal subgroups of the alternating and symmetric groups', *J. Algebra* **111** (1987), 365-383.

[L-P-S_2] M.W. Liebeck, C.E. Praeger and J. Saxl, 'On the O'Nan-Scott reduction theorem for finite primitive permuatation groups', *J. Australian Math. Soc.* **44** (1988), 389-396.

[L-P-S_3] M.W. Liebeck, C.E. Praeger and J. Saxl, 'The factorizations of the finite simple groups and their automorphism groups', preprint.

[L-S_1] M.W. Liebeck and J. Saxl, 'On the orders of maximal subgroups of the finite exceptional groups of Lie type', *Proc. London Math. Soc* (3) **55** (1987), 299-330.

[L-S_2] M.W. Liebeck and J. Saxl, 'Primitive permutation groups containing an element of large prime order', *J. London Math. Soc.* (2) **31** (1985), 237-249.

[L-Se] M.W. Liebeck and G.M. Seitz, 'Maximal subgroups of exceptional groups of Lie type: finite and algebraic', preprint.

[L-S-S] M.W. Liebeck, J. Saxl and G.M. Seitz, 'On the overgroups of irreducible subgroups of the finite classical groups', *Proc. London Math. Soc.* (3) **55** (1987), 507-537.

[Lin] S. Linton, 'The maximal subgroups of the Thompson group' *J. London Math. Soc.* (2) **39** (1989), 79-88.

[Ma] G. Malle, 'The maximal subgroups of the Ree groups ${}^2F_4(q)'$', to appear in *J. Algebra.*

[Mc] J. McLaughlin, 'Some subgroups generated by transvections', *Arch. Math.* **18** (1969), 108-115.

[Mi] E.T. Migliore, 'The determination of the maximal subgroups of $G_2(q)$, q odd', Ph.D.

Thesis, University of California at Santa Cruz, 1982.

[No] S.P. Norton, unpublished.

[N-W] S.P. Norton and R.A. Wilson 'The maximal subgroups of $F_4(2)$', submitted.

[Pa] R.A. Parker, 'A collection of modular character tables', University of Cambridge, 1989.

[Pas] D. Passman, *Permutation Groups*, Benjamin, New York, 1968.

[Pat] W.H. Patton, 'The minimum index for subgroups in some classical groups: a generalization of a theorem of Galois', Ph.D. Thesis, U. of Illinois at Chicago Circle, 1972.

[P-T] N.T. Petrov and K.B. Tchakerian, 'The maximal subgroups of $G_2(4)$', *J. Algebra* **76** (1982), 171-185.

[Po] H. Pollatsek, 'First cohomology groups of some linear groups over fields of characteristic two', *Illinois J. Math.* **15** (1971), 393-417.

[Ru] P. Ruffini, *Teoria generale delle equazioni in cui si dimonstra impossibile delle equazioni generli di grado superiore al quarto*, 2 vols. (Bologna 1799).

[Sch$_1$] I. Schur, 'Untersuchungen über die Darstellung der enlichen Gruppen durch gebrochene lineare Substitutionen', *J. Reine Angew. Math.* **132** (1907), 85-137.

[Sch$_2$] I. Schur, 'Über die Darstellung der symmetrischen und der alternierenden Gruppe durch gebrochene lineare Substitutionen', *J. Reine Angew. Math.* **139** (1911), 155-250.

[Sc] L.L. Scott, 'Representations in characteristic p', in *Proc. Symp. Pure Math.* **37** (1980), 319-331.

[Se$_1$] G.M. Seitz, 'The maximal subgroups of classical algebraic groups', *Mem. Amer. Math. Soc.* **67**, No. 36 (1987).

[Se$_2$] G.M. Seitz, 'Representations and maximal subgroups of finite groups of Lie type', *Geom. Ded.* **25** (1988), 391-406.

[Se$_3$] G.M. Seitz, 'Cross-characteristic embeddings of finite groups of Lie type', to appear in *Proc. London Math. Soc.*

[St$_1$] R. Steinberg, *Lectures on Chevalley Groups,* Yale University Mathematics Department, 1968.

[St$_2$] R. Steinberg, 'Endomorphisms of linear algebraic groups', *Mem. Amer. Math. Soc.* **80** (1968).

[St$_3$] R. Steinberg, 'Representations of algebraic groups', *Nagoya Math. J.* **22** (1963), 33-56.

[St$_4$] R. Steinberg, 'Variations on a theme of Chevalley', *Pacific J. Math.* **9** (1959), 875-891.

[Su$_1$] M. Suzuki, 'On a class of doubly transitive groups', *Ann. of Math.* **75** (1962),

105-145.

[Su$_2$] M. Suzuki, *Group Theory* II, Springer, Berlin-Heidelberg-New York, 1982.

[Tc] K.B. Tchackerian, 'The maximal subgroups of the Tits simple group', *Compt. Rend. Acad. Bulg. Sci.* **34** (1981), 1637.

[Te] D. Testerman, 'Irreducible subgroups of exceptional algebraic groups', *Mem. Amer. Math. Soc.* **75**, No. 390 (1988).

[Wa$_1$] A. Wagner, 'The faithful linear representations of least degree of S_n and A_n over a field of characteristic 2', *Math. Zeit.* **151** (1976), 127-137.

[Wa$_2$] A. Wagner, 'The faithful linear representations of least degree of S_n and A_n over a field of odd characteristic', *Math. Zeit.* **154** (1977), 103-114.

[Wa$_3$] A. Wagner, 'An observation on the degrees of projective representations of the symmetric and alternating groups over an arbitrary field', *Arch. Math.* **29** (1977), 583-589.

[Wie] H. Wielandt, *Finite Permutation Groups*, Academic Press, New York-London, 1964.

[Wi$_1$] R.A. Wilson, 'Maximal subgroups of sporadic simple groups', in *Proceedings of Groups — St. Andrews, 1985*, C.M. Campbell and L. Robertson, eds., London Mathematical Society Lecture Notes Series **121**, Cambridge University Press (1986), 352-358.

[Wi$_2$] R.A. Wilson, 'The quaternionic lattice for $2G_2(4)$ and its maximal subgroups', *J. Algebra* **77** (1982), 449-466.

[Wi$_3$] R.A. Wilson, 'The geometry and maximal subgroups of the simple groups of A. Rudvalis and J. Tits', *Proc. London Math. Soc.* (3) **48** (1984), 533-563.

[Wi$_4$] R.A. Wilson, 'The complex Leech lattice and maximal subgroups of the Suzuki group', *J. Algebra* **84** (1983), 151-188.

[Wi$_5$] R.A. Wilson, 'Maximal subgroups of automorphism groups of simple groups', *J. London Math. Soc.* (2) **32** (1985), 460-466.

[Wi$_6$] R.A. Wilson 'Maximal subgroups of some sporadic simple groups', Ph.D. Thesis, University of Cambridge, 1982.

[Zs] K. Zsigmondy, 'Zur Theorie der Potenzreste', *Monatsh. für Math. u. Phys.* **3** (1892), 265-284.

Index of Notation

Index

Printed in the United States
By Bookmasters